PAUL M. BINGHAM and JOANNE SOUZA

DEATH FROM A DISTANCE

and the

BIRTH OF A HUMANE UNIVERSE

HUMAN EVOLUTION, BEHAVIOR, HISTORY, AND YOUR FUTURE

ISBN: 1-4392-5412-5
ISBN-13: 9781439254127

Visit www.booksurge.com to order additional copies.

Praise for

Death from a Distance and the Birth of a Humane Universe

"...I should point out that the manuscript is magnificent! I am amazed that someone classified as having interests only inside cells can be as wise [] about ecology and evolution. I also think your language skills are terrific..."

– George C. Williams, 1999 Crafoord Laureate, Evolutionary Biology

"This wide-ranging and provocative book is overflowing with ideas and insights into what it means to be human as well as how and why we came to be the way we are. Not a typical book on human evolution; it is, to use the authors' phrase, "a theory of everything."

– Prof. John Fleagle, MacArthur Fellow, Distinguished Professor, Dept of Anatomical Sciences, Stony Brook University

"This book is an astonishingly wide-ranging and provocative overview of evolution, human origins and social organization. Breathtaking in its scope, and ranging from the earliest prehistoric period, through the course of the evolution of life on Earth, Death from a Distance presents a startlingly original thesis grounded in contemporary evolutionary theory. It will challenge countless

well-entrenched theories about who we are as a species, where we came from and where we are going. Their arguments about the evolution of human sexual mores, language, the role of projectile weaponry in human evolution, and their predictions about the future course of human society are sure to ignite controversy. Yet, the aplomb with which Bingham and Souza present their arguments will give readers a clear appreciation why their undergraduate course is one of the most popular at Stony Brook University."

– Prof. John J. Shea, Dept. of Anthropology, Stony Brook University

"It is not often that readers are offered a new theory of human existence encompassing our origins, our unique properties as biological creatures, the agricultural revolution, and the rise of modern states, but this book promises just that. The authors argue that the first humans of two million years ago evolved the capacity to throw stones accurately and thus kill conspecifics from a distance. This capacity for "law enforcement" allowed unrelated individuals to cooperate in ever larger aggregates, thus forever altering the social environment and paving the way for democratization on a global scale. Such bold claims require strong support. The authors are diligent in building their foundation, even if some of their premises can be challenged. This book is surely provocative. Reading it carefully is well worth the effort."

– Prof. G. Philip Rightmire, Research Associate, Dept of Anthropology, Harvard University; Distinguished Professor, Binghamton University

"With "elite throwing" as a unifying theme, Bingham and Souza launch an epic journey through two million years of human evolutionary history. Following lucid explanations of scientific method and the evolutionary process, the authors present a theory explaining the emergence of unique human behaviors, meticulously documenting the evidence. The book is so engagingly written that readers with all backgrounds will be drawn to critically examining the new perspectives on current and past human behaviors."

– Mary W. Marzke, Professor emerita, School of Human Evolution and Social Change, Arizona State University

⚬∽⚬

Death from a Distance and the Birth of a Humane Universe

Human evolution, behavior, history, and your future

Humans communicate, cooperate, reason, coerce, and influence in new ways that separate us from all other species. We are unique. How does this knowledge give us clues into our evolutionary origins and hope for the future? Paul M. Bingham and Joanne Souza address these questions and more in a remarkable examination of the biological, behavioral, and historical evolution of our species. *Death from a Distance and the Birth of a Humane Universe* reads like a gripping novel, while delivering an answer to Darwin's unanswered question how did humans become unique? The authors, one a molecular and evolutionary biologist and the other a research psychologist, dedicated to the evolutionary logic of human social behavior, have taken us beyond the fundamental concepts of biology into a theory that merges the natural and social sciences. Their work opens an entirely new way of looking at science and the human future.

❦

To :

GEORGE C. WILLIAMS

One of the great scientists of the
20th Century. His astonishing courage, thoughtfulness
and integrity stand as a beacon to everyone who
wishes to understand.

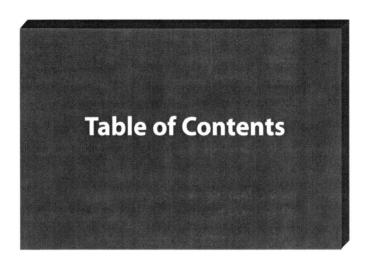

Table of Contents

Fifth Interlude: 537

*Never again - Pathological deployment of human coercive threat
and the brutalization of the powerless*

Postscript: 617

*Meaning and materialism – fulfilling our hearts
and minds*

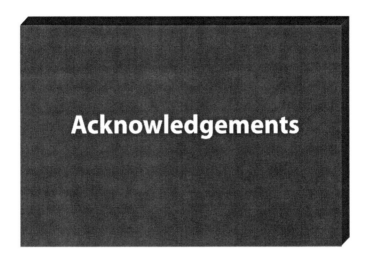

Acknowledgements

Many people have contributed to our work in diverse and vital ways. We cannot possibly name them all. Instead, we call attention to a few who made especially recent and salient contributions. To all our teachers and colleagues over the years who are not named, know that we are, nonetheless, very well aware of our debts to you.

Most importantly, we acknowledge Profs. Daijiro Okada, Zuzana Zachar, and the late Clifford A. Souza, Jr. as collaborators, sounding boards, colleagues, and critics. Their contributions touch every page of this book. They have enriched our lives enormously.

Profs. John Shea, Jack Stern, John Fleagle, and George Williams played crucial roles as informants and critics of our early work. Their clear-headed, expert support of our original endeavors was important to making this book a reality. Also extraordinarily valuable was the input of Profs. Mary Marzke, Rob Blumenschine, Philip Rightmire, and Leslie Aiello.

We note that none of these professional colleagues necessarily agrees with all the conclusions we draw. If some of our claims ultimately prove to be wrong, we, not they, will be entirely responsible.

Joanne wishes to particularly acknowledge Stephen, Courtney, Chris, and Clifford Souza; Josephine, Cesare, John, and Joseph Monastra; and Susan DiSario for their candid input, their testing of the theoretical work with their varied life experiences, and their critical eyes and ears as to clarification of our communication. Their patience with the unending discussions over the last seven years is deeply appreciated. She would also like to mention the late Paul Murray, her childhood horseback riding instructor, who introduced her early in life to the value of evolutionary origins when attempting to understand all animal social behavior.

Paul wishes to thank Maria Trinkle for her important help at one stage of this project.

We also thank the Two Elizabeths, Liz Dunn and Betty Ryan, members of the 111th PVI, for their generous help in understanding the vast extra-academic study of the American Civil War.

Equally important to this project has been the tremendously rich input from the approximately four thousand Stony Brook University students to whom we have taught this content over the last decade. Their passionate engagement and relentless questioning have improved our understanding (and this manuscript) in myriad ways. As students, we teach more than we know. As teachers, we learn more than we often recognize. This book is a monument to how scientifically productive and creative the so-inadequately-named student-teacher relationship can actually be. Our teaching assistants over the years have played a crucial part in helping us overcome the language barriers inherent in any educational enterprise. Our heartfelt thanks goes out to all of them.

Among our student colleagues, several have played important roles in sharpening our understanding. We are grateful to Arnav Shah, Michael Zannettis, Artem Sunik, Roman Spektor, Konstantin Peysin, Stephanie Beirle, and Katie Ching for these rich interactions.

A special acknowledgement also goes to the Round Tablers at Stony Brook University. They are a rich collaborative group of adult professionals who continue to learn and contribute to the scientific enterprise. In particular, we would like to mention Joel and Rhoda Spinner, Merton Reichler, and the late Prof. Egon Neuberger and his wife, Florence.

A very talented artist, Amy Radican, took the time and effort to create many of the more sophisticated drawings in this book. We appreciate her creativity and commitment to the science. Author photos are courtesy of Christina Luciw.

We would also like to acknowledge Todd Rothman, a very talented person in many regards. His brilliant influence as a fellow educator and a collaborator has energized us throughout this project.

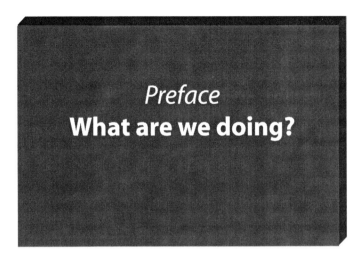

Preface
What are we doing?

An invitation

We have had the enormous good fortune to spend the last several decades on a scientific journey that has profoundly changed us. We will claim that this journey has taught us many vital new lessons about what it means to be human. Nothing in life looks the same to us as it did before. We read the newspaper; think about history and the human future; listen to the global economic and political dialog; or sense our own private thoughts with new eyes, minds, and ears.

This shift in perspective is a little unsettling at first. Very little about being human, about how we got here or about what the future may hold for us, is actually as we have always thought. Many things about ourselves are not as we have been taught to believe. But at the same time, this new way of looking at things is richly humane. It lets us understand ourselves and one another as we never have before. It invites a new mutual respect, even an admiration for our species. Most importantly of all, this new perspective provides us with insights into the human condition that give us unprecedented power over our common destiny.

This book offers you the opportunity to take the same journey. If you complete the trip, you also will be changed. As the story unfolds, you will sometimes be disturbed at first. Indeed, you will occasionally even find your ethical sense profoundly offended; but, be patient. Authentic self-understanding is resonant and rewarding. The ultimate destination is illuminating and ethically empowering. We will argue that we have greater potential to bring into existence a humane, wealthy, peaceful pan-global human world than we ever knew.

The form of our journey is to build, test, and explore a new scientific theory of what it means to be human—how we got here; why our ethical, political, parental, and sexual minds are the way they are; why our history has played out as it has; why our world is now as it is. For a few readers a *scientific* journey will sound dry and lifeless or daunting and intimidating.* The truth is that science is merely the ancient, universal human approach to understanding the world. We have two million years of practice and evolved skill at it. We do it very well. It electrifies us when it works.

We will seek to introduce you to that same scintillating experience of insight we had when we first followed this path. This trip was fun in the most cosmic sense. We look forward to sharing this deep pleasure with you. If we are successful, this book will begin a rich, new, life-long adventure for you, as the original journey did for us.

What this book is about

Humans are unique. We are utterly different than any other animal on Earth. In fact, we are *so* different from other creatures that it took us a very long time to realize that we were, after all, merely one of the hundreds of thousands of animal species who have lived on this planet over the last billion years.

We are not objective about how strange we are. For example, a typical non-human animal might cooperate with a few other adults. In stunning contrast, humans can cooperate on scales of states, empires, continents, and even the entire globe involving hundreds of millions of individuals. Aristotle famously recognized this special property and designated us the "political" animal. No other animal has ever achieved such a high level of cooperation among unrelated (non-kin) individuals.

As dramatic as our social cooperation is, it appears to be just the tip of a very large iceberg. We also possess individual capabilities such as speech, sophisticated cognitive abilities, and technological virtuosity utterly beyond anything previously seen on our planet.

This complicated suite of features becomes even more remarkable when we realize that its elements probably all evolved "explosively." These properties apparently emerged as a package extremely rapidly either at the same time as or immediately after the evolutionary origin

---

* You were almost certainly taught science badly, a tragically common experience in most contemporary cultures.

of humans. Our first uniquely human ancestors acquired their unique-ness overnight on an evolutionary time scale.*

All of these things about us, collectively, present us with the "human uniqueness problem." This book is about a rich, rigorous scientific expla-nation of our unprecedented properties—a solution to the human uniqueness problem.

Why should anyone other than a few professional scientists care about such things? In fact, the answers to essentially every question we might ask about ourselves, our friends and family, or the larger human world and its future depends, absolutely and in every detail, on our solu-tion to the human uniqueness problem.

For example, any of us might wonder *Why do my siblings, my children, or my mate behave as they do? Why are violence and chronic poverty so pervasive, yet people in some places live in peace and prosperity? Why does the economy that supports me and my family work as it does (for better and, often, for worse)? Are we doomed to endless cycles of flowering of cultures followed by their decay and collapse? Why are my sexual impulses so complex? Which features of human societies are inevitable and which are malleable local quirks?*

Throughout this book, we will argue that the answers to all these and many other questions come *directly* from our properties as unique biological animals. Moreover, we will find that all our unprecedented properties are products of our evolutionary history in a way we have not understood before. Indeed, all our uniquely human individual features apparently have a single underlying origin, emerging from our biologically novel *social* cooperation. Unexpectedly rich insights of many different kinds will emerge from this deceptively simple conclusion.

What we are doing

The foundational scientific work on which this book is based began with decades of thought and analysis by the first author, Paul. However, science only takes full form in social dialog. This dialog has gone on for the last fifteen years with many colleagues and with over four thou-sand students at Stony Brook University. This ongoing conversation has profoundly shaped what you will read. The second author, Joanne, has not only contributed centrally to the research presented here as a colleague over the last seven years, she has also helped manage and

---

*   A few specialists may think this picture of the evolution of human properties is an over-simplification. We urge them to be patient. We will defend this claim later.

organize this dialog with rich wisdom and brilliant insight (Paul's words), based on her own parallel decades-long scientific quest into human behavior. This book is a result of our pooled research, resources, education, and experience. As well, the gifted economist and game theorist, Daijiro Okada, has been a vital collaborator throughout the last nine years in the development of the foundational theory underlying this book. You will meet him in a later chapter.

New science is always more important for what it starts than for what it finishes. Making new progress accessible to those who will go on to use it is necessary to turn private insight into science. We have invested much effort over the last seven years learning to tell this story. We have been blessed by valuable criticism from professional colleagues as well as from many non-professional readers (see Acknowledgements).

Both of us want you to know, here at the beginning, that we are each parents who love our children, citizens of a modern democracy who cherish political equality and freedom, and citizens of the world who believe in a common, peaceful, just, transparent pan-human future. Later in the book, some readers will wonder whether our personal politics might be influencing our scientific conclusions. Only testing the science against the empirical evidence can meet this concern to everyone's satisfaction (Introduction). However, we note that our personal politics might be described as *centrist* and *pragmatic* (Chapter 17). We are equally and deeply skeptical of all political entrepreneurs, *rightist*, *centrist*, *leftist*, *anarchist*, *authoritarian*, *theistic*, or *secularist*. Our political faith is in the pan-human *wise crowd* (Chapter 10), whenever and wherever it is able to choose and manage its political entrepreneurs.

Finally, when we occasionally speak in a terse, analytical voice about humans, it is not out of condescension or disrespect. Detachment is merely the perspective we must all occasionally take, knowingly, to have the best chance at self-understanding and, thus, at leaving a better world for our children and grandchildren. At the end of our shared journey, we will be able to look one another in the eye with new, mutual knowledge and candor—even love.

About the authors

<u>Paul M. Bingham</u>: I have been intensely interested in science and the human condition since childhood. I began college as a social science major, but quickly moved to biology as the path to progress at that historical moment. This was in 1970 and the molecular revolution in biology was in full throat. I was in a position to contribute to its unfold-

ing.* I have had the good fortune to go on to work on important problems in this area, including the treatment of cancer. This wonderful part
of my life has played out in parallel to the work described here.

When I came to Harvard in 1975 to work on a PhD in Biochemistry
and Molecular Biology, another revolution was just beginning. During
my first year in Cambridge, Richard Dawkins's *The Selfish Gene* and Ed
Wilson's *Sociobiology* were published. I was working in the Biological
Laboratories and a stone's throw away, across Divinity Avenue in the
Museum of Comparative Zoology Ed Wilson, Bob Trivers, and others were
creating the foundation on which Joanne, Daijiro Okada and I would
later build. They were extending a process begun in the 1960s by our
Stony Brook colleague George Williams, Bill Hamilton, and others. The
result was a clear understanding of *non-human* animal social behavior.
This underpinning (Chapters 1-3) would ultimately be essential to our
quest to understand the evolutionary logic of *human* social behavior
(Chapters 4-17).

This rich intellectual ferment in the Cambridge of the late 1970s
empowered new purpose in my life-long quest. I have spent many free
moments in the ensuing thirty-four years thinking about the science
of being human. By the early 1990s I had explored and rejected every
earlier answer to the human uniqueness question. The new answer we
describe in this manuscript was beginning to crystallize. This answer ultimately took full form in active collaboration with Joanne and Daijiro. In a
real sense, this book is the fulfillment of one of the larger purposes of my
life, a purpose that has been a constant companion and a consuming
passion since my earliest memories.

<u>Joanne Souza</u>: I have been interested in all animal behavior since
early childhood. Throughout my life, I have spent many hours reading
about and studying animals, observing their behavior, and wondering
why they behave as they do in the present. I have also spent a good
deal of time comparing humans to other animals and have been
intrigued about our obvious uniqueness. In early childhood, I learned
to ride horses from a retired cavalryman who advised me to learn about
the evolutionary history of horses if I truly wanted to understand their
current behavior. I transferred that advice to human behavior as I grew
to be an adult and became interested in psychology and education.

---

*   Among other things, I contributed to the discovery of the P element transposon
and the use of this class of elements to investigate and manage the genomes of multicellular organisms; the role of these elements in genetics and mutation; alternative pre-
mRNA splicing; nuclear substructure; and, most recently, energy metabolism in cancer
and its chemotherapy.

I spent the first half of my working life in the business world. This work included acting as an industry consultant in health and education at AT&T during and immediately after divestiture, and later as a senior executive for a business of my own. I acquired extensive experience with human individual and institutional beliefs, behavior, and learning. However, my passion for understanding the underlying reasons for the differences between humans and other animals never left me. I could *describe* many behaviors, but did not *understand* the behavior, nor could I understand why we make decisions as we do. Why do we often succumb to social pressures when such actions sometimes degrade our physical and psychological health? Why do we stay in jobs we do not enjoy? Why, when learning seems so natural and enjoyable to us, do our children come to dislike school or consider it a necessary evil on the path to independence? Why do some see change as an opportunity, love the new learning, and thrive under unstable circumstances while others feel threatened, withdraw, and experience negative health repercussions?

I finally returned to academia, received a psychology degree summa cum laude, followed by six years of intensive post-graduate work in human evolutionary psychology and theories of history, and earned a Masters in research and evaluative psychology from Walden University. In the process, I have used my communications and technology knowledge, along with my education in human behavior to teach thousands of students. When I met Paul, we seemed to be looking for the same answers from different directions. We combined forces and this book is the result. I finally feel that I have made substantial progress in my life-long quest and wish to share these insights and scientific findings in hope of creating a better future for our children.

෧෧

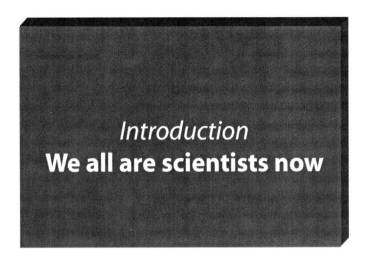

Introduction
We all are scientists now

***Old [scientists] never change their minds,
they just die.***

– Max Planck, Nobel Laureate, Chemistry

Knowledge progresses, funeral by funeral.

– Paul Samuelson, Nobel Laureate, Economics

Sciences mature in stages. At first, we collect isolated observations, looking for patterns that seem to be shared by superficially diverse events. Next, we develop theories that account simply for these shared patterns. These theories *unify* large portions of the world, letting us comprehend and use the patterns and similarities we can now *predict*. For example, we possess theories that unify our knowledge of all trees, all water molecules, or all galaxies.

The biological sciences underwent this kind of unification on a vast scale over the last two generations. As little as sixty years ago, biology was a loose collection of local theories under names like zoology, genetics, physiology, and cell biology. The molecular revolution of the last two generations has produced a powerful and general theory (Chapters 1 and 2). This new perspective pulls all the superficially diverse, older, local theories into a single whole that we can now call simply *biology*.

The incredible elegance and power of this unification of biology is one of the great achievements of the human knowledge enterprise. This achievement does not merely give us electrifying new insight; it also gives us vast new power to improve our everyday lives pragmatically, giving us better medical and environmental science, for example. In fact, many of our persistent problems result *not* from a lack of good *natural* science, but rather from our inability to muster the *social* wherewithal to apply what we know about the material world with real wisdom. We need better social science to go with our beautiful natural science.

In this book, we will argue that we now stand at the inception of another, comparably great unification. The social sciences of the present moment are very much like biology sixty years ago, with areas like sociology, economics, history, psychology, and anthropology all having different local theories, and different specialist jargons. We will try to convince you that we can now develop a simple, powerful theory that unifies all the social sciences into a single coherent whole, something we might someday come to call *human science*. The beauty, elegance, and practical power of this effort, if it is ultimately successful, will be at least as great as for the unification of biology.

Some brave first steps have been taken toward a general scientific theory of what it is to be human. Richard Alexander's *Biology of Moral Systems* was especially important. Also valuable were Matt Ridley's *Origins of Virtue*, Jared Diamond's *Guns, Germs and Steel*, Ed Wilson's *Sociobiology*, Robert Wright's *The Moral Animal*, and the well-done parts of traditional evolutionary psychology. Moreover, some brilliant work in economics, history, and political science have likewise begun to address this pressing need.

We will pick up where these early attempts left off. And we will go much, much further. We are going to argue that we have a very complete new theory of all of human existence. This new theory gives us unprecedented insight into ourselves and the human condition—past, present, future. It also brings all six billion of us actively inside the scientific enterprise.

The difference between scientific belief and all other beliefs is that everything we think we know in science is held perpetually in doubt. Nothing in science is accepted on the grounds of authority, revelation, faith, or longevity. Heresy is not hostile to science; it is the very process of science. *Creative destruction* of earlier belief is vital. This clearing away of dead wood is the way we rescue things of lasting value from the past as the foundation for the future.

When our beliefs become conventional wisdom, no longer susceptible to doubt and creative destruction, they are also no longer science.
To accept settled insight *prima facie* is to destroy what we imagine we
protect—to render it a moribund hulk retreating into oblivion. There
is much heresy in this book. The tool of authentic doubt by the global
community of all people will ultimately tell us whether it is helpful in
creative destruction and useful in new building.

For academics, this theory will be among the fundamental tools of
all future social scientists, if we are right. If we are wrong, the skeptical
young-of-mind (not the calcified old-at-heart) will show us where we
went off.

Theory and doubt – the stuff of science

The goal of this book is to build and test a new theory of how we,
as humans, became so different than all other animals. It is important
to be specific about what a scientific theory is. The word is sometimes
used incorrectly in the popular press. Theory does *not* mean a guess
or a conjecture about how something in the world works. The term
for such an initial guess is *hypothesis*. A theory is a hypothesis that has
survived many rounds of testing against the evidence of the world. A
theory is something in which we can have some real, if still provisional,
confidence. A theory becomes accepted (though never *ever* freed from
perennial doubt) only after years of extensive testing.

An example illustrates vividly the distinction between initial
hypothesis and increasingly established theory. One of the many ways
in which humans are unique is that we are the only animals to have
left footprints on the Moon. Stepping onto the Moon required us to
use the physical laws of rocket propulsion. These laws are theories, not
hypotheses. These very special footprints are one small part of a vast
body of evidence showing that our original hypotheses about physical
laws have matured into good theories.

This example also illustrates what we said in the Preface. All of us,
not just a few scientists, should care about good theories. They give us
the power to affect the world. In fact, they are the *only* source of our
capacity to affect the world on our *uniquely human* scale. *Without* good
theories, we stumble around ineffectually, having only a tiny repertoire
of skills that we learn accidentally in our local environments, the kinds of
skills non-human animals have. *With* good theories, we can leave footprints on the Moon or shape a new human future.

Of course, we now confront an obvious question. Exactly how do we decide when a hypothesis advances to the status of a theory? The details of this process really matter. Two major steps are involved.

First, the theory must make sense. It must be logically coherent, in more formal language. However, making sense is *necessary*, but it is not *sufficient*. Theories that appear to be logically coherent (at least in a narrow sense) can still be wrong. Also, logical coherence is more elusive than we sometimes think. Specious arguments are everywhere. We need more.

Second, the more we need is the answer to the question *How well does the theory work to account for events in the world?* This account-ability to the evidence is vital to the process. It is the practical way in which we doubt our theories. This practical doubting (testing) is the use we make of all the individual insights about the world we extract and examine. It allows us to discard specious arguments and confused hypotheses. It also empowers us to refine hypotheses that are useful but not yet complete.

Thus, the most important tool of science is not a microscope, tele-scope, test tube, or pipette. It is **doubt**. By holding everything we think we know perpetually in doubt we have the best chance of discov-ering how the world actually works. Ironically, the demagogue who is completely confident she (or he) is right is certain to be wrong, while the scientist (professional or not) who holds everything she thinks she knows in doubt has the only real chance the universe offers to see the truth. Theories that survive our doubting day-in and day-out become powerful tools, indeed.

What does a scientific theory look like? It turns out, in practice, when theories are good, they do not *merely* account for events in some local part of the world. They do so with style, *simply and economically*. Such simple, economical theories are sometimes said to be parsimonious. Such parsimonious, but effective theories are also elegant, in the sense of being beautiful. We get the same kick out of a good theory that a connoisseur gets out of great cooking or a dedicated basketball fan gets out of a back door cut or alley-oop dunk.

This love of elegant theory is not arbitrary. It reflects something deeper about the world. Elegant theories simply work much better than inelegant ones when it comes to "putting footprints on the Moon." In other words, the sense of beauty possessed by our (biological) minds was probably shaped by natural selection to help us recognize valu-able things in the world such as good food and good theories. In fact, forming good theories about the world has probably been a human trick for close to two million years (Chapter 10).

Because good theories are the only tools we have to understand the world and only elegant theories turn out to be good, we can make a strong statement. If we do not have an elegant, parsimonious theory for some part of the world (human uniqueness, for example), then we **do not understand** that part of the world—period, no exceptions.

Though a good deal of effort has gone into some brave attempts, a successful (elegant, parsimonious) theory of human uniqueness has been elusive. *However,* we will argue that the theory developed in this book is the first such successful theory. This new theory apparently accounts simply and directly for every important feature of the human story from our origins two million years ago through the present moment.

Testing such a theory by subjecting it to systematic doubt will turn out to be *relatively* easy and we hope you will join us in doing this. In the first several chapters of this book, we will develop the theory in sufficient detail so that our claims are clear and simply defined. This part of the book will also greatly enrich our intuition about how biological creatures (including us) look at the world.

Using these insights, we will proceed to build our theory of human uniqueness. Then, throughout the remainder of the book, we will explore how well our theory can predict the events of our evolution and history. We will ask whether the most interesting things about us (speech and big smart brains, for example) and about our history (the origin of the first humans, the agricultural revolutions, and the rise of the state, for example) are accounted for parsimoniously. Having taken this journey with us, you will have what you need to construct your own tests.

If our new theory is much better at surviving all these tests than any competing theory, then we can begin to consider it correct.[*]

The new human scientists

For most of the approximately four hundred years of the Scientific Revolution, there were good reasons why testing new theories was often the business of professional specialists.[†] Theory testing required hard-to-obtain information about relatively inaccessible parts of the world (the precise orbits of the moons of Jupiter or the growth habits of super-small microbes, for example). Only people with years of special-

---

[*]   Actually, of course, it is taken to be the most likely among available theories. No theory in science is ever taken to be final. It can always be displaced in the future by better theory—perpetual doubt, again.

[†]   Though brilliant amateurs like Ben Franklin have always been important, as well.

ized training, using expensive hardware and complex techniques, could collect this kind of information.

In contrast, the theory we will explore here is *not* about orbital mechanics or microbiology. It is about humans. Thus, we can test the theory by asking whether it predicts how we and the other people all around us look, behave, and feel. Each one of us is *already* a trained specialist in collecting this kind of information. The only hardware we need is our senses and minds. The only training we need is the ability to think patiently, critically, and skeptically. When you have completed the journey through this book, you will find yourself testing the theory every time you read a newspaper, listen to someone speak, sense your own inner self, or watch your fellow humans going about their lives.

The existence of the theory we will explore here means that *everyone* on the planet can now be a functioning scientist, a part of the scientific enterprise. Every human on the Earth today can become an *active participant* (not just an observer or consumer) in the scientific enterprise.

Debates, arguments, and discussions of the merits of the evidence supporting or refuting a theory are the way the theory is evaluated. This is the social process of organized doubt. However, debates about the particular theory we will be exploring will not be restricted to a few trained specialists. They will involve anyone in the world who cares to participate, whether it be in a sidewalk café in Paris, a diner in rural North America, a restaurant in Tokyo, a beach in Lebanon, or a savanna in Kenya*. These global discussions will be the final arbiter of the value of this theory. For the first time in human history, all of us, collectively, will make the decision about whether we believe a specific answer to an important scientific question—the human uniqueness question.

In participating in this process, we will find that we are not just testing a theory. All of us also will be building, revising, and improving the theory as we go. If the theory is ultimately successful, it will be because many contributed to this process. In the long run (the only time scale that matters in science), the value of good science is determined by what it subsequently gives birth to.

The practicalities of testing theories

There are a few practical issues you should have in mind as you evaluate everyday tests of the theory we will explore.

---

* You may wish to join one of these conversations at our website, www.deathfromadistance.com.

The appearance of "apparently": From here on, we will use the word apparently quite a bit. This word has a slightly different meaning in scientific writing than in everyday conversation. Scientists use *apparently* like prosecuting attorneys use *allegedly*. When we say something is apparently true, we will mean we think it is very probably true. However, there remains at least some disagreement among some scientists about whether the matter is likely to be established. Since this book deals with work at the active forefront of contemporary social science and biology, obviously, this use of apparently will often be necessary.

Predictive capacity: When a theory accounts for a specific fact about the world, we will say that it *predicts* that fact. This wording convention in science applies both to events in the future and to past events. In this sense, we will ask whether our theory predicts human language and big human brains, for example.

Empirical evidence: Scientists have an inclusive, generic term for the facts about the world a theory predicts (or fails to predict). These facts are said to be *empirical* evidence for or against the theory. We will use this term often.

Testability (falsifiability): If our tests of theories are to be useful and fair, our predictions cannot be trivial. We must predict things in such detail that it is clear when our predictions are wrong. Predictions that have this useful, clear quality are said to be *testable*. Scientists will sometimes say that such well-crafted predictions are *falsifiable*, meaning that if they are false, we can readily see that they are false.

We will look at many falsifiable predictions of our theory. You should be very aggressive, attentive, and doubtful at these moments. Ask how well or poorly our predictions agree with what you actually see in the world.

Natural experiments: One good way to test theories is to go into a laboratory. In this highly controlled environment we can carry out experiments that are carefully designed to give us (or fail to give us) specific facts about the world predicted by our theory. This method lets us make many different tests—quickly rejecting bad theories and supporting good ones. Much good science in areas like chemistry, cell biology, or experimental psychology is done this way.

However, this approach will not work when the events in the world our theory explains occur on the time scales of millennia such as the events of human evolution and history. Instead, we will have to do the best we can with events that have already occurred (and are occurring) in the world around us. Some of these earlier events left sufficient records that they are useful to us. Such useful events in the world are sometimes called *natural experiments*. What makes a good natural experiment will become more obvious as we go along.

Quantity and quality of evidence: In order for natural experiments to do us any good, we need many of them. The result of a single experiment will rarely convince us to accept or reject a theory. For example, any individual experiment might give us the answer we predict, but for some spurious, unrelated reason of which we are unaware.

However, if we continue to get the answers we expect from many different natural experiments, unrecognized alternative explanations become more and more unlikely until, eventually, we can dismiss them. Fortunately, the natural world and the two-million-year human story present us with many useful natural experiments. We will be able to ask whether our theory is successful in its predictions over and over in different circumstances.

These multiple natural experiments will come in two basic forms. On the one hand, our theory *inescapably* predicts that the two-million-year human story is not a huge sequence of historical accidents. Rather, it is made up of many repetitions of the same underlying fundamental process. For example, the origin of humans around two million years ago, the behaviorally modern human revolution about fifty thousand years ago, and the various agricultural revolutions around the world between approximately eleven thousand and one thousand years ago all look like very different phenomena. *However*, our theory predicts that they are *not* different. Instead, they are just local repetitions of the same underlying process at different scales. The details of this particular prediction will be very useful to us.

On the other hand, we will also make use of many different pair-wise comparisons of humans to non-human animals. Our theory has a good deal to say about how human and non-human animals will be the same and how they will be different. This comparing and contrasting applies to not just obvious things like language, for example. We can also compare our sexual, family, and political behaviors. (Yes, non-human animals have politics, just not on the scale that attracted Aristotle's attention.)

Using incomplete evidence: It is very useful to have a good deal of information about the results of many natural experiments. But a problem is created by the circumstances of collection of some of the older evidence. People originally gathered this information who did not know about the new theory we would be exploring here, many years later. They had no way of anticipating what details we would want them to attend to at the time. Occasionally, they threw out the baby and kept the bath water.*

In addition to this collection problem, the records themselves are often *inherently* fragmentary. Think of fossil or archaeological remains, for example. The upshot of these problems is that our empirical evidence is always incomplete and partially degraded.

This is a serious problem, but not a fatal one. To deal with it, we must ask two questions. One question is if the quality of the fit between the available empirical evidence and the predictions of our theory correlates with the reliability of the evidence itself. Evidence of high quality (good data) should produce good fits with prediction.

We will argue throughout the book that the fit with the predictions of our theory is excellent when the evidence available to us is detailed and reliable. However, we will also find a few partial or ambiguous fits. As expected, if our theory is good, we will find that these ambiguous fits are apparently attributable to incomplete empirical data. We can still make good use of such incomplete data. We can ask whether these fragments of information, *in aggregate,* support the theory. An individual case of partial fit with degraded information is only weak evidence; however, *many* separate cases of such fits become, collectively, very powerful evidence.

We will argue that we have many strong fits to detailed data and these are compelling. However, we will also argue that we have numerous partial fits to less complete data. We will suggest that, *in aggregate*, all of these fits are very convincing. It will remain to you the reader to decide whether you agree.

To professional academics

The internal logic of the new theory we develop here imposes specific interpretations on many kinds of evidence, from paleoanthropology and archaeology to psychology, linguistics, history, and economics. Important support for the theory comes from its unification of earlier, smaller local theories into a larger coherent whole. In effect, this general theory *predicts* much good earlier local theory.

---

* In spite of this problem, we will find that these earlier investigators often performed brilliantly, showing a profound intuition about what was important. Our quest here would be hopeless without this remarkable earlier work—this human empirical virtuosity. The inescapably limited vision of earlier workers will sometimes confine us, nonetheless.

However, in other cases, the new theory implies interpretations that are new—sometimes subtly, sometimes drastically. Such novel perspectives can be jarring, even off-putting to specialists. However, they also hold the possibility of fundamentally new insight. It will be important for academic specialists who wish to engage the theory authentically not to be reflexively skeptical merely because the theory contradicts isolated pieces of narrow, disciplinary, conventional wisdom. Confront the theory in its entirety.

Finally, in view of the vast scope of this project, occasional errors of fact and detail in this first presentation seem inevitable. We invite our colleagues to help us find and correct these for the future. However, we also emphasize that we do not believe that any such errors as might remain are likely to be sufficient to falsify the larger claims we make here. Of course, skeptics are always free to try to show otherwise.

৩

Chapter 1
The journey we will take

What does the nine-month human pregnancy have to do with Wall Street? What does the history of archaic states (like the Roman Empire) have to do with daycare centers? What does your sexual behavior have to do with Amazonian rain forest cultures? What does DNA sequence information have to do with the economy? What does wildebeest biology have to do with human political systems? What does a Homo erectus fossil have to do with baseball? What do handguns have to do with the Scientific Revolution? Why do our minds cause us to say that a person who beats another to death behaved "like an animal"? What does human evolution have to do with war?

The answer to each of these questions is EVERYTHING. Our social, sexual, political, cultural, and economic lives are also our biological lives. Humans are absolutely unique among all Earth's creatures. Our existence presents what scientists sometimes call the "human uniqueness problem." Nevertheless our uniqueness emerges directly and simply from our biology. We are going to explore a fundamentally new scientific theory about exactly how and why this is.

Our journey will commence at the birth of the solar system about five billion years ago. After briefly exploring only the few things we need to know about the emergence of organisms in general, our careful attention will turn to ourselves as a very specific new kind of organism. This exploration of ourselves will begin with the rise of the first humans around two million years ago. We will then turn to the essential features of our two-million-year history, from the emergence of language, our unique sexuality, and our powerful minds to events like the invention of agriculture and the rise of the first states. Our quest will not stop until we stand in the present instant looking forward into the human future. We will attempt to explain all the important things encountered along this journey as a single, coherent whole with no missing pieces, no magic, no hand waving.

Once this tour is completed, we will argue that we can comprehend the place of humans in the world with an entirely new totality. A far better understanding of our origins, our properties, our history, and the contemporary world will be our prize.

Once we grasp the concept of our real biology, we will comprehend ourselves with a stunning new simplicity. Arguably, we will understand ourselves for the first time. Much of what our exploration uncovers will be surprising. Though we will not always be pleased at first glance by what confronts us, we will ultimately find a new, deeper respect for human life, a vastly enlarged hold on our common humanity, and a realistic hope for a more humane future.

At the end of the book, it will be up to you to decide if this ambitious purpose has been fulfilled.

What is our theory?

Before we outline the theory that will (ostensibly) give us this new level of insight, be aware of two things.

First, you will, most likely, not fully understand the theory initially. Its reach is wide and far. Though the theory is ultimately simple, full assimilation of its insights and implications can only emerge gradually as we proceed.

Second, aspects of this picture will seem ethically disturbing at first. Be patient. Followed to its logical conclusion, our theory is richly humane. It puts our common humanity and our need for an ethical vision of our lives on a firmer footing than we have ever had before.

What is our theory of human uniqueness? It emerges from the fact that all biological creatures have what we can call *conflicts of interest.** What this phrase means is that all organisms have an incentive to compete with one another for access to the scarce and crucial assets we each must have from the world to survive and reproduce. These conflicts of interest normally limit social cooperation between

---

*   These conflicts of interest result directly from the fundamental physical and chemical nature of all biological creatures, as we will see in Chapter 2.

non-human animals to very close kin—parents, offspring, and siblings. Animals compete intensely with all *other* (non-kin) members of their species almost all the time.

It is helpful to visualize *conflicts of interest* from a simple, everyday human point of view. We all have conflicts of interest with merchants who sell us things we need, like food, for example. If we could take food without paying, we could get more food or more of other things we might also need. On the other hand, the merchant will not be able to buy the things she (or he) needs for herself and her family if we shop lift. Our interests and the merchant's interests are in conflict. (We rarely think consciously about these facts, but they exist nonetheless.)

Of course, we have a legal system, complete with law enforcement, that usually prevents us from stealing the merchant's food. If we steal, we will be prosecuted and incarcerated, suffering a much larger cost than would be justified by the food we might realistically be able to steal. We anticipate all this, though usually not consciously, and we are thereby prevented from stealing. As a result, food stores work, the merchant survives, and we eat. This legal system will be very important to us in a minute, but we need to understand three other things beforehand.

First, you might be thinking something along these lines. We are aware that, if we steal, the merchant will go out of business and we will not be able to obtain food from her in the future. Thus, even if we did not have a legal system, we still would not steal. Stealing would ultimately be futile.

But, for a moment, imagine there is no legal system. Now, suppose there are just a few individuals who choose to steal in spite of its being ultimately self-defeating. They will have more food now and do better than we do. Soon others will see that the thieves are doing better and join them. Ultimately, even you and we would be forced to steal to feed our children. We would have to seize some of the food before it all is stolen by those already choosing stealing over buying. In short order, everyone is forced to steal, the merchant goes out of business, and we are on our own for food. The fact that this outcome is stupid and self-defeating has nothing to do with it. This is precisely the way non-human animals live, and it is exactly their inability to prevent "stealing" (to control conflicts of interest) that forces them to live this way.

Second, you might be thinking that these rules for a trip to the grocery store might be interesting, but they are local and trivial. Surely, they cannot be the basis of a vast theory of everything it means to be human. If you think this, you are perfectly wrong. The laws of inertia pervade everything about our physical world, from a dribbled basketball, to a landing passenger aircraft, to the path of the Earth as it orbits the Sun. Likewise, conflicts of interest pervade every crevice of every event in the social lives of every creature, always. Everything about the social lives of human and non-human animals is completely determined by conflicts of interest and their immediate implications, no exceptions. Conflicts of interest are the central *force of nature* throughout the social world.

Third, alternatively, you might be thinking, "rubbish, social behavior at the grocery store doesn't work this way. People would not put up with stealing even if there wasn't a formal legal system." You are exactly right. **People** usually do not put up with stealing and other such behaviors, formal legal system or not. But non-human animals do. In other words, humans usually control and suppress conflicts of interest and non-human animals almost never do.

The question is why? The answer is simple—cost. It is too expensive for individual non-human animals to participate in control of conflicts of interest. In contrast, for humans, law enforcement is cheap and a good investment. It has become an innate natural behavior for us, one we usually take unconsciously for granted.

How did this difference between humans and non-human animals come to be? The ancestors of the first humans evolved, inadvertently, the capacity to kill or injure *conspecifics* (members of their own species, fellow humans) from a substantial distance. These ancestors could kill *remotely*, from many body diameters away. This ability arose, in turn, from the evolution of human virtuosity at accurate, high-momentum throwing. No previous animal could reliably kill or injure conspecifics remotely.

This novel physical virtuosity at throwing probably evolved at first as part of a local professional hunting or scavenging adaptation.* However, this elite throwing had unexpected, revolutionary implications for the evolutionary future of this new animal.[1] This unprecedented remote killing capability permitted multiple individuals to simultaneously project violence at vastly lower cost than in any previously existing creature.†

---

*   We will explore this in Chapter 7. For now, it is sufficient to recognize that humans throw the way a cheetah runs or a dolphin swims—with elite skill. We are "born to throw." Elite human skill at throwing has been recognized by many authors, from Darwin to William Calvin; however, before now, we have not had a sensible theory about why human throwing might be central and important.

†   The capacity to project violence cheaply also means that the *threat* of violence can be projected believably or credibly. In practice, day-to-day, it is threat that is more commonly employed than actual violence.

As we will see in Chapter 5, these reductions in cost are huge and highly significant. Thus, a radically new adaptive opportunity was *inadvertently* created. For the first time in the history of our planet, an animal came into existence who could suppress the conflicts of interest between non-kin (unrelated) conspecifics at low cost. "Law enforcement," thus, became a self-interested behavior for the original human individuals.

For the first time, natural selection could now "reward" individuals who actively suppressed conflicts of interest **in** others and responded to this suppression **from** others.* Not putting up with thieves became biologically adaptive. This drastically altered social environment is the one we have inherited and know so well. We will refer, collectively, to all the many new social behaviors that evolved as a result of this environment as **kinship-independent social cooperation**.

To be very clear about what are we saying, return to the merchant selling us food. We are saying the cost of law enforcement that prevents thieves from bringing down our local economic system is, in fact, the very thing that determines whether we have that economic system in the first place. Ancestral humans evolved a new ability that brought these enforcement costs down drastically, and uniquely human economic systems (kinship-independent social cooperation) became possible. Of course, we mostly do not throw stones to enforce the law these days. We prefer newer projectile weapons when we can get them. But the underlying principle used today remains precisely the same as it was in the ancestral environment.

When we go shopping or act as merchants, we are engaging in a behavior that is both ancient (probably about two million years old) and

---

* *Reward* is a metaphor. Natural selection is a blind mechanical in-the-moment process, not a conscious forward-looking one. Of course, we are referring to natural selection and evolution here, not how you and I appear to make decisions in the present. The important terms are as follows. *Selection* for a behavior results when individuals who have the behavior leave more offspring who, in turn, inherit that behavior. Over many generations, this selection causes such a beneficial behavior to become more common until it is characteristic of most or all animals in a population. We refer to a behavior that is beneficial in this way as an *adaptive* behavior. Such behaviors are elements of *adaptations* (which also include adaptive changes in physiology and anatomy). By definition, a behavior is said to be adaptive because it increases the likelihood of survival and reproduction of individuals in a local environment who display that behavior.

Adaptive sophistication refers to the relative competitive competence and ability of an animal or a person. For example, a person with access to complex tools like a computer or an assault rifle has a substantially higher level of adaptive sophistication than an illiterate ancestral human hunter whose most complex tool was a bow.

We will enrich our understanding of biological evolution, selection, and adaptation in Chapter 2 and beyond.

uniquely human. This novel way of getting along in the world is possible precisely and solely because we can afford not to put up with thieves. We can afford to control conflicts of interest. *Everything* else that distinguishes us from other animals flows from this utterly simple trick of enforcing social cooperation in spite of conflicts of interest.

At this early point in our journey together, many readers will be thinking that this picture is surely either incomplete or utterly incorrect outright. If you are thinking either of these things, you are almost certainly wrong. And in being wrong, you are closing off all possibility of understanding what it means to be human, how we got here, and why our history looks as it does. Our task throughout the rest of the book will be to try to show you that this simple picture of humans is not only correct and complete, but also startlingly, pervasively powerful.

This simple picture of human uniqueness will bear enormous fruit. Individual adaptations to coercively enforced kinship-independent social cooperation consist of **the entire suite** of traits we think of as uniquely human. These traits include complex language, large brains/powerful minds, our unique sexual and child-rearing behaviors, and our elaborate ethical and political sense. Just exactly how these human traits emerge simply from access to cheap coercion will not be obvious yet, but it will become clear in later chapters.

But there is more. Not only do we have a theory of our individual human properties and their evolution, we have a powerful new theory of history. We can outline this new way of understanding how human history apparently works for you right now if you consider a few more details.

First, the larger a cooperative collection of humans is, the more culturally transmitted expertise it can store. Larger social aggregates also can support much more individual specialization, permitting more new expertise to be discovered. As a result of effects like these, the scope of human capability (adaptive sophistication) is entirely dependent on the scale of our social cooperation. We will find that the major advances throughout our two-million-year history all resulted from increases in the scale of our uniquely human social cooperation.

Second, conflicts of interest are to social behavior as gravity is to astronomy. They are the central fact of our social existence, at every scale, large or small. For example, different nations have conflicts of interest with one another just like non-kin individuals do. Cooperation between either individuals or nations (or collections of people of any size) requires management of conflicts of interest *on the scale in question*. Thus, the scale of our social cooperation and, therefore, our adaptive sophistication, are determined by the scale on which we can manage the conflict of interest problem.

Third, the weapons that would make law enforcement feasible (cost-effective) in a local neighborhood (a police handgun, for example) are not the same as the weapons that allow practical law enforcement among nations (cruise missiles, for example). Thus, the scale of human social cooperation can increase over time, *but only* as new coercive technologies are invented that permit *cost-effective* law enforcement (control of conflicts of interest) on the new scale in question. Thus, adaptive progress will always await and flow from the introduction of new coercive technologies.

We will argue that this simple causal chain will prove to be a startlingly powerful and complete theory of history. All the important features of our two-million-year human journey are consequences of this single causal process.

Let us return to our ancestors and look at how these fundamental facts about the world have apparently determined the course of human history.

The original increase in adaptive sophistication from elite throwing in the first human ancestors produced biological evolution of new capacities. It also ultimately allowed diverse technical innovations. Eventually, among these innovations were new means for suppressing conflicts of interest on ever-larger scales with new weapons. As creatures now highly adapted to the coercive suppression of conflicts of interest, humans inevitably exploited these new technical means in pursuit of individual self-interest, precisely analogously to the smaller scale cooperative behaviors of their ancestors (originally sustained by elite throwing).

This pursuit, in turn, inevitably produced ever-larger cooperative social units, ultimately including enormous numbers of individuals. The resulting increases in scale of human social cooperation produced new adaptive revolutions.* This process was inherently autocatalytic. New adaptive sophistication produced further improvements in coercive technology, producing still further adaptive revolutions.

---

* An *adaptive revolution* refers to the relatively abrupt acquisition of dramatically increased adaptive sophistication. We will see in later chapters that the important features of human history can be described as a series of adaptive revolutions. For now, what matters is that human adaptive sophistication is limited almost exclusively by the scale of our kinship-independent social cooperation. So, if the scale of this social cooperation increases, our adaptive sophistication will increase, producing a new adaptive revolution.

This ongoing process eventually became relatively rapid by the standards of traditional biological evolution. Moreover, it produced a long sequence of ever more sophisticated adaptive revolutions in the two million years since the evolution of the first humans.

These revolutions represent all the major transitions in human history including the behaviorally modern human revolution, agricultural revolutions, the rise of the archaic and modern states, and the currently ongoing consolidation of pan-global human cooperation. Each of these adaptive revolutions has precisely the same underlying logic on our new theory—the self-interested application of *relatively* inexpensive coercion resulting in sustained social cooperation. This fundamental logic is merely applied at ever-larger scales with each historical transition.

The spectacular adaptive abilities conferred upon us by the huge scale of our contemporary social cooperation are the primary reasons that contemporary humans seem so totally different than non-human animals. While it is true that you and we are individually smarter than non-human animals, we are not nearly as individually smart as our personal conceit tempts us to believe. Rather, it is our capacity for cooperation with huge numbers of other people that really lifts us above the rest of the biological world, as we will explain in later chapters.

The fact that historical human adaptive revolutions are by-products of the *self-interested* application of coercive power has other important implications. First, human societies do not necessarily serve the interests of *all* their members equally. Rather, they are expected (inevitably) to serve the interests of individuals in proportion to the coercive power they exercise. Depending on historical happenstance and the properties of technologies, coercive power can be widely, democratically distributed or concentrated in the hands of a few. The resulting human societies reflect these distinct distributions of coercive power in predictable ways.

Second, interest groups within a local human society who hold decisive coercive power will resist the access to new coercive technologies by other disenfranchised individuals. The struggles to acquire and deploy new technologies for asserting coercive self-interest in the face of older entrenched interests are variously lengthy and chronic or cataclysmic and violent.

The "ideological" rationales (economic, religious, political, or ethnic) for these struggles prove mostly to be persiflage generated by our evolved human ethical/political psychologies. This ongoing competition for naked coercive dominance is the real story of history and a game our ancestors had no choice but to play. These struggles, together with the growth of knowledge/expertise with changes in the scale of social cooperation are the actual, substantive processes of historical change. These two processes (struggle for coercive dominance and accrual of knowledge) are the sources of essentially all the rich local color and superficial (false) appearance of complexity in the flow of human history.

You may want to reject these strong, simple claims about the nature of human history here at the beginning. If so, our task through the rest of the book will be to challenge your skepticism. As well, you may find this view of history grim and foreboding. However, we will see in later chapters that there is an alternative to this perspective. Coercive dominance by a democratized global *coalition of the whole* of humanity is probably an achievable outcome. Such a world holds immense promise both materially and ethically.

In summary, our theory is apparently complete. It is a theory of human origins and of our unique properties as biological creatures. Moreover, it is also apparently a theory of human history and social organization of unprecedented economy and scope.*

The rest of the book will build this theory in detail, test the theory against the evidence from many different areas of the knowledge enterprise, and explore its numerous and diverse implications. Scientifically sophisticated readers who are prepared to dive into constructing the new theory can proceed directly to Chapter 2. The remaining sections of this chapter are for readers who still have some intellectual or ethical concerns about our objectives or methods. If you are not comfortable with the natural scientist's definition of *reductionism*, for example, you may want to read the rest of this chapter.

How big is our problem?

We have set ambitious goals. In order to achieve them, we need a clearer understanding of the uniqueness of humans. We really are very different than other animals. For example, we all know that Dr. Doolittle conversing with the non-human animals is a fantasy. Humans talk, but animals do not seem to communicate extensively in this way. Moreover, we travel in space; other animals do not. We are

---

* It will even emerge that features of the contemporary economic crisis (2008/2009) are predictable on our theory, as are possible steps to reduce such problems in the future (Chapter 17).

also aware that humans seem to be much smarter than non-human animals. Being smart and talking are important, but they are just two of our many unique features. We often do not fully appreciate just *how* different we are. We lack perspective because we live *inside* a completely human world, in a human monoculture, so to speak. But think about the following things.

Chimps are our closest living relatives. They are among the animals most like us, **yet** there is a vast chasm between their properties and ours. The epitome of chimp engineering is to strip a twig and stick it into a hole in a termite mound, fishing for a few insects to eat. Humans can engineer enormous sets of tools allowing us to fly directly from New York to Tokyo in a few hours, visualize structures ten thousand times smaller than a hair, or peer billions of years back into the life of the universe.

A talented chimp can break open nuts with a stone. Humans can mimic the nuclear fusion reactions that power the Sun, setting off titanic thermonuclear blasts capable of vaporizing whole cities.

Chimps have several calls including a pant hoot. Humans can sing opera, recite Shakespeare, or write the Gettysburg address.

Chimps are defeated if they have to count to thirty accurately. Humans can count the number of water molecules in an ice cube— about 54,827,952,000,000,000,000,000.

Our differences from other animals matter. They give us near total dominance of the biological world. Consider two examples among millions. Tens of millions of buffalo once roamed the American Great Plains, eating the grasses that grew wild there. The grasses evolved to prosper in the presence of the buffalo and vice versa. Humans drove the buffalo from the face of the continent and replaced *their* grasses with *our* grasses—a world-feeding, continent-sized field of wheat and corn. Buffalo shaped their grasses inadvertently. We shaped ours by conscious design, and our grasses can only survive with our constant, planned, mindful supervision.

Some animals live on the arctic ice, others in the sub-arctic tundra, and others in the forests and grasslands of temperate zones. Still other animals live in tropical and subtropical deserts, savannas, and rain forests. Humans live in **all** these places and the other animals in *each* of them survive (or die) mostly at our pleasure.

We need to account for all of these differences and all of these consequences. We need a *theory of human uniqueness*. Since Darwin, many attempts, some clever and brave, have been made to produce such a theory. All have failed. These earlier theories and the reasons for their failures are too many to review in detail. But two examples are illuminating.

It has been proposed that complex language distinguishes humans from the other animals. On this view, all our other remarkable properties result from the fact that we can talk to each other. This theory fails for several very good reasons. For one, it merely restates the question. If language distinguishes us from other animals, why are we the only animals to speak? For another, we can speak to one another for mutual benefit, but we can also use language to deceive and manipulate one another. Our dominance over other animals arguably comes from our capacity to cooperate. Why do we use language (often) to cooperate and not (always) merely to manipulate? A good theory must answer these questions. Ours will.

Alternatively, it has been proposed that our large brains and powerful minds provide the essential difference between us and other animals. On this view, language and all our other unique properties are somehow made possible by our unique minds. This approach suffers from precisely the same problems as the language first hypothesis. It restates the question of why are we so smart, rather than answering it. Likewise, it fails to account for why we also use our powerful minds to cooperate (often) with non-kin and not (always) merely to compete and manipulate.

We can, in fact, make bad theories like these appear to work. However, making them work requires that we make additional *ad hoc* assumptions. Such inelegance is not the sign of a good theory (Introduction). We will argue in later chapters that these complicated, inelegant theories are clearly wrong. In addition to the problems already mentioned, we will find that these and other earlier theories misidentify effects as causes.

Our challenge is to do better. We can. And the rewards will be substantial.[2]

Success will also bring us some indirect and unexpected rewards

We have forecast that a good theory of human uniqueness will give us unprecedented understanding of the human world. However, by being able to "subtract" these uniquely human parts of our nature, we will also discover how we are *like* other animals. These similarities are many, profound, and pervasive. These universal animal properties are just as important as our uniquely human features to authentic self-understanding. Our universal animal bits have been fairly well understood by biologists for several decades; however, a new, clear understanding of how our universal animal parts interact with our uniquely human parts will enrich us deeply.

Consider the following details of our existential state. They emerge at the interface between our uniquely human and universal animals parts.

Young adults give birth to infants and spend a large fraction of the remainder of their lives providing everything these youngsters need to survive and grow to adulthood. The first generation of adults dies and the youngsters-grown-to-adults carry out the cycle again, as do their offspring and theirs. Within four or five generations, their descendents often do not even know their names, let alone who they were and what they did. Remarkably, rather than considering this utterly futile, we often consider having and raising children one of life's deepest satisfactions.

We fall in love. If our feelings are reciprocated, we are euphoric. Life is redolent with purpose and transcendent joy. If our feelings are not reciprocated, we are crushed, temporarily destroyed. Life seems value-less. We are the same person in both situations.

A man has sex with his mate. Though she might prefer otherwise, he finds it difficult to have sex with her twice within an hour. However, once, he has sex with both his wife and her girlfriend together. He finds that he is ready to have sex with her friend within seconds of his orgasm with his mate. His mate is annoyed and he is mystified and pleasantly surprised. On another occasion, he and a male friend cooperate to have sex with his mate. He is able to have sex with her over and over in a period of an hour, interspersed each time with his friend having sex with her. Again, he is mystified. This time his mate is pleasantly surprised.

Our theory will give new perspective to these and many other things about us.

Is that all there is? The nature of reductionist, materialist explanation

What does it mean to construct a scientific theory of human uniqueness? We explored some facets of this meaning above and in the Introduction. However, it is now time to add a new element to our picture of good theories. The universe turns out to be organized in a very special way. We have learned to recognize this organization over the approximately four hundred years of the Scientific Revolution. This organization is not merely *reflected* in scientific theories; it is the very thing that makes science possible. The existence of this property of the universe also is a thing of immense beauty.

We can understand this property of the universe by defining a widely used phrase—*levels of complexity*. Though slightly fuzzy, this phrase is precise enough to be useful. We will take it to mean that the universe is organized hierarchically and that each level in the hierarchy has properties best described at that level.

An excellent analogy to what we mean by hierarchical organization is written language. When we write, we assemble elements of the first hierarchical level or the first level of complexity, letters, to create elements of the second level, words. We then assemble words into elements of the third level, phrases, and, in turn, assemble phrases into fourth level elements, clauses and sentences. Fourth level elements are assembled into fifth level elements, paragraphs, which are assembled, finally, into sixth level elements, documents like the Gettysburg address or this book.

Several things are important to notice about written language if it is to serve as a useful analogy. First, each level of complexity or organization in the hierarchy emerges from assembling combinations of elements from the level below. This feature or property is called **combinatoriality**.

Second, combinatoriality allows tremendous quantitative accumulation (accretion) of complexity at each level. For example, a hardcopy dictionary on our desk has about one hundred forty-five thousand entries. Given that this outdated 1996 unabridged dictionary lacks such important words as *muggle, cell phone, truthiness,* and *blowback,* this is a conservative estimate of the number of English words.

Nevertheless, this enormous collection of words (a level of hierarchy or complexity) is assembled from a mere twenty-six letters—the elements of the next level down. Of course, these many words can be assembled into an effectively infinite number of phrases and sentences. For example, some of the sentences in this book appear here for the first and last time in the history of the universe. More important still is the infinite variety of a written literary tradition. Documents from Hobbes, Faulkner, Shakespeare, Snoop Dogg, or Stephen Colbert are each utterly different compositions.

Third, all this complexity is assembled according to a few simple rules making up the standards of spelling, grammar, and clear writing.

It is important to recognize that the hierarchical organization of written language gives us profound complexity, BUT it does so simply and transparently.

The hierarchical organization of the natural universe gives us precisely these same things—profound complexity emerging with simple transparency.

Before looking at the universe more carefully, a word of caution. The analogy with written language is incomplete in one important way. For example, letters are "designed" to be assembled into words; words are designed to be assembled into phrases; and so on. From the bottom of the hierarchy to the top, written language is (ostensibly) an integrated whole.

In contrast, the levels of the hierarchical organization of the physical world have no such larger purpose. Each level merely arises because it does or it can. Each level has its own internal logic and its own absence of purpose unrelated to the absence of purpose of levels above or below.* As long as we keep this crucial difference in mind, our analogy with written language can serve us well.

Exactly how are the levels of complexity of our world organized? If we were subatomic particle physicists, we might be interested in the properties of individual electrons, protons, and neutrons. If we were atomic physicists, we might be interested in the properties of individual atoms made up of combinations of copies of protons, neutrons, and electrons. In other words, atoms are assembled combinatorially. If we were chemists, we might be interested in the properties of atoms as they react to form chemical bonds with one another producing molecules made up of multiple atoms—combinatoriality, again. If we were biochemists, we might be interested in the interactions between large molecules, each made up of many atoms.

Subatomic particles, atoms, and molecules each constitute a different level of complexity. Each level is assembled from the level below, analogous to assembling letters into words into phrases and so on.

Moreover, we can and will extend this hierarchy upward through a number of additional steps, ultimately arriving at human societies. These societies will represent a new level of complexity above individual animals and non-human animal societies. As we discuss these issues, we will sometimes use *level of organization* or *organizational level* as synonyms for *level of complexity*.

Defining levels of complexity is not merely a descriptive or rhetorical convenience. When properly done, this parsing of the world apparently reflects how causality is actually organized in the physical universe.[3] Thus, when we explain the properties of any level of complexity in the universe (for example, molecules) we invoke the properties of the level of complexity immediately below (atoms) and so on and so on. When

---

*   This word of caution requires another. For reasons we will discuss in Chapter 2 and beyond, we humans habitually see "purpose" in the organization of the physical universe. This is an illusion produced by our "purposeful" evolved minds. It is not a fact about the universe.

we go through this process, we are constructing a **reductionist explanation** of the level of complexity in question.[4] We will use reductionism and reductionist explanation in this specific sense throughout this book. Our goal will be to construct a reductionist explanation of humans and of human uniqueness.

Reductionist explanations are actually far richer and more powerful than our brief discussion so far would suggest. Indeed, these explanations are the source of all the spectacular successes and profound intellectual beauty of the scientific enterprise.[5]

Reductionist explanation **IS** the scientific enterprise. All the tremendous successes of science, **without exception**, resulted from the discovery and development of reductionist explanations for ever-increasing portions of the universe and its components. Indeed, many scientists (authors included) would argue that the phrase *reductionist explanation* is redundant. We argue that *all* valid explanations are reductionist in the sense above. Equivalently, the phrase *non-reductionist explanation* is an oxymoron, on this view. *Non-reductionist* might be applied to a **description** or a **tautological statement**, but **never** to an authentic explanation.*

It is very important to notice one other implication of this story. Not only must scientific explanations be reductionist, they also must be simple and transparent. The simplicity of reductionist explanation results from the fact that the properties of one level of complexity results from a *small subset* of the properties of the level below. For example, the properties of atoms result from a small subset of the properties of subatomic particles. Likewise, animal societies result from a small subset of animal behaviors. Reductionist explanations, therefore, remain simple, no matter how complicated the level of complexity they explain.

This requisite simplicity is sometimes called the *front-of-the-tee-shirt rule*. It states that a mature scientific theory should *not only* explain a big chunk of the world (an entire level of complexity, usually) *but also*, it should be simple enough to be written on the front of a tee shirt.

A great example of the front-of-the-tee-shirt rule is Newtonian mechanics—three laws of motion and a single law of gravitation. Each can be written as a simple algebraic equation and all four equations fit easily on the front of a tee shirt. If what purports to be a scientific

---

* For example, to say that the wetness of water somehow "emerges holistically" from many water molecules rather than from the aggregate consequences of the individual molecules is a non-reductionist (and non-useful) explanation (Chapter 2). To say that fire results from a fire spirit being chased from the fuel is, likewise, a non-reductionist (non-useful) explanation.

theory is too complicated for the front of a tee shirt, it is probably wrong outright or, at least, unfinished.

Thus, the point and the beauty of science is simplification. It allows us to find and understand the fundamental simplicity that underlies the superficial, misleading appearance of overwhelming complexity we experience when we first confront some novel level of complexity in the universe. For example, many features of the complicated night sky are explained by the Newtonian mechanics we can write on the front of a tee shirt.

We will find this transcendental beauty and profound understanding yet again, as we build our reductionist theory of human uniqueness. So when a scientist answers the question about reductionism posed in the title of this section "Is that all there is?", she/he will answer, "Yes, but it is a great deal, indeed."

Fear and a theory of humanness

Our use above of the title of Peggy Lee's famous song *Is that all there is?* in relation to reductionism has an additional purpose. It captures the reaction many people (including some scientists) have when first confronted with the prospect of a reductionist theory of humans and human uniqueness. We tend to feel that such a theory would diminish us, make humans somehow less important, less remarkable.

Let us call this the *first fear* of a science of humanness. We have lived with a good reductionist theory of human origins for years. We can assure you that the opposite is true. A clear understanding of ourselves and of our origins only enhances our respect for and wonder at our common humanity. This understanding will emerge for you as you progress through this book.

Another, darker sense of fear is also impeding our progress. This is the fear that an objective understanding of ourselves and our history might somehow undermine our capacity to make humane ethical judgments. The definition of right and wrong might be taken out of our hands. We might somehow be forced to agree that "nature red in tooth and claw" is all there is. This fear is salient to us as we emerge from the 20th Century with its vast atrocities.

This *second fear* is also misplaced. While science is not a direct source of ethical judgments, it can inform them. A clearer understanding of our place in the universe will refine our grasp of our ethical frame of reference. More importantly, a complete biological theory of humanness will define the origins of our worst atrocities much more clearly, arming us to effectively confront and prevent similar events in the future. We will also see that the central point of uniquely human (and humane) cooperation is to raise us above and beyond the endemic violence of the non-human biological world.

Real knowledge is opportunity, here as everywhere.

∽

Chapter 2
We know what life is –
a special case of chemistry

This chapter will summarize what kinds of physical and chemical things organisms actually are. Common properties are shared by all organisms—including human organisms—that we will need to understand.

The selected details we examine will all be very familiar to readers knowledgeable about biology. However, even biological sophisticates tend to forget the material nature of organisms when we think about ourselves. It is important to be reminded.

For non-biologists, this elementary information may seem both difficult and esoteric at first glance. It is neither. It is simple to grasp if you invest a little effort and thought. Moreover, it is central and vital. We cannot understand human uniqueness without this foundation.

Our goal is to define the reductionist, materialist explanation of the **individual** *organism. This picture will show us the important properties of the level of complexity immediately below the level of social interactions* **between** *individuals. We will come to the powerful, but surprisingly simple logic of these social interactions, in turn, in Chapter 3. In Chapter 4, we will see how this logic of social behavior applies specifically to us as humans.*

Ours is a "chemical" world

Ours is a most unusual planet. The contemporary Earth (Mark II) and our Moon apparently arose from a spectacularly improbable collision between Earth Mark I and a planet sometimes referred to as Orpheus. Together with the unusual chemistry of our solar system, this collision created a planet that may be unlike any world anywhere in our galaxy or even in the entire universe.[1]

This incredible story has two implications. First, we probably own the galaxy and, possibly, the entire universe. Understanding ourselves takes on a new urgency in view of this insight. Second, all Earth's organisms, including you and the authors, are the products of this unusual planet. Our task, at the moment, is to take only from this story those details we need to understand the unique status of humans among Earth's progeny.

We are chemical systems, chemical children of the improbable chemical and physical system, Earth Mark II and its Moon. To understand this story we need to recall briefly a few of the details of what we mean by the word *chemical*.

Most of us know that chemistry emerges from the combination of three sub-atomic particles to form what we call atoms. The diversity of atoms in our world is produced in the same way the diversity of English words is produced with an alphabet of merely twenty-six letters—combinatoriality (Chapter 1). Specific combinations of protons, neutrons, and electrons generate specific atoms with specific and predictable chemical properties. More than one hundred types of atoms are known, each with its own set of chemical properties. The same simple combinatoriality produces them all.

Of course, if our focus were the reductionist explanation of chemistry, the many interesting details of this story would be important to us. However, our focus is on the next-highest level of complexity, organisms as a specific class of chemical system. Thus, only a small subset of the properties of the chemical level of complexity need concern us.

First, the chemical properties of atoms result from the interactions of the outer layer of atoms (their outer electron shells) with one another. These properties of atoms allow them to bond with one another by sharing electrons.

Second, many atoms can simultaneously form bonds with two or more other atoms, allowing the formation of large chains and complex networks of atoms. When this happens, stable large aggregates of atoms called molecules can be produced.

Third, molecules can interact reversibly with one another through weak chemical bonding between combinations of their atoms, allowing one molecule to influence the behavior of another on a hit-and-run basis. The rules for these interactions also emerge from the behavior of electrons and are reasonably well understood.

Most atoms in our everyday world are in molecules (or molecule-like solids). The details of sharing electrons within these molecules determine how they absorb or transmit or reflect light, how easily the solids they form bend or break, whether these breaks form smooth or rough surfaces, how well or poorly these solids conduct heat or light, whether copies of the molecules dissolve in the air or in saliva and interact with our smell or taste receptors, and so on and so on. All the properties of the world we experience are inherently chemical.

Do not be distracted by the seemingly **endlessly diverse, yet mundane** character of the "stuff" of our world—the smell of coal tar, the hardness of ice, or the color of dirt, for example. Our understanding of these things is reductionism at its best, and most elegantly beautiful. The way our physical world looks, tastes, feels, and smells is **entirely determined** by the properties of its component molecules, whose properties are **entirely determined** by their component atoms, whose properties are **entirely determined** by their component subatomic particles.[2]

Moreover, at each of these levels, complexity emerges simply, combinatorially. All the properties of ice or dirt or organisms are ultimately and completely determined by the properties of combinations of multiple copies of *only three subatomic particles*.

Pause and reflect on this for moment. Allow yourself to "get it." Even to professional scientists long inured to them, these facts about our world are utterly shocking, and perfectly satisfying.

The "youth" of Earth Mark II – chemistry on a "gifted" planet

If life is just a particular case of chemistry, just exactly how did it arise on Earth Mark II? In fact, we have known the answer in a very general sense since Darwin and many investigators have contributed important additional insights in the ensuing century and a half. However, the molecular revolution in biology over the last two generations has spectacularly improved our understanding of the *details* of this story.[3]

During the gestation and birth of our particular solar system, heavy atoms were relatively abundant (including carbon, nitrogen, and oxygen among others). Radiation from various sources, including the dying stars that ejected the matter into the cloud that would ultimately form our solar system, drove many chemical reactions in this material. Atoms reacted to form molecules of many, many types.

All this chemistry produced enormous quantities of what we will call *small molecules*. These molecules contained atoms numbering from two to around twenty or thirty in the cases that will concern us. More-over, some of these small molecules were produced over and over.

Once this material rained onto the surface of the young Earth Mark II (probably mostly from the comets that also delivered some of the water in our oceans), it formed a massive chemical system undergoing more and more chemical reactions.

There were probably so many of these molecules during the youth of our planet that they turned the ocean into a soup containing the water-soluble members of this group. Moreover, a truly stupendous "oil slick," containing the less water-soluble (oil-like) small molecules, covered this soup. This pan-global oil slick was probably many feet deep!

The young Earth Mark II resembled a planet-sized French onion soup.* This planetary soup contained uncountable billions of copies of the most frequently cooked up small molecules.

Of all the billions and billions and billions of chemical reactions of the young Earth, almost all are completely irrelevant to us here. However, a tiny, tiny fraction produced a very special class of molecules. These molecules were *polymers*—many-mers. They were long, linear strings of several similar (but subtly different) small molecules, like pop-beads of different colors making up a long linear chain. Each bead is referred to as a monomer—a one-mer. These molecular beads were joined to one another by a single chemical bond between one atom in one of them and another atom in the other.

The monomers of these polymers were among the more abundant of the small molecules present in the original planetary soup. In fact, many trillions of trillions of polymers of this general type would have formed and been destroyed in the volcanic heat and solar radiation churning the oceans of the young Earth Mark II. But, again, almost none of these polymers matter to us except one (literally one!). This one we will call the Universal Parent.

The Universal Parent was different from all the billions of billions of other polymers in the planetary French onion soup of the young Earth Mark II, in **only one** respect. Its chemical structure allowed it to fold up into a small molecular "machine" that could recruit from the surrounding soup other copies of the monomers that made it up and assembled them into new copies of itself.†

---

*   Actually, this analogy is a little wide of the mark gastronomically. In reality, the toxic emanations from all of this cookery would probably have killed us rapidly. The young Earth Mark II was the ultimate "Super Fund site."

†   Two pieces in the popular periodical *Scientific American* summarize some of our current picture in accessible form. These are R.E. Dickerson's 1978 piece (volume 239, p. 70) and Tom Cech's 1989 piece on RNA "organisms" (volume 255, p. 64).

In fact, the Universal Parent probably did this by treating its own sequence of monomers as a guide or template to making new copies. In practice, a growing linear copy of the Parent was aligned with the sequence of the Parent molecule itself and each new monomer was added to the growing copy by virtue of its interaction with the corresponding monomer on the Parent polymer. We will call this particular kind of machine a *polymerase* because it promotes the formation of new polymers. Notice that the sequence of the new polymers produced by this polymerase is determined or controlled by the sequence of the parental polymer that is being copied.*

This original Universal Parent molecule would eventually go the way of all flesh. It would have "died" by being destroyed in some random chemical event—cooked when it wandered too near a super hot deep-ocean volcanic vent, perhaps. However, as long as it sent off at least two copies of itself (molecular driftwood in a planetary ocean) before its demise, it was "alive."

The Universal Parent was able to do this through the simple chemistry we alluded to earlier. The details of this chemistry are sufficiently complex that only a very, very rare polymer has the structure to fold into a machine that will carry out the necessary chemical step, as we said. However, once this molecule is accidently formed it will take over the world, inevitably. Indeed, it will leave uncountable descendents over billions of years, including maple trees, race horses, the HIV virus, and us.[4]

The Universal Parent will make copies of itself. These copies will inevitably contain errors occasionally. Most of these errors are irrelevant, but a tiny few improve the performance of the Universal Parent, making it better. These "better" derivatives will make new copies of themselves more efficiently (by definition), taking over Earth's oceans at the expense of less efficiently replicating sequences that will lose out.

The process we have just described is referred to as *natural selection*. It is sometimes also called *Darwinian selection* (after Charles Darwin, of course). Since the new versions of a replicating molecule produced by this process are different from earlier ones, change produced by natural selection is also often called *evolution*.

Over hundreds of millions of years, many, many cycles of evolution by natural selection produced increasingly complex descendents of the original Universal Parent molecule—ultimately, four billion years later,

---

*   Readers sophisticated in how molecular copying works in contemporary organisms will recognize that this description is over-simplified in several respects. However, it captures the parts of the process that are important for us here.

including you and us. A few additional details of these later-version chemical systems (like us) are important to our quest.

The chemistry of Earth Mark II grows up – complex chemical systems are vehicles for replication of design information (organisms)

The billions and billions of cycles of natural selection since the first Universal Parent approximately four billion years ago have produced the complex chemical systems we think of as *organisms*. The original Universal Parent arose by accident and would have been very inefficient (by our standards) at making new copies of itself. However, natural selection would have relentlessly improved these capabilities in the descendents of the Universal Parent. By looking at the chemistry or molecular biology of contemporary organisms, we can infer a great deal about how this process actually happened.

First, the original Universal Parent molecule probably recruited Partner polymer molecules forming multipart machines, through the action of natural selection. These Partner machines would have promoted other chemical reactions useful to the Universal Parent and to other Partner polymers. For example, one Partner molecular machine might have promoted a chemical reaction that produced new copies of the monomer components of the polymers from other small molecules in the planetary French onion soup.

This process of adding useful new Partners would eventually produce a relatively sophisticated "team" of molecules. We can call it the First Team. The First Team would have taken over the Earth's oceans by natural selection, displacing descendents of earlier, simpler teams.[*,5]

Second, because the members of the First Team were all copied or replicated by the same chemistry as the parent, their structure was constrained. They could only use copyable monomers. These are monomers that could be recognized by the copying process the Universal Parent polymerase was designed (by natural selection) to carry out. This property severely restricted the range of functions these molecules could take on.

---

[*]   In fact, formation of these teams of molecules creates a conflict of interest problem (Chapters 1 and 3). The policing of these conflicts of interest evolved early in the history of Earth's first organisms, but the details of these molecular law enforcement mechanisms will not concern us here. See Bingham, 1997.

Thus, a new trick was ultimately "developed" by the blind process of natural selection. Descendents of the First Team "learned" to make tools from other, very different kinds of monomer units. These new Derivative Tools were not copied directly by the Universal Parent polymerase. They were built secondarily by some of the First Team polymers.

Building these new Derivative Tools required the solution of some formidable chemical engineering problems. However, the "cooperation" of the multiple different members of the First Team—modified by natural selection—made this result possible. Indeed, the solution invented by natural selection was beautifully elegant. We know because you and we have inherited this solution. Its processes are still going on in our bodies as we speak, building the new copies of Derivative Tools that we need to stay alive from moment to moment.[6]

Third, after invention of superior Derivative Tools, continuing rounds of natural selection increasingly reshaped the descendents of the First Universal Parent Team and its set of structurally similar Partner polymers. Now most of the Partner polymers no longer acted directly to carry out machine functions necessary to replicate themselves. Rather, they merely encoded the instructions for building the new Derivative Tools. These new Tools took over the various necessary machine functions, and they did them much better. These new tools, thus, carried out nearly every chemical process directly necessary for replication of the Team. Ultimately, even the polymerase copying or replication function of the original Universal Parent, itself, was taken over by one of these newer Tools.

The descendents of the First Team polymers were now reshaped by natural selection to act virtually *exclusively* as the reservoir for storage and replication of the instructions for building the new Derivative Tools. We will call this descendant team the Mature Team.

This picture puts us in the position to define a term that will be very useful from here on. We will call the instructions for building the advanced Derivative Tools *design information*. Again, this design information is encoded in the sequence of monomers in the Mature Team polymers.

Now we are ready to give an initial reductionist definition of a *living* or *biological* organism. It is a chemical system (descended from the Universal Parent and the First Team through the Mature Team) that consists of **design information** chemically encoded in the sequence of monomers in a specific class of polymers. This encoded information has two properties, and two properties *only*. It can be chemically replicated and it chemically produces chemical **tools** that assist in that replication. For reasons that will become clearer in a moment, it will be useful to call the physical object consisting of the design information plus all its tools a *vehicle*.*

That is it. This is a complete description of every organism that ever lived on Earth. including you and us. We all are descendents of the original Mature Team. We have inherited the properties of the Team, including its design information and its tools. We all are vehicles built by design information.

When we look in the mirror, we see eyes, a nose, skin, hair. All of these parts are mostly made up of different sets of Derivative Tools encoded by the remote descendents of the original Mature Team that make up our human design information. These Derivative Tools are, in fact, **protein** molecules. Each polymeric piece of design information (descended from a member of the Mature Team) building such a tool is a segment of a polymer (**DNA**) making up a human **gene**.

We are built by design information encoded in approximately twenty-three thousand such design-information-encoding polymers (genes) making up the *human genome*.[†]

Each of these genes or pieces of design information is still copied today by a process that makes occasional errors. In contemporary organisms, these errors are referred to as *mutations*. Mutations introduce new changes called *variation* into the copies of design information in a population of organisms. Natural selection acts on this variation to produce evolution in contemporary organisms just as it did long ago in the case of the Universal Parent, the First Team, and the Mature Team. Some versions of the variable design information replicate themselves better and take over their world while others are lost in this inevitable competition, now and always.[7] All organisms alive today (including us) have been shaped by about four billion years of this process.

What are we seeing in the mirror? Combinatoriality and complexity

When we look at ourselves in the mirror, we seem to see so much more than just an elaborate chemical system. We see a *person*—something living, feeling, thinking, believing, hoping. We feel that we cannot

---

*   To borrow Richard Dawkins' pithy term from his seminal 1976 book *The Selfish Gene*. Also see Dennett (1995) for a lucid, engaging discussion of these fundamental issues.

†   See *The Molecular Biology of the Cell*, 5th ed. (Alberts et al., 2008) for a good basic description of these molecular details of contemporary organisms—the surviving descendents of the original Mature Team.

merely be a set of chemical processes shaped by blind natural selection. Yet, we have an overwhelming body of evidence that such a chemical system is just exactly *and completely* what we are. There really is not any more doubt about this fact than there is about the claim that the Earth is (roughly) spherical rather than flat.

So why are our intuitive impressions of ourselves so different and so much more richly evocative than our abstract, reductionist understanding of what we are? It is impossible to understand the evolution of organisms, including the evolution of their social behavior and the *unique* social behavior of human animals, without first answering this question. Our goal in this section is to find this answer and to begin to gain the intuition we need to be able to look at organisms (including ourselves) as chemical systems in a crowded world.

One of the reasons this picture is so intuitively challenging is that the complexity of organisms results from many layers of combinatoriality (Chapter 1). Small molecules make large, but simple linear polymers called macromolecules. Small numbers of these macromolecules make molecular machines and functional bits of cells (sometimes called organelles). Small numbers of machines and organelles make cells. Moreover, we can make a number of different kinds of cells by using somewhat different subsets of the molecular machines our design information can make. In turn, small numbers of different kinds of cells combine to make functional tissues, muscle fiber, or lung epithelium, for example. A few tissues make an organ and a few organs make a *system* like the digestive or circulatory systems. A set of organ systems makes a functional organism. That is it, that is all there is.

Organisms look complicated to us for the same reason a Maserati streaking along the autobahn looks complicated if we are not mechanically sophisticated. If we do not understand how spark plugs, fuel injectors, transmissions, and so on are assembled beneath the skin of the Maserati, the contraption looks like magic. Of course, it is not; it is just mechanics. Likewise, beneath the skin of an organism, it is just chemistry.

Needing time to get used to ourselves as chemical systems that have grown complex through multiple layers of combinatoriality produced by the blind, material process of natural selection is one large part of why our reductionist view of ourselves seems subjectively inadequate. We will gradually overcome this problem as we proceed through the book. However, there is a second hurdle to our intuitive understanding of reductionist interpretations of organisms. The following section will let us begin to deal with this problem.

Matter and mind – how our nervous systems "understand" the world

Hidden layers of combinatoriality make it difficult for us to understand what we are. However, the way our minds interact with the world is another reason we have trouble accepting that maple trees and human beings are simply chemical systems. In fact, we encounter this problem every time we undertake a reductionist explanation of *any* part of the world. So let us first understand how our minds see the simple chemical and physical world around us and then return to what we see and how we feel when we gaze into our own eyes in the mirror.

Think of a piece of charcoal and a segment of copper wire. Visualize how they feel when you bend or crush them with your hands. Imagine the dry flat taste of the charcoal and the sharp metallic tang of the copper wire. The appearance and behavior of these two objects are very different. We know intellectually that they are both just solids containing many different copies of the same single simple atom chemically bonded in enormous arrays—carbon and copper atoms, respectively. Their differences result from the way electrons are shared between atoms of the non-metal, carbon, and the metal, copper. However, we have the strong intuitive feeling that this explanation is shallow, incomplete and, perhaps, even trivial. It seems to fail to capture what we most intensely experience about these two objects. It lacks subjective juice, it does not give us the look and feel, the essence of copper or charcoal.

Ironically (and beautifully), there is an excellent reductionist explanation for why reductionist explanations do not give us complete subjective satisfaction. Our minds are biological (chemical) devices designed by natural selection to interact with the world in ways that contribute to our individual survival and reproduction—ultimately, for the "purpose" of getting our design information replicated. Thus, one of the most important things our minds do is to give us the greatest possible capacity to discriminate between different useful (or dangerous) parts of the world.

However subtle the differences between two different physical substances might be in some objective, cosmic sense, our minds are designed to detect and amplify those differences, if they matter. Our minds have super *image analysis/contrast enhancement* capability, so to speak. They present a subjective world of sharp contrasts to us, because that is what works best from the point of view of natural selection.

When we construct detached, analytical, reductionist explanations of this contrasty subjective world, these explanations seem pallid by comparison. Thus, our subjective dissatisfaction with reductionist explanations is not due to their inadequacy. Rather, this unease comes from the distortion and exaggeration our subjective minds impose on an otherwise simple world.

Another example that might help is water. Individual water molecules consist of two copies of the hydrogen atom and one of oxygen. Strong chemical bonds hold together these three atoms in a water molecule formed by robust sharing of electrons.

Because of the way electrons are distributed between these oxygen and hydrogen atoms *within* a water molecule, water molecules also form weak, transient chemical bonds with other water molecules—the hydrogen of one water molecule forming a weak bond with the oxygen of another.[*]

If we remove enough heat energy (molecular motion) from a collection of water molecules, the molecules stick together stably through these weak interactions, forming an orderly lattice we experience as ice.

If we add back a little heat energy to the ice, the molecules move a little more rapidly. At any moment in time, some of them are stuck together in tiny ice-like aggregates of a few molecules and others are broken loose from their neighbors. These little moving aggregates form and break very rapidly, exchanging members like the people moving between the small groups in a large, complex square dance. This produces what we experience as liquid water.

As we add yet a little more heat energy, individual water molecules move still more violently, breaking completely free of one another. This is steam.

These are the element of the reductionist account of water. How does this picture square with our subjective experience? It is sometimes argued that this reductionist explanation of water must not be complete because its fails to explain things like the wetness of water—the way it looks and feels when we wash our hands, for example. Such supposedly unexplained properties are said to be *emergent*, to evade reductionist explanation.

At first glance, this objection to reductionism sounds right. Thinking of water molecules forming highly dynamic little aggregates forever trading members seems a long way from the look and feel of liquid water on our skin. However, before we take this objection to reductionism very seriously, stop and remember what our minds are designed to do. They

---

[*]   These weak bonds are different than the strong bonds *within* the water molecule. These weak bonds are a little like the static electric bonds that holds your socks to your shirts as they come out of the drier.

are designed to present a sharply defined picture of what is important to our (chemical) survival.

With this in mind, notice water's importance. It is an essential component of our world. We need water as the solvent in which almost all our internal chemistry occurs. Indeed, when you step on the bathroom scale, most of what you are weighing is all this necessary water throughout the cells and tissues in your body.

In addition to this internal role, water is an extremely important element of the external world navigated by our bodies. For example, water has a very high heat capacity; so it is important for us to drink it when we are over-heated and to avoid it when we are cold.

Thus, our minds are designed to generate a highly engaging internal image of water—to produce a multimodal sensory blitz when confronted with water, particularly liquid water.

The existence of this rich subjective cocktail of image and sensation produced by the chemical device of our minds hardly constitutes meaningful evidence that a reductionist explanation of water's properties is inadequate.

When you hear arguments in the future about things like the supposed *emergence* of the *wetness* of water (and the ostensible incompleteness of reductionist explanation), ask the following question. *Do the supposedly emergent properties pop up precisely at the interface between the world and the human nervous system?* We find that they usually do and we predict you will too. Objections to reductionist explanation based on this interface are quite unreliable and very likely wrong.[*]

Now we can return to our reductionist explanation of the more complex physical objects represented by organisms. Other organisms (potential competitors, cooperators, mates, predators, prey, or parasites) are among the most important objects in our individual personal universes—as chemical systems (organisms) surviving in a crowded, competitive, *Malthusian* world.[†]

---

[*]   Of course, our argument here leaves us with the ultimate obligation to construct a fully reductionist explanation of the functioning of the subjective minds producing these intense subjective images of the external world. At the moment, reductionist explanation of all the properties of our minds is still incomplete; however, we know enough to believe that one will ultimately be forthcoming. Moreover, we know a number of useful things about minds that we will return to in later chapters.

[†]   The crowded, competitive worlds that organisms inevitably occupy are called *Malthusian*, after Thomas Robert Malthus, the 18th Century thinker who first popularized the universal tendency of organisms to overgrow their living space over time. Notice that each organism produces copies of itself (offspring). *Each* of these offspring produces multiple copies of themselves, and so on. Thus, biological organisms grow in numbers

Thus, we expect our minds, as shaped by natural selection, to construct rich and highly evocative subjective pictures of these other organisms and of our own bodies.

This feeling of recognizing some unexplained aliveness in ourselves and other organisms is just like the feeling of the wetness of water. It is a subjective effect produced by our minds—minds well designed by natural selection to contend with our world.

Information and purpose in our world

An organism is a chemical system, built by design information and capable of replicating that design information. It will be convenient to continue to use Richard Dawkins' term for this kind of chemical system. We will call it a *vehicle*. This term evokes just exactly what we need to keep in mind about organisms.

Vehicles that are alive today for us to find are the surviving descendents of those ancestral vehicles who were successful in leaving progeny in their crowded, Malthusian world. How do we expect such "winning" vehicles to behave? Such a successful vehicle will be good at making new copies of itself. Indeed, that is the reason (the **sole** reason) it wins this competition. There is no other criterion for victory in this arena than existence—and no ticket to individual existence in a Malthusian world other than successful replication by one's immediate ancestors. It could not be any simpler.

Thus, a surviving vehicle is inevitably "designed" by blind natural selection, acting on the design information in its ancestors to make copies of itself very efficiently. As a matter of fact, chemical vehicles are shaped by natural selection so that they appear to have one "purpose"— making new copies of themselves. This is true of all organisms, even a virus with no mind to feel purpose. Our use of the word purpose for any vehicle's behavior is, thus, strictly metaphorical. Purpose in this sense is entirely self-referential and internal to chemical replication in a Malthusian world.

Nevertheless, this metaphor of purpose is a perfectly complete and accurate *de facto* description of the *behavior* of biological vehicles. It is so powerful that we will use the word *purpose* in this sense—never

---

explosively over multiple generations. They always fill their world to its limits to sustain them and this will happen rapidly, no matter how small the organism or how big the world.

forgetting, however, that it is just a metaphor, another subjective illusion like the wetness of water, strictly speaking.*

We can generalize this argument. A living organism is a chemical vehicle built by chemically encoded design information. The sole purpose of the vehicle is chemically replicating that design information.

This is an incredible level of insight into the biological world. We really do understand just exactly what organisms are and what they are not. We know why they exist and what their properties are expected to be.

As we mentioned, fully assimilating this dramatically simple, powerful insight takes time. If this picture of a living organism is not already familiar to you, be patient. You will gain deeper insight gradually as we proceed. Indeed, even if this level of understanding of biology *is* familiar, the relentless subjective minutia of our everyday lives pulls our focus away from this insight. We are constantly seduced and harassed into experiencing the world *only* through our richly evocative subjective minds rather than *also* through our detached analytical minds.

We will return throughout the book to strengthen and enrich the analytical view of ourselves. True self-understanding absolutely demands it. We will find over and over that this view contains the power to enrich our comprehension of what it means to be alive and *what it means to be human*.

Keep your eye on the information

Consider the following facts about us. First, the design information that built you and us was inherited as a single copy of each human gene from our mothers and a second copy from our fathers, as we will see in more detail in Chapter 3. These molecules then made many billions of copies of themselves as we grew and developed—one copy in each of the cells making up our bodies. Moreover, design information molecules in our bodies are constantly vulnerable to chemical damage of many sorts such as UV damage from sunlight on our skin, to name one of many. These molecules require constant repair to remain intact.†

---

*   The astute reader will wonder how we know that our own subjective mental sense of purpose as human beings is not just as illusory and metaphorical as that of the original First Team. This is a most interesting question, indeed. We will return to it in later chapters.

†   This design information is stored in the sequence of monomers in the polymeric molecule DNA in contemporary organisms like you and us. Moreover, DNA molecules are *double stranded*. This redundancy allows repair of one damaged copy of the design

This repair involves replacing damaged polymer segments with new ones. Material excised from a damaged segment of design information is recycled, ultimately being released into the surrounding environment (in our urine, for example).

The upshot of all of this is that the original DNA molecules we received from our parents at the moment of fertilization are long gone. The particular atoms that made up those specific molecules are now spread around the world. Some of these original atoms, no doubt, are floating in the ocean and the atmosphere. Others, might, perchance, be a part of a rabbit, an oak tree, the Prime Minister of England or the Secretary General of the UN.

We have **not** inherited atoms and molecules from our parents. Rather, we have inherited information. It just happens to be encoded in disposable molecules.

Second, we each have many memories from our decades of life to date. For example, Paul remembers being drenched by water as a four-year-old in the process of discovering the principle of the siphon (using a hose and a rain barrel). You have your own set of such memories. If we could go back and magically label each of the atoms in the body of the four-year-old Paul, we would now find that very few of them remain. They have all been dispersed around the planet after being exhaled or excreted from his body. They have all been replaced by the molecules from food Paul has eaten during the relentless repairing of damaged molecules and cells in his body over the last fifty odd years.

Yet…the memories remain. Memories are information stored (somehow) in our nervous systems. This system is repaired and maintained in such a way that information is retained while physical substance is replaced.

Both of these examples illustrate something very important. Our minds are designed to keep us alive in the physical world. So the material substance making up our bodies at any particular moment is highly salient and precious to us. We must protect this body if we are to survive from moment to moment.

This creates another of the subjective illusions to which our biological (chemical) minds are so prone. We see ourselves as these physical objects, our momentary bodies. In fact, however, what makes you *you* and us *us* is **not** this transient physical substance. Rather, we are really the information *encoded in* the parts of our physical bodies. In a very fundamental sense, we are informational objects, **not** material objects.

---

information (one of its strands) by templating that repair on the remaining, undamaged copy (the other strand).

This is a profound fact about biology to which we will return again and again. BIOLOGY IS ABOUT INFORMATION—KEEP YOUR EYE ON THE INFORMATION.

How does all of this insight help us?

We apparently really do understand what organisms are. They are a specific class of chemical systems. Organisms are chemical vehicles built by chemically encoded design information for the "purpose" of chemically replicating that design information in a Malthusian world. This insight is one of the most monumental achievements of the human knowledge enterprise. It is staggering in its power and elegant simplicity, in its parsimony.

So what? Our purpose in this book is not to understand how all organisms are alike. Rather, it is to understand how one of them, we humans, got to be so different. Does this reductionist picture of organisms help us?

It does—in spades. We will see later that the fundamental thing that makes humans different from all other organisms, the thing that ultimately produces the entire suite of unique human properties, is how multiple individual non-kin human vehicles interact with one another. In other words, what is most fundamentally important and unique about us is our social behavior. But, why is the social cooperation of animals even an issue?

The picture we have just built of organisms lets us answer this essential question, right here and right now. Consider how two different vehicles who are members of the same species are expected to behave toward one another—two lions, for example. They live in a crowded, Malthusian world. They compete for the limiting resources they both need for successful replication—the *same* limiting resources, *inevitably*. What should their attitudes toward one another be? Hostile. The life or replication of one will often come at the expense of the other.

Our reductionist picture thus predicts competition—sometimes even fierce conflict—between members of the same species. As we will shortly see, this really is how most organisms behave most of the time, when they are not simply avoiding one another out of *anticipation* of hostility.

However, there are exceptions. A very select subset of members of the same species sometimes fails to compete. They may even actively cooperate. This cooperation between different individuals allows a new level of functional adaptive sophistication very analogous to the increased adaptive sophistication the Universal Parent achieved by cooperating with various Partner molecules.

Thus, social cooperation between animals is a new level of complexity in the biological world and it is very important. To proceed from here we will need to understand when, where, and how non-human animals cooperate or do not cooperate. In Chapter 3, we will find that there are some beautifully powerful and simple answers to these questions. Moreover, in Chapter 4, we will find that we humans also often (but not always) play by these same universal animal social rules.

With these insights in hand, we will find (in Chapter 5 and beyond) that we have the final pieces we need to understand how humans went on to build something **new**. We humans added a fundamental new level to the rules of animal social cooperation. We retained the patterns of social cooperation that non-human animals have **and** we added, *on top* of these, a fundamentally new human pattern. This additional pattern of social cooperation, in turn, revolutionized the way we human vehicles pursued self-interest in a Malthusian world. This new human trick changed the biological world of Earth forever. This human *revolution* will be our focus from Chapter 5 throughout the remainder of the book.

❧

Chapter 3
The only way to win is not to play – the problem with cooperation between organisms

In this chapter, we will complete our exploration of those few specific details from 20th Century biology that form the foundation for our theory of human uniqueness. Though these insights are longstanding, we will look at them from a specific perspective that will be vital to our goal.

In Chapter 2, we were mostly concerned with organisms as individuals—*what their bodies are made of, how their design information is encoded, the physical basis of the variation between them, and so on. Now we step up to the next level of complexity in the biological world, the interactions* between *individual animals or their* social *behavior. As we examine these behaviors, it will be crucial to remember one particular insight.* **Natural selection builds the brains and behaviors of animals just as surely as it builds their bodies.** *We cannot understand a bird in flight without understanding* **both** *the aerodynamic properties of its wing* **and** *the mind controlling that wing.*

Brains and the minds they encode are just one more class of tools built by design information. Brains are crafted from combinations of different cell types expressing different subsets of protein molecules (Derivative Tools; Chapter 2) interacting in brain tissue to carry out mental functions (Chapter 10).[*]

---

[*]   The facts that brains learn from experience and receive culturally transmitted information does not change these fundamental claims, as we will see in later chapters.

We will find that the social behavior of animals, just like the properties of their physical bodies, is remarkably predictable when we view these animals as vehicles built by design information for the purpose of replicating that design information. This insight is one of the great intellectual achievements of our species.

Non-human animals generally cooperate socially with only a few very particular other animals. Specifically, they cooperate selectively with close genetic kin, like parents, siblings, and offspring. This property evolves in a straightforward way. Close kin individuals are guaranteed to share identical pieces of design information. After all, they inherit these pieces from their recent common ancestors (parents, for example) that make them close kin. Thus, when design information builds an animal to cooperate with (to help) close kin, that information is likely to be helping other copies of itself.

This pattern of evolved behavior is called <u>kinship-dependent</u> social cooperation. When natural selection produces such social behaviors by exerting selection on the design information shared by close kin, this process is called <u>kin-selection</u> and we say that such behaviors are <u>kin-selected</u>.

In defining the logic of this cooperation between close kin, we will also be implicitly defining the circumstances when animals **do not** cooperate—and when they actively compete. We will find that **non-kin** conspecifics have active **conflicts of interest.*** This pattern of competitive behavior evolves just as simply as kin cooperation. The potentially different repertoires of design information distinguishing any two non-kin individuals are competing to survive and reproduce in the crowded social world these individuals inevitably occupy. The specific kinds of design information winning this competition survive for us to find in the world. Ancestral design information building organisms that did not compete with non-kin has been lost.

We can **only** understand the social behavior of **any** animal, including our own uniquely human social behavior, by understanding conflicts of interest and how they are managed. Once we understand how conflicts of interest between non-kin are managed, we can apparently predict essentially everything else about an animal's social evolution and behavior. It is this management of conflicts of interests that allows the evolution of the better angels of our human nature—our novel human social cooperation.

Below, we will discuss how non-human animals actually behave in their natural environments. This tour will bring conflicts of interest into

---

*   Strictly speaking, non-kin members of the same species will behave *as if* they have conflicts of interest, but we need not articulate this complexity most of the time. These conflicts of interest are the same as those from our everyday human economic lives (Chapter 1).

*sharp focus by showing how they emerge simply and directly from natural selection. We will sometimes find our moral sensibilities disturbed—even shocked or outraged—by what non-human animals do. This reaction is an expression of our **uniquely human** ethical sense. Non-human animals behave by some very different rules than we do. If we are to understand how **our** rules became so different, we must first grasp the rules of non-human animals. We must not let our human ethical sense stand in our way.*

*It will be a thing of great beauty when a clear understanding of the sometimes brutal, **inhumane** behavior of non-human animals leads us to an understanding of our own unique, common humanity.*

Morning on the savanna

Minnie, a tiny young lion cub, lies artificially still in the tall grass. For reasons she does not understand, she feels deeply uneasy. Minnie has not seen her father for several days. Then, just last night, her mother, Jane, brought her here. Jane indicated with her body and vocalizations that Minnie needed to stay very still. It has been hours since her mother left. Minnie wants to meep softly to call her mother, but she does not dare.

Suddenly, Minnie hears a large animal coming slowly through the grass. At first, she assumes and hopes it is her mother. Scent soon tells her otherwise. It is a male lion, a stranger. He must be a large and vigorous individual judging by the power and low pitch of his rumbling subvocalizations. We will call this new adult male Arthur. Minnie is utterly terrified.

Arthur springs over the last bit of grass into the tiny clear spot where Minnie lies. In an instant, Arthur uses his powerful jaws to grip Minnie across the back of the neck. With a blindingly quick shake, Arthur snaps Minnie's neck. Her death is instantaneous. Her fear and her life are over.

Minnie has several half-siblings, children of her father and her mother's sisters, hiding elsewhere in the grass. Arthur is relentless. He systematically hunts down every one of these youngsters and dispatches them. By noon, none are left alive. It is systematic carnage.

For the few days leading up to Minnie's murder, Arthur had been busy. First, he found and killed the adult male, Franklin, Minnie's father. This confrontation was violent and Arthur received several deep claw wounds that took many days to heal in the humid tropical heat. On several occasions, Arthur fell feverish, but he ultimately struggled back to health.

By today, the day Minnie and the other cubs were killed, he had recovered. Once Arthur is satisfied that no more youngsters remain, he prowls the territory that Jane and her sisters have long occupied. These adult females are the heart of this pride of lions. Though they have lost this year's cubs, they are still in possession of this territory and of their older offspring. Arthur will now have to deal with them.

As he roams and searches, Arthur passes Franklin's decomposing body, now heavily scavenged by vultures and the other denizens of the savanna. Eventually, Arthur finds Jane and all her adult sisters. Surely now a massive conflagration will ensue. No. In fact, Jane and her sisters quietly tolerate Arthur, no longer harassing him.

Indeed, after a short time, they warm up. They soon actively begin to seduce him, aggressively, shamelessly flirting and soliciting. For a few days, Arthur is so busy mating with all these females that he has trouble finding time for anything else. Soon all the females are pregnant.

After a few months, the new lion cubs sired by Arthur are born. We *humans* might think, *Surely this psychopathic killer will destroy them all*. Precisely to the contrary, Arthur proves to be a doting father, allowing the cubs to play all over and around him, nipping at his tail and ears and showing no respect. More importantly, he aggressively protects them from other adult males prowling around the margins of the pride's territory.

These first cubs grow to partial independence over about a year and Arthur, Jane, and her sisters produce another round of cubs. This cycle goes on for several years. Eventually, Arthur ages, losing a step. One day the inevitable happens. Arthur is killed by a new adult male, who also kills that last round of Arthur's cubs—those youngsters who have not yet been weaned.

This new male is soon sought out by Jane and her sisters. The pride goes on—year after year, male after male.

What just happened?

This short story is not really fiction. With a little simplification of a few details, it describes events that are very common in the animal world.[1] In fact, as many as 25 percent of all young lion cubs may perish in this way in some areas. This routine killing of very young lion cubs by adult males when they take over a new pride, as Arthur and his successor did, is given the specific technical name of *infanticide*.

Infanticide in the specific case of lions is well known. However, infanticide is also frequently seen in many different animals. Moreover, depending on the details of the animals' lives, adults of either sex kill youngsters.

What, exactly, is going on here? Who benefited from Arthur's infanticide? On the one hand, it is easy to say who *did not* benefit. Obviously, Minnie did not benefit. She lost her life. Did Jane or her adult sisters benefit? Certainly not! They lost their children.

We might be thinking, then, that only Arthur is left. Did Arthur benefit? No, he did not! When he fought with Franklin, he risked his life. For adult males lions, fighting is dangerous business. He took a large personal risk. He could have perished from the wounds these fights produced. It could easily have been Arthur's bones, rather than Franklin's, bleaching under the East African sun.

Indeed, Arthur's bones did join Franklin's when Arthur was killed, in turn, by his successor. Arthur could probably have lived to a ripe old age by just minding his own business and not getting mixed up in pride "politics."

Well then, we might also be thinking that lions *in general* somehow benefit from Arthur's behavior. Maybe this episode was somehow "for the good of the species." This is a very seductive line of thought. Some professional biologists accepted it during the first half of the 20th Century. We will return to this very interesting mistake in different guises a little later. For the moment, it is sufficient to get an initial feeling for why this intuitively appealing perspective is incorrect.

First, recall that one quarter of lion cubs are killed by the likes of Arthur. It is very difficult to see how this could represent a benefit to lions as a group. This behavior does not look like it is for the good of the species.

Second, we might be thinking that the cubs that Arthur killed were somehow inferior to the cubs he fathered in their place. His cubs may have somehow been fitter so that lions *as a species* were better off as a result of Arthur's infanticide. But stop and think for a moment. Yes, it is true, Arthur defeated Franklin. But Franklin defeated his predecessor and Arthur will be defeated by his successor, in turn. The outcomes of these conflicts are determined by the deterioration of the males with age (and some luck). They are generally not determined by differences in their overall value or fitness.

There is no reason at all to think that Franklin's dead cubs were not just as good as were Arthur's first generation of cubs who replaced them.

Arthur was not acting in the greater interests of lion-hood. He was not promoting the good of his species.

Wait, then! There's no one left. Did no one benefit? No, "someone" *did* benefit. "Who" was this beneficiary? **It was Arthur's *design information*!**

This insight is straightforward, once we see it. It is ***absolutely essential*** for us to understand this point. Consider the alternatives to Arthur's behavior. First, he might elect not to fight with Franklin in the first place and just continue to live among the bachelor males in the territories surrounding those of the prides. He, as an *individual*, does better this way. He has no fights, no wounds, and lives a long and tranquil life. As the classical Greek saying goes, he could have chosen "a long life of tranquility rather than a short life of glory."

In contrast, his design information does badly if he chooses tranquility. He makes no new copies of his own personal design information in the form of cubs he fathers. Thus, by fighting with Franklin, he risks his individual, personal life. But, if his risks pay off, his design information "wins." If Arthur's personal design information is to be replicated, he has no alternative but to fight with Franklin.

But what about Minnie and the other cubs? Why does Arthur kill them? Again, consider Arthur's own personal design information. If he allows Minnie and her infant peers to live, the females will continue to nurse them for months to come. While nursing, the females will not ovulate. Thus, they will not mate and produce any new offspring until Minnie and the other youngsters are weaned. In contrast, when Arthur kills Minnie and the other cubs, the females are suddenly bereft of nursing infants. The cub's absence triggers a response in the adult females' bodies, originally designed to cope with the death of cubs by disease or accident, perhaps. They now will ovulate and become sexually receptive again within a few days.

Our uniquely human ethical sense predicts, incorrectly, that the adult females will seek vengeance on Arthur. However, retaliation would not improve the replication of the adult females' design information at this point and they do not, in fact, behave that way.

Now, we can put these details in perspective from the point of view of Arthur's personal design information. The first round of his new cubs born to Jane and her sisters contain copies of his design information. Had he waited for Minnie's generation to grow to independence, these copies of his design information would never have been born.

Infanticide gives his design information extra copies it would not otherwise generate. By forcing the issue, infanticidal males get extra rounds of cubs (on average) generation after generation. This is just natural selection increasing the frequency of infanticidal design information.

Arthur's behavior here appears highly "purposeful." Its purpose is to produce extra copies of his own personal design information. He need not be consciously aware of this larger, ultimate purpose. But, it is the source of his behavior nonetheless.

For the moment, it is sufficient to notice that Arthur is behaving as we would expect of a vehicle built by his own personal design information, for the purpose of replicating that personal design information. Indeed, his behavior makes **no sense** on any other interpretation we have.

Evolutionary biologist Theodosius Dobzhansky famously remarked that "nothing in biology makes sense, **except** in light of evolution."[2] Infanticide illustrates this crucial point *perfectly*.

The natural selection of "purpose" – Part 1

Natural selection is the only possible source of Arthur's purposeful behavior *if* the picture we built in Chapter 2 is correct. Natural selection is a physical process just like the bouncing of a ball. We do not say that a ball bounces as it does to fulfill some larger plan. It bounces merely as a by-product of the physical laws of the universe. The morphology *and behavior* of animals are just like this bouncing ball. We look only to the internal workings of natural selection to comprehend them.

Understanding that natural selection is only about the physical replication of personal design information and not about some larger purpose, is difficult for us, intuitively. We humans subjectively feel that we plan our world with a sense of larger, extra-personal purpose in a million different ways. The spacing of the keys on our computer keyboards was designed by engineers we never met to match the placement of typing fingers *so that* we could write, *so that* others could read and learn, *so that*…and on and on.

Human artifacts are usually designed *as part* of some larger enterprise; therefore, we tend to see *all* designed objects in this way. But objects designed by natural selection are **not** designed *so that* something larger can occur or *for* some larger purpose. Rather, biological "artifacts" are only designed to assist in the immediate replication of the specific copies of design information that builds them.*

---

* This fact can become especially confusing to our subjective human sense of purpose under various specific circumstances. For example, the different pieces of design information that "cooperate" to build a vehicle (the vehicle's genome) appear to us to be

This view of natural selection is essential. Nothing in biology makes sense on close inspection *unless* we understand this fundamental fact about the world. If we fall back into our usual human habits of thought about design *for* some larger purpose, we are immediately doomed to confusion and incomprehension.

The natural selection of "purpose" – Part 2

Now back to Arthur's purposeful behavior. Natural selection shapes brains and behaviors just as it shapes bodies. There is every reason to believe that natural selection is, indeed, the source of Arthur's behavior. Consider ancient ancestral lions. The individual pieces of design information that built them varied from individual to individual. Different copies of the same piece of information are slightly different as a result of copying errors that occurred in the past.[*] The form of this variation is the existence of slightly different versions of each copy of the thousands of genes that encode the design information building a lion. These slightly different versions of genes are given a technical name. They are called different *alleles* of that gene.[†]

---

designed for the larger purposes of the vehicle. In a superficial sense, we can say they are. But in the more fundamental and very important sense, they are not. Each piece of design information is simply selected—through natural selection, a purely chemical and physical process—to replicate efficiently. Each piece evolved in a context where the best route to its ancestor's replication happened to be to contribute to the building of the vehicle. This produces an individual piece of design information that functions as it would if its "purpose" were to help build the vehicle. (This requires "law enforcement" within the genome. Conflicts of interest between different pieces of design information must be suppressed and managed for this cooperative strategy for replication to arise and prosper. See Bingham, 1997.)

Another way of saying this is that each piece of the design information in our genomes is exploiting every other piece in order to get itself replicated. From the perspective of this particular piece of design information, all the other pieces of design information (and our bodies and minds) are just tools to get its own "personal" replication done.

[*]   These copying errors are the inevitable consequence of a simple physical law—the Second Law of Thermodynamics (also see Chapter 2).

[†]   As we all know, each different piece of design information is commonly referred to as a *gene* by contemporary biologists. *Alleles* are different versions of the same gene. The existence of multiple different alleles for many of the genes in our genome constitutes variation—sometimes explicitly called *genetic variation* or *allelic variation*.

Different randomly arising alleles and combinations of alleles at multiple genes build brains that produce different behaviors. Any new behaviors arising from new allele combinations will usually be bad from the point of view of successful replication. A **random** new internal arrangement for a device like a mind/brain is very, very unlikely to work better than a previously existing, orderly and functional arrangement produced by earlier natural selection. We can see this point by asking what happens if we bang on our TV with a hammer. Once in a very long while, this method improves the performance of the TV, but almost always the TV gets worse.

The slightly altered versions of design information that produces such "bad" behaviors will be lost soon after they randomly arise because the vehicles they produce will perform badly, generally failing to leave offspring. However, a very rare few of these randomly arising design information variants can have more interesting properties.

First, for example, once in a while, one of these randomly arising variants in design information in our ancestral lions built a male that protected his individual life by avoiding conflict. This particular version of lion design information would have lived a nice, long life in that individual, but the design information producing the behavior perished with him at the end of his natural life. He left no copies of his "peaceful" design information in the form of offspring for us to find here millennia later. He was not one of Arthur's ancestors. His design information vanished from the Malthusian world. As "happy" and "well-adjusted" as its vehicle was, this was bad information from the perspective of natural selection.

Second, another rare, random ancestral design information variant happened to have built males that fought with other adult males to take over lion prides. This *fighter* version of lion design information *did* leave copies of itself in the form of offspring. These males were among Arthur's ancestors. He inherited their design information and hence the

---

When a piece of design information is copied over many generations it very rarely, but inevitably undergoes change from unavoidable errors in replication. Many of these changes have little or no effect on the function of the gene. They are silent changes in the sequence of monomers making up the polymer (DNA) in which design information is encoded (Chapter 2). Others produce the differences that make individuals distinguishable—by eye or hair color, for example.

behaviors produced by it. From the perspective of natural selection, this is "good" information.*

Finally, among these fighter adults, further random variation produced yet another new variant, *infanticidal-fighter* design information. Individuals with this new version of lion design information not only fought for pride ownership, they also killed the non-weaned young cubs they found in their new prides.

This version of lion information initially left *more* copies of itself in offspring at each generation than did the simple fighter version. This is because of the extra round of cubs infanticidal fighters leave, as we saw.

At each subsequent generation, the fraction of males built by infanticidal-fighter information increases at the expense of simple fighter information. Today's infanticidal males are the great-great... great-grandsons of the original infanticidal male, inheriting these pieces of ancestral design information. Ultimately, only the infanticidal-fighter version of lion information remained. In a Malthusian world, infanticidal-fighter information did best of all— a "triumph" of natural selection.[†]

In a very real sense, we now have everything we need to proceed to Chapter 4 and the first steps of building our theory of human uniqueness. Non-kin members of the same species have intense conflicts of interest and their social cooperation is limited by them. However, it will serve us well to get a deeper and broader grasp of these issues. The remainder to this chapter will give us this richer understanding.

---

* Obviously, "good" refers to the internal logic of the physical process of natural selection. Good has no ethical implication in the human sense.

† We notice that, once all adult males are infanticidal, individual males no longer get an extra round of cubs from infanticide. Their final round of cubs is killed by the next infanticidal male just as they killed the last male's final round. There is no net benefit compared to a *purely* non-infanticidal population. This is madness from the perspective of lions as a whole or even the local group or population of lions. Of course, this fact about the process is not relevant. Natural selection is a consequence of the blind, individual, head-to-head competition of different behavioral strategies, and infanticidal-fighter wins this competition.

Conspecific competition – a cosmic dilemma and its resolution

Most large organisms around today are a rather odd kind of vehicle. They do not make new copies of themselves alone. Instead, they replicate in cooperation with another vehicle. They reproduce **sexually**. It is not obvious exactly why sexual reproduction is so ubiquitous. Some smaller organisms are largely asexual (bacteria, for example) and often make copies of themselves directly, without sex.

What matters here are the un-sexy abstract, reproductive functions of sex and the kind of populations of organisms sexual reproduction produces. Sexual reproduction engenders an entity that biologists call a *species*. For our purposes, the simplest definition of species will be sufficient. A species is the entire set or population of organisms whose opposite-sex members can reproduce sexually in cooperation with one another.

All humans belong to the same species because any fertile human female from anywhere in the world can mate with any fertile human male from any other place and produce fertile human offspring. African lions are a species, as are all the North American gray squirrels nesting in the trees around us, all the white tailed deer, all the dogs of the world, and so on and so on for essentially every organism on Earth that we can see with our unaided eyes.

Members of the same species are called *conspecifics*, simply meaning *same species*. Thus, we are conspecifics and we are much more like one another than we are like the oak trees or the squirrels in our back yards.

Our concern is how members of the same species will behave toward one another in a Malthusian world. This question presents a crucial dilemma.

First, since conspecific individuals are very much like one another, they are very likely to compete for the same limiting resources in their crowded, Malthusian world. As fellow humans with similar needs, we might compete for the last quart of milk at the grocery store in the face of an approaching hurricane. However, we would usually not compete with the squirrels in our back yards over the acorns the hurricane brings down.

In fact, in general, the most important and persistent competitors any individual organism has in its life are other conspecifics. Thus, individuals who compete successfully with other conspecifics will usually do better (their design information will do better) in a Malthusian world.

Our male lions, Arthur and Franklin, are vivid examples of the evolved consequences of this competition.

Second, another conspecific competitor can sometimes be built by the same versions of the species' design information as oneself. Thus, if we were to compete with this other vehicle built by exactly the same information, we would be shooting ourselves (our design information, actually) in the foot, so to speak. This would be the very quintessence of "robbing Peter to pay Paul."

This problem sounds like a classic damned-if-you-do-and-damned-if-you-don't situation. It is…almost. One particular strategy is expected to win over all the others in the face of this dilemma *if* our fundamental picture of organisms is correct.

Originally, three people primarily worked out the resolution of this dilemma. Two of these were John Maynard Smith and the authors' Stony Brook colleague, George C. Williams. The third giant of this story is the late William D. "Bill" Hamilton whom Paul also had the good fortune to know. Bill was an exuberant personality with mind racing, eyes flashing, long hair flying about. He also had the characteristic that all the giants of science have. He relentlessly asked the simplest questions and insisted on getting answers that made sense—answers that could be tested against the facts of the world. No other scruple is relevant—only the truth. Bill (and George, John, Isaac Newton, Charles Darwin, and so on) was always the first to yell out that *the emperor has no clothes* and was the last to stop trying to find him something to wear.

Bill could look at a stunted evergreen tree, the genital swelling on a female chimp, or the colors of fall foliage, all with the same intensity and imagination. He examined everything with a flinty, critical eye that simultaneously managed to be as open and wondering as a gifted young child's.

Bill's relentless search for insight took him to every corner of the world, including to the tropical rainforests where he contracted the malaria that finally killed him. The world is a poorer place for his passing. But he has left us something immortal—insight.

Bill, George, John and others showed that the logic of natural selection acting on design information building sexual vehicles resolves the conspecific competition dilemma by building individuals who cooperate very specifically with only very close kin (Box 3.1).[3] It is convenient to think of this as follows. An animal helping very close kin is unambiguously assisting in the replication of other copies of some of its own personal design information, inevitably shared by close kin. As described in Box 3.1, we will call individuals who preferentially help very close kin, Family Helpers. From now on, we will call individuals built by Family Helper design information *kinship-dependent cooperators*. We will see that most animals in the real world (including the you and us, in the narrow context of our families) are just exactly kinship-dependent cooperators.

Individual vehicles and "groups"

This brings us to the next crucial piece of the puzzle, an insight originally generated by our friend and Stony Brook colleague, George C. Williams. George is a strikingly tall, dignified man with a close-cropped beard. The adjective *Lincolnesque* has been used over and over for George because it is so utterly perfect. In superficial personality, he is rather different than Bill Hamilton. George is quietly thoughtful, compared to "wild" Bill's extravagance.

But he also shares important, unique skills with Bill and science's other giants. He has courage and relentless intellectual honesty. He is not willing to call the game to protect feelings or prerogatives, that is, not until the answers are in.

He also has the priceless gift of asking just the right *simple* question. As we know from personal experience, these simple questions can be electrifying, instantly plunging a room full of over-loquacious professional academics into intense, silent reflection.

A lifetime of asking such simple, penetrating questions has allowed George to help shape our modern understanding of how evolution by natural selection works.

Let us look at one of George's revolutionary contributions by examining kinship-dependent cooperators (Family Helpers; Box 3.1) a little more closely. What happens when everyone in a local population comes, over time, to be built by Family Helper design information? Every individual is now a kinship-dependent cooperator. Does the logic of this behavior not break down? Under these conditions, even when an individual competes with non-kin (distant relatives), its design information is competing with other copies of itself.

This is logically perverse, but exactly correct. Remarkably, this paradox does not destroy the logic of kinship-dependent cooperation. Natural selection is just a physical process that "works how it works." There is no larger plan and no foresight to the process. Natural selection is blind, purposeless, amoral. Whatever is produced by the internal logic of natural selection acting "in the moment" on the design information that builds vehicles is what we get—"good" or "evil," paradoxical or not, even "insane" or not.

A thought experiment will add clarity to this crucial point. Suppose we were omnipotent, so that we could engineer a population of animals who behaved in any way we wanted. We start by making a population of individuals who are all perfectly identical genetically—no variation. Moreover, these individuals are all built by an artificial version of design information that causes them to cooperate with all conspecifics they meet, no matter who they are, kin or not. Let us call this Indiscriminant Helper information.

Box 3.1: Design information's winning strategy – conspecific cooperation between close kin

The world of biological organisms is almost always crowded with conspecifics. The biological world is Malthusian. We begin by imaging an animal species where all individual members just compete like crazy with *every* other conspecific individual they meet, without exception. This means that the design information that builds these animals, builds this particular pattern of behavior. We will call this particular version of design information Indiscriminant Competitor.

Other versions of information that build animals that do not compete at all (call this information Pacifist) will lose out to Indiscriminant Competitor information as we have discussed. Indiscriminant Competitor will come to rule this particular world at this particular moment.

Now consider a new version of the species' design information—produced purely at random by errors in replication as always. This new version of design information builds individuals that still compete like crazy with other conspecifics, *with one exception.*

This new information builds individual vehicles who refrain from intense competition, when they can, with those few conspecific organisms who are very close kin. We will define kinship a little more precisely in Box 3.2 and the text. For the present, just think of it in our everyday sense of the term where close kin are parents, siblings, and offspring.

We will call this new version of design information Family Tolerator. When Family Tolerator first comes into existence, it will be rare among all the other copies of this species' design information. Thus, when it builds individuals who compete with other individuals who are not close kin it will almost always be competing with Indiscriminant Competitor information, **not** with Family Tolerator information.

Consider what happens when an individual built by Family Tolerator avoids competing with very close kin. For simplicity, consider siblings. Each of two siblings is likely to have inherited Family Tolerator information from one of their parents (Box 3.2). When these two siblings compete with distant kin and avoid competition with close kin, two good things happen. Family Tolerator competes with Indiscriminant Competitor but avoids competing with the other copy of Family Tolerator in its sibling.

Natural selection will shape this competition-avoidance to extend to parents and offspring as well as siblings—in other words, all close kin. Family Tolerator design information will certainly do better than Indiscriminant Competitor information over many generations as a result of this pattern of behavior in the animals (vehicles) it builds.

In fact, we predict yet another new version to arise at random and come to prominence through natural selection. We will call this last new version Family Helper information. Family Helper builds animals who do not merely avoid competing with close kin, they *actively cooperate* with these same close kin. This cooperation allows productive combinatoriality between the assets and abilities of close relatives (Chapter 2). A set of multiple individuals can do things none of them could accomplish alone. Thus, Family Helper alleles can do better than non-cooperating Indiscriminant Competitors or even mere Family Tolerator individuals. Again, Family Helper will ultimately displace both Indiscriminant Competitor and Family Tolerator because of the extra benefits individuals built by Family Helper design information give one another through their cooperation.

Now compare a separate population of Family Helpers to this population of Indiscriminant Helpers. At first, members of the Indiscriminant Helper population will be at least as well off as the members of the Family Helper population. In fact, they will probably be *better* off because they have more opportunities for productive cooperation. We might think that this is the end of it. Once we have *universal* Indiscriminant Helper design information, it will displace Family Helper information because the pure Indiscriminant Helper population is better off.

If you thought this, you would be very wrong. But take heart, a generation of professional biologists made exactly the same mistake. It is called the *group selection fallacy*.[4] One of George Williams' many contributions was to decisively reveal and debunk this fallacy in his famous 1966 book *Adaptation and Natural Selection*.

One particular example of the group selection fallacy concerns us here. This version assumes that *because* individuals who are members of a *group* of non-kin individuals behaving in some specific way would do better in principle than individuals who are members of a second group who behave differently, *individuals* will evolve the behavior pattern of the first group. This is logically incorrect (again, as George showed). What matters is *not* an imaginary uniform group, but actual local populations (groups) with lots of behavioral variation. In these populations, multiple distinct individual strategies come into head-to-head competition. The outcome of this individual competition determines which behaviors evolve and which do not. It will be vital for us not to mistakenly focus on groups as the unit of natural selection as we build our theory of human uniqueness in later chapters. Box 3.3 is an opportunity to think this through in more detail. [5]

Box 3.2: The details of sex and relatedness.

Kin-selection in sexual animals requires that individuals "play the odds" in a very particular way. Sexual kin share some pieces of design information with one another, but not others. Moreover, there is no way to know which individual pieces are identical and which are not. However, natural selection has a very precise "knowledge" of the *probability* that a particular piece is shared. Thus, animal social behavior is shaped to take this probability into account. In effect, in social interactions, an animal looks at a kin individual as a discounted version of itself. The factor measuring this discount is called genetic *relatedness*.

When talking about genetic relatedness, the two different copies of a gene (a particular piece of design information) in a *diploid* sexual individual (see main text below) will always be considered different alleles (whether the DNA sequence of the two copies differs or not).

Now, consider any two genes or different pieces of design information. We will call them *gene 1* and *gene 2*. For the moment, label the two alleles of each of these genes in a specific diploid individual A and B. Thus, such an individual has the 1A and 1B alleles of gene 1 and the 2A and 2B alleles of gene 2.

When this individual makes one gamete (a sperm or an egg), that gamete might receive 1A and 2B. A second gamete might have 1A and 2A—one allele the same as the first gamete, the other different. Of course, the 1B, 2B and 1B, 2A combinations will also occur. Now average this process over the roughly twenty-three thousand different genes that build a mammal like us. To an excellent approximation, *one half* of the genes in one gamete will be the same allele (A for A or B for B) and the other half will be opposite alleles (A for B or B for A). This is what it means when we say any two gametes are 50 percent related. Half of their design information is identical by virtue of having been inherited from a very recent common ancestor.

Now extend this to first cousins. Their parents include a pair of siblings (50 percent related) and (usually) two unrelated parents. Thus, we might imagine at first glance that first cousins are 25 percent related, the average of their parents' relatedness. Not so. There is an addition factor of two. When the sibling parents of the two cousins choose single pieces of design information from their diploid complement they have a 50 percent chance of choosing the same shared piece when they make a new gamete. For example, both siblings might have genotype 1A and 1B as the two alleles at gene 1. This 100 percent identity at the 1 locus offsets 0 percent identity at other loci (one might be 2A and 2A and the other 2B and 2B, for example). But when each sibling parent makes a gamete there is only a 50 percent chance that both will choose 1A (or 1B) rather than one choosing 1B and the other 1A. Thus, their gametes will be only 25 percent related, not 50 percent related. As a result, first cousins have the relatedness of the average of one 25 percent related gamete and another roughly 0 percent related gamete. They are only 12.5 percent related. Genetic related drops *very rapidly* as pedigree relationship becomes more distant.

Sex and kinship

The word sex is very evocative to us for the perfectly good reason that we are sexually reproducing vehicles. Our design information insists that we feel this way. For now, though, we are only concerned with how sexual reproduction affects the kinship recognized by Family Helper information (Box 3.1).

No matter how good it may feel, sexual reproduction is still just reproduction. It is merely a method whereby vehicles make new copies of the design information that builds them. These copies are in the form of offspring vehicles (Chapter 2). This fundamental logic is *indistinguishable* in sexual and non-sexual organisms. Only the details are different.

Figures 3.1 to 3.3 in the online endnotes provide a detailed discussion of sex and kinship for those who want to understand its features more deeply[6]. However, for us, only a few details are crucial.

First, sexually reproducing organisms are built by *two different* copies of each piece of design information. The technical term for this is *diploid* meaning "two-ploid," or having two-copies of the species' genome.

Second, as most readers will know, in preparation for sexual reproduction, adult organisms make sex cells. In the case of animals, these are called eggs when produced by females and sperm when produced by males. Collectively, eggs and sperm are called *gametes* and their production is called *gametogenesis*.

During gametogenesis, each piece of design information is present in two distinct copies in the parent, as we said. She or he chooses one of these two copies to put into a gamete (her egg or his sperm). This "choice" is the result of a purely random chemical (molecular) process occurring in the ovary or testis of the animal. Moreover, for each different piece of design information, this random choice is made independently.*

Thus, each gamete now has *only one* copy of each piece of design information. The technical term for this one-copy state is *haploid* or "one-ploid."

Third, two gametes, an egg and a sperm, join at fertilization to produce a new fertilized egg that will grow into an offspring, a new vehicle. This offspring will have each of the two haploid complements of design information (called haploid genomes) contributed by its respective parents. As a result, the offspring will be diploid again, just as its parents were.

Again, when these offspring reach adulthood they will make haploid gametes of their own from the diploid complements of design information they have. These offspring will be the grandchildren of the first set of adults and so on and so on.

---

* This is a slight over-simplification, given the existence of what geneticists call *linkage*. However, that small complication need not concern us here.

Fourth, as a result of these processes, any two eggs or sperm produced by a specific individual will be 50 percent related in the sense of the genetic kinship that natural selection "sees." (Another useful phrasing is to say that the relatedness of two siblings is 50 percent.) This 50 percent related condition arises in the simple way described in Box 3.2.

Fifth, if the same two parents produce two offspring, each of the eggs and each of the sperm that fuse to produce these two siblings will be 50 percent related and the siblings will, thus, be 50 percent related to one another.

Sixth, how does this relatedness change along the animal's family tree? The crucial point is that it drops drastically with more distant relatives. We just saw that siblings are 50 percent related as is a parent to its offspring (and vice versa). In contrast, two first cousins are only 12.5 percent related and two second cousins are only 3.125 percent related (Box 3.2).

Is this quantitative detail really important? It is. Organisms that are built by unambiguously shared design information (close kin) will have been shaped by natural selection to behave as if they have common or confluent interests. They will be built by Family Helper information designed to assist other copies of their personal design information in close kin (Box 3.1). Such close kin include parents, siblings, and offspring, and essentially no one else. In contrast, organisms will behave as if they have conflicts of interest with every conspecific that is not close kin.[7]

Bill Hamilton drove this vital point home by constructing his famous mathematical expression of this effect. In the following section, we can see what Bill did.

Hamilton's Law and kin-selected cooperation

Bill Hamilton's Law (or Rule) is written below. For a behavior by one animal generating a benefit to a second animal to evolve on the basis of kin-selection, the cost (C) of that behavior to the first animal must be less than the benefit (B) to the second animal *discounted* by the probability that each piece of design information building the first animal is identical in the second, receiving animal by virtue of being inherited from a recent common ancestor. This probability is the genetic relatedness (r) we have discussed. This compact mathematical expression certainly fulfills the front-of-the-tee-shirt rule for simplicity in good scientific theory (Chapter 1). It is a remarkable intellectual achievement.

Hamilton's Law: Br > C

It is illuminating to think through how and why this law should apply to animal social cooperation as a route to intuitive mastery of the conspecific cooperation (and competition) problem. A useful place to begin is to consider Family Helper information (Box 3.1) in a sexual species. When a specific, individual piece of design information "looks" out at another creature to see how likely that other individual is to have been built by another copy of itself, it must consider recent shared ancestors. It can then *discount* the certainty of its being present in the other individual accordingly (Figure 3.1). This discount determines when a *self-sacrificial* (from the first vehicle's point of view) behavior is "sensible" (from the point of view of all the copies of this personal piece of design information potentially shared by the two vehicles). This is what Hamilton's Law expresses, with relatedness (r) being the factor of discounting. "I am worth two siblings or eight cousins," to paraphrase J.B.S. Haldane's famous intuitive statement of what came to be known as Hamilton's Law.

Box 3.3: The group selection fallacy and kinship

Suppose we were completely omnipotent. Imagine two artificial populations, one pure Indiscriminant Helper and one pure Family Helper, the strategies we have already discussed in the text. We keep our two different populations apart and we repeal the Second Law of Thermodynamics so that no new variation arises. In this very artificial case, group selection would not necessarily be a fallacy. The Indiscriminant Helper population would be better off and would remain so, forever. Keep in mind, this is NOT the real world.

However, suppose we are almost omnipotent, but we cannot repeal the Second Law. Now what happens? Consider the Indiscriminant Helper population. Inevitable random variation in design information will produce many different new behavior patterns in vehicles. Once in a very long while one of these will be the Family Helper pattern. What happens to these rare, new copies of variant information? The Family Helper information "wins," ultimately taking over the population and driving Indiscriminant Helper information to extinction.

Here is how this works. When a Family Helper meets a non-kin Indiscriminant Cooperator, the Family Helper accepts cooperation offered by the Indiscriminant Helper individual, but does not reciprocate. This produces extra resources for the replication of Family Helper information. Family wins, Indiscriminant loses. However, when a Family Helper meets a close kin (usually Family Helper) individual, the two cooperate, generating extra resources for the replication of the Family Helper design information they both share. Thus, Family on Family does as well as Indiscriminant on Indiscriminant, but in Family on Indiscriminant, Family trumps. In the head-to-head, in-the-moment world of natural selection, Family wins.

Inevitably, over time, Family Helper information displaces Indiscriminant Helper information. Indiscriminant is not stable in the real world, but, Family is.

Our thought experiment shows us something remarkable. The Family Helper population of design information derived by natural selection from the Indiscriminant Helper population could, in principle, still be perfectly uniform. (Suppose, for example, that we omnipotent creatures ran an infinite number of such local experiments and examined only those where the very first new variation that arose produced Family Helper behavior.) In principle, every copy of design information in this thought population can be perfectly identical to every other.

In other words, genetically identical individuals will still be built to distinguish close kin from non-kin even when both sets of individuals are indistinguishable by any objective criterion other than the circumstances of their immediate ancestry. Such individuals will be competing with identical copies of their design information in distant kin at the same time they are cooperating with still other identical copies of their design information in close kin.

In a detached, "rational" sense, this is perfectly mad—the grossest nonsense. However, as we have seen, this does not matter. The equilibrium behavior that survives natural selection for head-to-head, in-the-moment competition is Family Helper. This is all that matters—sense or nonsense, sanity or insanity, notwithstanding.

Is Family Helper information really the winner in all circumstances? Apparently in all non-human circumstances, it is.

Our thought experiment has shown how design information that builds vehicles who engage in kinship-dependent social cooperation wins out over all other kinds of design information. If our picture is right, this should be how animals actually behave. We will see in the Empirical Test sections in the main text and the online endnotes that this is what animals do.

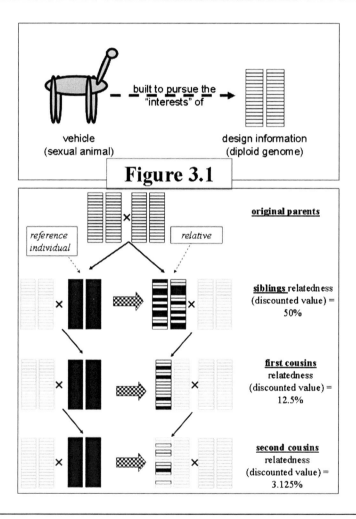

Figure 3.1: The TOP BOX illustrates the diploid genome that builds a sexual vehicle to pursue the design information's interests (text). The BOTTOM BOX examines a reference individual at each of three successive generations. Each individual is symbolized by its design information. The design information of the reference individual is solid black. When "planning" its social behavior,

this reference individual "looks" (block arrow) at the indicated relative at each generation as a version of itself discounted by its relatedness to that individual. This relatedness is the fraction of identical design information the two relatives inherit from the original parents at the top of the box (also see Box 3.2). Hamilton's Law (text) is the mathematical statement of this principle of social behavior. Design information from mates of the offspring generations (unrelated) is indicated with light gray lines.

The crucial point is how quickly relatedness drops at each generation. In practice, non-human animals rarely recognize individuals more distant than half siblings (25 percent related) as kin.

Empirical tests - The rules in science

When we talked about chemistry and early biological evolution in Chapter 2, we were talking about relatively well-established basic science. Thus, we did not stop to look at the evidence that our claims were correct.*

However, we are now approaching the questions of how humans behave and how non-human animal behavior helps us understand human behavior. This behavior is not only our real concern here; it is also newer science. It is science that is still "in play". These are hypotheses and young theories still being argued.

Thus, we return to the issues we discussed in the Introduction and Chapter 1. We need to ask how well our theories account for or predict the world they are supposed to explain.

The rules for this kind of testing of hypothesis and theory are very specific. They are a little different than the rules for many other parts of our world for very important reasons. We talked about this briefly in the Introduction, but we now need a few more details.

Think for a moment about how people sometimes argue on the floor of the US House of Representatives or on the political campaign trail. They often are very selective about the evidence they present in support of their views. For example, politicians are not above hailing the private benefits of a tax cut while failing to mention the resulting collapse of public infrastructure it produces. Alternatively, politicians are capable of arguing for a tax increase in the *public interest*, which is actually designed to fund payoffs to political supporters.

---

* However, skeptical individuals are *always* free to engage in such doubt, even of "settled" science.

What all of these and many other sleazy political arguments have in common is what a professional scientist would call *data selection*. The easiest way for us to fool ourselves (or others) is to go looking for empirical evidence (data) that fits a theory we like and sweep everything we do not like under the rug. Because this is such a bad way to gain insight, it is against the rules in scientific argument. We are required to consider **all** the evidence.

This presents a practical problem. In a finite argument, we can only actively discuss a limited amount of evidence. Thus, in practice, this consider-all-the-evidence rule is implemented in a particular way. We are required to actively look for evidence that contradicts our hypothesis. We must try to "break our theories on the facts." We look harder for information that proves our theory wrong, than for information that seems to support the theory.

We should continue to search until one of two things happens. Either, we find evidence that breaks our theory (such evidence demonstrates a theory to be wrong and is given the somewhat ostentatious technical name of *disconfirming* evidence) or, alternatively, we continue to search for disconfirming evidence, but we fail to find any. Instead, we find only consistent or confirming evidence.

At this point, after we have searched long and far enough that we come to believe (always tentatively) that there is no disconfirming evidence to be found, we can take the position the theory is right.[*]

We are aware of no disconfirming evidence for the central role of kin-selection in shaping non-human animal social behavior[†]—both cooperation and competition. Below is one simple example of confirming observations. Other examples can be found in the online endnotes.[8]

Empirical tests -The mammalian life style

Our daily lives are filled with mundane matters that we accept without interest or question. They are just "facts of life" or the "way the world is." For example, grass is green, the sky is blue, and mothers usually love their children.

One of the simplest, but deepest pleasures in science is grasping that these mundane matters are actually fascinating empirical evidence

---

[*]   Of course, again, this belief in the correctness of a theory can *always* be changed later by a new, good case of disconfirming evidence.

[†]   A few additional effects are added on to kin-selection in shaping non-human animal social behavior and we will return to these later.

susceptible to reductionist explanation. When this happens to us, we are elevated. We cease to be helpless spectators to a bewildering world and acquire the potential to become cognizant participants in a knowable one.[9]

Just as with green grass and blue skies, *mother love* is amenable to simple, reductionist explanation by kin-selection theory. First, we usually visualize mothers as warm and fuzzy, that is, we are usually thinking about mammals. We might think of humans specifically, of course, and we will return to them in the next chapter. But we might also think of a mother deer nursing and protecting her fawn, a mother grizzly teaching her cubs to fish for salmon, or a mother cheetah training her cubs to chase down a gazelle.

Now, think about how mammalian reproduction works. The fetus grows in the female. When the offspring is born, its mother (its maternity) is not in doubt. Until the recent invention of *in vitro* fertilization, if a mammalian youngster popped out of your vagina, 50 percent of its design information was yours, beyond all question.

However, the female who gives birth to this youngster might have mated with one male or dozens or hundreds before becoming pregnant. Thus, in contrast to *maternity*, mammalian *paternity* is *always* uncertain.[*]

So, what do we expect in view of kin-selection? Mammalian mothers should, under these circumstances, *always* be more solicitous of their offspring than mammalian fathers. In effect, the relatedness (r) in Hamilton's Law is lower, on average, for *ostensible* fathers than for *unambiguous* mothers. In species where female promiscuity is the rule and paternity is always *very uncertain*, fathers should be *very* much less solicitous. In species where the male has higher confidence in paternity, a father should be more solicitous of his young, but still less so than the mother. This should be true everywhere and always in mammals. It appears to be.

Indeed, for the typical mammal, the male's contribution to the young might consist of sperm donation and (sometimes) subsequent protection *from other, infanticidal males*! This was pretty much the sum total of the roles of our lion fathers, Franklin and Arthur, for example.

Thus, the parenting behavior of non-human mammals is a strong piece of empirical evidence supporting kin-selection as the source of animal social behavior.

---

[*]   Reliable molecular paternity tests are too new to have affected any of the evolved behavioral properties of organisms we are talking about here, of course.

What about humans? Human males are sometimes accused of being just as unfatherly as Franklin and Arthur. While this is no doubt true of some individuals, it is generally not the most common pattern. Does this human pattern disconfirm our theory?

Actually, it does not. And the reasons it does not are illuminating. We encounter here a pattern that we will see over and over in later chapters. Recall that our largest objective is to understand human uniqueness—how we are different than other animals. However, we humans are descended from non-human ancestors. Thus, we expect our behavior to be some kind of mixture of non-human behaviors inherited and retained from ancient ancestors and new, uniquely human behaviors, just as our bodies are mixtures of uniquely human properties (our speech-enabled vocal apparatus, for example) and ancient, retained properties (our very mammalian hair, for example).

When we look at parenting by human males, we see both ancient, non-human components and new, uniquely human components. The uniquely human component is a little subtle and we will leave it for Chapters 6 and 8 when we discuss uniquely human aspects of sexuality and child rearing.

However, the ancient, non-human component to human male parenting is more straightforward. Human male contributions to child rearing in our culture usually occur only when the human male in question has relatively high certainty of paternity. In other words, this part of the human pattern is inconsistent with kin-selection theory.

Moreover, even here, human mothers commonly take a greater role in the day-to-day rearing of the very young than do human fathers. For example, we, the authors, love our children very much, but their relationships with their mothers (Joanne or Zuzana) are just a little different than their relationships with their fathers (Paul or Cliff).

This is all just exactly what kin-selection theory predicts since paternity certainty was *never* complete in the *ancestral* human environment, nor, indeed, in our present world.

Finally, one of the best tests of a theory is how well it can turn the mundane *background* of our world into sharply focused *foreground*. Animal and human child-rearing most emphatically provides this kind of satisfying empirical test of kin-selection theory (also see Chapters 4, 6, and 8). This kin-selected pattern of non-human (and some human) social behavior will be the foundation on which our new theory of human *kinship-independent* social cooperation will be built.

Proximate and ultimate causation

We have been talking about animals as if they were behaving to serve the "purposes" of their design information. Do we mean that the animals are thinking about their design information and calculating what they should do? Of course, we do not. To understand what our theory predicts about what animals (or humans) are thinking, we need to stop and look a little more carefully at how brains control behavior.

Consider your own behavior for a moment. Say you eat a sandwich for lunch. Now, your friend asks you why you did that. How would you reply? Often, you would reply that you were *hungry*.

Hunger is another of those mundane facts of life. What is its reductionist explanation? Consider what is called the Second Law of Thermodynamics. We are highly orderly chemical systems. According to this Second Law we need the continuous input of matter and energy to maintain the evolved orderly state of being "alive."*

Thus, in one sense, the correct answer to the question of why you ate the sandwich is, "I am a non-equilibrium thermodynamic system in need of a new input of matter and energy." Of course, almost no one ever gives this answer. Indeed, no one *could* have known to give this answer, *ever*, before the 19th Century development of thermodynamics.

Our minds are devices shaped by natural selection to sustain us as a vehicle for the design information that, in turn, builds our minds. Our minds have an internal logic of their own. We do not understand all its details, but the mind creates a kind of virtual world in which it represents the outside world in a way that lets us successfully navigate that world as vehicles.

Among the properties of this internal virtual world are motivational rules, an appropriately vague term for something we only dimly understand as of yet. We subjectively desire and avoid things according to these rules. Moreover, we can learn to associate novel external circumstances with satisfaction of these desires and aversions.

Be this messiness as it may, the bottom line is simple. Our minds are devices that control our behavior based on our subjective motivations. These motivations are arranged to control the behavior of the vehicle so that the need of vehicles (of its design information) are met. These subjective motivations do not represent those needs in any direct or objective sense.

---

* More formally, we all are what are called *non-equilibrium thermodynamic systems* as a result of these properties.

Thus, we make an ***absolutely crucial distinction*** when we discuss the causation of animal behavior. We will say that the minds that motivate a behavior are the ***proximate cause*** of that behavior. In contrast, when we discuss the underlying evolutionary rational for a behavior, we will refer to these rationales as the ***ultimate cause*** of that behavior.*

Consider, again, your sandwich for lunch. The *proximate cause* of your behavior was that you "felt hungry." The *ultimate cause* of your behavior is that natural selection shaped your mind to cause you to behave ***as if*** you understood your needs as a non-equilibrium thermodynamic system.

Another evocative and illuminating example is our sexual behavior. Its *ultimate cause* is that sex is necessary for us to replicate our design information. Thus, natural selection has shaped us to pursue sex when it potentially furthers that objective.† Of course, the *proximate cause* of our behavior is subjective feelings like love and lust. Almost *no one* is ever thinking explicitly about replicating his or her design information during sex.

We will use this insight over and over. Evolution by natural selection acting on design information is the *ultimate cause* of essentially all animal behavior under normal circumstances. This ultimate level of causation is what will usually interest us. Natural selection acts by producing specific mental or psychological devices that control and influence behavior in furtherance of *ultimate* goals. These psychological devices are the *proximate* cause of behavior and we will sometimes refer to them as *proximate* devices.

Minds as proximate devices – the nature of "as if" explanations

Now we face a semantic problem. Minds are proximate devices directly causing behavior. But we are more interested here in the ultimate causation of behavior. We need a way to talk about this clearly and efficiently without sacrificing factual accuracy.

---

* The terminology for this essential distinction between levels of causation is adapted from Ernst Mayr. Ernst shared the Crafoord Prize for evolutionary biology from the Swedish Royal Academy with our Stony Brook colleague George Williams and John Maynard Smith.

† More precisely, we pursue sex under conditions where it might have fulfilled reproductive purpose in our ancestors. If our contemporary environment is novel from an evolutionary point of view—a brothel, for example (Chapter 8)—our behaviors might not fulfill realistic contemporary reproductive function.

An easy way to do this is to describe behavior *as if* it was directly produced by its ultimate causation—that is as fulfilling the evolutionary "objectives" of the design information building the brain that directly controls the behavior.

From the point of view of immediate, proximate causation of behavior, such explanations are metaphorical. However, this metaphor is so accurate in so many different circumstances that we can very often use it with no loss of meaning or content.

In fact, we have already been doing this just a little bit when we discussed things like infanticide. We know virtually nothing whatsoever about what a male lion is actually thinking and feeling when he goes about his kin-selected behaviors. And we do not need to know. We can understand and predict behavior remarkably well while ignoring questions about subjective minds entirely.

The subjective mental states of animals represent an interesting little technical or mechanistic question. But such questions are quite secondary and relatively unimportant if we wish to understand why animals really do what they do at the most fundamental level.

This point is **crucial** to everything else we will do throughout the rest of the book. Moreover, it is profoundly enlightening—but a little shocking—when we reflect on its *inevitable* implications for us.

Our minds are proximate devices just like the minds of any other animal. Thus, when we ask ourselves why we do what we do, the subjective answer is usually *unimportant*, even *uninteresting*.* This subjective answer is just a description of the output of a quirky proximate device—our minds. It says virtually nothing about what we really care about—the *ultimate* cause of our uniquely human behaviors.

This property applies not just to explaining our feeding or sexual behaviors. It applies to *all* our behaviors whether they be social, economic, political, ethical, religious, and (most emphatically, yes) even our scientific behaviors.

This is a well-established and long-understood fact about the world for people who have invested the necessary time to think about the problem. But, if this conclusion is new to you, it should profoundly reshape your view of the human world. If it does not, either it is not new after all or you are missing the point.

---

* It is equally important to recognize that our *proximate* judgments and feelings about life and the world can richly inform the esthetic and ethical basis of a humane future. We will return to this subject later. For now, if we fail to understand the ultimate sources of our proximate feelings and judgments, we cannot distinguish between those that serve our mutually enriching common human purpose and those that serve potentially toxic, narrow self-interest.

If we really want to understand ourselves and our world, we have to outgrow our provincial, self-referential mental world. As people who have spent years looking at the world from this other perspective, we can assure you that this adjustment is worth the effort. It provides stunning new levels of self-understanding. You will never look at your world, yourself, or other people in the same way again.

Moreover, and perhaps unexpectedly, this perspective turns out to be very humane. We ultimately find a new sense of acceptance and tolerance for ourselves and for others here. We will return to these vital ethical and existential issues in Chapter 17 and the Postscript.

"As if" and ultimate causation, "interests," and "conflicts of interest"

In order to be able to talk clearly about our uniquely human social adaptation in later chapters, we need to put *conflicts of interest* into our newly sophisticated understanding of the evolution of behavior. We said earlier that vehicles appear purposeful and that this purpose is to replicate their design information. In other words, they appear to be agents of purposeful design information.

We can also say that design information has *interests*. It "wishes" to replicate successfully. Moreover, we can say that vehicles behave *as if* they are aware of those interests and are agents obligated to pursue those interests. Again, these are NOT real statements about design information having mental states—a preposterous suggestion. Rather they are extremely precise metaphorical descriptions of the ultimate logic of social behavior.

Of course, in a Malthusian world, the *interests* of one vehicle often cannot be fulfilled without interfering with the interests of another vehicle. Franklin's and Arthur's interests could not simultaneously be satisfied. Under these conditions, we can very realistically (but metaphorically) say that the interests of non-kin vehicles (their design information) were in conflict.

Thus, the social cooperation problem can be restated in a useful way. The many pieces of design information building a vehicle have *conflicts of interest* with the pieces of design information building another, non-kin vehicle. When only non-kin individuals are involved, the problem is simple. Individual conspecific vehicles have pervasive conflicts of interest. These conflicts of interest prevent social cooperation most of the time. These conflicts of interest represent the ultimate level of causation if we wish to understand social behavior.

The problem with non-kin cooperation – the only way to win is not to play

Social cooperation between vehicles in a Malthusian world is limited by non-kin conflicts of interest. As a result, the only way for animals to win at the game of cooperation with non-kin conspecifics is not to play most of the time. This is exactly what they do.

Consider an illuminating example. Gazelles on the East African savanna suffer predation by cheetahs. It is common for a nursing mother to be killed, leaving a calf that will starve without mother's milk. It is also common for a nursing mother to lose a calf. She is still lactating, but has no calf to nurse and must wait until next year to try to reproduce again. We might think that the childless mother and the motherless child could join forces allowing the calf to survive. Gazelles virtually never do this.

Childless mothers do not nurse and soon stop lactating. Mother-less children (calves) starve. To us as humans, this seems grotesque. It is precisely, however, what we expect if non-human animal cooperation is limited to close kin. Non-kin cooperation essentially never occurs.

However, non-kin cooperation can occur under one very special set of conditions. If non-kin conflicts of interest can be managed or suppressed in some way, cooperation between non-kin conspecifics should be possible.

For example, suppose that there was some way for the childless mother gazelle to assure that the non-kin calf she nursed would not compete with her or her offspring in the future. Moreover, suppose she could be assured that the nursed orphan calf would grow up to nurse other orphaned calves including descendents of our nursing mother here. Under these conditions, adoptive nursing of non-kin infants might produce higher survival and design information building this behavior might ultimately come to characterize the gazelle population.

But notice what is required. Some way must exist to assure that this non-kin nursing does not inflame future non-kin conflicts of interest. This is a good trick if we can figure out how to do it.

In fact, this trick was apparently figured out on a grand scale once and only once in the four billion year history of the Earth. This unprecedented social strategy was developed by a primate in East Africa around two million years ago. The descendents of this primate went on to inherit this trick and, as a result, to rule the world and to walk on the Moon.

We are the descendents of this primate and the inheritors of its very special social trick. Everything uniquely human about us flows directly from this new social strategy. It is our history, it is our destiny.

We will introduce this very special, uniquely human, adaptive trick in Chapter 5. But it will help us to first look a little more carefully at a few of the ways in which we are *not* unique—the ways in which we are like other animals. Humans have learned (evolved) to manage non-kin conflicts of interest, but not to eliminate them. Thus, we show *some* kin-selected behaviors just like all other animals and *fail to show others*. Understanding these behaviors (Chapter 4), will helps us greatly in understanding how we *added* cooperation that is independent of genetic kinship *on top of* our kin-selected social behaviors.

৩

Chapter 4
The facts of life –
reproduction and kin-
selected behavior
in humans

*In this chapter we will explore some important direct evidence for a crucial fact about us. Humans are different from other animals in a fundamental way. We engage in substantial social cooperation **independently** of close genetic kinship. Moreover, we apparently achieve this unprecedented cooperation by **actively suppressing** important kin-selected **competitive** behaviors common to non-human animals.*

*Our uniquely human pattern of social cooperation independent of close genetic kinship might seem to contradict the lessons from Chapter 3 about conflicts of interest. However, we will find that it does not. Humans are still replicating chemical vehicles subject to natural selection in a crowded Malthusian world. We have merely **added** the new ability to cooperate independently of kinship to compete more effectively **under very particular circumstances**. More specifically, we evolved a new way of controlling or managing the conflict of interest problem that all biological vehicles always face.*

In spite of this novel human pattern of cooperation, many of our social behaviors (everyday behaviors we take for granted) are, in fact, kin-selected, just as they are in non-human animals. We will explore these kin-selected behaviors in this chapter. Comprehension of our kin-selected behaviors will give us important insights into some of our most intense private concerns and deepest emotional feelings.

With this picture of our kin-selected behaviors in hand, we will be positioned to see that our unique, kinship-independent social cooperation does *not occur **instead** of kin cooperation. We are no more free of kin-specific behavior than any other animal. Rather, our kinship-independent social cooperation is **added on top of** the kin-selected social cooperation we share with all the other animals.[*,1] Specific subsets of kin-selected behaviors are highly modified or **quantitatively** attenuated in us, though they are **qualitatively** similar to the corresponding behaviors in non-human animals.*

Finally, looking at our kin-selected behaviors will give us a chance to practice our skills at looking at ourselves with detachment—as just another animal. This capacity for self-objectification will serve us well throughout the rest of our journey.

*Some of the examples of human behaviors in this chapter are shocking and repulsive to us. We cannot let this human ethical reaction deflect us. Looking at our behaviors as animals will ultimately elevate us, not degrade us. In the end, this understanding will give us a clear view of the biological origin of the better angels of our human nature—**of our common humanity**.*

Strategic infanticide in humans – stepchildren

In Chapter 3, we saw adult lions like Arthur systematically killing non-kin cubs in support of the interests of their own close kin cubs, built by copies of their personal design information. We humans also engage in this pattern of strategic infanticide, but with a crucial difference. Consider a child in a household with one natural parent and a stepparent. The child's natural parent is committing resources to the child. From the stepparent's point of view, these resources are being diverted to non-kin. They are taken from any other close kin children the stepparent might have, either from a previous mate or with her/his contemporary mate. Eliminating this competition to the stepparent's own offspring (or potential offspring) by killing the stepchild should be a "good" adaptive idea, just as it was for Arthur in Chapter 3. Yet, rather than killing stepchildren, most human stepparents help raise them—often responsibly and supportively.

---

*   As we will see in Chapters 6 and beyond, some elements of our kin-selected reproductive behavior have been modified as a consequence of our kinship-independent social cooperation. In contrast to these uniquely human sexual and child-rearing behaviors, the facets of our reproductive behavior that we share with all animals (our focus in this chapter) have been well understood by professional biologists for some time. Thus, we will look at them only briefly. For readers interested in exploring this older story and fascinating subject in more detail, a series of excellent books is available as listed in online Endnote 1 to this chapter.

We must immediately ask whether kin-selection theory is thus disconfirmed (falsified) by this quirky human behavior. Perhaps humans really play by some *entirely* different set of rules than non-human animals. Actually, we do not play by different rules in the larger sense. Understanding step parenting in humans will bring us much closer to understanding how our universal animal (kin-selected) behaviors are affected by our uniquely human social behavior.

The most common form of adoption of non-kin youngsters in humans occurs when death or divorce ends a marriage with young children and the surviving biological parent remarries. This often creates households with a stepparent. In this situation we can ask whether stepchildren are more likely to be killed by their stepparent than by their natural biological parent.

The answer is very dramatic, in two ways. First, human children are about **eighty times** more likely to be killed by a stepparent than by a natural parent!!* This is a huge difference and it is exactly the kind of effect that simple kin-selection theory predicts. In this sense, humans are capable of behaving just like Arthur, *selectively* killing non-kin youngsters competing with their own kin youngsters.

Of course, this differential treatment of natural and stepchildren is intuitively accessible to us. It is part of our shared picture of the human world. We possess cross-culturally universal stories and myths based on this differential relationship. For Western Europeans and North Americans, *Cinderella* with her "evil stepmother" is one of our versions of this universal story.

But interestingly and secondly, human adults in modern societies like ours kill their stepchildren **much, much less** frequently than do non-human animals. Indeed, many stepparent/stepchild relationships are warm and nurturing. Non-kin infanticide is the **rule** for many non-human animals, it is the **exception** in humans. Roughly 25 percent of *all* lion "children" perish at the hands of their "stepfathers" (like Arthur) whereas step parental infanticide in humans is hundreds of times more rare than this in most contemporary human societies!![2]

Thus, the kin-selected *patterns* in our infanticidal behavior are just like non-human animals, *but* we are very much less likely to carry out infanticide.

We need to know where this large quantitative difference between us and non-human animals comes from. We can understand one source of this difference from our own experiences. Infanticide comes with

---

*   These data and insights are from the book *Homicide* by Margo Wilson and Martin Daly whom you will meet in a moment.

a huge off-setting cost. Other humans (including the legal system in contemporary states) will impose an enormous cost on us if we commit infanticide. This consequence changes the cost/benefit logic of infanticide in humans compared to non-human animals.*

Family life – how we grow up

Many of us live in a world of electronic communication and computation. Moreover, enormous, integrated cities supporting international travel networks allow us to consume tropical fruit during our local winters and enjoy a thousand other amenities. These and the many other features of contemporary life were unknown to our great-grandparents, never mind our Paleolithic ancestors.

We might look at our environment and wonder whether we live in a profoundly novel world compared to our ancestors. In fact, the evidence indicates that our world is really not very different than the ancestral human world (even a million years ago) in many ways **that matter.**† In this world of supersonic jets, perishable food grown in Guatemala and consumed in Groton, iPods, and cell phones, humans still show (some or all of) the kin-selected behaviors that all other animals do and that our pre-modern ancestors did.

Our individual experiences and memories are one valuable source of evidence that this is the right way to look at our behavior. Indeed, we now encounter another case of the beauty of science as it takes the mundane facts of our lives and turns them into processes we understand, with depth and clarity.

First, a typical human child grows up with the support of his/her mother and father, with exceptions, of course. In fact, at some level, we intuitively recognize this fact as an evolutionary product when we refer to the adults that raise most children as their biological or natural parents. They are *just* that—*biological* vehicles, in the sense of the special case of chemistry from Chapter 2.

The task of child-rearing involves substantial investment of resources (a *cost*) by one creature (a parent) generating a *benefit* for another creature (the child). We would expect natural selection to produce this

---

*   Notice that we are talking here about *ultimate* causation, not about what individual stepparents might be thinking with their *proximate* minds (Chapters 3 and 10).

†   A vital implication of this different-but-the-same view of the contemporary human world is that the evolved *proximate* psychological devices (Chapter 3) we inherit from our remote ancestors should still "work" in our world—at least often enough for us to detect their effects.

behavior only when the two individuals are close genetic kin (Chapter 3). The "natural" human child-rearing patterns that produce the world we occupy are just exactly what kin-selection theory predicts they should be. As we would expect, this pattern is universal. For the most part, everywhere in the world, close kin rear human offspring.*

Second, human maternity is always certain but paternity never is.[3] Thus, when a human pair-bond (marriage) breaks up, we expect mothers to be more likely to take primary responsibility for young children. In most cultures, this is what we see. A mother's commitment to very young offspring is (on average) greater than the father's.

Third, when neither parent can care for young children, we predict that this task should usually fall to close kin of the children, specifically older siblings, aunts, uncles, grandparents. Indeed, since maternity is certain, we expect that close kin of the mother are more likely to step in than close kin of the father. Again, the behavior we observe is precisely as kin-selection theory predicts.

Fourth, our public economic and political behavior involves cooperation between large numbers of individuals, most of whom are not close kin. This is our uniquely human *kinship-independent social cooperation*, again. Can we understand how this *public* kinship-independent cooperation relates to our *private* kin-selected behaviors? One requirement is clear. We cannot let kin-selected behaviors *subvert* kinship-independent social cooperation.

Thus, we expect our public behavior to be governed by rules that preclude such subversion. As we would predict, public institutions typically have nepotism rules. For example, in the United States, each government official is hemmed around with various layers of regulations that make it difficult for her or him to hire siblings, children, nieces, or nephews selectively for public jobs. In the absence of kin-selected behaviors in humans, nepotism laws would be as unnecessary as laws against flapping our arms and flying away.

Fifth, is it easy to think of objections to the claim that our reproductive behaviors are subject to kin-selection. For example, as we have already discussed, we occasionally raise non-kin infants. This turns out to be uniquely human behavior (Chapter 8). Non-human animals essentially never do this in their natural habitats. What is important to us at the moment is that even in the *unique* human animal, this raising of non-kin youngsters is unusual. Most human children are *not* reared by adoptive parents. Statistically, adoption of non-kin infants is relatively rare behavior even in humans.

---

*   With some interesting variations we will encounter in Chapters 6 and 8.

This evidence will take us a little further. Humans have sometimes experimented with childrearing that was *systematically* done by non-kin. Some kibbutzim in Israel early in the 20th Century are famous examples. Children were weaned early and moved into institutional dormitories where they were cared for by all the adults and/or by designated professionals. In principle, they were to be unaware of genetic kinship and to be raised as equal members of a collective.

In practice, such systems have always proven unstable. Parents seek out their own offspring, give them selective benefits, and ultimately subvert the entire cooperative enterprise. Of course, this is precisely what kin-selection theory predicts.

Aldous Huxley* wrote a famous science fiction novel early in the 20th Century entitled *Brave New World*. The book depended on collective, non-kin childrearing to paint an emotionally bleak picture of a hyper-collective future. Children were conceived, gestated, and born in laboratory containers and raised by anonymous bureaucrats.

Kin-selection theory tells us where the shock value of Huxley's brave new world comes from. Our evolved minds (proximate devices) have been designed by natural selection to look at childrearing as a natural function and prerogative of close kin. Anything that upsets that pattern feels very wrong to us.

Men and women really are different – sexual desire and jealousy

Not only is our behavior toward offspring kin-selected, but our mating behavior to produce offspring is also plainly kin-selected, as well. Psychologist David Buss and his colleagues have done outstanding work on human sexual behavior.[†,4] Some of the rich fruits from this tree are very useful examples of our kin-selected behavior.

It is easier for us to picture our evolved sexual behavior intuitively if we set it in the contemporary context. We can picture a mated pair—call them husband and wife for book keeping purposes. The couple shares a home somewhere in North America. Now we picture the potential positions they are in as self-interested vehicles. (They are not *consciously* thinking about self-interest most of the time, of course.)

The husband mates with his wife. However, if he can also mate with other women, he can potentially have additional offspring. This fact is especially true if these other women are mated to other men who

---

*   Aldous was the grandson of Thomas Huxley, Darwin's well-known "bulldog."
†   Much more about our sexual behavior in Chapter 8.

might provide resources for raising any of these "external" children he fathers.

Of course kin-selection theory predicts a strict limitation on this male strategy. Men will generally resist contributing to rearing children fathered by other non-kin males. Thus, if a husband is to achieve this extra reproductive success, his external mating will need to be hidden *from the spouses of the females he mates with*. It is less important that his external mating be hidden from his own spouse. She is not in danger of giving birth to children who are not her own or of directly contributing resources to the rearing of her husband's extracurricular non-kin youngsters.

The wife's position is different than the husbands. The number of children she can give birth to is not increased by the number of males she mates with—one sperm donor is as good as two or a dozen for this purpose. Thus, at first we might think that the wife has little incentive for external mating. This intuition is incorrect.

Giving birth is only the beginning of successful reproduction for a human. The offspring will need years of resources provided by parents (including fathers) to grow and learn optimally, as we all know from personal experience. These resources are called *parental investment*. Moreover, in the competitive social world of humans, this child can also benefit from such investments by adult friends, uncles, and so on.

A wife can obtain resources of this form by mating with other men. If these other men believe they *might* have fathered some of her offspring, kin-selection theory predicts that they are more likely to contribute resources to those offspring. Of course, her own husband, as a self-interested vehicle, would be just as *unwilling* to contribute resources to these same children. So, the wife's external mating must be hidden *from her spouse*. It is relatively less important that they be hidden from the spouses of men with whom she mates. These other women are not in danger of rearing a non-kin youngster as a result of their husband's external mating.

Of course, all wives of other men have this same motivational structure. This is why a husband's original external mating is sometimes possible.

Buss, a gifted public speaker, brought down the house with a version of the following story at a meeting of the Human Biology and Evolution Society a few years back. The story captures the essence of this kind of mating behavior.

A young man, Benjamin, approaches his father, Jason, with the news that he is in love with the girl next door, Heather. He wants to marry her. His father is quiet for a moment and then, with difficulty, says to his son, "Your mother and I were having some problems when we were young and I had an affair with Heather's mother. You can't marry Heather. She may be your half sister. Please don't tell your mother."

Benjamin is crushed. He goes away to college for a year and gradually recovers. The following summer he is staying with his parents and he meets a girl from across town, Jennifer. Benjamin and Jennifer fall in love. Benjamin feels renewed and he comes to his father with the great news.

His father is clearly deeply troubled. Jason says, "As you know, your mother and I were having some problems when we were young and I had an affair with Jennifer's mother. You can't marry her. She may be your half sister. Please don't tell your mother."

Benjamin is outraged. Against his father's wishes, he tells his mother everything. Unexpectedly, she smiles with equanimity. She says to her son, "Don't be concerned. Marry the girl. Jason isn't your father."

This story all sounds a little more like a soap opera than the daily lives of most of us. However, this scenario was probably much more common in the ancestral human world than it is in ours (Chapter 8). Our contemporary human design information was shaped by two million years of evolution in a mating environment that held the kinds of opportunities (and risks) we just described for ancestral Janices and Jasons. Buss' data support this picture.[*]

Mating preferences are illuminating. Certain preferences in mates are shared by both women and men, including health, vigor, and intelligence, among others. However, it is easier for us to measure *differences* than to try to understand similarities without any standard of comparison. For males, some of the most important criteria for mates are youth and vigor, because these characteristics determine the likelihood (especially in a relatively difficult ancestral environment) that the female will survive childbirth and live long enough to be a successful mother. The ability to contribute material resources to the raising of offspring, while valuable, is slightly less important in a mate from the point of view of an ancestral male.

The other major feature of a female's attractiveness as a mate to an ancestral male is the likelihood that the children she has will be his.

---

[*]   Buss (1994) and (2003).

Any children of hers that are not his are a huge reproductive penalty for his design information. He would be lavishing resources in the form of paternal investment on non-kin youngsters. Thus, females who are conspicuously promiscuous are much less attractive as mates than females who appear relatively chaste and sexually selective.

Females also show the mate preferences we expect. Given that her nominal mate need not even be the genetic father of her offspring (if she is sufficiently discreet) a human female's primary concern is access to resources. In our contemporary world these symptoms are wealth and status.* In contrast to her husband's attitudes toward her extra-marital trysts, she is less concerned about his dalliances as long as they do not divert substantial resources from her own offspring. His fathering a child by another married woman carries only this (smaller) risk of modest resource loss.

Buss' studies are clear. Women, in general, are more concerned with wealth and status in potential male mates than men are in their female mates. Conversely, men, in general, are more concerned with youth, vigor, and sexual fidelity in their mates than are women in their mates.

The feeling of jealousy is also an interesting window here. Men tend to be intensely jealous of extramarital sex by their female mates and relatively less concerned with their emotional connections to others. Women, on average, are a little less concerned with extramarital sexual encounters by their husbands, but a little more concerned with emotional attachments (that might translate into resources transferred outside the home).

These are all kin-selected tastes and traits we would expect to evolve in the ancestral human mating environment.

Two sexes, two ancestral adaptive strategies, two modern behavior patterns

As we just saw, the sexual appetites of men and women have almost entirely different "purposes." **For men**, the primary purpose of sex is the potential for more offspring—the direct chance to make new copies of their design information. If they can achieve these matings under the appropriate conditions, the resources to raise the resulting offspring might be mostly provided by other men.

For women, in contrast, the actual production of a fetus is not a challenge. Sex for this purpose is widely available to them. Thus, the act of sex

---

*   The social attributes that correlate with male access to resources are also important, of course.

has a different primary purpose for women. Making new copies of their design information is a given. The problem is getting resources to rear those copies (offspring) to successful adulthood. An efficient approach for the ancestral female would likely have been to use male access to mating opportunities as a tool to obtain those resources.

We need to master this perspective. When a woman and a man make love, they are not really participating in a shared experience in the "ultimate" sense. Each is manipulating the other (generally unconsciously) in her/his own interests. Their longer-term interests are sometimes confluent, but often in conflict. In view of this insight, it is remarkable that men and women ever manage to really understand and enjoy one another as sex partners—as, in fact, they sometimes do.

This story will prove to have other rich properties (Chapter 8). For now, we need to think about why pair-bonded couples often do comparatively well in partnering for reproduction. Most do not reproduce like Jason and Janice from Buss' story even though two non-kin mates certainly have conflicts of interest. It will be fruitful for us to remember that the members of this couple are surrounded by many others who take an interest in what they do. For example, our contemporary world includes divorce attorneys and family courts.

Before leaving this subject of the kin-selected behaviors we inherit from our ancient human ancestors, contemplating jealousy is illuminating. Almost every one of us has felt this emotion. It can be incredibly painful—among the very worst emotional pains we ever feel. Adjectives like devastating and crushing come to mind. This mundane fact of our lives is predicted by kin-selection theory. Jealousy is the fear of the loss of either a genetic mate (males) or resources (females), both lethal threats to ancestral human reproduction. Our proximate minds are well designed to recognize this threat.

Men and women really are different – prostitution and erotica

Any mating opportunity has the potential to yield additional offspring for an already mated male, but not for an already mated female. Thus, we expect mated males to be generally more motivated to casual promiscuity than mated females. The sex industry strongly verifies our expectation. For example, the vast majority of customers of prostitution are male. (We are defining prostitution in the legal sense here.) The large majority of prostitutes are female and even those prostitutes who are male mostly serve male (not female) clients.

The fundamental asymmetry in the "purposes" of sex for female and male vehicles is retained in prostitution. Males get (ostensibly, at least) an opportunity to produce an offspring, while females (prostitutes) obtain needed resources (usually money) in exchange. The human pattern of prostitution emerges as one of the many predictable behaviors produced by our inherited, kin-selected sexual minds in the contemporary world.[*]

We expect the erotica to which men and women respond to differ substantially given the fundamentally different purpose that sex has for each. It does, as we are all aware. Erotica for each sex is a massive industry, though each is almost completely different from the other.

Hugh Hefner became a wealthy man in the 20th Century as a result of selling images depicting nubile, healthy young females. These are precisely the most desirable properties in external female sex partners for men. Hefner's and other magazines were purchased by men in substantial numbers. The corresponding magazines (ostensibly) for women (*Playgirl*, for instance) were less successful.

Conversely, we expect females to be a little less concerned about vigorous youth and more concerned about resources (wealth and social status) and commitment (to resource provision) in their mates. Thus, sexual excitement is much less about naked young bodies and much more about context. There is a large industry that sells highly sexualized fantasies built on these desires and marketed almost exclusively to women. These books of erotica go by the trade name romance novels.

We might doubt the symmetry of these two publishing enterprises—visual pornography and romance novels. But if we look at the paperback bestseller list in the Sunday New York Times Book Review section we find that three to six of the top twenty sellers are romance novels—week in, week out, year in, year out. Moreover, the best selling fiction author in all of human history is probably not one of the authors we first think of—Tom Clancy, Shakespeare, or J.K. Rowling. Rather, it is reputed to be the late Dame Barbara Cartland. This famous author of romance novels apparently sold over one *billion* books.

Again, mundane facts of our world are interpretable on basic kin-selection theory. Our sexual minds are what we expect of proximate devices designed by natural selection in the competitive, Malthusian world of our ancestors.

---

[*]   In practice, contemporary prostitution is also controlled by elite power (generally male pimps and organized criminals). We will have more to say about elite power in human social behavior later (Chapters 13-17).

The minds of men and women – proximate and ultimate causation, again

We behave as sexual animals and as parents—as vehicles built by our design information for the "purpose" of replicating that design information. To grasp the power of this view completely, we need to focus once again on the vital distinction between *ultimate* and *proximate* causation of behavior (Chapter 3).

When we talk about the "calculations" that design information makes in influencing our behavior as its vehicle, we are, of course, *not* talking about this calculation *ever* being *explicitly* made *anywhere*. In spite of this crucial fact, these imaginary calculations, nevertheless, describe the *ultimate* strategic logic of our behavior.

Design information itself is not conscious and can make no calculations in any literal sense, whatsoever. Rather, it is merely shaped by natural selection to build vehicles, including their minds. Moreover, these minds-in-vehicles do not make calculations about the "interests" of design information either. Rather, these minds are designed to make simpler calculations in response to immediate signals in their local environments that cause them to behave *as if* they were making the relevant *ultimate* calculations.

These local *as if* calculations are the *proximate* causes of our behavior. However, they can have a quirky historical origin that is only obliquely related to ultimate causation. They commonly do not directly report to us about ultimate causation at all. In other words, the ultimate cause of our behavior is almost always entirely hidden from our conscious minds

To get this fact in clear focus, we can recall the case from Chapter 3. We eat because our minds cause us to feel hungry—*proximate* causation. However, our minds produce this feeling or desire in us because we are shaped by natural selection to contend with the fact that we are non-equilibrium thermodynamic systems who must have continuous inputs of energy to survive—*ultimate* causation.

Equivalently, our desires, such as lust, longing, love, parental affection are the *proximate* causes of our reproductive behaviors. These behaviors are most emphatically *not* produced by our conscious (or even unconscious) calculations about the interests of our design information.

Ultimate causation is a beautifully powerful concept for predicting our eating or sexual behaviors. However, it can sometimes be a terrible, unreliable tool for predicting how our minds think and feel about these behaviors.

This is a profound lesson that we must **never** forget. It will be relevant to *every page of this book,* from here to the very end. If we want to truly understand and predict our behavior, *ultimate* causation is what matters. In contrast, *proximate* causation—what our minds seem to be feeling and thinking as we carry out a behavior—is not directly important to understanding our behavior. These proximate feelings *might* give us some indirect insight from time to time but, just as commonly, they *actively mislead* us. We must always keep our focus on ultimate causation. The idiosyncrasies of proximate causation—of the quirky local calculations of our minds*—cannot be allowed to distract us.

Our *ethical reactions* to some human behaviors are among the strongest outputs of our *proximate* psychologies. Keeping these in check will be vital if we wish to achieve a detached, analytical understanding of the origins and causes of those behaviors. Only by taking this step can we keep our focus on the *ultimate* causes of human behaviors. We will find that many of those behaviors are actually intensely mercenary in their service to the interests of design information. These objectives are commonly hidden from us even as we actively try to look into our own minds. We will remind ourselves of this vital point from time to time.

What more can we predict? Kin-selection and conspecific conflicts of interest

Just *exactly* what other behaviors do we expect kin-selection to produce? This question is slightly more complicated than it appears at first glance and *this small complication is important for us.*

In the online endnotes to Chapter 3, we discuss infanticide by field voles (small rodents). During breeding seasons in their very crowded nesting areas, adult voles circulate throughout their territories, killing pups in the nests of others when the parents are away. The adults (both sexes) have an interest in killing the offspring of surrounding adults because those non-kin infants will grow up to compete with their own kin offspring.† They kill infants, but only when the parents of those other youngsters are away from the nest. If the parents are home, the nest is safe.

---

* We are *not* saying that our proximate feelings are unimportant to our everyday enjoyment of life or our sense of fairness about the world. They are a part of being human. Thus, they are and should be an important consideration as we decide how we want to build the human future (Chapter 17 and Postscript).

† Again, actually their design information has an "interest" (Chapter 3).

Now we can contrast this behavior with the lions in Chapter 3. Franklin was "home," but Arthur still fought him to the death then killed his offspring anyway.

What is the difference in these two cases? First, we must notice what the difference is *not*. In each case, a parent is on the scene with a large incentive to protect its (kin) offspring. The violent confrontation with this adult will inevitably entail a risk (a likely cost) to the would-be infanticidal (non-kin) adult. Arthur sustained potentially fatal injuries in his struggle with Franklin and he could have lost that struggle.

If the *costs* of infanticide are the same in both cases, exactly why do the two situations play out so differently? The *benefits* of infanticide are very different.

In the lion case, the costs of violent conflict are offset by a huge potential reproductive advantage. Arthur was able to father several litters of cubs who were his own (50 percent related).

In contrast, in the vole case, this same large cost would be paid in return for a smaller, more indirect benefit. For example, when a male vole commits infanticide he will not usually obtain any extra mating opportunities. Rather, he is merely eliminating some youngsters who *might* otherwise grow to adulthood and *might* come into future competition with him or his offspring in the crowded Malthusian world of these animals. If this smaller benefit can be had cheaply, it is worth having, to be sure—but it is *not* worth risking their lives.

So, we expect natural selection to design two different behaviors in the lions and the voles and that is what we see.

This is a complication in understanding the behaviors we actually find in the world. It illustrates one way in which kin-selected behaviors emerge (or do not emerge). Kin-selected behaviors can arise when their net benefits according to Hamilton's Law are positive, of course, but this is merely *necessary*, not *sufficient*. *It is also necessary that we consider the reactions of non-kin conspecifics to such behaviors as part of the cost.* We can call these the *costs of opposition*. A behavior favored by Hamilton's Law, *considered in isolation*, would still *not* evolve if the costs of opposition are large enough to affect Hamilton's inequality (Chapter 3). For example, the divorce attorneys and family courts of Jason and Janice's world impose costs on cheating spouses.

The behavior of our voles helps us think about this. The best *imaginable* behavior for them would be to kill all other field voles except their mates and offspring and to resist the attempts of all other voles to do the same thing. The best *actual surviving* behavior is something rather different. Since all other voles would violently resist such genocidal behavior, the best remaining strategy is to engage in this behavior only when its costs are sufficiently low. Attacking helpless youngsters but not mature adults is a workable strategy.

In other words, we see what we expect in non-human animals once we understand all the issues. When we ask what behaviors kin-selection *actually* produces in the *human* case, we will find many things that we expect by extrapolating from non-human animals. *However*, we will also find that humans sometimes do not display (or display rarely) some kin-selected behaviors that non-human animals show frequently and conspicuously. More specifically, we will find that we need to include the reactions of other non-kin humans in our predictions in a way that is much more important in humans than in any non-human animals. Understanding this difference will help us immensely in our larger quest to build a new theory of ourselves.

"Good data" – homicide and kin-selection

Some of the empirical evidence (data) about human behavior we discussed above was relatively simple to obtain. It is stored in the life experiences of each of us. However, we need more if we are to build confidence in our understanding of the quirky way kin-selected behaviors evolved in humans.

This is a difficult task for many different reasons. For example, we lie to one another and even to ourselves. Consider how you would answer a set of questions that required you to reveal publicly every detail of your sexual feelings toward others. Or suppose you were required to be fully candid about your opinions regarding social friends and professional colleagues. *Could* you answer these questions completely—and even if you could, *would* you? As well, much of our behavior is controlled by impulses to which we have, at best, only partially conscious access. Even if we wanted to, we could not report them accurately (Chapter 10).

Can we get reliable information about our impulses and behaviors (ostensibly shaped by kin-selection) in spite of this problem? Margo Wilson and Martin Daly recognized that there is a very good way to access this information by looking at homicide statistics. Their approach is brilliantly simple.[*]

In modern nation-states (like Canada and the United States where Daly and Wilson work), governments insist on the fullest possible infor-

---

[*]   Indeed, we already used a little of Daly and Wilson's data when we talked about step parental infanticide near the beginning of this chapter.

mation about homicides.* Contrary to what we might think from skewed media coverage, most homicides in most places are solved.† We usually know who killed who.

Daly and Wilson used homicide statistics to ask the kinds of simple, yet penetrating, questions that are the life-blood of good science. Their book *Homicide* is one of the master works of 20th Century social science. We recommend it in the strongest possible terms to every reader of this book. A small sample of its insight will help us here. We will find a highly illuminating pattern in Daly and Wilson's observations.

Strategic infanticide – Susan Smith and the killing of our own children

Human parents sometimes kill their own (kin) offspring. At first glance, this behavior seems like an even more egregious disconfirmation of kin-selection theory than non-kin adoption. Actually it is not. In fact, it is just an extension to humans of the same logic we saw in our cannibalistic infanticidal mouse in the online endnotes to Chapter 3. The essential specifics are these. Offspring are only 50 percent related to parents. They have only a partial (though substantial) confluence of interest (Chapter 3). If killing (even eating) kin offspring has a sufficiently high future payoff and a sufficiently low cost, it will happen. Non-human animals commonly abandon, kill, or even consume their own youngsters under the right strategic circumstances. They are apparently generally serving the interests of their design information when they behave this way.

To understand better how this logic applies to humans, we have to consider a few details of how human parents invest in their offspring. First, for a vehicle to be successful, it is not necessary that all its potential offspring survive, merely that a sufficiently large number of them do.

Second, how many offspring can actually be successfully produced (reared to reproductive maturity) is limited by access to parental

---

*   We will return in Chapters 13 to 17 to the reductionist explanation of exactly why "the state" is so interested in this information.

†   Clearing rates for homicides have gone down somewhat over the last decade because of the rising relative importance of gang- and drug-related anonymous killings. However, this does not impact the quality of the data Daly and Wilson use.

investment. This includes one's own investment and the investment of mates.*

Third, the cumulative parental investment made in a child is dependent on the age of the child. The older the child, the more the parent has already invested and the less future investment is required. Such a child is more valuable as a vehicle than a younger, more needy offspring.

Daly and Wilson use a simple measurement of the relevant variable here. They call it *reproductive value* (Figure 4.1). From the point of view of the design information that builds a parent (vehicle), the parent's own body and that of any of the parent's offspring has a value at any moment in its life that corresponds to its likely *future* contribution to replication of this design information.

Two things cause this reproductive value to change over the life of the individual. One is infant mortality. In the ancestral environment, roughly half of human infants probably died before reaching adulthood. This attrition resulted from things like childhood diseases (that either kill or immunize) and the risks of death from injury or starvation before one accumulates adult skills. Thus, from the point of view of an ancestral parent's design information, a newborn is only half as valuable as an older adolescent offspring about to enter reproductive adulthood.

Thus, if killing a newborn now increases the likelihood of the future successful production of a mature adult offspring, killing the newborn "makes sense."

The other factor that influences reproductive value is parental aging. From the point of view of a parent's design information, the value of the parent's own body *at any moment in time* is proportional to the expected *future* reproductive output of that body. So a young adult parental body is more valuable than an older one. Thus, when a parent (its design information) makes a decision about the value of a child, this value is normalized to the future reproductive output of the parent.

This last point is slightly subtle. We can enlarge our understanding of it as follows. For a young parent, a newborn is a comparatively small fraction of its remaining potential reproductive output. Sacrificing such a newborn now under circumstances that might improve the future reproductive output of the adult can have a high value. For an older adult, this same decision to kill a newborn means risking a much larger fraction of the adult's total *remaining reproductive output* to enrich *a much shorter remaining period* of future reproduction. A "good" decision for a young parent to kill its own newborn can be a "bad" decision

---

*   This also includes social investment from non-parents in the human case (Chapters 6 and 8), but we can ignore this part of the story for the moment.

for an older parent. As always, we are considering *ultimate* causation of behavior here—*not* the idiosyncratic conscious calculations in the minds of the specific people involved.

Now we can understand how natural selection acting on design information can build human parents who will sometimes kill their own (kin) children. Sometimes the killing of a child (one copy of design information) can actually improve the chances of producing other copies in other offspring in the future, at least in the ancestral environment that shaped our minds. The design information that behaved this way in the ancestral past survived preferentially.

Does this picture actually predict what Daly and Wilson find in the homicide data? It does. They observe exactly what kin-selection theory predicts. Young parents (high *future* reproductive value) are much more likely to kill their own offspring than are older parents. Further, very young infants (lower ancestral reproductive value) are much more likely to be killed by their natural parents than are older children (higher reproductive value).

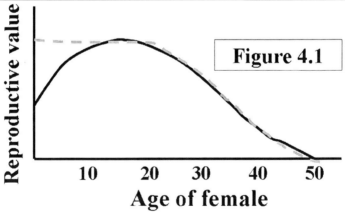

Reproductive value of an individual as a function of age in the ancestral (solid black) and modern (dashed gray) worlds

Figure 4.1

Reproductive value

10   20   30   40   50

Age of female

Figure 4.1: This figure plots reproductive value as a function of the age of a female. (Male reproductive value is similar in its overall properties.) Notice that these plots reflect the number of likely *future* offspring a surviving individual will have at each age as she progresses through her life. Our evolved minds are shaped by the reproductive values of individuals in the ancestral past (solid black line), not by their new reproductive values in the context of our very recent contemporary world (dashed gray line). Data are from *Homicide*, Daly and Wilson (see text).

Daly and Wilson also use various data from people living in environments other than modern nation-states to argue that these patterns are in fact, quite universal—as kin-selection theory also predicts.

A famous example illustrates this pattern. In the American South in the early 1990s, a young woman named Susan Smith reported to the police that her car containing her two young children had been hijacked. She went on national television to make a tearful plea for the safe return of her boys, Michael and Alex.

Police investigation ultimately revealed that Ms. Smith had driven her car into a nearby lake with the boys still in the car, belted in place in the rear seat. Maternity is never in doubt. Ms. Smith had killed her own (kin) children.

The details of this case are archetypal and illuminating. Ms. Smith was young and estranged from her husband. Neither she nor her husband was wealthy. The likely future success of any of her children (Michael, Alex, or new children) could be enhanced by obtaining a new mate who could provide more resources.

Ms. Smith had been dating an excellent candidate for this new source of resources, a wealthy local eligible bachelor. However, he had broken off their relationship, at least in part because of her children. Acquiring the valuable assets for possible future reproduction represented by this new male was the potential benefit. Eliminating her young (low reproductive value) children early in her own potentially long reproductive life was the (low) cost.

In an ancestral environment, this was a choice that might well have been worth making. Ms. Smith made that choice. Had she evaded detection she might, in fact, have had other future children with better "prospects."

The upshot of Daly and Wilson's analysis of homicide data is that killing of children by their natural parents, ironically, represents some of the strongest evidence *for* kin-selection theory.

We also confront, yet again, what we saw with step parental infanticide. Humans execute strategic infanticide of their own kin offspring, just like non-human animals do. However, we apparently do so much, much less frequently than non-human animals. Moreover, the reason for this lower frequency appears to be countervailing social reactions from other humans. We see competitive human kin-selected behaviors occurring, *but* strongly attenuated, apparently by the *cost of opposition* by non-kin conspecifics. This is a recurring pattern in homicide data.[5]

We are peeking through a window onto human uniqueness.

Adult homicide, a young adult male business

The basic asymmetry in the roles of males and females in the reproduction of mammals lets us make some useful predictions. The female provides a large parental investment unique to her in the form of incubation of the fetus inside her body and nursing the newborn. This asymmetry has a profound effect on the optimal strategy for design information when it is building a female vehicle and when it is building a male vehicle.*

What kinds of male and female vehicles should human design information build? One easy way for us to approach this question is to consider a simple implication of male/female asymmetry. When a male mates with multiple females, each mating increases his potential reproductive output (increases the number of copies of design information he can make). In contrast, for a female, mating with multiple males rather than a single male has little or no direct effect on her reproductive output. She cannot be "a little" pregnant.†

Male vehicles that compete successfully with other male vehicles for access to mating opportunities with females can leave more copies of their design information. Males will fight over females. In contrast, females who compete for access to males are often wasting their time. From the narrow perspective of actual mating and fertilization, males will come to them if they simply wait. Moreover, female competition for access to resources from multiple males must be discreet given its fundamental logic, as we saw above. Thus, overt physical violence (inherently public) is a relatively useless strategy for females to deploy against one another in this particular context.

In the complex social world of humans, things may seem a little more complicated than this, as indeed they are, but only *a little*.‡ The

---

* We have only tiny amounts of *sex-specific* design information—negligible from our point of view here. The design information that builds a male, for example, came from both parents, half from his mother and half from his father. But the half from his father was originally half from *his* mother, and so on and so on. Essentially every bit of design information that builds a male was building a female, one, two, or three generations ago—and vice versa for a female.

More generally, essentially every piece of human design information has spent just exactly half of the last two million years building females bodies and minds and the other half building male bodies and minds. Thus, design information evolves to "know" which sex it is building and to build bodies and minds accordingly.

† As we saw above, the beneficial effects of multiple female mating are restricted to resource access—a smaller return.

‡ Chapters 6 and 8.

underlying mammalian pattern is still very much a part of our lives. Human design information has been shaped to make many of the same male/female-specific choices as the design information of any other mammal. Only the superficial "rules of engagement" are human-specific.

This is a strong, categorical statement. We make it because the empirical evidence is clear. Daly and Wilson's analyses provide excellent illustration. For example, in the ancestral human environment, who do we expect to be overtly competing most aggressively for mating opportunities? Males who do not yet have mates, of course. Who are such males? They are generally *young* males. Thus, what kinds of males do we expect human design information to build? Males who overtly compete with one another intensely, especially when they are young and un-mated.

What does this predict about homicide data? We should expect young, unmarried males to commit most homicides and we should expect males to be the primary targets. This is precisely what Daly and Wilson find.

First, males are nearly ten times more likely to commit homicides than females, overall!! This fact is quite amazing in its own right.

Second, males are two to three times more likely to be the victims of homicide than females.

Third, young males entering adult reproductive years (ages twenty to twenty-four) are roughly five times more likely to commit homicides than older, established males (ages fifty to fifty-four).

Fourth, if homicides often grow out of competitions for mates (directly and indirectly, consciously and unconsciously) we expect unmarried males to commit more homicides and to be homicide victims more frequently. As we previously stated, they are—about two to three times more likely.

Fifth and finally, there is another wrinkle to this evidence with a lesson for us. We said that females can wait for males to come to them. Though this is true, not all males are of equal "value." Thus, it is in the interest of females to wait for males who can contribute the most paternal investment to offspring. Males who have resources are more attractive to females than males who do not.

This leads to a simple prediction. Unemployed males (intrinsically less attractive to females as a result) should compete more fiercely with one another for mates than economically successful males. The homicide data confirm this picture. Unemployed males during peak reproductive years (ages twenty-five to forty-four) are about five times more likely to commit homicides than are employed males of the same age.

Of course, as readers of newspapers, most of us would not need the homicide statistics to tell us these things. We are aware that homicide really is a young male business. Another mundane fact of life becomes understandable and predictable to us.

Moreover, as we saw with the other kin-selected homicidal behaviors above, this one is much more rare in humans than in other animals. A very large fraction (possibly a *majority*) of adult male lions perishes as a result of doing battle with other male lions. This is a very high frequency of "leocide." In contrast, the risk of homicide for human males is much, much lower, at least in the context of the modern state.

Again, the properties of our societies suggest that the cost of opposition by conspecifics is responsible for this much lower human homicide rate. Manslaughter and murder are serious felonies in contemporary states and there are police, prosecutors, and courts designed to impose very large costs on those who commit them.

One last issue before we leave this subject. There is a large racial difference in homicide risk in the United States. This has lead to the suggestion of race-specific proclivities to violence. This is a continuing question. However, if we subtract away the effects of age and marital/economic status, a large fraction of these race effects goes away.

The biology of good and evil

Some non-human animals probably have a rudimentary ethical sense.[6] However, humans seem to have this particular sense in spades. Indeed, after our directly kin-selected emotions (love, lust, jealousy, and parental affection) our feelings of moral outrage and guilt are probably the most intense emotions we experience.

In order to approach our own behavior objectively, we have been holding this uniquely human ethical sense in check from the beginning of the chapter. It is now time to bring it back into the picture. Seeing how our ethical feelings relate to kin-selected human behaviors and to the *cost of opposition* is extremely illuminating.

Think about the specific events of this chapter that provoked your ethical sense—Janice and Jason, the infanticidal stepparent, the corrupt nepotist, Susan Smith. In each case, what did the person do to become a target of our moral outrage? They simply executed a behavior that had the potential to improve their chances of producing extra copies of their own personal design information—extra kin offspring.

They were obeying Hamilton's Law but they were doing so in a way that entailed substantial costs to others. Moreover, those others (the surrounding human social enterprise) were able to impose costs of opposition, which made these behaviors (usually) unfavorable from an adaptive point of view. Thus, these behaviors are much less common in humans than in non-human animals.

Importantly, these behaviors provoke **moral outrage** in us and, thereby, cause us to participate (directly or indirectly) in imposing *costs of opposition* on the offenders. Notice *also* that these behaviors are precisely those that would provoke **guilt** in us if we were to engage in them, leading most of us, most of the time, to avoid them.

These facts are *not* coincidences. Our uniquely human ethical sense is designed by natural selection to do *precisely* these things. Our ethical sense is apparently a proximate psychological mechanism designed to suppress those kin-selected behaviors in others that impose unacceptable costs on us.

We can make an even stronger statement—one that we can fruitfully contemplate at length. The world is often portrayed as being a struggle between good and evil. We now have the first glimpse of the fact that good and evil have a straightforward (uniquely human) biological definition.

Evil consists of the pursuit of the interests of personal design information that are in conflict with the interests of the design information of the non-kin humans who are members of our own cooperative social set.

Good can also include the pursuit of interests of design information. However, it entails pursuing those interests in ways that do not conflict with the interests of the non-kin design information in other members of our social set. This happens in two ways.

First, we can pursue self-interest in ways that also actively serve the interests of the design information in non-kin members of our social set. This is our "good" *public* behavior—our *economic* behavior. We can earn a living by being a good person as an economic actor. For example we can be a good teacher, a good physician, a good carpenter, and so on.

Second, we can pursue self-interest in ways that do not directly contribute to non-kin interests, but do not conflict with them too violently either. This is our "good" *private* behavior. For example, if we raise our children in a way that does not prey on other parents and that teaches these children to be good public persons, we are said to be good parents.

This is a little tedious and clinical. But we must not be distracted. We are encountering a profoundly important fact about the world. We will ultimately find ourselves in the position to establish a *biology of good and evil.* Good and evil are not properties of the universe. Rather, they are idiosyncratic, internally self-referential properties of the uniquely human social world. We can make a strong case that the conflict between good and evil is just the conflict between our interests as participants in (uniquely human) kinship-independent social cooperation and our interests as kin-specific cooperators (universal animal behavior).

We can enrich this insight by recalling that non-human animal social behavior involves directly pursuing the interests of the individual animal's personal design information. And, when a human is evil or destructively pursuing the interests of her/his design information, we often say that person is "behaving like an animal." This choice of words is not an accident. It is *not* a metaphor.

We will find ourselves returning to this picture of our ethical sense many times.

Two final questions. First, *why do only we humans look at the world this way?* It seems to work so well for us—we walk on the Moon and rule the world. Secondly, *why did it take hundreds of millions of years for the first "moral animal," us, to arise?*

We will argue that there is an utterly simple, elegant answer to this profound, central question. We turn our attention to it in the next chapter.

᧳

Chapter 5
Death from a distance – the evolutionary logic of kinship-independent social cooperation

This chapter's title sounds foreboding. In one way it really is. We will discover for the first time some unexpected things about what it means to be human **at the most fundamental level.** *In fact, the single most fundamental thing about us, from which everything else flows, is a unique capacity to carry out (and threaten) violence against one another* **inexpensively.** *The costs of coercive violence against conspecifics can be inherently low for an animal that can kill from a distance. We are that animal. At first glance, this does not sound like a very elevating view of our humanity.*

*In another way, though, any foreboding is utterly misplaced. Our unique capacity for inexpensive violence against our fellow humans is precisely what makes possible our uniquely human ethical sense with all its humane potential for our world. Our unprecedented relationship to coercive threat is the basis—***the complete and indispensable basis***—for uniquely human social cooperation* **independent of genetic kinship.**[*] *This capacity for inexpensive coercion permits us to manage the conflicts of interest that almost always prevent cooperation between non-kin members of the same species in non-human animals (Chapters 3 and 4).*

At this point, you might be thinking that this is mildly interesting, but not what you signed on for. It does not seem to have anything to do with what we really care about such as our economic system, present day politics and war, education, the quality of our lives, and so on. Likewise, this claim

---

[*] The theory described in this chapter is built on earlier work published in the professional scientific literature (Bingham, 1999, 2000; Okada & Bingham, 2008).

does not seem to have anything to do with the specifics of our uniqueness, our human speech, our brilliant minds, our conquest of the planet and all the rest.

If these are your thoughts, you could not possibly be more wrong. We will see in later chapters that all the things we think of as making us unique among animals are apparently just consequences (effects) of this single cause—the control of conflicts of interest **as a consequence, a by-product** *of our use of inexpensive coercion in pursuit of individual self-interest. Put differently, we display a special pattern of social coopera-tion and this pattern is made possible, for the first time in the history of the Earth, by our unprecedented access to individually inexpensive coercive threat.*

There is a subtle confusion we need to be vigilant against from here on. Humans have become a completely new kind of animal, never seen before on Earth. However, the process that produced us is **not** *new. All humans, including us today, are just as completely the products of natural selection as any other non-human organism. Though we are novel crea-tures, there is no magic, no new forces, no anomalies in our evolutionary history. Our lack of objectivity about ourselves will seduce us to think that at some moment in our history, the rules suddenly changed for us. They did not, not ever.* As we proceed, try to see how the processes that produced us are quite mundane. This perspective is a useful antidote to self-defeating conceit.*

At a few points, this chapter may seem a little abstract, possibly even bloodless. Be patient, the insight we build here will explode into life in later chapters. It will, quite literally, let us truly understand our common humanity for the very first time.

We might choose to think that this picture of the ultimate cause of our humanity is appalling, if true. It is almost certainly true, but it is not appalling. Understanding exactly why we say this with such confidence will require the journey of the remainder of the book. But we can give you a first sense of this right here. For example, our deployment of inexpensive coercive threat in pursuit of self-interest is what enables us to impose the high costs of opposition that make murder and infanticide so rare in humans (Chapter 4). More generally, our unique access to individually inexpensive coercion makes possible all of our most noble endeavors, including our proximate desires to contribute to the greater good. Indeed, our use of coercion is what liberates the "better angels of our nature."

---

*   In spite of this, our growing knowledge of our origins and status in the universe will give us many rich opportunities to make more humane choices than our blindly ignorant ancestors would have made (Chapter 17 and Postscript).

Non-human animals and conflicts of interest - *Inter*specific cooperation

We forecasted that the uniquely human evolution of *kinship-independent* social cooperation is the thing we need to understand. No other animal cooperates in this way on a scale that even remotely approaches humans. We argued in Chapter 3 that *conspecific* conflicts of interest are, in fact, the barrier to the evolution of kinship-independent social cooperation. If this conclusion really is correct *and complete*, the existence of this barrier should be the *sole* problem limiting the evolution of kinship-independent conspecific social cooperation. That would mean that solving the conflict of interest problem would immediately allow the evolution of cooperation between non-kin individuals.

One way to see if this picture is accurate is by looking at cooperation between members of different non-human species—*interspecific cooperation*. This kind of cooperation should be possible because members of different species can have different needs. Having very different needs means that what is a *limiting resource* for one of them need not be for the other. Individuals of two very different species do not necessarily compete significantly with one another.

Thus, individual members of very different species will sometimes find (by the blind processes of natural selection) that their conflicts of interest are few and mild. It follows that they can evolve cooperation. *Again, this will be true* **if** *individual conflicts of interest really are the nub of the issue.*

In fact, this expectation is often fulfilled when we look out at the world of non-human organisms. Cooperation between individuals from very different species is widely observed. It is just as *common* as cooperation between non-kin members of the *same* species is *uncommon*. Here are some examples to remind us.

For many flowering plants, reproduction occurs when an animal, like a bee or a humming bird lands on the male parts of the flower of one plant. This animal "inadvertently" picks up pollen (the plant's sperm) on its body. It then moves on to another plant of the same species, sometimes landing on the female parts and leaving behind some of the pollen from the first plant. *Voila!* The two plants have successfully mated.

What concerns us is the ultimate causation of these reproductive practices. They represent cooperation, a business deal. Specifically, our bee is self-interested and is looking for nectar to make honey to feed its nest-mates. Our hummingbird is also self-interested. It is looking for sugar-rich plant nectar to consume on the spot to fuel its frantically high-energy lifestyle.

Likewise, the plants are self-interested. They want the animals to act as devices for transporting pollen. They do not have any use for the nectar they produce other than seducing the animals to do this work for them—a fair *market exchange* of value between a plant and an animal.*

Another remarkable class of examples of cooperation between ecologically different organisms enriches this picture. Ants are famously social creatures.[1] Their cooperation, as members of the same species, will be very illuminating to us in a few moments. For now, let us just accept this conspecific cooperation and focus instead on the cooperative relationships of members of some ant colonies with other organisms.

Farming leaf-cutter ants of the South American rain forests and savannas excavate enormous underground tunnel systems. These subterranean nest complexes can house millions of colony members. Worker ants from these colonies spend all day going out to cut the leaves of surrounding trees into pieces and bring them back into the underground nest. Other colony members process these plant cuttings into food for a fungus that the ants cultivate and care for. As the fungus crop matures, the ants feed on it, saving spores to plant new crops. Fungus gardens at different stages of maturity are spread throughout the nest complex, providing the ants with a stable food supply.

This relationship is so ancient and highly evolved that these ants can no longer survive without the fungus nor the fungus without the ants. Again, this cooperative business deal between ants and fungus (very different species with few conflicts of interest) is just exactly the kind of thing we would predict—*if conflicts of interest are the* **sole** *impediment to cooperation*.

Another example of ant business deals with other species involves "dairy farming." These ants keep a kind of bug called an aphid in their nest along with their own colony members and offspring. They protect and rear young aphids. The adult aphids go out (with ant protectors) and feed on plant sap. They are highly adapted to using this food source. The feces of the aphids form as a droplet at the back tips of their abdomens where they eventually drop as honey dew. However, this material still contains a lot of the sugars and amino acids from the original plant sap. The aphid's digestive system does not remove every last bit of nutritional value. The ants feed on this waste, imbibing the droplets directly from the tip of aphid's abdomen.

---

* Note that both participants in this deal have ways of policing one another. The bees will not come to flowers that do not produce good nectar and the flowers can place their pollen so the birds and the bees cannot realistically evade it. Moreover, members of neither species are in the position to actively force the other to provide its part of the reciprocal deal without payment.

When a particular plant ceases to be a good source of food, the ants gather up their own youngsters and the aphids' and set off for greener pastures. The ants and aphids exchange food for protection and transportation. These species are highly adapted to one another. Without conflicts of interest to stand in the way, interspecific cooperation can be a very effective approach to making a living.[2,3]

The crucial point for us is that members of the **same** species could take advantage of the many benefits of cooperation, like ants cooperating with aphids, *if* they could get around the conflicts of interest that prevent them from doing so.

Non-human animals and conflicts of interest - *Intra*specific cooperation in ants and bees

Farming or cattle-herding ants and the colonies of bees sending out legions of pollinators have another lesson to teach us, a lesson that is central to our understanding of uniquely human social cooperation. If conflicts of interest between ecologically similar individuals are the decisive barrier to cooperation, these conflicts should usually be worst, most intense between members of the *same species* (Chapter 3).

However, ants and bees live in enormous cooperative colonies of conspecifics. Does this fact tell us that the conflict of interest problem is less important than we thought? No, it does not. Ed Wilson famously referred to these colonies as *insect societies*, analogizing these colonies with human societies[*]. We argue that Wilson's analogy is both more profound than he knew *and* also a little misleading.

Wilson's analogy is misleading because ant and bee colonies are less like human societies and more like human nuclear families. All the millions of individuals who make the colonies up are usually, mostly very close genetic kin. In fact, most bee and ant colonies are actually massive sisterhoods. Virtually all of their members are offspring of a queen and are sisters and half sisters. Most of the (female) members of the colony work to raise their sisters (and a few brothers) who are close kin. This behavior is quite analogous to raising their own offspring from the point of view of genetic design information.[†]

---

[*]   Wilson (1971).

[†]   Why bee and ant societies do not have many regular adult male members apparently has to do with the consequences of their chromosomal sex-determination system and is not directly important for us here.

So, part of the explanation of massive bee and ant colonies is the kin-selection we explored in Chapter 3. However, there is another, quite remarkable part to this story. This part will show us how the society analogy for insects is profound. To appreciate this other part we need to look at a couple of details of the logic of bee and ant reproduction.

Natural selection acts through differential replication of design information (Chapter 2). Moreover, what matters for the evolution of most animal social cooperation is the probability that a piece of design information in one animal that "looks out" at another conspecific animal, sees another copy of itself (identical by virtue of having been inherited by both animals from a very recent common ancestor).[*]

Design information that builds animal behaviors to help other conspecific animals based on this probability of identity will do better for the simple, obvious reasons we saw in Chapter 3. This probability, again, is given the formal term *relatedness*.

Because of the details of the particular chromosomal inheritance system in bees and ants, females are more closely related to their sisters than to their daughters. Thus, workers raising their sisters are easy to understand as a product of kin-selection.

However, the rarer males produced by the colony are a different and more interesting story.[†] The queen, whose eggs hatch into all the workers of the colony, must have a male mate—called a drone—to produce these worker females. This mating with queens is essentially the sole purpose of drones. They are generally born, leave the colony in search of a new young queen, mate with her, and die—a short, happy (?) life. The queen stores the drones' sperm in a special organ inside her body, for years in some cases, and dispenses it one sperm cell at a time to produce fertilized eggs, worker females, as she wishes.

In bees and ants (again, because of their special chromosome system), females are 50 percent related to their own sons, but only 25 percent related to their brothers. Moreover, it is possible for colony worker females to lay eggs that will develop into males even though they have not mated. Male bees and ants are produced by unfertilized eggs, virgin birth, so to speak. This is a very different system than in a mammal like us, of course.

---

[*]   See Chapter 3 and Figure 3.1 for a discussion of this highly accurate metaphorical description of kin-selected social behavior.

[†]   This story is actually several different stories, depending on which ant species we are talking about. We will be looking at only one of these stories here.

Thus, raising brothers (25 percent related) instead of sons (50 percent related) appears to contradict kin-selection theory.* This pattern looks like a violation of Hamilton's Law. Therefore, when colony females raise brothers, they appear to be cooperating in violation of kin-selected interest. They are displaying a (very limited) form of kinship-independent social cooperation.

How does this happen? Let us go back to the queens when they first establish these colonies. They usually mate promiscuously at the beginning of their reproductive lives. Thus, the colony females—produced from the queen's fertilized eggs—are a mixture of full sisters (having the same drone father) and half sisters (having different fathers). Moreover, when a worker ant produces a son (lays an unfertilized egg), this son is 37.5 percent related to his mother's full sisters (to his ant aunts). Thus, these full sisters should agree to raise their sons together (50 percent and 37.5 percent related) rather than raising their brothers (25 percent), given Hamilton's Law.

However, the presence of half sisters creates a new problem. The sons of half sisters (half nephews) are now only 12.5 percent related to their mother's half sisters. Moreover, individual colony females *cannot* tell which of the other colony females are full sisters and which are half sisters. They all look too much alike. Finally, there are more half sisters around than full sisters.

What is a worker female to do? Her best bet is to produce her own sons (if she can). Her second best bet is to help her full sisters raise her full nephews (but she cannot recognize these). Her third best bet is to raise her brothers. Her *least* favorite choice is raising her half nephews.†

How is her dilemma resolved? Here is where the story gets very interesting for us. Each worker eats her sister's (mostly half-sister's) male eggs (her least favorite choice). If they survived, these males would compete for colony resources that would otherwise go to males who are more closely related to our worker. Moreover, she would otherwise be raising half-nephew males who would leave the colony and compete for future mating opportunities with males more closely related to her. So, when she eats these (mostly) half-nephew eggs, she is pursuing individual self-interest (actually the interest of her design information, of course).

---

* These relatedness values are also different than for humans yet again, because of the different chromosome system in bees and ants. This chromosome system is also the reason ant and bee females can give virgin birth to sons.

† All of these statements reflect the worker female's *ultimate* motivation, of course. Ultimate causation determines how she will behave. We know nothing about what she is actually thinking, and we do not need to (Chapter 3).

Thus, she may try to lay a few eggs of her own—her favorite choice. But this behavior mostly fails, because her sisters (mostly half sisters) feel toward her eggs just exactly as she feels toward theirs and they eat hers. So she "settles" for her best **remaining** option. She rears her brothers.

Put differently, she lays a few of her own male eggs. But she invests only a little in this high-return, but high-risk approach. This behavior is a little like us buying an occasional lottery ticket. She puts most of her effort into the somewhat lower return, but also much lower risk activity of raising her brothers—a "solid investment."

Let us take stock. From the point of view of an individual female, her best option is closed off because of the infanticide perpetrated by her sisters. Her second best option is not viable because she cannot distinguish her full nephews from her half nephews. So, she settles for her third best option.[*,4]

There is another way to look at the logic of these ant behaviors, a way that is extremely useful in understanding uniquely human social cooperation. Each colony female carries out self-interested violence (infanticide) to suppress the self-interested behavior of her half-sisters (raising their sons, her half nephews). However, she is simultaneously a target of the analogous violence coming back at her from her sisters. Thus, her behavior is changed by the resulting natural selection. She is adapted to spend most of her effort raising her third-best choice (brothers) because design information that builds this pattern of behavior does best *under these circumstances*.

The fact that the presence of her sisters' violence (and threat of violence) changes her best reproductive behavior means that we can call this violence **coercive**, of course. Each colony female is actively coerced to pursue some behavior that is less desirable (less efficiently self-interested in an immediate, short-term sense) than another behavior she would pursue in the absence of that coercive threat. Moreover, this active coercion is individually self-interested. It is the product of natural selection acting on design information *building the coercing individual*.

In fact, natural selection is creating individual behaviors that, themselves, become sources of future natural selection on other conspecific animals. **Effective pursuit of self-interest is very strongly influenced by what other self-interested conspecifics are doing.**

---

[*]   Our worker female could leave the colony and attempt to raise her own sons (50 percent related). But the likelihood that she would fail in this dangerous undertaking is probably greater than 50 percent. Thus, staying home to raise her brothers (25 percent related) is a net positive from the point of view of her design information.

Finally, because each individual worker female is forced to raise her brothers rather than competing with the queen to raise the worker's own sons, the colony is very stable. There are no revolutions in this society. It can grow very large and continue to exist for years. Thus, individuals who are members of this highly cooperative endeavor will often do better than individuals who try to live outside of it or who live in smaller, unstable colonies where workers compete with the queen to raise the worker's sons rather than their brothers.

But this *group advantage*—the advantage of being the member of a large stable colony—is **not sufficient** to produce this cooperative behavior (brother rearing). *First*, immediately self-interested son-rearing has to be coercively suppressed, *then* this cooperation can emerge.

Group selection (again)

We talked about group selection theories of social cooperation in Chapter 3. Such theories are usually wrong and our bees and ants help us increase our grasp of why this is.

Large cooperative colonies can out-compete smaller or less cooperative colonies. Thus, we might be tempted to say that this advantage of the cooperative group (the colony) represents the selection for the cooperative behaviors that make the colony possible. However, when we make this suggestion, we are committing the group selection fallacy.[5]

The group advantage of being a colony member results from coercive suppression of *individual* conflicts of interest as we just discussed. Moreover, this suppression of conflicts of interest is, itself, the consequence of *individually self-interested* actions (policing by eating half-sister-laid eggs).

It is useful to state this conclusion from a different point of view. If, for some reason, worker policing were not possible, stable, brother-raising colonies would never come into existence in the first place or would be immediately destroyed by workers cheating on the system if they did. Never mind that such colonies would be more well-adapted "groups" if they *did* exist. Group advantage almost never determines evolutionary outcome, even if those group selected outcomes might be better. Only head-to-head, in-the-moment competition of *individual* strategies can determine what social behaviors evolve under almost all circumstances.

This point is very, very important. If you feel you understand the fallacy of group selection arguments, we can proceed. If you still find group selectionist thinking attractive look at online Endnotes 6 and 7 for further clarification. We really must leave the powerful illusion of group selection behind if we are to understand how humans came to be the unique animal we are.

A pivotal moment – what is a true society?

Our bees and ants, using worker policing to create the circumstances for cooperation, carry an essential insight for us. *Under the right circumstances*, individuals sometimes cooperate at variance with their *optimal,* **short-term** self-interests (ants raising brothers instead of sons, for example). When they do this, they are still pursuing self-interest, but doing so more obliquely, indirectly. We said previously that Ed Wilson's society analogy for insect colonies was right in one profound sense. This pattern of coerced cooperation against short-term self-interest is that sense.

Societies, in our strict sense of the word, reflect this cooperation **independently of close genetic kinship**. Moreover, we will find that worker policing is paradigmatic of a **far more general insight**. Kinship-independent social cooperation is utterly and unavoidably dependent on the presence of coercive violence (and threat) deployed in pursuit of short-term individual self-interest. This self-interested threat sustains kinship-independent social cooperation **as a by-product**. The fact that this kinship-independent social cooperation can be highly adaptive is gravy, not the ultimate cause of the evolution of that cooperation.

In such societies, behaviors that would otherwise emerge from the straightforward, short-term functioning of kin-selection instead become untenable because of coercion by self-interested others. As a consequence, **alternative** approaches to self-interest now become the best available option. Thus, kinship-independent social cooperation emerges from the direct and immediate pursuit of individual self-interest, not from group selection.

It is just that simple, we will argue. Everything else we think of as constituting any animal society, *sensu stricto*, including human societies, is merely an effect of this fundamental underlying causal process.

This insight about the world is crucial.[8] It will ultimately give us the power to understand how humans first evolved two million years ago; why we look, act, and feel as we do; why we have the properties we have; and why our two-million-year history through the present instant has played out as it has. This insight is the secret to an unprecedented level of human self-understanding. It gives us a level of knowledge about ourselves that our grandparents could not even remotely have imagined.

A day in the life*

[W]hen strangers attacked some residents on their way to work, stripping them of money, tools and even clothes, the response was swift and deadly.

Two of the muggers escaped, but a third [] was marooned at the center of a growing mob of local residents, who rained blows, kicks and stones on him until he was dead.

They hanged the body from the bridge[], evidence of this [town's] growing recourse to street justice.

"We didn't decide to kill him," said a 36 year old carpenter. "It was something that just happened. But afterward, everybody said, "At least this way they won't mess with us again."

The details of why [the mugger] was killed varied with each witness.

"For two months they were holding up four or five people a day. They would take their things and run."

"This was like the sixth time that they had ambushed people going to work. People got tired of it."

"I don't feel there can be any judgment, [] because it was the entire community, not one person. When they screamed for help I ran, and many others did too. First there were 50 people, and then more than 200."

"Everyone agreed that it was well we did it[]. The feeling of the people was "Let's unite." Listen, we were the ones who were violated. If the community did not help us, it would have been we who died."

The police investigating the [killing of the mugger]…were handed a statement with 150 signatures claiming responsibility.

[T]he lynchings had begun with two in 1994. In 1995 there was one every other month. By this spring [1996] they were averaging one a week.

These real events have much to teach us. Embedded in the details of this story are unexpectedly rich clues about what it means to be human and why humans are different than other animals. We will understand why even better as we proceed through the book, but we can get a clear first glimpse here and now.

---

* Excerpts from a story in the *New York Times*, May 13, 1996, p. A6, by Diana Jean Schemo.

First, this killing of a mugger occurred in El Encanto, one of the many poor suburbs of Caracas, Venezuela, where the formal state police presence is weak. We are looking at something humans do when there is no state coercive authority around them. Of course, the state is a recent invention (Chapters 13-15). Our ancestors lived without it throughout most of our evolutionary history.

Second, the people who were the initial victims of the muggers were in the course of participating in public human economic activity. They were on their way to work and were carrying valuable objects (tools, money, etc.).

Third, the muggers would have continued to obtain self-interested benefits from later acts of theft if their victims had not stopped them. Indeed, they would have ultimately been able to steal most of the proceeds of their present and future victims' work. The muggers would have done better—as self-interested individuals—than the workers. If muggers were left unchecked, workers would have eventually been forced to become muggers and the entire basis of the local economic system would have collapsed.

Fourth, however, their victims were able to recover and/or forestall their personal individual losses through the use of violence. Their self-interested behavior created a *cost of opposition* (Chapter 4) for the muggers. For one of the muggers, the cost was his life. Under these conditions, the costs of being a mugger came to outweigh its potential benefits.

Being a mugger was a poorer individually self-interested choice than being a worker in El Encanto *that* day. Moreover, imposing these costs was self-interested for the workers. It allowed them to protect resources they already possessed—to pre-empt a loss. Thus, the self-interested behavior of *other* individuals determines which choice is the best available self-interested option for *each* individual.[9]

This scenario is remarkably analogous to our worker bees or ants rearing their brothers rather than their sons because the self-interested behavior of their sisters forces them to make this choice.

Fifth, the targets and potential targets of the attempted mugging did not merely kill the mugger. They hung his body from a footbridge in very public display. This is a clear message—a threat. It is intended to influence the future behavior of other would-be muggers. This is **coercive threat** in its most naked form.

Sixth, intense human reactions are revealed by the self-descriptions of members of the attacking crowd. This is our evolved, uniquely human ethical psychology (a proximate mechanism) in play (Chapters 3 and 10).

Remember this fact particularly. We will return to it many times, and each time it will teach us something new, something more.

Finally, we will see in a moment that humans have access to means of coercive violence against other humans that make the costs of the self-interested coercive violence at El Encanto (and everywhere) much cheaper than they are for any non-human animal. This access to *cheap* coercion will be vital. Before we explore this property more carefully, it is important to understand clearly the *consequences* of this unprecedented access to cheap coercive threat.

Another pivotal moment – what is a *human* society?

Like large stable ant colonies, vast human societies can confer enormous advantages on their members. Our societies rule the planet; isolated individual humans cannot compete or even survive. We are tempted to think that this dominance of our societies is the ultimate cause of their existence, of their evolution. When we think this way, we commit the group selection fallacy.

The cooperative behavior creating our societies, our public economic activity, arises not from group selection but from pursuit of individual self-interest. Moreover, this cooperation is a viable, individually self-interested strategy *only* when an alternative strategy—theft—is rendered unattractive by the threat of coercive violence.

You might be thinking, *Wait! This is a ridiculously jaundiced view of human cooperation. It mixes things up. Most people are decent and hardworking. This is the real reason the human economic system works, not the threat of coercive violence. Coercive violence is a secondary, marginal part of our society.*

If you have this view, it is perfectly wrong, as we will argue in a moment and throughout the rest of the book.[*] The evidence that this view is incorrect is really quite overwhelming. In spite of this evidence, we all still tend to feel that coercive violence is not really very important. This view is an illusion produced by our (proximate) minds because of the way our uniquely human ethical sense works in detail (Chapter 10).

It is, in fact, true that most people are good, decent, and hard working. But it only takes a few cheaters (muggers, extortionists, etc.) who are not good *and who are not the targets of coercion* to change the rules. These few who are not good do better under these conditions. Their behavior spreads to others (Chapter 1).

---

[*] By the way, the authors' subjective intuition about this part of being human was just like yours and just as misleading.

For example, formerly "good" parents find themselves losing out to cheaters in the competition for the resources their children need. They have no choice but to become cheaters themselves. These new cheaters shift the balance even further toward cheating as the best available strategy. It becomes ever harder to play the good role and ever more necessary to play the role of cheater if one's children are to have enough to eat.

Ultimately, everyone becomes a cheater. In practice, now, everyone is forced to find resources on her/his own (with help from close kin, of course) *unless coercion is available to prevent this relentless chain of events.*

Again, you might be thinking, *Wait! This is crazy. This 'pure cheater' population is worse off than the original 'mostly good' population because they are not cooperating. They are all on their own.*

If you are thinking this, you are quite right about the status of individuals in the universal-cheater population. They **are** worse off. But you are quite wrong if you think this means that cheaters will not take over the population in the absence of coercion. They will. This is just how natural selection works—no big picture, no eye to overall efficiency, just winners and losers in **head-to-head**, **in-the-moment** competition between **individuals.**[*] Natural selection produces the outcomes it produces—sane or insane, logical or illogical.

Of course, universal-cheater human societies do not usually arise or they do not last long, if they do. But the reason is NOT because they work poorly (though, again, they do). Rather, the reason is that humans use the self-interested application of their capacity for cheap coercive violence in a way that makes cheating individually unattractive—leaving "good" economic behavior as the best *remaining* option. It is our highly evolved "human nature" (Chapter 7).

We will see much more evidence in future chapters that this fundamental claim is right. But, here is one last piece for now. Consider the world in which most of us live. Coercive violence does not seem to play a very large role. But think for a moment. For example, who comes when you dial 911 (in the US) or when a bobby blows a whistle (in the UK)? People with guns. Are those guns loaded? Would policing work if they were not? We all know the answers to these questions if we are honest with ourselves. Economic systems (cooperation) are only sustainable through continuous *credible* coercive threat. One last question, *Are the*

---

[*]   In the case we just described, natural selection was operating on *culturally transmitted* design information (Chapter 10), but the logic is the same as selection acting on genetic design information in this case.

people with guns who show up when we dial 911 acting altruistically or are they paid for their efforts?

The theory we briefly outlined in Chapter 1 can now be stated in more general, simple terms:[10] **Humans are the first animals in the history of the Earth to have access to individually inexpensive violence (and threat) against conspecific adults. Our self-interested deployment of this coercive threat has the effect of suppressing non-kin conflicts of interest on a large scale. This self-interested suppression of conflicts of interest produces our vast cooperative societies as a by-product.**

Put differently, humans play the same game as our brother-raising, nephew-killing bees and ants, but with an unprecedented scope. We engage in self-interested behaviors that result in suppression of conflicts of interest, not just occasionally, here and there, as non-human animals do, but consistently, every moment, and on a truly massive scale.

That is it. That is all there is.[11] Everything important about our origins, our unique properties, and our history—each salient detail of all the uniquely human empirical evidence—represents one of two things.

First, some facts about us are part of the explanation of how we came to possess the various unprecedented coercive means that empower our self-interested suppression of conflicts of interest. We have relied on different means at different times and therein lies a very important story we will explore in Chapters 11 to 16.

Second, all the other uniquely human facts about us are simply *implications* and *outcomes* of our self-interested use of these means of coercion. All our uniquely human properties and our history are **effects** of this single primary **cause**. Self-interested use of appropriate means of coercion *yields* kinship-independent social cooperation as a by-product. We carry many deep and rich adaptations to this use of force and to the individual opportunities within the cooperative societies it sustains. Our large brains and powerful minds, our complex language, and our profound sense of purpose and justice are examples of such adaptations. They are **all** just consequences and details of our adaptation to kinship-independent social cooperation spun off from our self-interested use of coercive threat.

If this is the most general statement of the solution to the human uniqueness problem, it leads to the next question, *How did our original human ancestors come to be the first animals in the history of the Earth to have the ability to execute violence against one another cheaply—to possess the means for self-interested suppression of conflicts of interest on a massive scale?*

To understand the answer to this question a little more background is required. We will look first at the logic of coercive violence in non-human animals, then at the uniquely human approach to this adaptive behavior.

The costs of coercive violence and the killing strategies of animals, the non-human case

The key to understanding the unique human capacity to engage in coercive behaviors that result in suppression of conflicts of interest lies in the **costs of coercion**. Specifically, for coercion to be an individually self-interested behavior, its immediate costs must be **less than** its immediate benefits. If its costs are **greater than** its benefits, coercion becomes a self-defeating behavior under almost all circumstances. Animals who avoid it will do better than animals who do it.

Consider what must happen if kinship-independent cooperation is to become a pervasive by-product, vast in scale. Conflicts of interest between healthy, full-grown adults must be suppressed or managed so that the selfish strategies these conflicts of interest would produce are no longer the best available short-term options. Recall what we have seen about violence between adults in non-human animals.

First, in the case of adult male lions (Chapter 3), violence is extremely risky, often involving death or serious injury. Given this, we would expect such huge risks to be taken only when the potential adaptive benefits are comparably enormous. This is, in fact, what we see. Adult males take these risks only when control of a pride of females is at stake (a large, new reproductive opportunity).

Second, recall our field voles from Chapter 4 and the online endnotes to Chapter 3. They commit infanticide on non-kin pups (a small, indirect adaptive benefit to their close kin offspring) only when the parents of the pups are not around to resist. The costs of direct adult-on-adult violence are too high to risk for a small benefit.

Third, our policing bees and ants teach us the same lesson. The nest members do not kill or threaten an adult worker female who attempts to lay an egg. Rather, they kill the helpless egg itself, apparently when the egg's mother is elsewhere. Again, this pattern of coercive violence indicates that adult-on-adult violence is too expensive to be a good adaptive individual choice for non-human animals under most circumstances.

These are representative examples from a large body of evidence about how adult-on-adult non-human violence works. The pattern they reveal is predictable. Consider our lions again. When two adult lions fight, each has about a fifty-fifty chance of being injured or killed. Indeed, this is a minimal estimate. It could even transpire that both males are seriously—even fatally—wounded in a one-on-one fight. High cost is easy to see here.

Of course, we are also concerned with *social* coercion. This type of coercion will not generally involve conflicts of interest between two isolated individuals. Rather, it will often involve cases where a number of individuals are involved. Examples like the El Canto episode where many cooperative individuals are coercing a few cheaters are telling.

Though is takes a few minutes of tedious attention to detail, it will be very well worth the effort to understand how this works.

Let us look at many-on-one coercive violence in lions. This violence would seem to be cheaper for each individual on the many side than would participation in a one-on-one fight. It is. Given that the target individual will move around aggressively in attempting to evade and counter-attack, at each moment in time only one or a very few of the many can actually engage the single target. Indeed, to a good approximation, only one attacker can effectively engage the target individual at any one time.* Does this not reduce the costs of violence to each individual member of a group of <u>five</u> individuals attacking <u>one</u> individual, say? It does. But by how much? If they trade off, tag-team style, during the conflict, each takes about 1/5 of the risk or the total cost over the entire duration of the fight. Thus, the cost to the target individual should be about five times as high as the risk/cost to each of the attacking five.†

Is this reduction in cost enough to make social coercion cost-effective? In most cases, it is not. For example, the five attacking individuals might be a set of females competing with other females for territory or some resource like food or water. Alternatively, they might be males competing for access to mating opportunities with females.

How much of the disputed resource will each member of the five get? What fraction of the food or the mating opportunities will they

---

* This is especially true when we recall that exchanges of violence will often involve kin on both sides. Thus, five-to-one will more often be ten-to-two (or fifteen-on-three) with the close kin two standing back-to-back. Under these conditions, neither of the two can be effectively flanked and each will be engaged by roughly only one attacker at a time.

† Notice that the costs of violence result mostly from receiving violence, not projecting it. For example, the cost of firing a gun is much lower than the cost of being hit by gunfire or the cost of swiping with a claw is much lower than having one's flesh torn by a swiping claw.

obtain? The answer, of course, is about 1/5. The individual being coerced will possess a resource that can be shared by the five attempting to drive him (or her) from the field. If we normalize the value of the resource this way, we say that there is *one unit* of the resource at stake. Thus, if this individual is expelled by coercion and the five share the resource, each of the five gets roughly 1/5 unit.

This quantitative argument is a little subtle. Here is another way to see the point. Think about how hard the target will be willing to fight to resist coercion by the five. How much cost is he willing to incur before it is no longer sensible to keep fighting. He will fight up to the point where the cost he incurs is equal to the likely benefit of the resource he is fighting to retain, but no further.* We are free to use whatever units we want to measure these costs, like choosing either inches or meters to measure distance. So, we can call the costs that the target of coercion is willing to incur *one unit of cost* in pursuit of *one unit of benefit*.

Now suppose that the target fights until he has absorbed this full unit of cost. How much cost will he inflict on each of his five attackers during this time? For the moment we assume that all adults of the same species have similar fighting abilities. Thus, the five will have collaborated to inflict one unit of cost during the same time interval when they *collectively* received one unit of cost. Thus, *each* individual member of the five will absorb 1/5 unit of cost in anticipation of 1/5 unit of benefit.

We could think that this still might be a good idea even though it looks like a wash, 1/5 unit of cost for 1/5 unit of benefit. However, now differences in fighting ability become important. We assumed that each of the six individuals involved above were equally good fighters. But this is rarely true. In the real world, some animals are bigger and stronger than others. Thus, when the target is bigger and stronger, he will be dishing out more risk to the five than he is absorbing. But the bigger target will still stay and fight as long as his costs are less than the value of the resource he is fighting to retain.

By the time the five have scared him away, each of the five will have absorbed more cost than the resource is worth to each of them because of the superior fighting ability of the target. This is not a sensible behavior for the five and would never evolve by natural selection.

---

* His mind will not be consciously aware of these considerations, of course. But his mind will have been shaped by natural selection so that he behaves *as if* he knew. When we approach things this way, we are using the *rational actor* fiction—the pretense that individuals have conscious access to complete information about the implications of their action. This works because natural selection produces animals whose minds (proximate devices) often cause them to behave as if they had access to complete information in the specific circumstances that resemble the *ancestral past*.

One last piece of this picture is needed before we can see how all of this works. Real animals rarely actually fight intensely. Intense fights usually occur only when neither party has any better alternative.

More commonly, each individual has the choice to fight or withdraw and live to fight another day. What does natural selection produce in such circumstances? The animals size up their situation. They may even rough one another up just a little to test who is strongest.

But once this ritualized assessment of relative fighting ability has been done, the weaker individual withdraws.[12] She/he would do worse to stay and fight. The stronger individual stays. She/he would do worse to withdraw.

In other words, the cost/benefit logic we went through above does not necessarily reflect the costs animals **actually** pay. Rather, it reflects the costs they *would, in principle, be willing to pay*. All animals in this interaction have minds that are designed (by natural selection) to understand this cost/benefit logic (not consciously, again) and fight or withdraw on the basis of that understanding. There is no bluffing here. Only real coercive threat and its real costs matter.

What is the upshot of all of this for social cooperation between non-kin? Big, strong individuals do best if they just sit around and wait for non-kin others to generate some resource. They can then take what they want by dint of projecting superior coercive threat.

Of course, it is not in the interests of others to allow themselves to be parasitized this way. But they cannot cost-effectively deal with this exploitation by overt counter-threat the way the El Encanto humans did. Rather, they try not to be around the big guys. For example, they might try to hunt only when away by themselves and so on.

More generally, there is virtually no basis here for the evolution of cooperation between non-kin adults. Adult-on-adult coercion is just too expensive for all but the strongest and the strongest can often better use their advantage to steal than to enforce cooperation. Each individual does his/her best to pursue self-interest without helping other, non-kin individuals. All other behaviors do worse under natural selection.

This lion story, in fact, is applicable to all non-human animals, with only relatively minor modifications that need not concern us at the moment. Cooperative coercive violence against conspecifics is not usually sensible.

Now we can turn our attention to how this logic was changed in a profound way for the first time in the history of our planet in the immediate ancestors of the first humans about two million years ago.

Another pivotal moment – the costs of coercive violence and the killing strategies of animals, the *uniquely human* case

There is one way—and, apparently, *only* one way—to change the logic of social coercion substantially among non-kin adult animals. The answer could not be simpler. The universal cost-benefit logic of social coercion in non-human animals can only be changed in an animal *that can kill adult conspecifics from a distance*.

It will be convenient for us to designate non-human animals who kill up close (with "tooth and claw"), like our lions above, as **proximal killers.** Analogously, we will call animals that can kill from a distance of many body diameters away from their target, **remote killers.** For clarity and simplicity we can think of remote killing as consisting of gunfire for now. We will return shortly to how our human ancestors actually first acquired the ability to kill remotely approximately two million years ago.

We begin with the details of coercive violence that are the same for proximal and remote killers, then turn to the *differences*. As we saw, when five lions attack a single target, the risk each of them takes (his costs) is proportional to 1/5. Moreover, if there are ten attacking individuals targeting a single individual, each attacker receives 1/10 of the cost/risk and so on. This is because each receives 1/5 or 1/10 of the return fire from the target. Thus, in general, **n** attackers each absorb 1/**n**th of the cost.

This is true for **both** proximal killers and remote killers. Again, think of gunfire. A target can only fire her/his gun at one attacker at a time. This is the part of coercion that is similar for both kinds of animals. But, it is the **differences that are important**.

When multiple lions attack a single target, they get in one another's way. The consequence of this is that the target is receiving violence from only one or a very few attackers at any instant in time. *In contrast*, remote killing animals do not interfere with one another in this way. Again, think of guns. The members of a firing squad do not interfere with one another. Each can fire simultaneously at the target.

What does this mean for the single target in a gunfight under attack by five individuals? He experiences five times—**n** times, in the general case—as much risk per unit time as the single target in the lion case. This means that he will use up the costs of fighting he is willing to pay five times—**n** times—faster.

What does this mean, in turn, for the five remote killing attackers? They will have to fight for an interval of time that is five times shorter, before the target runs away, than would be true for our five lions. Because they are taking return fire at the same rate per unit time as for the lions, *but* for a five-fold shorter time, each attacker will experience a five times lower risk than for five proximal killers like the lions.

This logic is crucial to us, so we should be carefully precise and specific. Each of five remote killing attackers, on average, receives 1/5 of the violence (gunfire here) emanating from the single target just like the proximal killers. But they do so for 1/5 as much time. In contrast, the target still receives the same amount of cost as for the proximally killing lions from above (5/5 or *1 unit* here). He just accumulates this cost in 1/5 the time (five times faster).

Now consider the **ratio** of these rates. The remote killing attacker absorbs 1/25 as much cost as the target rather that the 1/5 as much as the target for proximally killing attackers like our lions above.

More generally, *n* remote killing attackers engaging a single target experience $1/n^2$ as much risk as the target experiences. This is a very large effect. For example, if there are ten attackers and one target, each attacker experiences one hundred times less risk than the target does. If one hundred attackers engage a single target, each attacker experiences ten thousand times less risk than the target! [13]

This very large effect is **absolutely crucial** to the evolution of kinship-independent cooperation as a by-product of self-interested coercion. Remember what happened with proximal killers. Five individuals might have shared the costs of coercive violence, 1/5 unit each, but then they also had to share the benefits likewise, 1/5 unit each.

Contrast this with remote killers. Five of them also share the benefits—1/5 as much as the target individual, (*1/5 unit* as we defined it). **But**, these remote killing attackers each only have to pay 1/25 unit of cost to obtain this benefit. This approach is now very "profitable" from an evolutionary point of view.

When the target of coercive threat is a thief attempting to steal a cooperatively generated asset, this self-interested coercion has the side-effect of rendering the original cooperative behavior adaptive. Thus, in contrast to proximal killers, social coercion can be a very good idea for remote killers.

Coercive violence can be highly adaptive for self-interested *remote killing* individuals. In turn, because coercive violence can be an individually adaptive behavior, individuals do it and other individuals must evolve to adjust their behaviors in the face of this coercion, just as our ants evolved to adjust their behavior in the face of coercion (worker policing). But unlike our ants, remote killers can *directly* coerce adults. Ongoing coercion of adults makes active, pervasive cooperation between all non-kin members of a local social environment possible.

It is impossible to overstate the importance of remote killing. It changes the rules for the evolution of social behavior, utterly. Remote killing makes the evolution of kinship-independent conspecific social cooperation possible. Indeed**, it makes evolution of this cooperation inevitable**![14]

This is the core of our theory. Humans are unique because they are the first animals in the history of planet Earth to be able to kill adult conspecifics remotely. Everything else about us is simply a consequence of this single fact.

What is the most direct prediction of this theory? The only animal on the planet that has evolved kinship-independent social cooperation should also be the only animal that ever evolved the capacity to kill adult conspecifics from a distance.

This is a strong prediction, very **testable/falsifiable**. And this prediction is fulfilled. Our earliest human ancestors did not have guns as in our example above, however. So how did they manage to kill adult conspecifics from a distance? You and we have inherited this unprecedented remote killing capacity. It is our capacity to throw with elite human skill, skills that we use when we play American baseball or throw a stone at a strange growling dog.

This skill seems so mundane that it is hard for us to grasp how important and how revolutionary it is. Two things are crucial to understand.

First, our throwing skills truly are unique. We throw the way a cheetah runs or a dolphin swims—with utterly elite command and skill that no other animal has.*

Second, these unique, elite skills really are lethal. A small group of humans armed with baseball-size stones can easily kill another human. Indeed, death by stoning was long a popular form of extreme social ostracism. It is mentioned dozens of times in the Bible, for example. Indeed, it is still in use in a few cultures today.

Our theory is as simple as it could possibly be. Individually self-interested remote killing animals can afford the costs of self-interested coercion against conspecific adults. A by-product of this coercive threat is the suppression of non-kin conflicts of interest. Proximally killing animals cannot afford these costs and rarely cooperate with non-kin. As a result,

_____

* We have video clips of chimp throwing that we play for students. These inevitably produce laughter at the comic ineptitude (by elite human standards) of chimp throwing.

the first remote killing animal that evolved on planet Earth should have quickly gone on to evolve kinship-independent social cooperation.

Such remote killing animals did evolve and we are their descendents. Their original weapons were apparently thrown stones. Of course, now the question becomes how and why did humans first evolve elite throwing. This question also has a simple, straightforward answer that we will explore in Chapter 7. For now, it is important to look a little more carefully at the logic of the social cooperation that evolves among remote killing animals.

The precise logic of the evolution of kinship-independent social cooperation in a remote killing animal

It will be very important at every step through the rest of the book to understand the exact logic of self-interested deployment of remote killing capabilities. This logic allows suppression of conflicts of interest and vastly enlarged social cooperation as a by-product.[15] We will find that grasping the rules of this process apparently gives us everything we need to comprehend what it really means to be human. We will also see that once our ancestors evolved this species-typical trick, their descendents (including us) exploited the adaptive opportunities it provided continuously throughout all of human history.

Consider what happens when a set of non-kin proximally killing individuals cooperate.[*] The members of this set generate some benefit from their cooperation. It might be meat from a cooperative hunting or scavenging episode, for example.

The strong, large individuals stand back during the hunt, stepping in to take food for themselves at the end. This is their best available option. Under these conditions, less dominant animals who hunt cooperatively are parasitized by non-kin and do worse than animals who find ways to hunt alone or exclusively with close kin.

In contrast, in a remote killing animal, a large, strong individual finds it impractical to parasitize a set of self-defending individuals. Even if this large individual can fight twice as well as the average of his/her targets (a large asymmetry), this advantage is swamped by the $1/n^2$ effect we explored above. Figure 5.1 illustrates this effect graphically.

---

* More realistically—in early human ancestors—this set would contain some close kin and some more distant kin. However, this does not need to concern us here. What is at issue is what extra cooperation (beyond that produced by kin-selection) inexpensive coercion would allow these ancestral humans to evolve.

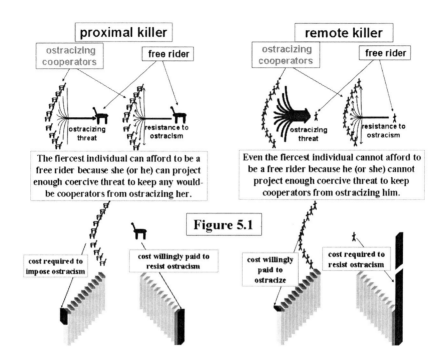

Figure 5.1: TOP LEFT: Shown are the relative rates at which multiple *proximally-killing* would-be ostracizing cooperators and one especially fierce individual free rider would exchange costs (risk of injury/death) during an episode of coercive violence.

BOTTOM LEFT: Each light gray column plus dark gray extension represents the yield to an individual ostracizing cooperator of a cooperative episode. The dark extension represents the loss each cooperator would experience if a single free rider were to share in the proceeds of cooperation. The free rider's share (the sum of the gray extensions at left) is indicated as the dark gray column at right. If the fierce free rider absorbs as much risk as is cost-effective (black column at right) the ostracizing cooperators will have absorbed much more than is cost-effective. Thus, ostracizing the free rider is an irrational behavior, one that will not be produced by natural selection.

TOP RIGHT: Shown are the relative rates at which multiple *remotely killing* ostracizing cooperators and one individual free rider would exchange costs (risk of injury/death) during an episode of coercive violence. Notice how much more cost remote killing ostracizers can inflict on a target.

BOTTOM RIGHT: Each light gray column plus dark gray extension at left represents the yield to an individual ostracizing cooperator of a cooperative episode. The gray extension represents the loss each cooperator would experience if a single free rider were to share in the proceeds of cooperation. The free rider's share (the sum of the gray extensions at left) is indicated as the dark gray column at right. If each ostracizing cooperator were to absorb a cost equal to her/his expected return from ostracizing the free rider (equal to the gray extension) she/he would inflict a massive cost on the free rider symbolized by the black column at right. For any realistically fierce free rider (text) it is rational for the ostracizing cooperators to persist in projecting threat to ostracize and it is irrational for the free rider to resist ostracism. As a result, preemptive ostracism is strategically viable for these remote killers. Natural selection will produce it and cooperation between conspecifics with conflicts of interest can evolve as a result (text).

Now consider each individual's best strategy (her/his most adaptive behavior) as a member of a remote killing species. She (or he) can forage and hunt alone. Or she can forage cooperatively and more efficiently with others including non-kin (combinatoriality, again; Chapter 2). Moreover, if she makes the second (cooperative) choice she and her collaborators are in the position to cheaply drive away anyone else who did not contribute to the cooperative effort. Indeed, it is in their head-to-head, in-the-moment individual interest to do so.

Notice, when she does this she is using coercion to suppress conflicts of interest.* In effect, her self-interested decision to recapture stolen goods (or pre-empt their theft) results in coercive *enforcement* of the kinship-independent social cooperation that generated those goods.

Her choice to enforce cooperation is self-interested **only if** others make the same choice. Of course, they have the same incentive she has. Moreover, if benefit accrues only to those who stay and fight to drive the parasitic individual away (rather than leaving the field), coercion becomes the best self-interested strategy.†

What will the psychological mechanisms producing this behavior "feel" like? We can ask the question just this way because this animal is us. We have all subjectively experienced these mechanisms in action.

First, these animals should have a strong tendency to do things cooperatively. Indeed, they might even feel lost and purposeless unless they were a "part of something larger than themselves." **We do.**

Second, they should have a very strongly developed sense of the proper distribution of the fruits of cooperation. We might even say that they should have an intense sense of "justice" and "fair play." **We do.**

---

* Of course, she is not consciously aware that this is what she is doing—proximate mechanisms and ultimate causation, again.

† This is a vital detail. Under any other circumstances, the choice of foisting the costs of "law enforcement" onto other cooperators can become the best available individual option, and coercive suppression of conflicts of interest can break down. Most human cooperative enterprises are apparently assembled so that they have this payoff structure, however indirectly.

Third, they should have psychological devices that spur them to action when cheaters try to parasitize their cooperation enterprises. *We have precisely such a mechanism.* We call it moral outrage.

Lastly, they should have a mechanism that causes them to avoid selfish behaviors that are likely to make them targets of the (potentially lethal) coercive violence of their remote killing conspecifics. *We have just such a mechanism.* We have all felt its bite. It is guilt.

Box 5.1: Great collaborators and new clarity – Daijiro Okada and the game theory of human kinship-independent social cooperation

One of the greatest pleasures from science and one of the most productive approaches to *doing* science lies in having brilliant collaborators. Many people contributed in diverse ways to making this book possible. Non-kin cooperation is what we do, after all. But, one of the most important events in the subsequent development of this work since the original publication of the theory by Paul in 1999 was meeting the gifted economist, Daijiro Okada.

To understand why this collaboration has been so important, reflect, for a moment, on what the theory is saying. Uniquely human social cooperation emerges from suppression of conflicts of interest as a result of the self-interested exercise of coercive power. In other words, our complex *cooperative* behavior results from the *immediate, direct pursuit of individual self-interest* making cooperation our best available option.

What is "best," from an evolutionary point of view, for each individual to be doing is dependent on what everyone else is doing, and vice versa. Situations like this, involving the interactions of many individuals, can be complicated and confusing. A branch of social science with the technical name *game theory* turns out to give us some important tools that help.

The inventor of game theory, John von Neumann, is a fascinating character. He was a member of Princeton's Institute for Advanced Study (where Daijiro also spent the 2007-2008 academic year), the home of diverse luminaries over the years, including Albert Einstein.

von Neumann was a brilliant polymath, contributing to the Manhattan project and the invention of the modern computer. However, von Neumann is most interesting to us here because of his invention of game theory in his 1944 book with Oskar Morgenstern, *Theory of Games and Economic Behavior.*

Their approach was called game theory because of its initial foundation in trying to predict the behavior of actors in "games," whose rules were simple enough to be easily modeled mathematically. Moreover, this initial foray stimulated a generation of thoughtful people to analyze new situations and develop new tools. (One of the most famous of these was John Nash, of course, played by Russell Crowe in the popular romantic film, *A Beautiful Mind.*)

Moreover, we now recognize that, as game theory develops its tools, those tools should be applicable not merely to human economic actors, but to any self-interested actor (vehicle) in a world with conflicts of interest (a Malthusian world), which is to say, they should be applicable to *all* of biology.

One of Daijiro's many vital skills is mastery of game theory. Moreover, his intellectual courage and openness to new challenges are extraordinary. Daijiro and Paul were able to give the fundamental logic of the theory a clearer, firmer foundation in their 2008 game theory paper (Okada and Bingham). Indeed, some of the important wrinkles in how the theory was described through the earlier parts of this chapter are due to this crucial collaboration. The core theory of this chapter is not just ours. Important parts of it are also Daijiro's.

Some unexpectedly general implications

We now have the fundamentals of a theory of the emergence of human social cooperation. This discovery alone would be worth all the work that was required to find it. But, this theory turns out to be far more general and valuable than just these fundamental elements.

Our claim is that the evolution of remote killing produced a new evolutionary pattern. Remote killing yielded a new adaptive opportunity, the capacity for individuals to project coercive threat in pursuit of self-interest and cooperate in response to this coercion from others. Moreover, we claim that this novel pattern of natural selection produced all the traits we think of as being uniquely human. These traits include complex language and cognitive virtuosity, for example. Our powerful minds arise, in part, from our access to the vastly enlarged amounts of *culturally transmitted* information to which we have access. The members of our uniquely large cooperative social aggregates are the source of all this extra information (Chapter 10) and these aggregates can exist only because of self-interested policing. Likewise, our complex language is part of our adaptation to the exchange of these very large amounts of information with many individuals, in spite of conflicts of interest that would otherwise cause us to lie and misinform (Chapter 9).

Moreover, some of our uniquely human traits are direct elements of the adaptation to enforcing such cooperation, itself. These include our elaborate ethical sense, including guilt and moral outrage, as we said.

Finally, individual human animals produced by this selection, with all their predictable traits, continued to pursue self-interest, resulting in suppression of conflicts of interest throughout their history. Their cognitive virtuosity repeatedly gave them access to small technical innovations of many sorts over the two million years of their history. A few of these technical innovations bestowed upon them coercive technologies of ever-larger scales. These new coercive capabilities, deployed in pursuit of self-interest, enabled social cooperation on correspondingly larger scales. The results of these various increases in the scale of uniquely human social cooperation will prove to be the central events of our vast history (Chapters 11-17).

How do we begin?

Now that we have defined the basics of our theory of human uniqueness, the next logical step might seem to be to travel back in time (using the fossil record) and look at the moment of our origins to see if this event occurred as our theory predicts that it must have. However, there is one more piece of the puzzle we must have *first*. We need to understand just exactly how our uniquely enlarged human brains came to be.

To get this new puzzle piece in hand, we will look at one of the earliest adaptive consequences of the emergence of kinship-independent social cooperation in the first ancestral humans—the evolution of our uniquely human reproductive strategy, **kinship-independent social breeding**. Chapter 6 will introduce us to this remarkable human property.

Our uniquely large human brains can only be produced and maintained through a wholesale reorganization of human lives, from conception through birth, infancy and childhood. These changes, in turn, created both youngsters and mothers who require massive new levels of social support, levels far beyond those available to any non-human mother or offspring. Only uniquely human social cooperation can provide this support, cooperation sustained through self-interested coercion by remote killing animals on our theory. In other words, our brains (and bodies) have been redesigned from top to bottom in ways that are only made possible because of vast resources generated by kinship-independent social cooperation, and our theory predicts how this should come to be through the process of natural selection.

This insight allows us to exploit uniquely human brain expansion, revealed by fossil skulls, to date the initial evolution of our unprecedented social cooperation. In Chapter 7, we will use this capability to carry out a rich test of our theory using the fossils of the first humans and their immediate ancestors.

෧෧

Chapter 6
It takes a village to raise a human child

Cooking is Jefferson's greatest skill. He spent several years training with a master chef and he now runs his own small neighborhood restaurant. It is a local favorite, and many of his customers are also friends. Today, though, he is a little distracted. He just received a long-anticipated phone call and his wife Camille dominates his mind. She will deliver their first child very soon.

Camille loves teaching. She spent her 20s acquiring the skills and the experience necessary to help children through their grade school years. She knows what they need to grow in skill and confidence. Camille took time off to prepare for her first child, and the delivery is imminent. Because she is such a valued employee and because of legislation supporting her family rights, her job will be waiting for her when she returns.

Camille consulted with her friend and obstetrician, Janine, throughout her pregnancy. She is in qualified hands and as comfortable as a first time mother can be. The baby soon to emerge from her body has become so demanding that all her energies are required.

Janine organized the final details of her day as she dried her hair and dressed. As an obstetrician at a major teaching hospital, her time was intensively scheduled, with teaching, overseeing interns, and patient care. Her skilled time was valuable and her hospital sought to maximize its return on her. One of the deliveries she expected today was a personal friend, Camille. Camille was having some problems, but Janine thought they could be managed. She is in a very optimistic frame of mind.

Camille enters the hospital around 3 PM. Her contractions had begun. By 6 PM—with Jefferson at the bedside—her labor has become more intense. The contractions grow stronger through the evening and night. By early the next morning—after twelve hours of crushing struggle—Camille finally pushed out her son, Jermaine.

She elected natural childbirth. With all its benefits for Jermaine, it takes a terrible toll on her. She is beyond tired. She is bone-sapped, jelly-muscled, mind-blinded tired. Jermaine was laid briefly across her sweat-wet breast. In spite of her fatigue, she is elated beyond mere words. No other moment of satisfaction—professional, sensory, sexual—would ever match this one. As Jermaine is taken away to be warmed and wrapped, she falls into a deep sleep.

For Jefferson, the worry of the day before has vanished. In the weeks preceding delivery, Camille's blood pressure had spiked dangerously. If she had not gone into labor when she did, her life would have been at risk. Janine's long experience and knowledge would have been necessary to take steps to save her. Luckily, she was now through that danger. It was time to look ahead.

Jefferson now stepped into his next role. He had felt marginal and irrelevant, as Camille had struggled through labor and delivery. Now he could contribute.

The physicians at the hospital would see mother and son through the next several days. Camille would leave the hospital with her surgical sutures (from the labial tears of her son's passing) melting away in pink, well-healing wounds. Jermaine would be nursing lustily, obliviously.

But Camille would need Jefferson's help. He would need to shop for her and transport her until she felt comfortable to drive. He would need to watch over her. Janine had described the symptoms of post-partum depression and a few others "tells" that Jefferson should watch for. He would take her to their warm, clean, and comfortable home, built and maintained by so many others, but purchased with funds he and Camille had earned. He would bring her the food, the diapers, the extra blankets, the entertainment, the reading…everything she would need, he thought. Camille's sisters and several friends would be standing by, ready to assist as needed.

Over the next several months, Camille and Jefferson found their lives changed utterly. But they noticed this change only in brief moments stolen for reflection. Mostly, the changes were unconscious, natural. They had all the resources they needed to support Camille and Jermaine through this vulnerable time and all the advice they would require as Jermaine surprised and challenged them.

The next year settled into a pattern. Camille arranged her schedule to be able to nurse Jermaine during the day while he was cared for, together with other infants, by day-care professionals.

Around his first birthday, Jermaine began working in earnest on his walking skills with all the child-proofing problems this posed around the house. His mobility and curiosity greatly exceeded his judgment. He needed constant supervision.

Over the next several years, Jermaine emerged as a fully articulate child mastering that most mysterious and wonderful of human skills, language. Jefferson, Camille, day-care workers, Camille's sisters, and others spent endless hours coaching, teasing, playing, and helping Jermaine's mastery grow.

By two and a half years of age, Jermaine's teachers worked hard to help him learn as his age-sensitive skills flashed into existence, one after the other. Camille had long since ceased to nurse him. But he received the best in nutrition (with supplements galore) that Jefferson and Camille could identify, with the help of Manfred, Jermaine's pediatrician. Indeed, Jermaine was soon able to consume Jefferson's more easily chewed and digested adult concoctions.

Formal school was an adjustment for Jermaine. He did not always care for the constraints on his time and behavior. However, he adapted and did what his teachers asked of him. He performed well when he cared enough to focus. Year after year, his teachers prepared him with reading, writing, arithmetic, and all the complex academic skills he needed. They ultimately helped him with many life skills, as well.

As Jermaine grew toward adulthood, he took a keen pleasure in music. Benton, his junior high school music teacher, recognized authentic talent in him and helped develop these natural abilities.

Jermaine grew under the tutelage of Benton's successors in high school and college. He emerged as a first-class cello player even studying briefly with Yo Yo Ma. He ultimately matured into the first chair in the local philharmonic orchestra, supplementing his income by teaching gifted youngsters about his beloved music.

But today's rehearsal was a struggle for Jermaine. He was a little distracted—a lot actually. His spouse was soon to go into labor and things, some very new things, would be expected of him….

An introduction to one very strange animal

The brief vignette above reminds us of something we normally pass over thoughtlessly, unconsciously. A huge number of people play crucial roles in our being born and developing over many years into competent, functioning adults. Indeed, we will argue that our basic biology—the bodies and minds built by human genetic information—has been radically redesigned in response to vast levels of social support. Just as with us today, many helpers were required for the survival and growth of our ancient ancestors a million or more years before formal professions like pediatrician or butcher were invented.

Understanding this socially driven evolutionary (genetic) redesign and the uniquely human biology it produces is a wonderful window on humanness. We have evolved to be biologically dependent on our massive social cooperation, on the human village, for our development and survival. This is one of the features that define us, distinguishing us from all other animals.

Comparing the lives of a human family like Camille, Jefferson, and Jermaine with those of our closest living relative, the chimpanzee gives us clear insight. In fact, we can compare ourselves to two of Jane Goodall's famous East African chimps, a mature female (Flo) who gave birth to a new youngster (Flint). We begin with a quick high altitude reconnaissance flight over these two families.

First, human Camille is a significantly larger animal than chimp Flo. The average size of human females who grew up in conditions like our ancestors is about 50 kilograms (kg)(110 pounds). In contrast, Flo is likely to weigh around 35 kg (77 pounds). Camille is around 30 percent larger than chimp females.[1] See Figure 6.1 for the ranges of chimp and human weights.

Second, human pregnancy takes approximately nine months. In fact, Jermaine was born at thirty-eight weeks of gestation, right on the human average. We do not know Flint's gestation time exactly, but for chimps in general it is about thirty-three weeks. Gestation is a full month longer in humans. More importantly, humans invest so much nutrition and energy in our fetuses that they are *very large* at birth. A typical human newborn might weigh around seven pounds (~3.4 kg), while a typical chimp newborn might weigh closer to four pounds (~1.75 kg).

Thus, though a typical human mother might be 30 percent larger than a typical chimp mother, the human mother's newborn is closer to 75 percent larger.[2] This difference is even more dramatic than these weights suggest. All the great apes (including us) have similarly sized birth canals in the female pelvic bone. As a result, most of the great apes (except us) have similarly sized newborns even though these newborns may grow into very heavy adult gorillas or relatively light adult chimps. When we compare ourselves to all the other apes we see how oversized we really are as newborns! (See Panel B of Figure 6.2.)

Our newborn brains are even more oversized than our bodies. Jermaine's brain at birth is about 300 percent larger than a newborn chimp's brain and much larger than any other great ape newborn's brain (Figure 6.2, panel C).

Third, within a few weeks of his birth, the chimp infant, Flint, could grasp his mother's rich fur coat, hanging under her as she moved around their rainforest home. He could support his own head so that it did not bounce around dangerously. His skull was relatively hard and resistant to compression. His brain had grown rapidly during fetal life. Now it continued to grow, but at a slower rate characteristic of infant chimps.

In contrast, Jermaine could not hold up his head for many weeks after he was born. Moreover, the bone plates of his skull were still not firmly connected to form a solid structure at birth. He had the human *soft spot* on the top of his head. This allowed his oversized skull to flex and bend as he passed through Camille's still ape-sized birth canal. Finally, Jermaine's brain did not slow its rapid growth nearly as soon after birth as Flint's did. Instead his brain continued its very fast *fetal* growth rate for nearly a year.

By the time Jermaine was around ten to twelve months old, his brain ceased its very fast fetal rate of growth, settling into the slower infant rate. But Jermaine already possessed the first element of a uniquely human legacy. His brain at one year was *far larger* than Flint's.

Fourth, both Flo and Camille will nurse their young sons. Flo will nurse Flint for four to five years as his infant chimp brain grows, weaning him toward the end of this period of post-birth brain growth. Flint's brain growth will slow and ultimately stop by the time he is weaned.

Camille will nurse Jermaine for less than a year. But this short nursing time is special. Camille has access to modern formula suitable to replace mother's milk (ostensibly). For nearly two million years, our female ancestors would probably have nursed their babies much longer than Camille did. However, even this longer nursing period was much shorter than Flo's. Camille's ancient female ancestors would have weaned their youngsters at around three years of age, effectively cutting them off from mother's milk earlier than chimp mothers. (We will discuss exactly why they did this in a moment.) In spite of this aggressive early weaning, Jermaine (like his ancient human ancestors) will continue to grow his brain for three more years, through about six years of age.

Jermaine is apparently able to continue to grow his brain after weaning, unlike Flint, because of the high quality food he has access to *other than mother's milk*. Jermaine's ancient human ancestors also must have had this advantage. Flint does not have access to such high quality food and his ability to grow his brain is limited by his access to Flo's rich milk.

So, the way we wean our infants (when and how we are able to wean) is another important difference between us and other animals. Infant brain growth in humans continues longer and remains faster than in chimps (*again, our closest living relatives*) or in any other ape for that matter.

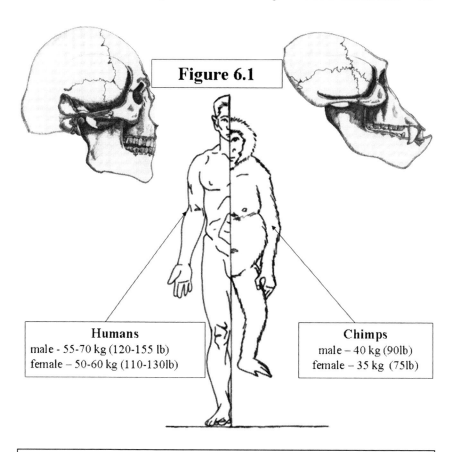

Humans
male - 55-70 kg (120-155 lb)
female – 50-60 kg (110-130lb)

Chimps
male – 40 kg (90lb)
female – 35 kg (75lb)

Figure 6.1: TOP: Comparison of modern human and chimp adult skulls. Notice how much larger the human brain case is than the chimp brain case. The brain **fills** the brain case. Also notice how much larger and heavier the chimp's teeth and jaws are than the human's. This difference reflects the different diets humans and chimps have evolved to eat. Skulls redrawn from various sources including Johanson and Edgar (1996).

BOTTOM: Comparison of the body silhouettes of a chimp and human. The longer human legs will interest us in Chapter 7. For now, two things are important. First, notice the more "pot bellied" shape of the chimp's body. Like the teeth and jaws above, this reflects the difference in the chimp's diet from our own. Second, notice that humans can be more than 50 percent heavier than chimps. (The human sizes in the figure refer to lean humans growing up in somewhat nutritionally limited primitive conditions. This is to make the comparison with chimp weights more realistic. Those of us growing up in very well fed modern societies are often substantially larger still.)

Drawings by Amy Radican.

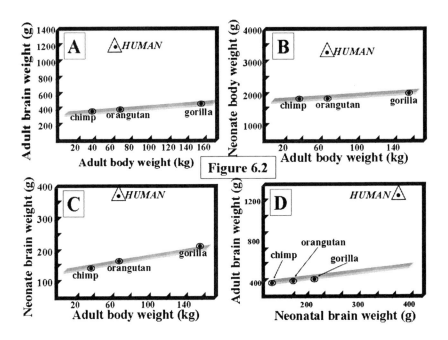

Figure 6.2

Figure 6.2: Shown are four different plots of body and brain sizes for humans and three other great apes. Notice how profoundly different humans are. These large differences arise from evolutionary changes in human life history, affecting how we grow both before and after birth (text).

PANEL A shows that the human brain is much larger than the brains of any other ape, even apes with much larger body sizes. [All size are in weights, measured in grams (gm) or thousands of grams (kg).]

PANELS B AND C show that we begin to build our oversized human brain by growing very large overall while still in our mother's womb. Our brains and our bodies are much larger at birth than the other apes. For example, consider the proverbial '700 pound gorilla'. The gorilla grows to be almost three times bigger than we do at adulthood, yet a human newborn (and her brain) is 1.5 times *bigger* than a gorilla newborn. Growing to be this large while still inside our mother imposes a large extra burden on her. She can support this extra burden only with uniquely human social support, with the help of the human village (text).

PANEL D shows that we continue to grow our brains more extensively after birth than the other apes, as well. If we grew our brains in proportion to their size at birth (already large, as we saw), humans would fall on the gray line. We do not. We are far above this line. This additional extra growing of our brains after birth requires much extra care and nutrition, which can only be provided with the help of the human village (text).

See Bogin (1999) for a review of human and ape life histories.

Unique human *life history* (all the sundry details of our fetal life, childhood, adolescence, and adulthood) is a result of natural selection acting on the design information building our human ancestors in their specific adaptive context. These selective pressures produced Jermaine, who will be a significantly larger animal than Flint (Figure 6.1). Even more importantly, all the differences in the first six years of Jermaine's and Flint's lives bestowed on Jermaine a brain that is an incredible **three-and-a-half times bigger** than Flint's (Figures 6.1 and 6.2). Jermaine will read Shakespeare, understand the calculus, and play Bach with aching passion and subtlety. Flint will learn to find ripe fruit and build a leafy nest for sleeping.

We are not merely different from chimps in our life history. We are different from any other animal that ever lived on earth. Moreover, the large brains enabling our unprecedented cognitive virtuosity are inextricably intertwined with this unique life history.* Understanding all of these connections will give us some useful, unexpected insight.[3] Our immediate goal is to grasp exactly why and how this dramatic feature of our humanness came to be. There is very good reason to believe that our unprecedented life history is a biological adaptation to our unique social lives.†

What are big human brains good for? A first look

We need to deepen our understanding of *how* our ancestors had their life histories altered by natural selection, culminating in human babies and human children. It is useful to begin by asking *why* natural selection produced our unique pattern of development. This *why* question actually has two parts. Part one is the *why could* question. Why could humans change their life history in ways that no other animal ever did before?

Part two is the *why did* question. Why did natural selection produce our massively enlarged brains, through alteration of our life history? The meaty details of the answer to this second *why did* question will come in Chapters 9 and 10, when we turn to the evolutionary origins of human

---

* The relationship between brain size and "smartness" is important, but subtle (Chapter 10).

† The distinction between *biological* and *social* is quite artificial, most emphatically so in humans. Biologically we are social creatures and our social behavior is deeply biological.

intellectual virtuosity and complex language. However, we already have enough background to understand the gist of the answer.

Specifically, one of the crucial sources of the information that allows a brain to produce "good" behaviors is culturally transmitted information.* This is information we learn from and transmit to others. This transmission happens when a cheetah mother teaches her cubs to hunt or a human teaches his or her students how to play the cello or to fly a plane. This information is *not* encoded in DNA sequence. It is encoded in the minds of individuals and is replicated by transmission from one mind to another.

Using culturally transmitted information is, in fact, a property of many animals. However, the *amount* of this information we humans have access to is vastly larger than for any other animal. We have expanded enormously on an old adaptive trick inherited from our mammalian ancestors. Our access to all this *additional* information is ultimately the reason we are so much smarter than other animals.† Storing and processing all this extra culturally transmitted information is apparently the primary adaptive purpose of the human version of enlarged brains. Tuck this insight away for future test against the evidence. All we need at the moment is to know that an answer to the **why did** question will eventually arrive. We can now turn to the crucial **why could** question, followed by the **how** question.

Why could human life history undergo such massive redesign?

Our **why could** question actually has a simple answer. In non-human animals, adult helpers in addition to the parents sometimes contribute to rearing the young. This is called *social breeding*. In the cases of birds and mammals, these extra adult helpers are often older siblings of the newborns.‡ Of course, the social hymenopterans (bees and ants) we

---

*   "Good" behaviors are adaptive. These behaviors are beneficial to the replication of the design information of the vehicle that evinces those behaviors.

†   We will see in Chapters 9 and 10 that human expansion of our repertoire of culturally transmitted information apparently has a straightforward source. Specifically, because we can suppress non-kin conflicts of interest, we can expand the number of individuals from whom we can acquire information without unacceptable risks of being manipulated to do things that are not in our adaptive interest. We manage conflicts of interest and, thus, enforce the broad sharing of reliable information.

‡   See online Endnote 8 to Chapter 3, for example.

met in Chapter 5 are also rather dramatic examples of social breeding. Millions of females (colony workers) assist in the rearing of their mother's (the queen's) offspring (the workers' sisters and brothers).

Social breeding in non-human animals has an essentially universal property.[4] The extra adult helpers (non-parents) are close genetic kin of the youngsters they raise. The reason this is true is obvious. Non-kin have conflicts of interest. Any random genetic variation that produced individuals who helped to raise non-kin offspring would be rapidly eliminated by natural selection (Chapter 3).[*]

This pattern of close kin helping is *kinship-<u>dependent</u> social breeding.*

However, there can be an exception to the kinship rule for social breeding. *If* an animal evolved control of non-kin conflicts of interest, that animal would be able, in turn, to evolve social breeding that was not dependent on close genetic kinship. This is the strategy we will call *kinship-<u>independent</u> social breeding.*

We argue that humans are just exactly such an animal. We are the first animal in the history of the Earth to be able to cost-effectively manage conflicts of interest on a large scale (Chapter 5). We thus predict exactly what we see. Humans are Earth's first large-scale kinship-independent social breeders. This uniquely human adaptive strategy is the answer to our **why could** question, we claim. All the changes in human life history that allow us to build our big brains require a new level of social support for the offspring and the parents.[†] The African proverb it *takes a village to raise a child* describes human kinship-independent social breeding perfectly.

Remember the story of Jermaine's life or your own. Jermaine was not related to Jefferson's employees or customers, to the students Camille taught, to Janine, to his daycare caregivers, to Manfred, to Benton, to Yo Yo Ma, to the members of his orchestra, to the women and men who grew his food and built his home, and so on and so on. Yet all these people were crucial to his successful growth into a uniquely human adult and a full-fledged member of a species truly standing astride its world.

---

* In a transiently uncrowded environment, active conflicts of interest might be temporarily unimportant. However, just as importantly, there is no positive selection *for* helping non-kin even here (Chapter 3). The simple physical principles of the Second Law (Chapter 2) tell us that no significantly complex trait can evolve without active positive selection.

† Notice that the only way for a mammal to obtain this extensive *new, extra* support is by recruiting non-kin individuals. Unlike bees and ants, mammals produce too few offspring to generate a large band of close kin helpers.

Our theory predicts that our ancestors could first evolve this novel adaptive strategy because they were able to ostracize (threaten or kill) individuals who *failed to contribute* to this unique new breeding system. Their access to inexpensive coercive violence (and the credible threat of violence) made this possible.* Again, our ancestors would not have been *consciously* aware of this logic. However, we can now begin to see it clearly.

Over time, natural selection produced *proximate* psychological devices that caused early humans to behave *as if* they were aware of this logic. These devices would have included such things as a sense of "fair play" for distribution of food and resources to offspring and an "ethical feeling" that adults who were neglectful or infanticidal toward one's own kin youngsters were "evil." We have all experienced ethical feelings like these and the desire to react coercively (even violently) to them (Chapter 10).

We are the planet's first true kinship-independent social breeders. This is an extremely effective strategy and it is important to grasp fully the adaptive power of this approach to breeding. The colonial ants and bees help us to appreciate this power. These social insects achieve relatively impressive local ecological dominance. Other insects, even some mammals, fear them and live in their shadows.

They achieve this dominant status by virtue of a tiny amount (by human standards) of suppression of conflicts of interest. Individually self-interested worker coercion, through infanticide of worker-laid eggs, permits the evolution of large stable colonies (Chapter 5). These are generally NOT fully kinship-independent breeding systems. Instead, their small coercive acts simply expand kinship-*dependent* social breeding by extending it to a slightly lower level of kinship (brothers rather than sons; Chapter 5). However, even this small move *toward* kinship-independent social breeding has made them enormously successful creatures. It remained for humans to invent kinship-independent social breeding *in earnest* and to put ant ecological dominance to shame.

We are now prepared to return to our *how* question in more detail. Exactly how does kinship-independent social cooperation produce the large-brained, cognitively gifted, ecologically dominant animal that we are?

_____

*   As we saw in Chapter 5, the ultimate logic of ostracism is that the costs of driving out (or killing) free riders must be less than the benefits from preempting losses to these social cheaters in a remote killing animal. Benefits must accrue preferentially to self-interested individuals who actually engage in ostracism. For example, those actively ostracizing a non-cooperative individual might also take possession of their target's assets such as tools, clothing, mates, etc.

Extended human gestation – Flo can go it alone, but Camille cannot

As many readers will know from direct experience, the last month of human pregnancy is extremely demanding. For many women it requires extended bed rest and avoidance of strenuous physical activity. Recall the remarkable size of the full-term human fetus in relation to its mother's body (Figures 6.2 and 6.3).

The large human fetus pushes against the mother's major organ systems squashing them and reducing their efficiency. At the same time, it makes huge new demands on these very same organ systems. For example, the growing fetus compresses the mother's heart and lungs into the upper half of her chest cavity. Her liver is approximately where her heart used to be by the time she is ready to give birth. Yet, these self-same lungs and heart must supply large amounts of oxygen to the fetus (*in addition* to the mother's own needs) if the infant is to be healthy and fully developed.

Contrast this with the last month (the *eighth* month) of pregnancy for Flo, our chimp mother. She ceased to lactate and weaned her last offspring before becoming pregnant with Flint. Otherwise, she continues all her normal activities. Fetal Flint is not so large that he demands as much from her as ninth-month fetal Jermaine demanded of Camille. Flo is able to find the food and shelter she needs to support herself and the fetal Flint.

Thus, Flo (like most mammalian mothers) does not need much direct social support to give birth to her offspring.[5]

The very human Camille is in a different position. For a moment, we need to ignore a few of the novelties of our modern environment. So, we will consider an ancestral Camille—a human woman living sometime between the origins of the first human around two million years ago and the rise of the first large permanent towns around ten thousand years ago.

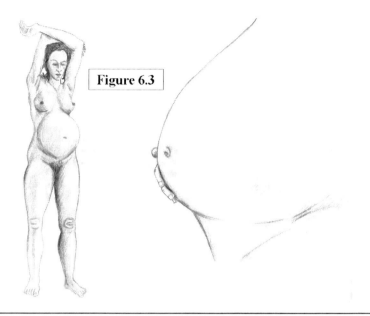

Figure 6.3: This figure illustrates how large the human fetus becomes during the last month of pregnancy. *Drawings by Amy Radican.*

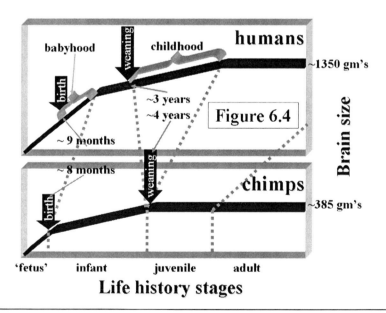

Figure 6.4: The life histories of chimps and humans are compared by plotting brain size as a function of developmental stage. The black line showing brain size is steeper where the brain is growing faster and becomes horizontal where brain growth slows substantially or stops completely.

First, notice that humans extend gestation a full month longer than chimps. Moreover, more resources are invested in the fetus. These changes produce a larger brain size at birth (Figures 6.2 and 6.3).

Second, humans continue the rapid fetus-like rate of brain growth well after birth. Chimps grow their brains at a more sedate infant rate at this time. This is one of the ways we continue to build our oversized brains after birth (Figure 6.2).

Third, humans continue infant brain growth for several years after weaning whereas chimps do not. This is another one of the ways we continue to build our oversized brains after birth (Figure 6.2).

The consequence of all these changes in early human life history is the production of an adult human with a brain around three and one-half times larger than an adult chimp. Moreover, the human adult has spent six extra years acquiring a vast amount of additional culturally transmitted information and an array of advanced technical skills (text; Chapters 9 and 10). All of these changes in life history are entirely dependent on the material and informational support of the human village, the unit of human kinship-independent social breeding. [Note that the weight units used here, gm's, are roughly equivalent to volume units, cubic centimeters or cc's.]

Suppose our ancestral Camille had no social support. What would she *lose* that she otherwise would *have* as a human?

Of course, the ancestral Camille would not have a job or an income to lose as our contemporary Camille does. These are very recent improvements in our system of kinship-independent social cooperation (Chapters 11-16). But the ancestral Camille would nevertheless lose economic support provided by her interactions with surrounding non-kin humans, the members of her "village" of cooperating non-kin adults.[*,6] Consider her predicament under these circumstances.

She would not have a home to live in unless she built it herself. This is a tall order, especially during the ninth month of pregnancy. Of course, this shelter will have to be well made because ancestral Camille will be naked except for clothes she made herself. There will be no one to make clothing for her.

---

* We use the word village metaphorically for any unit of human kinship-independent social cooperation. It can apply to many different detailed arrangements; however, human villages will share three properties, always and everywhere. First, they will include multiple adults of both sexes who are *not* close kin. Second, the village adults (including non-kin) will cooperate in ways that support many or all the members of the village. Third, some or all adult members of the human village will always be armed, permitting the projection of coercive threat that ultimately allows the village's cooperation. This armament might consist of throwing stones, bows and arrows, or sophisticated handguns; but it will always be present *without exception* (Chapter 5).

Likewise, no one would be foraging and hunting the food she eats. She would need to find all her food (while building and maintaining shelter and making clothes), even during her ninth month. Moreover, the extraordinary size of the fetal ancestral Jermaine means that she would need more, richer food than Flo.

Ancestral Camille would not have an obstetrician or its important pre-modern equivalent, a midwife with helpers. So she would have to give birth alone—dangerous, difficult *human* childbirth. In the shelter she built, nourished by the food she gathered, she must survive childbirth alone, pass the placenta, avoid bleeding to death from the inevitable labial tearing produced by the passing of a large human fetus, catch the newborn, warm him, and begin nursing him, all alone.

These tasks are, of course, not quite impossible for Camille alone. They are, nonetheless, risky and difficult. Natural selection plays the odds, nearly flawlessly (speaking metaphorically). Increased risks like these without corresponding benefits do not work. Without social support, natural selection does not bet on a needy animal like us.

Extended human gestation requires expanded social support. This is the first part of the answer to our **how** question. There is more.

Giving birth to a fetus – human "babyhood" requires extensive social support

Our newborns Flint and Jermaine, fellow apes, have just come into the world. Now what?[7]

Consider Flint. He is smaller at birth than Jermaine. Moreover, within days of birth, his brain ceases to grow at the rapid fetal rate. It slows to the infant rate (Figure 6.4). This change in rate has two important consequences.

First, Flint does not need as much food each day as he would need if he were bigger and growing his brain faster. Flo need not produce as many calories in milk as the ancestral Camille does. Flo can survive on a poorer diet.

Second, for reasons we can guess at (but do not fully understand) the sophistication of the behaviors a brain produces is apparently limited by how fast the brain grows. A very rapidly growing brain adds new brain tissue quickly, but this tissue does not perform well right away.* Thus, if Flint were growing his brain faster, his behavior would be less sophisticated than it is.

---

* This tissue requires "expert wiring" before it can be fully functional (Chapter 10).

In other words, a human newborn grows its brain more rapidly *not* because that newborn is immediately smarter. Indeed, it is "dumber" than it would otherwise be during this phase of its life, precisely because of its rapid brain growth. Instead, the human infant grows its bigger brain as an investment in the future, in order to grow into a smarter adult.

A chimp infant that made this investment (more rapid brain growth, poorer infant behavioral performance) would create a new problem for his mother. The newborn Flint might not be able to hang onto Flo's coat when she moved. He might not be able to sit upright on a tree branch where he was left for (comparative) safety while Flo foraged for food alone. Indeed, he might be in constant danger of simply flopping over and fracturing his skull or falling face down in a mud puddle and drowning.

Thus, Flo would be saddled with an infant that *simultaneously* required her to consume more and better food *and* to provide more attention. She would be forced to spend more time foraging alone for food at the same time she was also being required to spend more time watching over and protecting the less capable redesigned chimp infant. Such an insanely demanding infant is poorly designed for the chimp social environment and its design information would be eliminated by natural selection.

In contrast, consider Camille. First, she herself is bigger than Flo. Her own body requires more calories each day. Moreover, Jermaine needs lots of calories to sustain his large body and faster (fetal) rate of brain growth. Camille must find the food (and rest) to sustain herself and the necessarily high rate of lactation to sustain Jermaine.[8]

While providing for both herself and her child, Camille is afflicted with a very "slow learner." For example, the chimp Flint will be walking by around five months of age. Jermaine will take more than twice as long to walk proficiently. Indeed, at five months Jermaine will still be struggling to sit up without toppling over.

Thus, Jermaine seems to be precisely the poorly designed, insanely demanding infant we just said Flo *could not* raise. Yet Camille *can*.

How does Camille meet this challenge? The answer is obvious to us by now. She does not attempt to meet the huge demands of a human baby all by herself. That option would not work for her any more than it would for the solitary Flo.

Instead, for example, she may "forage" in the human cooperative enterprise (the village) while entrusting the moment-to-moment care of Jermaine to others (sometimes non-kin babysitters). Alternatively, others may forage for her (again in the human village) while she cares for Jermaine during his vulnerable babyhood. We will come back in a moment to just who it is that is doing this foraging for her. It is a little more interesting than we might think.

For now, two things are important. First, the strategies available to Camille to give her baby the best chance to survive depend on her being embedded in the human village—our kinship-independent social breeding system. Without the village, there could be no human-like baby. It simply would not evolve.

Second, the word *baby* as we will use it here, does not apply to other animals. It applies to a fetus-outside-the-womb. Jermaine continues to grow his brain at the very rapid fetus-like pace for many months after birth, but the chimp infant, Flint, does not. Only humans have evolved this long new stage of life history. So when we talk about newborn *non-human* animals, we will not call them babies. We will call them by others names (newborns or neonates, for example).

Again, the uniquely demanding human *baby* cannot evolve or survive without the human kinship-independent breeding system.

Wait! – Doesn't a "husband" solve all these problems?

At this juncture, we might be nurturing a very interesting doubt. We might be thinking that it is true, our modern Camille makes extensive use of the contemporary human village (our economic system) to fulfill the very young Jermaine's uniquely human needs. BUT, would not the ancestral Jefferson's (husbands) have been sufficient to fulfill the uniquely human needs of ancestral Camille's and Jermaine's? Perhaps the title of this chapter is misleading. Instead, should it not be "It takes a *couple* to raise a human child"?

At first glance, this is not a bad hypothesis. For example, the kind of pair-bonding we do (with extensive male contributions to raising the young) is, in fact, virtually unique among the mammals. Indeed, a few anthropologists have suggested that human uniqueness results from human pair-bonding.

However, this hypothesis is almost certainly wrong, even actively misleading. First, though few other mammals pair-bond as humans do, many bird species raise offspring this way and, of course, they remain bird brains. There just is not any empirical hint among the birds that pair bonding is sufficient to produce an ecologically dominant (or large-brained) animal like us.

Second, a pair-bonded couple would certainly have advantages over an isolated single parent. However, this is small change in comparison to the enormous productivity of larger cooperative human enterprises. This big advantage does not merely arise from our huge contemporary economies. It also applies to the simplest uniquely human social aggregates. Even twenty to fifty adults can undertake massively more ambitious projects than a single couple, especially a single couple with demanding, dependent children.

Again, our uniquely human capacity to suppress conflicts of interest makes this social cooperation possible. Even our very first human ancestors would have had the capacity to suppress conflicts of interest on the scale of twenty to fifty individuals, as we will see in Chapter 7.

But, in spite of the obvious importance of the human village, do we not still have evidence that pair-bonding is important? Both human sexes *fall in love*. This behavior looks just exactly like the product of a proximate device designed to support pair-bonding. It almost certainly is exactly what it appears to be. But we will see in Chapter 8 that these matters are more complex than they seem.

For now, it is sufficient to keep the following in mind. Pair bonding does not imply sexual monogamy, *a priori*. In fact, as we are all aware, both sexes have much more complex and ambiguous sexual feelings and impulses than expected if pure pair-bonding were our exclusive adaptation.

Indeed, we will find in Chapter 8 that ancestral human sexual behavior probably cycled over time from fairly monogamous to rather promiscuous and back again repeatedly, depending on local context. Obviously, as soon as even a modest level of promiscuity occurred in our ancestors, kin-selection would no longer reward individual males who invested exclusively in the offspring of their nominal spouses.

However, as long as conflicts of interest between such males could be cost-effectively suppressed (as they could in our uniquely human ancestors), it was still adaptive for males to cooperate to raise the offspring of a promiscuous mating system, a subset of which are likely to be their own.

The crucial point for us at the moment is that kinship-independent social breeding was a fully workable strategy, whether ancestral sexual behavior was monogamous or promiscuous. The ancestral human village had the power to produce unique human offspring across a broad range of adult sexual behaviors. The resulting kinship-independent social breeding system, has served us (and human design information) well throughout our two-million-year evolution.[9]

The butcher, the baker, the candlestick maker – unique human childhood

After Jermaine emerges from the very rapid brain growth of baby-hood he still continues to grow his brain (though more slowly now) until he is around six years old (Figure 6.4). Flint will continue to grow his brain until about age five. Thus, he will stop earlier than Jermaine. More-over, Flint will grow his brain *even more slowly* than Jermaine during this period.

These differences in brain growth are the next part of the overall redesign of human life history to produce our big brains. What concerns us at the moment is how Flint and Jermaine will obtain the food they need to get through this final stage of brain growth. The differences between them have crucial implications.

We need two formal, technical terms. The first of these is *infancy* or the *infant* stage. When used in our specific technical sense, this term applies to the period immediately *after* rapid fetal or fetal-like brain growth. As we saw above, infancy for Flint begins almost immediately after birth. In contrast, for Jermaine, infancy begins around ten to twelve months of age (after the uniquely human stage of babyhood).

The second technical term we need is *childhood*. This is a word we all use in daily conversation. This common use is a little vague and vari-able. In contrast, here we will give this word a very precise meaning. For us, *childhood* will refer to the period after weaning (when mother's milk is withdrawn) but before the youngster has adult teeth and is able to process adult food effectively. Jermaine has a childhood. Flint *does not*!

By exploring how and why these two youngsters are so unlike one another at these stages, we gain further insight into our origins and evolution. The first big difference is the nature of the food that Jermaine will have access to, even before he is weaned. Humans (like most mammals) begin to take some solid food while they are still nursing. This is part of working up to adult food.

For human youngsters, this food must be of much higher quality than the adult food that the infant, Flint, will have to supplement moth-er's milk. Specifically, Jermaine has a smaller gut (gastrointestinal tract) than Flint (an issue we will return to in a moment). Moreover, Jermaine will use the energy he takes in less efficiently. He will spend a larger fraction of this energy in maintenance (including of his over-sized brain) and less on growing new tissue than Flint will. The upshot of these prob-lems is that Jermaine will need richer food (bone marrow, fat-rich meat, oil rich nuts, and starch-rich tubers, for example) and more of it. Moreover, this richer food is harder to obtain than the lower quality adult chimp food (leaves and fruit) that Flint will eat.

The second big difference is that Jermaine will be fully weaned much earlier than Flint. To understand the significance of this earlier weaning we need to look away again from our lives in the recently arising modern state. We must ask how most human infants are raised in more primitive conditions and were, presumably, raised in ancestral conditions. The answers here appear to be fairly clear. Ancestral humans would have been weaned around three years of age as opposed to roughly five years of age for Flint!

At first glance, this seems paradoxical, even perverse. The ancestral infant Jermaine must grow his brain faster and longer. In order to do this he needs larger amounts of extra rich foods. But he is deprived of the single richest food source he has (mother's milk) while this growth is still going on!

The reason he is "deprived" in this way is enlightening. To grasp what is happening here, one other detail of human reproductive physiology is important. We in modern states are so well fed that obesity is a much more common health problem than starvation. However, an ancestral human mother was in a very different situation. Very often her calorie budget was sharply limited (in spite of the support of the ancient village) and with it, her reproductive opportunities.

Lactation (production of milk) requires a large number of extra calories. When calories were limiting, the nursing mother would become more slender (even emaciated in extreme cases) under lactation's demands. So, if she could wean her infant she could divert calories to restoring body fat. This body fat matters. Her reproductive system was designed not to allow her ovaries to release new eggs until her fat reserves were sufficient to give her a good chance of a successful pregnancy.[*,10]

Thus, when an ancestral human mother weaned her infant earlier, she could have another child sooner. Over a lifetime this translated into extra children and, thus, extra copies of the mother's design information.

However, this strategy would have been utterly self-defeating if the prematurely weaned infant died or failed to develop into a healthy reproductive adult. This danger was acute for the ancestral Jermaine, as we have already seen. Remember that he needed rich food and lots of

---

* Contemporary human females can sometimes find themselves in calorie-restricted situations significant enough to impact their capacity to reproduce. This can happen as a consequence of severe poverty. It also sometimes happens in extreme athletes or in individuals with certain kinds of eating disorders.

it. All other things being equal, he was far more likely to starve without mother's milk than Flint.

We have now arrived at the crucial insight. All other things were *not* equal for the ancestral (or the modern) Jermaine and Flint. Flint must rely on mother's milk or the food he can gather on his own. Jermaine has access to vast resources generated by the human village. This village can forage, scavenge, and hunt with an economy of scale and a cooperative efficiency that no individual (or pair of individuals) could match. The ecologically dominant human village can also protect these rich assets from other animals in ways individual non-human parents cannot. Large quantities of rich, diverse foods for uniquely human "children" are a result.

Taking stock – the evolutionary logic of living *within* the human village

It is illuminating to pause and reflect. With the capacity to suppress conflicts of interest inexpensively came expanded, kinship-independent cooperation. This cooperation opened new reproductive strategies. In effect, the *best available self-interested option* for individuals within the human village was to contribute resources to the survival of all (or many) of the village's youngsters. Those who did not were ostracized—or worse.

In this context, in turn, another element of the best self-interested strategy for individual ancestral females was to capitalize on the enhanced resources of the village to shorten their inter-birth intervals. Females who engaged in this exploitation left more progeny and more copies of their design information. As a result, we modern humans are all the descendents of such ancestral females.

The enhanced resources of the village also provided new opportunities for the human infant. She (or he) could grow her brain faster and longer even when prematurely "deserted" (weaned) by her self-interested mother.

This is natural selection in its simplest and most transparent form. You and we (as parents and as infants and children) are its inevitable products.[11]

Human adolescence – lords of the planet become adults

Jermaine and Flint are now about six years old. Both have largely finished growing their brains and Jermaine is the proud owner of his much larger uniquely human brain.

We might imagine that they will now proceed to mature into adults similarly. But we would be wrong if we did. The details of this story are surprisingly rich and complex. They will turn out to be important to understanding uniquely human properties, including our powerful minds, elements of our sexual behavior, and our elaborate ethical sense. For now, we need only a few details.

At age six, a hypothetical Flint will enter a new *juvenile* phase of his life. (The real Flint died young by the ill fortune so common in wild creatures.) He will spend the next six years growing his body (remember, his brain is already full-sized) and acquiring experience (wiring up his brain). At the end of this phase, he will be a full-grown adult. A typical chimp can become a parent at around twelve years of age. In contrast, the modern Jermaine will become a parent much later in his life. Moreover, later parenthood would also have been true of ancestral humans. Judging from the ethnographic record and other kinds of evidence, humans typically became parents for the first time around eighteen to twenty-one years of age for at least the last five hundred thousand years of our history.[*,12]

Even after brain growth is complete, the ancestral Jermaine would have taken *another* twelve years—instead of Flint's six—to become a fully functional, reproductive adult! What is going on here?

Some of the details are a little complicated. We humans also grow our bodies from the age of six to the age of twelve but, we do it more slowly. This is our *juvenile* stage. Americans often describe humans of this stage as *grade school children* though we are using the word *child* with a different meaning than above.

Then around the age of twelve to thirteen we enter *puberty* and undergo a growth spurt. We acquire our adult height within a few short years of rapid growth. However, for various subtle reasons, most of us

---

* Since the very recent advent of the state, with its relatively high wealth (at least, for some), young people are sometimes much more well fed than was common in the ancestral environment. Under these conditions, pregnancy as early as twelve and well before eighteen becomes more physiologically possible. However, this possibility was probably very uncommon for most of our prehistoric ancestors.

are not fully functional (socially and reproductively) until around the ages of eighteen to twenty-two. This period between twelve and eighteen years of age is uniquely human in its extent and can be referred to as *adolescence*. In this case the technical term adolescence means almost exactly what we mean when we use the word in everyday conversation.[13]

For now, what matters is that chimps take six years after their brains are full-sized to grow their adult bodies and acquire adult expertise, while humans take twelve years! This enormous discrepancy apparently has one primary origin. A chimp like Flint will grow up mostly learning what a single adult, his mother, can teach him.* His mother's cultural legacy will serve Flint well. But it is only the amount of information a single individual can pass on in the time she has for this task.

In contrast, in the human village, the ancestral Jermaine (just like the modern Jermaine) will spend his extra six years of juvenile and adolescent development learning from many adults in addition to his mother.† Moreover, these adults, in turn, will have learned from many other adults during their own development. Thus, the stream of culturally transmitted information in humans is vastly larger than in non-human animals.

The expertise Flint acquires in six years from one person, his mother, will make him a successful denizen of the East African rainforest. The information that Jermaine acquires in twelve years from many people (each of whom, in turn, learned from many people) made him a lord of the planet.

Notice, again, how we think of humans during juvenile and adolescent development. We Americans call them "grade school" children, "junior high school" children, and "high school" students. Their primary business remains the one uniquely human thing it has always been. *They learn*, acquiring the vast store of information held by the human village. In so doing, they become uniquely subtle and sophisticated *human* animals.‡

----

*   Non-human animals can trust other animals as a source of culturally transmitted information only if they are very close kin (Chapters 9 and 10).

†   The human capacity to suppress non-kin conflicts of interest means that human youngsters can trust a much larger number of adults than non-human youngsters can (Chapters 9 and 10).

‡   As we will see in later chapters, the volume of culturally transmitted information has grown explosively with the rise of the modern state. Under these evolutionarily novel conditions, we have extended this period of learning into early adulthood, with *college students*.

A sleek, handsome youngster comes of age on a rich human diet

We have seen how the human village supports our growth to uniquely human adulthood. However, the human village also affects Jermaine's adult anatomy (and ours) in obvious and direct ways.[14] These anatomical adaptations will be most useful to us a little later.

The food produced (or gathered or hunted) by the human village is of higher quality than the food that an individual ape can normally find for itself. Human food is richer and denser in calories and key nutrients so that we need to eat and digest less of it. As a result, our gastrointestinal tract (our gut) has been streamlined. It is shorter than a chimp's. Our stomachs or waists are thinner because they need only accommodate this smaller gut (Figure 6.1).

This is reflected in our skeleton in both the pelvis and the rib cage. In a chimp, these flair outward to accommodate the pot bellied shape housing the large gut. In contrast, in us, these structures are narrower and more cylindrical. As we said, this slimming of the human physique apparently requires the high quality diet produced by the human village.

Second, the elements of the high quality human diet (soft meats or honey, for example) are easier to chew than most parts of ape diets. This apparently permitted our teeth and jaws to become smaller and lighter. This is extremely obvious when we compare a chimp and human skull in Figure 6.1. This gracilization of the human chewing apparatus is, again, apparently dependent on the high quality diet of the human village.[15]

Modern economies are massive kinship-independent social breeding systems – the "purpose" of Main Street and Wall Street is the production, care, and feeding of human vehicles

We now have a deeper understanding of the contemporary human condition. Our economic system is really just a massive kinship-independent social breeding system. It is a device for generating the resources to sustain human adults as they raise human offspring. We carry out our roles as economic actors so (relatively) effortlessly that most of us think of our daily world as naturally peaceful and relatively prosperous (usually). Nonetheless, a legal system (backed by armed police) is essential, everyday, to make this kinship-independent breeding system possible.

In other words, our contemporary economic system apparently reflects kinship-independent social cooperation sustained by coercive suppression of conflicts of interest. Our massively redesigned life history implies that this pattern of cooperation is ancient. On our theory, the only important detail that has changed over the two million years since the first humans is the *scale* of this novel adaptive trick (Chapters 11-16).

Your mind and ours were built by the genetic (and culturally transmitted) design information shaped by natural selection to recognize and exploit these *facts of life*. Everything we do, everything we think, is the product of this uniquely human mind.

First, remember what minds are. They are collections of proximate devices designed by natural selection to direct our behavior in ways that are adaptive (or, more precisely, that *were* adaptive in the ancestral past).

Second, these proximate devices are *as if* devices. That is, they cause us to behave as if we were looking out for the interests of our design information without ever being conscious of those interests. Thus, we have little or no direct access to the real (the ultimate) reasons we feel and act as we do.

Third, a crucial implication of this lack of conscious access, in turn, is that we really do not understand our daily lives at a very deep level. We feel that we pursue one thing or another; for example professional satisfaction, making a contribution, carrying out the will of God, or obtaining money, power, sex, love, fame, glory, and so on. In actual fact, these desires are just the self-referential products of various *tactical* proximate devices, designed to produce *strategic* outcomes beyond our conscious awareness.

Each of us is, in reality, a vehicle that has evolved to replicate its design information successfully, but in a way very different than any other vehicles that have ever lived on this planet. The best available self-interested strategy for us is to contribute to the cooperative human village to the extent demanded by the local human coercive environment. At the same time, we exploit that cooperative village to produce and raise our offspring vehicles (our children). We also project coercive threat in defense of our interests within this village and we are vigilant about access to the necessary power to do this.

The global economy (the global village) is produced by billions of self-interested human actions programmed by unconscious proximate devices designed by natural selection. All the massive infrastructure of Wall Street, Main Street, Pennsylvania Avenue, the Kremlin, Beijing's Forbidden City, and so on and so on is merely the ancestral human village writ large, and having the sole "purpose" of producing new human offspring and rearing them to adulthood.

This may seem like a rather harsh or sterile view of our world. It may appear to diminish many things we believe in and hold dear. Be patient. As we proceed through the remainder of the book, we will make precisely the *contrary* argument. Truly understanding our evolutionary history puts us in a position (for the first time in the two-million-year existence of our species) to comprehend fully the possibilities for the meaning and purpose of our humanness.

We will argue that this insight gives us a new opportunity to choose from our evolutionary (and historical) heritage those things we wish to retain. There will be many of these. Moreover, we will also be able to choose which pieces of our history we wish to leave behind. There will be a few of these also, including some of our more inhumane behaviors and institutions. With authentic self-understanding we can find a new maturity and a new set of choices—new possibilities to find joy and fulfillment in our lives (Chapter 17 and Postscript).

Bodies, brains, and birth canals – insight into the human fossil record

Our theory makes extensive, precise predictions about what we will see if we time travel back to the origin of humans (using the fossil record). Kinship-independent social cooperation should emerge rapidly after the evolution of the first capacity to kill adult conspecifics remotely, that is, the capacity to throw with elite human skill.

Reflect for a moment on what a powerful test of our theory this is. *Only one animal* (among millions of species over hundreds of millions of years) could kill adult conspecifics remotely. *That same animal* should also be *the first and only animal* to evolve expansive kinship-independent social cooperation. Moreover, this cooperation should *follow* (not *precede*) remote killing and it should do so rapidly (on an evolutionary time scale).

If any one of the details in the preceding paragraph is wrong, the theory is wrong—period, end of story.

In this chapter, we describe the last critical tool we need to ask whether the theory survives this first challenging test. We found that we can score kinship-independent social breeding by scoring the evolution of the unique human life history. Moreover, kinship-independent breeding is merely the reproductive consequence of kinship-independent social cooperation more generally. Thus, our life history reflects our unique human social adaptation directly and vividly. Finally, several of the crucial details of this redesigned human life history are plainly recorded in our skeletons, *and our skeletons fossilize.*

The fossilized skeletons of early human ancestors will show us the evolution of our large brains and redesigned bodies. Moreover, we can tell from fossil pelves (plural of pelvis) with their birth canals that these enlarged brains involved a lot of growth after birth, apparently supported by the ancestral human village. Finally, we found that our slender-waisted, small-faced physiques (and skeletons) probably likewise report to us about the rise of the uniquely human village with its rich diet.

Thus, we can score the emergence of uniquely human diet and life history in the fossil skeletons of our ancestors. *Equally importantly*, we will find that the evolution of elite human throwing also resulted in the redesign of parts of our skeleton. We can see this in fossils, as well.

We now have everything we need. Both kinship-independent social cooperation and remote killing are directly recorded in the fossilized skeletons of our ancestors. Armed with these insights we can now ask our questions—we can subject our theory to its first strong test in Chapter 7.

෨

Chapter 7
Throwing strikes on the village commons – the planet's first remote killer and first kinship-independent social cooperator

PLACE: The East African Savanna. TIME: About two million years ago, shortly before the evolution of the first recognizable members of the genus Homo, *the first "humans."*

Two primates roam the grasslands. One is the ancestor of the savanna baboons still alive today and the other is our ancestor. Our ancestor is bipedal (walking up on two legs as we do), while the ancestral baboons favor walking and running on all fours (quadrupedally). These differences will matter to the human story, though not for the reasons traditionally assumed. A day in the life of each of these animals will reveal something important about them and about us.

The baboon troop moves across the savanna, foraging as they go. Seeds, tubers, fruits, and nuts are all attractive foods. Insect larvae are a treat and a useful source of protein when they can be found.

This is a special time of the year. The rains came a little while back and the savanna is lush. Grass is not good food for the baboons, but professional herbivores like wildebeests, topis, and gazelles love it. All these great savanna grass eaters have timed the birth of their offspring for this season of plenty.

The resulting baby gazelles are of special interest to the baboons. The newborn gazelles are small enough for the baboons to handle without undue risk of being injured. Moreover, these youngsters still run relatively slowly and do not yet have much endurance. The best way for these slow baby gazelles to protect themselves is to lie still in the grass and hide. Their coats are dappled as camouflage to improve their chances of evading detection (natural selection, again). The baboons "know" this and they are on the lookout.

A large, adult baboon male has just captured a baby gazelle. With characteristic vigor he begins eating the abdominal organs. Shock and blood loss quickly end the prey's suffering. Other baboons gather round, but the male is robustly healthy, too formidable to challenge one-on-one. The male consumes the gazelle, leaving scraps that the other baboons scavenge for tiny bits of nourishment.

The second primate, our ancestor, also lives in a troop. We can call this animal the "accidental ape." This troop of bipeds moves across the savanna and has its own interest in the explosion of new baby gazelles. The accidental apes fan out as they approach the herd, encircling some of the females and their fawns. As the fawns lie still in the grass, the accidental apes sometimes get close enough to spook them. The slow fawns could be caught by the apes in the same manner as our baboons above. But, the accidental apes have a better trick. As the calf bolts, he runs directly toward another individual in the circle of accidental apes. One of these other apes grabs him and immediately breaks his neck with a single, violent shake.

The accidental apes have yet another, even better trick. These apes are bipeds. Thus, their hands and arms are free allowing them to evolve the capacity to throw rocks with extremely great speed and accuracy.* They have learned to carry a few baseball-sized stones with them when hunting. As the speedy adult gazelles bolt, the accidental apes throw at the vulnerable knee joints of their forelegs. Sometimes the apes miss, sometimes they do not. When an adult gazelle is hit, the foreleg collapses and the animal tumbles down. The apes immediately storm the downed animal, quickly clubbing her to death with large sticks some of them carry.

But this still is not the end of the better tricks the accidental apes possess. They have one more. The apes gather around the dead fawns and adult gazelles and begin to feed. Larger individuals might like to seize more of the prey and consume most of the meat for themselves. However, they are surrounded by other apes who will all throw stones at them if they attempt to do so. Even a large individual is no match for a concerted barrage from all the other apes. Any individual who tries to monopolize the prey would become a target for numerous well-aimed, dangerous missiles.

As a result, the accidental apes can and do share their kills. Each individual needs to spend only a modest amount of effort in the original hunt and each is well fed. These accidental apes grow fat even when the baboons struggle.

---

* This is an example of evolutionary *historicity*. For a quadrupedal animal (like the baboon), natural selection would have to have the foresight to redesign its body into a form that could then, in turn, evolve elite throwing. Since natural selection has no foresight, the baboon is in a you-can't-get-there-from-here situation when it comes to evolving elite throwing.

Part I: An introduction to the evolution of the first humans[†]

An ape in the human village

How can these two animals—baboons and accidental apes—behave so differently? The answer makes perfect sense. And it is utterly profound.

Continue with our imaginary case from the opening vignette. Compare the situations of two unusually large individuals—one a baboon, the other an accidental ape. The large baboon could selfishly feed and wait for others to come to him to try actively to take away meat. The large baboon knows that such an approach would be too risky for other, smaller baboons. Such behavior is too likely to result in bodily harm for each of them for too little gain, a few bites of food. The smaller animals would have to fight him one-on-one. He would never be approached.

In remarkable contrast, for the large accidental ape individual, others did not need to approach him individually to obtain a piece of meat. They could stand off and together throw at him to chase him away or even kill him if he attempted to take an excessive share of the kill. This remote, synchronous killing capability meant that their individual costs of obtaining a share of the meat were much lower than for the baboons, even if the apes were individually smaller and weaker than the target individual.

This situation reflects the lower costs of coercive violence in remote killing animals (Chapter 5). In overview, multiple remote killers can all project threat at once, but multiple proximal killers—animals that kill with teeth, like the baboons—cannot. As a result, remote killers subdue their targets much more rapidly than proximal killers. The time they spend in harm's way is correspondingly reduced.

The best available self-interested choice for large baboon individuals and large accidental apes is completely different, as a result. For baboons, this best choice is to keep the food for themselves. Therefore, baboons usually hunt alone and eat alone. For the accidental apes, the

_____

[†]   Good general references on human fossil evidence include Leakey and Lewin (1992); Walker and Shipman (1996); Tattersal (1998); and Klein and Edgar (2002). We emphasize that these and other books give solid accounts of the empirical evidence, but (in our view) give no adequate theories of how humans got to be the unique animal we are. This missing theory is our goal here.

better choice is to surrender shares of the food to the other apes around. They hunt together and share their kills. Again, even a very large accidental ape will be subjected to dangerous violence from those around him if he does not share. The large baboon will not.

Likewise, for smaller, less fierce individuals, the best choices are *also* completely different in these two species. For the baboons, it makes no sense for smaller individuals to attempt to coerce or extort a share of the food from a large individual. Such coercion would require too much cost for too little benefit. Contributing to a cooperative hunt whose proceeds will be stolen by other non-kin individuals is foolish. Such cooperative behaviors will never evolve.

In stark contrast, for the accidental apes, it makes perfect sense for smaller individuals to coerce larger individuals to surrender a share of the proceeds of a hunt. This method of cheaper conjoint coercion now makes coercing a meal a good deal. Coercing apes receive the same modest benefit as the baboons *would have* received, but at a much lower cost.

Under these conditions, evolving cooperative behaviors is adaptive, sensible. Notice that the logic of this cooperation is not dependent on close genetic kinship. This scenario is just an example of our old friend—*kinship-independent social cooperation.*[*]

These distinct social logics ramify throughout the lives of these two animals. For example, forming a hunting party makes no sense to most of the baboons. The larger individuals would simply take the kill at the end. Thus, the smaller individual baboons' best choices are to hunt alone for high value foods, like meat, as they generally do.

Each smaller individual among the accidental ape pursues a diametrically opposite logic. Contributing to a cooperative hunting party increases the likelihood of flushing game and, *unlike the smaller baboons*, he will get a share. Any larger individual who attempts to monopolize the kill will be forced to disgorge the prey under very credible threats of violence. This capacity for inexpensive projection of threat by multiple remote-killing (elite throwing) individuals allows the apes to realistically pursue this tactic of *pre-emptive ostracism* (Chapter 5).

As a result of this new capability, many more accidental apes have access to more rich food, like meat, marrow, and brains, than baboons. Sharing this increased bounty is an unavoidable choice. Those who fail to share are ostracized or killed.

---

[*]   We will return later in the chapter to the details of how this cooperation probably evolved from pre-human kinship-dependent social cooperation.

The vital point – elite throwing produces a fundamentally new kind of animal

It is crucial that we recognized that the specific type of foraging, hunting, or scavenging does NOT matter in the story of the accidental apes above. RATHER, it is the availability of elite throwing that matters.Elite throwing allows inexpensive coercion, which, in turn, allows kinship-independent social cooperation to emerge and evolve.[*] Likewise, simple kin-selected cooperative hunting, as lion sisters might do, does not produce kinship-independent social cooperation, the resulting big brains, or other unique human features (Chapter 6).[†]

Further, it is the capacity for non-kin cooperation to be rendered the best available individual option (in many different activities of this

---

[*]   The possibility that human throwing might somehow be centrally important has been recognized since Darwin (see pages 49-51 in *Descent of Man*, 1871/1909). Barbara Isaac's 1987 paper is an excellent recent case. However, until now, we have not had a convincing theory for *why* throwing might matter. For example, Darwin thought that throwing might make the first primitive warfare possible. The needs of warfare might then select for improved social cooperation. This "conflict model" has been updated many times since Darwin (see Choi and Bowles, 2007, for a recent example). However, this idea is fundamentally flawed. Selection for better abilities at collective conflict as the basis of social evolution commits the group selection fallacy (Chapter 3). Individuals who hang back and do not engage in warfare do better than individuals who fight under biologically reasonable conditions.

In contrast, our theory puts organized human conflict in a realistic context. Warfare is like most other cooperative activities. It presents the conflict of interest problem and can only evolve *after* conflicts of interest are managed (Chapter 5). In other words, human warfare is just one of many *effects* of uniquely human social cooperation on our theory, *not* its *cause*.

Another proposal was that evolution of the capacity for hyper-fast controlled movement in elite throwing selected for rewiring of the human brain in ways that supported the subsequent evolution of uniquely human intelligence and/or speech (Calvin, 1983).This proposal is flawed in at least two crucial ways. First, many animals engage in species-typical hyper-fast movements. Think of a cheetah running across the uneven surface of the African savanna at 60 miles per hour, for example. Yet none of these other animals evolved uniquely human-like intellectual or linguistic abilities. Second, as we will see in Chapters 9 and 10, evolution of elite intelligence and human speech requires the solution of the conflict of interest problem. Simply having a fast mind can never fulfill this requirement.

[†]   Recall Hamilton's Law from Chapter 3. Kin-selected cooperative hunting involves less opportunity for adaptive cheating. Among the hunters, each of the cooperative kin is playing the odds that the benefits are going to other copies of their own personal design information. Of course, a bigger non-kin animal can still come and steal the kill and sometimes they do. Even a group of lion sisters together cannot inexpensively repulse a large male if he happens to show up after the feast has begun, for example.

species) that really matters to the evolution of the accidental ape. This cooperation reverberates through every facet of the accidental apes' lives. Reliable sharing, even among non-kin, applies not just to hunting baby gazelles, but also to everything accidental apes do that generates a *sharable* resource whose hording can be coercively pre-empted. Individuals who are adapted to contributing to and capitalizing on the benefits of this cooperation will do better when this coercive umbrella exists.

Thus, the evolution of inexpensive conjoint coercion will produce an adaptive revolution including many, diverse new behaviors. There are vast numbers of different scenarios for individual capitalization on cooperation sustained in this way. The origin of humans is the story of our evolution of the capacity to use all of these diverse tactics. These accidental apes (or actual animals much like them) are our very last pre-human ancestors. We evolved directly from them, by adapting to make use of the myriad opportunities for productive, enforceable cooperation remote killing made possible.

This picture predicts that our adaptations to kinship-independent social cooperation should be legion, as indeed they are. We will see in later chapters that individual humans are exquisitely designed for this cooperation, both generating and using its benefits and projecting the self-interested coercive threat that sustains it. Moreover, our history is a sequence of applications of this underlying adaptive strategy—over and over and at ever-larger scales through the present moment.

For now, only the most simple and direct ways for individuals to make use of this new cooperation need to concern us. For example, consider a reproductive female in the accidental ape troop. If she now has more reliable access to high-quality food from cooperative foraging, it is possible for extended gestation to evolve (Chapter 6). Offspring resulting from extended gestation, thus, have the opportunity to have slightly larger brains.* Better yet, nestled in the information-sharing cooperative human village, these brains will be adaptively useful, making their owners smarter (Chapters 6, 9, and 10). These smarter individuals are more likely, in turn, to survive to adulthood and leave more offspring, the original female's grandchildren. Her long-gestation design information will do better, generation after generation. In other words,

---

* Extended gestation is produced by genetic variation among alleles in the population that produces these alterations in life history (Chapters 2 and 3).

her lineage contains vehicles redesigned to replicate her design information better in this new socially supportive environment.*

Consider this same process from the perspective of a self-interested
youngster. He (or she) can afford to grow his brain longer as he has
more reliable access to rich food. If, as a result, he grows into a smarter,
healthier adult who leaves more offspring, his design information does
better, generation after generation.

What just happened? The reliable availability of socially provided
enriched nutrition and protection makes brain enlargement—through
life history redesign—possible and, indeed, a good adaptive idea
(Chapter 6). This enriched access to necessary foods and protection, in
turn, is a result of kinship-independent social cooperation, which is a
consequence of the access to the cheap coercion necessary to suppress
conflicts of interests between non-kin individuals.

Notice, again, that the only tool that is required is cheap coercion.
All the other implications and ramifications are merely effects of this
primary cause.

How this chapter is organized

How do we know that the scenario of the preceding section is likely
to be an accurate description of our early evolution? Therein resides a
magnificent evidence-filled story—the product of several centuries of
human investigation of our own ancient origins. In this chapter, we will
examine telling parts of the massive body of physical evidence from
this beautiful project, ranging from evidence for local climate change
to human fossils showing the redesign of the human body for elite
throwing. We will find that the scenario we just outlined accounts for the
diverse details of this rich body of evidence with unprecedented power
and completeness. This answer has the strong parsimony we seek and
demand of a good *reductionist,* scientific theory (Preface, Introduction,
and Chapter 1).

When we look back at our ancestors during a brief interval around
two million years ago, we will find extensive evidence that the evolution of elite human throwing was **followed rapidly** by the evolution
of kinship-independent social breeding. Fossil skulls will tell us about

---

*   In this and the following paragraphs, we are speaking loosely for semantic economy as if individuals (or their design information) were somehow consciously aware
of these new adaptive opportunities. Obviously, they are not. It is the random search
procedure of natural selection operating over many generations on variation among
copies of design information that produces these outcomes.

social cooperation by reporting brain expansion (Chapter 6). Fossil skeletons below the neck will tell us about elite throwing. We will find a bipedal, small brained, non-throwing ape-like animal ambling into the front end of this brief time window and a large-brained, powerful, elite-throwing human striding out the other end. These first humans are the direct ancestors of all of us. Their descendents will build all of human history.

In the following section (Part II), we will introduce and summarize the crucial punch lines from this strong body of evidence. In the final section of the chapter (Part III) we will explore the details of the human fossil record more completely for those who wish to grasp the remarkable particulars of this evidence.

Part II: The essential features of the evolution of the first humans*

Phylogenetic trees – how we can visualize the evolutionary history of animals

How does a new animal species come into existence on an evolutionary timescale?[†] Since the origin of the very first organisms over three and one-half billion years ago (Chapter 2), all new species on Earth always arise from previously existing species. The complete set of species alive on Earth at any moment results from a long history of two competing processes.

First, *speciation* is the generation of two or more new species from one ancestral species over time.[‡] Our understanding of some of the details of speciation is incomplete; however, we have a good general grasp of the process. The following realistic example will illustrate what we mean.

A single species might become separated into two non-overlapping populations by some geological event. For example, climate changed repeatedly in Africa over the last twenty million years and particularly over the last six million years.[1] These changes probably included tens of thousands of years of drier climate alternating with periods of wetter climate.

---

* Outstanding images of human fossils can be found in Johanson and Edgar (1996). For readers wishing to engage more technically advanced material, Aiello and Dean (1990) and Klein (1999) give excellent accounts of existing fossils. Vrba, et al., (1995) discusses African climate change during the era we are concerned with.

[†] We will use the phrases *evolutionary timescale* and *geological timescale* to refer to the tens of thousands or even millions of years that evolutionary changes sometimes require. These choices of words reflect the fact that our time intervals are sufficiently long that the surface of the Earth is being remodeled by various geological processes such as volcanism and tectonic plate movement (text).

[‡] It is important for us to recognize that new species often evolve in a small local sub-population of an existing species. Thus, a new, adaptively changed species in one location can co-exist with largely unchanged members of the ancestral species in other locations. Evolution of a new species need not imply the extinction of its immediate ancestral species.

When the climate dried out, a large continuous area of rainforest would be broken up into smaller chunks of forest separated by zones of grasslands (savanna) and desert. Such surviving fragments of a particular habitat are called *refugia* in the sense of refuges for their resident creatures. Likewise, climate drying will break formerly large expanses of savanna into smaller savanna refugia separated by stretches of desert.

It is important to consider what happens to the creatures trapped in these refugia over time. More specifically, we need to focus on the genetic design information that is building these refugia residents.

Remember that the pieces of design information making up each animal will often differ slightly from one another. For example, one copy of a particular piece of design information, a gene, might differ from another by one or two DNA bases. These differences might make the protein gene product (the tool) perform slightly differently in each case. We discussed genetic variation like this in Chapters 2 and 3.

Recall that we call such different versions of the same gene different *alleles*. When the different animals in one of our refugia populations have a number of different alleles at a gene, we say the population has a lot of *allelic variation* or just variation, for short.

Slight differences in performance of the protein gene products produced by different alleles might be valuable to the animal (and, thus adaptive) or these differences might disadvantage the animal or even not matter at all depending on the local environment in which the animal finds itself.* Likewise, a new variant allele in the sequence of *control elements* that determine when, where, and how much a protein gene product is made might sometimes be adaptively superior (or not)—again, depending on the animal's environment/context.

The animals in each of these isolated small refugia will reproduce, generation after generation. At each generation, events will occur that change the collection of alleles building the individual animals making up the local population.

These events can be *selective*. For example, consider two refugia that are small fragments of the same original large piece of forest. When they first form, these refugia contain members of the same animal species—a primate, say. However, perhaps because of accidents of formation and

---

*   Recall that the existence of allelic variation is inevitable because of the Second Law of Thermodynamics (Chapter 2). Each time a gene is copied, there is a tiny but significant chance that an error will be made, creating a new allele. Thus, when we look at the genes in a population of animals, we will almost always find variation between the alleles that are present. Moreover, new alleles are continuously being created by these inevitable changes. These changes are sometimes referred to as *mutations* (Chapter 2).

local soils, one refugium has lots of fruit trees whereas the other has few, but is rich in nut-producing trees.

Alleles that affected the feeding behaviors and digestive systems of the animals in these two refugia would be subjected to different selective forces. Animals carrying alleles that improved recovery of nutrition from fruits would do well in one refugium but poorly in the other. As a result, these fruit-eating alleles would become ever more frequent in one of the refugia, but not the other. The opposite would happen in the nut-rich refugium.

In addition to selective events, there can also be accidental or *stochastic* events. Stochastic events occur in many different ways. For example, if individual animals from the species in question are physically large in size and the species' refugium is small, there may only be a few thousand individuals alive at any moment in time. Suppose a flash flood drowns a hundred of these individuals, each containing her/his idiosyncratic sample of the variation in this population. The alleles present in subsequent generations will be changed significantly by this accidental drowning.

Many such stochastic events occur over long time scales. Their aggregate effect on a population and its gene pool is always significant and sometimes dramatic. For example, the coat color of some local populations of birds and mammals is probably the result of such stochastic events in some cases, but may be the result of selective events in others.

Thousands of generations of selective and stochastic events operate on many alleles of many different genes. These processes result in animals in our fruit-rich refugium that look and behave differently than animals in our nut-rich refugium even though both our fruits eaters and nut eaters originally descended from the same ancestral population.

Now suppose that the climate grows wetter again, perhaps 100,000 years after initially drying out. The isolated patches of rainforest expand and ultimately fuse into a single large forest again. Now the various refugia are no longer separated. Animals are free to move back and forth between them. Formerly separate descendents of a common ancestral species meet again.

When this occurs either of two things will happen. First, the animals from different refugia can find themselves *inter-fertile*. They are able to mate and produce viable, fertile offspring. Under these conditions, they remain members of the same species, as we will define the term. The genetic differences between the two formerly separated populations will be joined in a single new composite gene pool. Selective

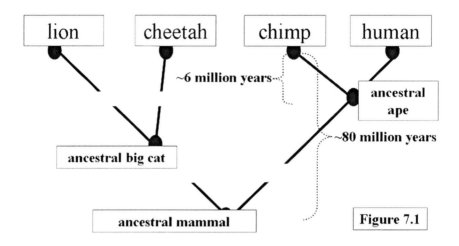

Figure 7.1

Figure 7.1: This is a phylogenetic tree showing the evolutionary relationships of four currently surviving animals along the top. (Trees can be as simple or as complex as we need them to be, showing many or a few species.) Each black circle represents a species, either living or extinct, at a particular moment in time. The lines connect species through their closest evolutionary relationships and the lengths of the lines can indicate the times separating the two species. The lines connect the descendent species above with their ancestors below. (The white interruptions in several lines show where the time scale has been compressed for convenience.) Thus, the branch points on the tree reflect speciation events. Moreover, the two species represented by the lines proceeding upward from each branch point will accumulate selective and stochastic changes (text) throughout the many generations represented by the lines, becoming more and more different over time.

For example, if you begin at the cheetah and follow the line down you come to a circle labeled *ancestral big cat*. If you begin at the lion and proceed down you come to this same ancestral big cat circle. Thus, the tree indicates that several million years in the past lions and cheetahs did not exist. Rather a single ancestral big cat species existed and this ancestral species underwent a speciation event. This event produced two new species whose living descendents include the lions and the cheetahs, respectively. Some of the traits the cheetahs and lions share (body plan, carnivore teeth, and so on) they inherited from this ancestor (their *last common ancestor*). In contrast, living chimps and humans did not arise from this ancestral big cat ancestor and do not share these "cat" traits.

Chimps and humans did not exist roughly six million years ago. Rather an ancestral ape existed which underwent speciation producing two species whose surviving descendents include chimps and us, respectively.

> Notice that around eighty million years ago, neither cats nor apes existed. Rather a single mammal ancestral to them all existed, possibly a small rodent-like animal. This animal underwent a speciation event producing two species. One went on to produce the ancestral big cat (and ultimately lions and cheetahs), while the other produced the ancestral ape (and ultimately chimps and humans). Thus, chimps, lions, humans, and cheetah share some properties inherited from this common ancestral mammal (hair, giving birth to live young, producing milk, and so on). Notice also that the ancestral mammal is the last common ancestor of the chimp and the lion, but *not* the last *common* ancestor of the cheetah and the lion. Moreover, the ancestral mammal *is* a common ancestor to the lion and the cheetah, for example, just not their *last* (or *most recent*) common ancestor, as we saw above.

and stochastic events will continue to act on this new, larger gene pool thereafter.

A good example of this first kind of outcome is the contemporary human population. Diverse, formerly isolated local populations (African, Asian, European, Native American, Polynesian, and so on) have accumulated differences in their pools of design information resulting in what we sometimes think of as *racial* traits. However, we are all inter-fertile, all members of the same species.

Second, the alternative outcome is that the descendents of a common ancestral species will no longer be able to interbreed successfully when rejoined as their refugia expand and fuse. For example, our fruit-feeders and nut-feeders may find that they can no longer mate to produce viable offspring. These two populations would then be two different species.

When two new species are produced from one ancestral species by speciation in this way, the new species are said to be *sibling species*. When two new sibling species meet one another again, they will usually be very similar to one another, like cheetahs and leopards, for example.[2]

In addition to speciation events, there are also *extinction* events. These occur when all the members of a species perish and the species ceases to exist. This can happen for many different reasons, including loss of habitat due to climate change or the presence of some lethal new predator or disease.

Speciation and extinction events apparently go on all the time. On a geological timescale, the Earth sometimes reaches a kind of equilibrium

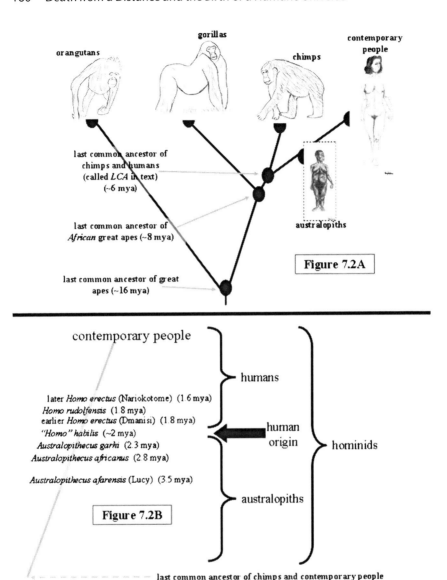

Figure 7.2A

Figure 7.2B

Figure 7.2: PANEL A: This panel depicts the evolutionary relationships between four surviving great ape species (including humans) along the top of the figure and an example of an extinct ancestral species (an australopith) lower down in the figure. The similarities between chimps and gorillas strongly suggest that the last common ancestors of African great apes and the last common ancestor of chimps and humans were more like chimps and gorillas than like humans. Further, chimps and gorillas like to walk on all fours, knuckle-walking on their hands. In contrast, humans do not walk this way and the fossil evidence indicates that walking up on two legs first evolved in our extinct australopith ancestors. [**mya = million years ago**. *Drawings of the person and the australopith are by Amy Radican, the others by P. Bingham.*]

PANEL B: This panel depicts a detailed view of the branch of the tree from Panel A leading directly to contemporary humans. Along this approximately six-million-year timeline are listed several key ancestors, now extinct (without the clutter of individual circles). In some cases, specific fossils or fossil sites are listed beside the species names (Lucy, Dmanisi, and Nariokotome; see text). Meanings of some useful generic terms (australopith, hominid, and human) are illustrated along the right edge of the panel. Older species (like the australopith, Lucy, for example) could be our direct ancestors or contemporary cousins of our direct ancestors. In either case, they are probably good exemplars of what our ancestors were like with respect to the properties that concern us here. For simplicity we will discuss them as direct ancestors in the text. Lastly, the quotes around *"Homo"* in *Homo habilis* reflect the fact that some fossils called by this name shared some properties with australopiths, including small brain size. Our theory probably accounts well for these "transitional" animals (text).

or steady state. New species are occasionally created by speciation; old ones can occasionally be destroyed by extinction.*

Given this historical sequence, an important question is how to organize the relationships between all the diverse species of organisms alive at any one moment in time. One useful way to visualize these relationships is to use a *phylogenetic tree*. Some properties of these trees are illustrated in Figure 7.1.

What is crucial to us at the moment is that currently surviving species are indicated as points along the horizon at the top of the tree (lions, cheetahs, chimps, and humans in Figure 7.1). Moreover, we are free to look at only the few species that happen to be of interest to us. Closely related species are each joined by a line projecting downward (back in time) to the most recent common ancestral species from which they originally descended. This ancestral species is thus represented as a point below the two contemporary species (Figure 7.1). The length of these lines is often chosen to be proportional to the time since the two contemporary species had this last common ancestor.

For example, chimpanzees are our closest *surviving* relatives. Around six million years ago, neither chimps nor humans existed. Rather,

---

* These processes are not always gradual, affecting a few species at a time. They can also be sudden and extensive, affecting many species at once. For example, uniquely human ecological dominance (and the resulting habitat destruction) has rapidly driven many thousands of species to extinction and continues to do so as we speak.

a single ape species would go on to be ancestral to both chimps and humans. This ape species underwent a speciation event producing two new species. The descendent species derived from each of these new sibling species underwent many subsequent adaptive and stochastic changes (and probably multiple speciation and extinction events) over the ensuing six million years. One of these two lines produced currently surviving chimps. The other six-million-year-old sibling species ultimately produced us, its sole *surviving* descendent.

We understand a very great deal, indeed, about *when and where* contemporary species, including us, came into existence. However, until now, we have had little understanding of *why and how* the uniquely human species came into existence. Our new theory ostensibly answers these how and why questions.

The following five subsections will introduce us to these answers in their most direct, fundamental form. The remainder of the chapter (Part III) will explore these answers in more detail for readers who wish deeper technical insight.

Cast of characters – our recent family tree

To understand our theory's answers to the fundamental questions of our origins approximately two million years ago, we need to know the players and put them in context. Figure 7.2 includes phylogenetic trees showing our relationship to our closest living relatives and some recent deceased (extinct) relatives.[3] (Notice that these are higher resolution, more detailed views of a part of the tree we saw in Figure 7.1.)

Sometime shortly after the speciation event, around six million years ago, that sundered the ancestors of chimps from the ancestors of humans, our remote human ancestors began walking bipedally, upright on two legs. All the descendents of these animals (including us) continued to walk this way. These bipedal animals are sometimes collectively called *hominids*.[*]

The six-million-year history of hominids can be broken into two very distinct phases that will be crucial to us. The first phase consists of an approximately four-million-year series of animals extending from about six million years ago to about two million years ago. These first-phase animals remained fundamentally ape-like, in spite of walking upright, and they retained ape-sized brains. As we saw in Chapter 6, this tells us that they probably possessed no new level of social cooperation beyond

---

[*]   Recently, some authors have switched to the term *hominins* in place of hominids. For continuity with the earlier literature, we will use the older term.

that seen in contemporary non-human apes. These first-phase animals are called *australopiths* for short, with a genus name *Australopithecus.** Several species of australopiths that are of particular interest to us are indicated in Figure 7.2B.[†]

Beginning the second phase of hominid evolution, the first humans arose. Their genus name is *Homo* (man) and we will usually refer to all members of our genus simply as *humans*. Humans clearly evolved from an australopith ancestor around two million years ago. Humans are easily distinguishable from their australopith ancestors in diverse ways (below). Most importantly for us, these earliest humans show significant brain expansion almost from the beginning. Their brain sizes indicate that expanded social cooperation is a fundamental feature of our genus *Homo* from its inception (Chapter 6).

Continents adrift and the Africa of our first human ancestors

The evolutionary origin of the first humans almost certainly played out in Africa. We are African apes who escaped our provincial heritage to colonize the wider world. Our theory makes very specific predictions about how these first African humans evolved. These predictions are quite unlike any earlier theory of human origins in ways that matter. It will serve us well to begin by placing this remarkable evolutionary story in context.

The world we know first-hand sits on top of the Earth's thin crust. Beneath this crust (beginning about forty kilometers or twenty-five miles down) is the mantle, extending thousands of miles toward the Earth's core. The mantle consists of rock at extremely high temperatures and under spectacular pressures.[‡] Most of the mantle is so hot it is molten (semi-liquid).

---

*　This slightly complex name is easy to remember. It translates "southern" (australo, as in Australia or southern land) "ape" (pith), referring to the discovery of the first fossils of these bipedal apes in southern Africa.

†　At least one species of non-human australopith managed to survive for almost one million years after the first humans arose. They were apparently able to avoid competing with early humans well enough to survive for a while. We will not be concerned with these animals here and they are not shown in Figure 7.2.

‡　Slow radioactive decay processes, over long periods of time, produce heat that accumulates to generate these tremendous temperatures.

Molten mantle is called magma. Magma does two things that matter to us here. First, it breaks through the crust at weak points to produce volcanoes, spewing magma as lava. Second, the surface of the Earth floats on this giant sea of molten magma. The magma sea has complex patterns of currents so that different parts of the crust are being pushed around, a little like big slabs of ice on a lake surface in the wind. Different large chunks or rafts of crust are forced in different directions, pushing against one another. For example, we are writing in North America. North America rides on one of these rafts of crust, South America rides on another and Africa on yet a third.*

Plates are pushed apart at seams in the Earth's crust where fresh magma comes to the surface. These seams are usually low points on the Earth's surface so that they are almost entirely under the sea, mostly in the deepest parts out in the middle of the oceans.

Where plates collide, one is pushed under the other and the top plate at this seam is pushed up producing high mountain ranges like the present day Himalayas, Rockies, and Andes, for example.

To get a feeling for the huge scale of these process consider the following. South America, Africa, North America, and Eurasia were joined together into a single large super-continent around two hundred million years ago. New seams in the Earth's crust opened and spread separating these four continents and, over time, producing the present day Atlantic Ocean.

More recently, South America and North America were separate until around three million years ago. Their plates were pushed together and ultimately the continents joined near the present day Isthmus of Panama.† Today a warm ocean current (the Gulf Stream) runs from the Caribbean up along eastern North America and across to northwestern Europe warming these areas as it goes. Without this current, northern Europe would be much colder. Scandinavia would probably be under the polar ice cap, for example. Three million years ago, the Caribbean did not exist and this Gulf current probably flowed from the Atlantic out into the Pacific through the space separating the North and South

---

*   The technical name for one of these rafts is a *tectonic plate*.

†   Indeed, at the time of joining, South America was populated by a collection of marsupial mammals, giant flightless predatory birds and other exotic animals. It had experienced several hundred million years of independent evolution producing a very different outcome. (This is quite analogous to Australia's separate evolution, producing the exotic animals like kangaroos, koalas, platypuses, and Tasmanian devils encountered by the first Europeans.) Most of the unusual South American creatures were driven to extinction by competition with the placental mammals we think of as conventional when these placentals surged south across the new land bridge at Panama.

American landmasses. Over the same period, when North and South America were joining, the Himalayas continued to be uplifted, altering global wind flow patterns.

Changes on this immense scale occur continually over geological time and climates of specific local places—like sub-Saharan Africa—are altered as a result. More specifically, the African climate has been drying for at least the last ten to twenty million years. Moreover, there is strong evidence for intense periods of drying around two to three million years ago.[4] Embedded within this overall drying has been a cyclical rhythm alternating periods of more extreme and less extreme African desiccation.[†]

This complex pattern of drying of the African climate has produced the Africa we know today. Ten to twenty million years ago, Africa was mostly rainforest. As Africa became dryer, there emerged substantial expanses of both desert, like the Sahara and the Kalahari, and grassland, like the enormous East African savannas. Moreover, the cyclicity of this process would have produced eras sometimes dryer and sometimes wetter than today, where pockets of savanna might have been separated by areas of rainforest or desert, for example.

This creation of massive and sometimes fragmented African savannas is probably important to us.[5] First, the early smaller scale expansion of grassland around five to seven million years ago may have had something to do (directly or indirectly) with the divergence of the hominid and chimp lineages (Figures 7.1 and 7.2). Second and more importantly, the subsequent dramatic expansion of the grasslands around two to three million years ago is very likely to have had something to do with the emergence of the first humans from our australopith ancestors. Remember that *Homo* first emerged around this time.

Why is grassland important here? Rainforest and grassland both have large masses of organisms (biomass). However, in a rainforest, most of this biomass is in plants, fundamentally because wood and tree leaves are relatively poor foods for animals. In contrast, grass is a relatively good food for the right kinds of plant-eating animals (herbivores). Thus, as quickly as grass grows it is eaten and turned into animal flesh.

The upshot of this is that there are more pounds of animal flesh per square kilometer on the savanna than in the rainforest. Apparently, in consequence of this, the savanna has supported a complex array of meat eating animals (carnivores) that prey on the herbivores.

[†]   These shorter-term rhythms apparently correspond to the cycles of polar ice sheet expansion (ice ages) and warmer interglacial periods. We are now living in the latest interglacial period. Judging from the record, our interglacial is likely to end soon and we will enter the next ice age.

Thus, the East African savannas have huge herds of gazelles, topis, wildebeests, and other herbivores preyed on by hunters and scavengers like lions, cheetahs, leopards, hyenas, wild dogs, and jackals, among others. It is of more than passing interest that a contemporary savanna primate, the baboon, still occasionally preys on these herbivores as we saw at the beginning of this chapter. It is this context in which the adaptive trick of being a human being evolved.

The groundwork for evolving humanness is laid

The expansion of the African savanna grasslands and their cyclical fragmentation suggests a number of obvious scenarios for the emergence of an elite-throwing, professional hominid hunter/scavenger near the beginning of the human story. We will have to await future evidence to see *exactly* how this happened in detail. However, it is illuminating to visualize specific scenarios like the following.[6]

A local australopith population around two and one half million years ago finds itself in an isolated savanna fragment (refugium). This fragment has a significant herbivore population, but, purely by accident, the large hunters and scavengers, like hyenas and lions, did not survive in this particular refugium. Only relatively lightweight, solitary hunters, like cheetahs or leopards, were present.

At first, the hominids in this refugium occasionally scavenged leftovers from cheetah kills and the incompletely eaten prey that leopards characteristically stash in trees.* Both the meat and the rich marrow in the long bones was valuable nutrition. Moreover, these hominids, like contemporary apes, were clever and observant. These australopiths also had the rudimentary ape ability to throw and club as contemporary chimps do. Though chimp (and presumably our early accidental ape) throwing is comically inept by our elite human standards, it is sufficiently proficient to be a little intimidating to animals who are close by or who might be surrounded by too many throwing chimps (or australopiths) to be able to keep track of all the incoming missiles.

These hominids soon learn that a cheetah who has fed to the point of significant satiation can be scared off the kill a little early if he is pelted with rocks. (Notice that such a satiated and solitary cheetah is also not much of a threat to the accidental apes.) Moreover, cheetah intimidation is more likely to be successful if multiple apes throw simul-

---

* These hominids might have been the immediate ancestors of our *accidental apes* from the beginning of the chapter.

taneously. When multiple accidental apes chase a cheetah from a kill, we call this *power scavenging*.

It will be important to think carefully about how throwing together (conjoint throwing) might most likely have evolved in these ancestral pre-humans. There are two likely factors.[7]

First would have been kin-selection for improved scavenging outcome. For example, early accidental ape troops might have been *matrilocal* like contemporary baboons (adult females stay in their birth troop and the males move on to other troops). If so, sets of adult females would have included close kin. Thus, sharing risks in cooperative hunting could have been directly self-interested, not presenting a serious conflict of interest (Chapter 3). Contemporary lion prides have this matrilocal structure and the closely related females sometimes hunt cooperatively, for example.

Second would have been *by-product mutualism*. One or two animals throwing at a cheetah might have had little effect. However, if three or four animals throw at once, the cheetah might be intimidated. Notice that there is no opportunity for one of the three or four accidental apes to free ride. If *any* individual does not throw in this small group, *no* one eats. The only available self-interested strategy is to throw together with others and scramble to get your share when the cheetah bolts.[*]

In this refugium, power scavenging proves to be particularly important. On the one hand, it is a useful way to supplement the diet. On the other, perhaps, this particular refugium was (by hypothesis) not particularly flush in some other rich dietary sources (fruits or nuts, for example).

Under this set of conditions, there would be intense selection for animals that can get more of their food by evolving to become more efficient and effective at this kind of power scavenging.

This local evolutionary process is expected to produce various effects. Of most concern to us, it would produce ever better aimed throwers, generation after generation culminating in hominids with the capacity to throw with elite skill—to "hit the strike zone with a fastball" in American baseball parlance.[†]

---

*   Under these conditions, unrelated australopiths can evolve conjoint throwing on a small scale, because there is no conflict of interest. However, notice that this strategy is very limited in its scope. As soon as there are more than enough individual throwers around to scare off the cheetah, there is an incentive to hang back (to free ride) and let others do the intimidation.

†   Though it need not concern us at the moment, these elite throwing animals might well have moved easily from power scavenging to professional hunting. For example, a

Power scavenging would also put a premium on the ability to process large amounts of meat, brains, and marrow efficiently. Traditional carnivores do this by evolving dental machinery with the appropriate properties.

This dental solution to the meat-eating problem would not have been immediately open to our early accidental apes. Their teeth were most emphatically not like carnivore teeth and wholesale redesign would have been required. However, a second strategy was open to them. They were the descendents of several million years of ancestors who used simple unmodified stone and wood tools to process food (as contemporary chimps still do).[8] They could use simple stone hammers to break open long bones and skulls to get at marrow and brains. Moreover, it is a small additional step to learn to smash stone to produce sharp shards (flakes) that will cut meat off bone. Indeed, contemporary chimps can master a related skill with a little encouragement.[9]

This scenario for the evolution of a scavenging/hunting australopith is expected to leave three traces in the record. First, because of the higher quality of a meat/marrow/brain-rich diet, the guts of these animals should grow smaller just as a meat-eating cheetah gut does, for example (Chapter 6). Thus, early hominid professional hunters/scavengers should undergo an anatomical transition from a pot-bellied ape to a svelte-waisted predator.[10] We should find evidence for this change in the shape of the waist through alterations in the rib cage and the pelvis in fossil remains.

Second, we should find simple stone flakes deliberately produced and used for butchering.

Third, we should find fossil prey bones with cut marks on them from stone flake de-fleshing and with percussion bulbs from breaking open long bones for marrow extraction.

In fact, we have extensive evidence for all three of these signs of early hominid scavenging, or hunting, in the fossil and archaeological records beginning around 2.3 to 2.5 million years ago and extending through the first clear humans around 1.8 million years ago (Part III of this chapter). Indeed, it has long been recognized that these clues probably reflect the importance of a scavenging/hunting adaptation to human origins. The problem has been to explain why hunting or scavenging in human ancestors produced a species that walks on the Moon, while hunting and scavenging in the big cats had no such outcome. Our new theory provides the first specific and satisfying answer to that question.

---

group of throwers could chase an herbivore into an ambush and other elite throwers could shatter a knee on its foreleg, making it easy to catch.

The vital point, again – elite throwing produces a fundamentally new kind of animal

Now that we have a clearer picture of how the immediate ancestors of the first human might have evolved, it is illuminating to return once again to the origins of these very first humans. Our accidental apes can now throw with elite skill allowing them to scavenge/hunt effectively. Let us be *absolutely clear* about why these animals are about to become unintentional evolutionary revolutionaries rather than just another member of the tiresomely large guild of African carnivores.

Consider what happens as these accidental apes engage in power scavenging. They will sometimes be in a situation where an opportunity presents itself for one of them to free ride, allowing his (or her) fellows to chase off the cheetah, for example.* Indeed, this would be the optimal self-interested choice, all other things being equal, and this is precisely the choice non-humans animals would be expected to make. However, for our accidental apes, this free-riding choice is preempted by self-interested others. Because they are remote killers, the surrounding others can cheaply keep this free rider from sharing in the trophy herbivore carcass and thereby keep a portion of the potential free rider's share for themselves (Chapter 5).

Individuals who exploit this option to ostracize will do better than those who tolerate free riding. They will have more food to eat in the short term. Moreover, in the longer term, by suppressing free riding, they "inadvertently" require everyone else to help with cheetah intimidation duties, thereby reducing their own risks and costs.

We expect these animals to evolve relatively quickly into something East African herbivores had never seen before—a scavenger who systematically traveled in packs including adults from multiple unrelated kinship groups. These packs (or something very much like them) were the first version of the uniquely human village, on our theory.

This unprecedented adaptive capability did not merely throw a chill into East Africa's herbivores. These accidental apes would inevitably have extended and expanded this new trick, discovering new ways to act cooperatively and making these new tactics stick as a consequence coercively expelling (or killing) individuals who attempted to free ride.

---

* This free riding would be most likely on the part of adults who were not close kin to most of the other adults around, of course. Thus, when this behavior is suppressed, non-kin cooperation is the consequence.

For example, merely chasing carnivores off a kill is just the beginning. Bringing food back to share with others not present (like pregnant females) is another cooperative activity that could have been coercively enforced.

Still another example is the sharing of information and insight. Individuals who consistently failed to lead the troop to some useful resource might ultimately have been shown the door—the original what-have-you-done-for-us-lately membership rule.

Most generally, these animals would have discovered—by random, Darwinian processes—a fundamentally new cooperative approach to each adaptive problem. Selection to exploit the new strategy of expanded cooperation to solve every adaptive problem would have been intense and the resulting evolutionary changes rapid.

For example, life history changes would have evolved quickly (Chapter 6). Likewise, the intellectual and emotional psychological devices to exploit the opportunities of kinship-independent social cooperation would also have evolved relatively rapidly as this socially imposed selection would have been intense (Chapter 10).

Stop and reflect on a crucial insight. The mundane processes of climate change apparently produced a seemingly prosaic change in lifestyle in a local australopith population. However, the quirky details of this new lifestyle had an unintended consequence. This evolutionary process created the first animal in the history of the planet with access not merely to a single new adaptive <u>tactic</u>, but rather to a **fundamentally new <u>strategic</u> approach to each and every adaptive problem, kinship-independent social cooperation**.

This animal would evolve to take advantage of this fundamentally new strategic approach to adaptation over and over. In so doing, these creatures would evolve into an unprecedented *kind* of animal. This animal's descendents (us) would come to "have dominion" over all the other animals and, indeed, over their planet.

With this insight in hand, we can ask how well (or poorly) the fossil record supports this picture.

Stones, bones, elite throwing and a crucial prediction

Our theory makes a crucial prediction about chronology. Elite throwing should have emerged first, **followed by** expanded kinship-independent social cooperation. Does the record of the emergence of early hominid scavenging/hunting support this chronology? We argue that it does.

Beginning around 2.3 to 2.5 million years ago, several very interesting things appear in the East African record.[11] We see stone flakes that are obviously produced deliberately. We can be very confident of this because of their precise shape and size. Moreover, these flakes do not merely occur, they become very numerous. Further, fossil prey bones appear with markings produced by cutting meat off them with stone flakes or smashing them open with tools for marrow extraction.

These remains are the evidence we expect for the evolution of australopith scavenging/hunting. These features of the record are relevant to our quest here *only* if the evolution of elite human throwing was associated with this early scavenging/hunting. There is considerable reason to believe it was.

First, the same archaeological deposits showing flakes and fossil hominid prey bones also show likely throwing stones. These stones (called *manuport*s in the technical literature) are often of the right size to be throwing stones. Moreover, they were clearly transported around the landscape by someone (hominids being the only likely candidate). Finally, though manuports are also the right size to be stone hammers (for breaking into marrow bones and so on), they sometimes occur in such large numbers that *ammunition* is a better interpretation for at least some of them rather than just crushing tools.

Second, the skeletal evidence from this transitional period suggests that elite throwing evolved during this interval and *preceded* overt expansion of kinship-independent social cooperation.*

For example, animals originally called *Homo habilis* (Figure 7.2; Part III) lived during this period, and appear to be contemporary with early simple flake tools in at least some places (including Olduvai Gorge in East Africa).† These animals have relatively small cranial volumes, suggesting limited expansion of kinship-independent social cooperation, but some details of their hands and feet suggest that they might have already begun to evolve elite human throwing.[12]

Further, a late australopith called *Australopithecus garhi*, living around 2.5 million years ago, is possibly informative (Figure 7.2; Part III). Fossils of

---

* This transitional period is from about 2.5 million years ago, when the early stone flakes and manuports first appear, until about 1.8 million ago, when fully human animals are clearly present.

† Olduvai Gorge is a large valley cut through ancient depositional layers (strata) by water flow (text). Is it one of most productive areas explored by the early pioneers investigating the fossil record of human origins, especially Lewis and Mary Leakey.

this animal are found in the same geologically deposited sedimentary layers (called strata) with flaked stone tools and fossil prey bones bearing the imprints of stone tool butchery. This suggests that *garhi* may have scavenged or hunted. *Australopithecus garhi* had a relatively small cranial volume, again, indicating no expanded social cooperation. However, a fragmentary (ostensibly) *garhi* skeleton shows arm-to-leg length proportions more similar to modern humans than the long-armed, short-legged physique of earlier australopiths (like the famous Lucy australopith fossil; Figure 7.2; Part III). This suggests *relatively* long legs and, possibly, that *garhi* showed early adaptations to human-like throwing.

Very recently, some exciting new evidence has emerged. The Dmanisi site in central Asia (Georgia) has yielded one of the richest collections of fossils of some of the very earliest members of our genus to leave Africa. This site is still being investigated, but it has already yielded vital insights. Most importantly, the morphology of multiple skulls indicates that these Dmanisi hominids are very early member of *Homo* (Figure 7.2). Yet, they show only modest, variable brain expansion. Moreover, several elements of the skeleton of these animals indicate that they had a human like, rather than australopith-like, body. As we will see in detail later (Part III), the human-like body has almost certainly been redesigned for elite throwing. Finally, these hominid fossils are found with fossil prey species (indicating hunting/scavenging) as well as a very large number of manuports (further suggesting elite throwing). This extensive, diverse set of remains at Dmanisi is remarkably similar to our theory's predictions for the signs our earliest human ancestors are expected to leave behind.

There is also an excellent African candidate for early stages clearly recognizable as the uniquely human life history redesign as evidenced by brain expansion. This candidate is *Homo rudolfensis.*[*]

The type skull for *Homo rudolfensis* is the famous KNM-ER 1470 skull (1470, for short). This 1470 has a cranial capacity of around 750 cubic centimeters (cc) and a second likely *Homo rudolfensis* skull has a capacity a little over 800cc.[13] This is larger than expected for an australopith, but not as large as later members of *Homo* (Figure 7.4 in Part III). Though the issue remains in some doubt, it is certainly conceivable that *Homo rudolfensis* brain volume reflects an early case of altered, human-like life history requiring some elementary form of the human village and producing an enlarged brain.

---

* This animal is sometimes renamed *Australopithecus rudolfensis* and this naming controversy highlights why it is a candidate transitional animal.

This becomes particularly interesting because two femurs were found nearby the 1470 skull and these could reflect the *Homo rudolfensis* skeleton.[14] These femurs are large and robust. This suggests that the body of *Homo rudolfensis* had already been redesigned to throw with something approaching elite human skill (Part III). Again, elite human throwing is already present at the earliest signs of brain expansion, as our theory requires.

Finally, we come to the earliest, highly complete skeleton of an unambiguous member of our genus, the famous Nariokotome fossil (Figures 7.2 and 7.3).[15] This remarkable fossil shows clear brain expansion (around 900cc; Part III). Moreover, we have a nearly complete skeleton of this young man and it is quite clear that his body is very human (and NOT australopith-like) including the thin waist of a carnivore. It also includes the extensive redesign of diverse skeletal features, apparently permitting elite human throwing (Part III).

Though we would like to have more fossils (and will watch new digs with great interest), the available evidence is already strong. We have reasonably high confidence that elite throwing evolved first in the hominid lineage, followed rapidly by expansion of kinship-independent social cooperation as indicated by brain expansion. These are precisely the predictions of our theory.

We knew going into this chapter that the only animal species in the history of the Earth to show extensive, kinship-independent social cooperation (us) was *also* apparently the only *contemporary* animal to be able to kill adult conspecifics from a distance. We now see that these two capabilities apparently arose over a very short period in our evolutionary history and in precisely the predicted sequence. These observations represent extremely strong empirical support for our theory.

For readers who wish to understand the quality of the fossil evidence in more detail, Part III is dedicated to an exploration of important elements of this large body of information. We will find there that the feet, hands, shoulders, and pelvis (hips) of the very first members of *Homo* are already radically redesigned to support elite throwing. This redesign apparently briefly precedes the uniquely human, cooperation-dependent life history redesign.

Readers who are comfortable with this overview of the evidence may proceed directly to Chapter 8. We will begin there to explore how our humanness and our history emerge from the simple adaptive strategy of sustaining kinship-independent social cooperation through self-interested projection of coercive threat. This exploration will reveal us to ourselves in ways that were never possible before.

Part III: The details of the human fossil record[*]

Fossils and volcanoes – how we can "see" our origins two million years ago?

The synopsis of our knowledge about human origins in Part II depended on enormous amounts of information. How can we possibly know these things? How can we look back at the brief time interval around two million years ago when the first humans evolved from ape-like australopith ancestors? Further, how can we know what the East African savanna looked like so long before any of us was there to see it? Brilliant people over the last several centuries have discovered just how to get the insights we need. This section describes the fruits of their combined, uniquely human genius.

Much of our bodies consist of soft tissues, like our hearts, livers, brains, and muscles. We do not collapse into an immobile mush because of the internal hard tissue of our skeletons. Moreover, the fact that our muscles attach to our skeletons and move us around when they contract, means that the precise shape of the bones (and the joints between bones) can tell us a good deal about just exactly how any particular animal is designed to move. The skeleton of a cheetah and a pig are very different, reflecting the different movement patterns for which they have been shaped by natural selection.[†]

---

[*]   For readers who wish to explore the extensive evidence more completely, selected references from the professional scientific literature are in the numbered online endnotes. As well, Agur, et al. (1999) is an outstanding medical anatomy textbook for help in getting a clearer picture of our bones and muscles. Aiello and Dean (1990) and Klein (1999) are solid technical introductions to the human fossil evidence. Vrba, et al. (1995) discusses the evidence for African climate change during the evolution of humans. References to important work on human throwing published in the scientific literature by Mary Marzke and her colleagues and by Holly Dunsworth, John Challis, and Alan Walker who can also be found in the online endnotes.

[†]   We tend not to be conscious of how much our movements are constrained by how our skeleton is built. We pay more attention to how we CAN move than to how we CAN-NOT. A simple way to get a better feeling for our lack of perspective about our movements is as follows. Hold out your hand, palm up. Place a baseball in your palm and grip it. Now take the ball out with your other hand, invert your hand (palm down), replace the ball (now on the back of your hand), and try to grip the ball backwards. Of course, you cannot. The main reason you cannot is that the bones and joints of your hand are not designed to permit this backward grip. What matters to us here is that examining the skeleton of a human hand (including a fossil skeleton) will tell us a great deal about the

When an animal dies in the wild, the soft tissues are usually quickly lost to scavengers and microbial decay. However, its bones decay much more slowly. Indeed, under the right conditions, bones do not decay at all. Instead, they can be slowly converted into rock-like fossils that can survive for hundreds of millions of years.[16] Our two-million-year-old ancestors occasionally left their fossilized skeletons for us to find.

Simply having fossil skeletons of animals that no longer exist can be helpful. But these fossils are much more useful if we know *when* the animal lived. For example, an ape-like animal that lived two million years ago might conceivably be a recent ancestor of ours, but one that lived twenty million years ago could only be a more distant relative.

More generally, if we have lots of animal fossils that are well dated and cover many millions of years, we may be able to infer the evolutionary processes producing these animals. We do have just exactly this information for many ape-like animals, including humans and our immediate ancestors.*

One very powerful way of obtaining ages for fossils is to take advantage of radioactive isotopes in the soil or rock containing the fossil, called the fossil's *matrix*. These isotopes allow the matrix to be dated and the fossil it contains will generally be the same age. There are different ways to do this kind of radiological dating. Each is useful on different kinds of samples or on samples of different ages. The following scenario illustrates one specific example of how we come to be able to do this.

Picture a large lake. Over many thousands of years, sediment brought into the lake by its feeder streams is laid down along the bottom of the lake. Dead animal bodies likewise wash into the lake and fall to the bottom. Sediment accumulates on top of the bones. The stream water also brings in dissolved minerals. These minerals slowly replace some

---

grips that the human hand will (and will not) do. Moreover, this is true for all the different moving parts of our skeletons.

*   Evolutionary relationships were originally deduced *solely* on the basis of this kind of dated morphological (anatomical) information. We now have extensive confirmation of these relationships from independent lines of evidence. For example, we can compare the DNA sequences encoding the genetic design information in surviving descendants of extinct animals (their descendants' genomes). When two living species have a very recent common ancestor there has been relatively little time for DNA sequence change (Chapter 2) and the sequences of their genomes will differ relatively little. When animals have not shared a common ancestor for a very long time, their genomes will have very different sequences. Intermediate levels of evolutionary relationship will produce intermediate levels of DNA sequence similarity. Based on these and other lines of evidence, we now have very high confidence in many of the important details of our evolutionary history.

of the different minerals making up the original bone, turning the bone into stone, a fossil. This replacement is atom-by-atom so that the fossil bone is often an exquisitely precise replica of the original bone.

New sediment is deposited, new animal bones arrive, then more sediment, then more bones—slowly, over thousands and thousands of years. This produces a layer-cake of sediment, with older sediments at the bottom and progressively newer ones toward the top.*

Now picture a volcano, perhaps fifty miles from the lake. It erupts violently once every ten thousand years or so. Each time it does, it blows volcanic dust high into the atmosphere. The heat of the eruption drives off all the volatile elements in the dust, including the elemental gas argon. A layer of this dust falls across the landscape and onto the surface of the lake after each eruption and settles into the sediment. Sediments that built up between eruptions can be thought of as the layers of a cake; the thin layers of volcanic dust can be thought of as "icing" periodically inserted between layers creating a large, many-layered cake.

Each layer of volcanic dust contains various radioactive elements. These radioactive atoms undergo decay events at very predictable rates. Some of these decays replenish specific volatile elements (including argon) that were lost in the eruption.[17] Millions of years later we can measure how much volatile argon has accumulated since the volcanic dust was originally laid down.† This will tell us how old this particular thin layer of icing is. From the dating of multiple different icing layers (called *volcanic tuffs*), we can reconstruct the dates of all the parts of the cake.

Usually we do not dig to the bottom of actual lakes to retrieve these cakes of layered strata. We do not have to. Over time, lakes sometimes dry out. A lake bed might even be pushed up by the folding of the Earth's crust as another effect of volcanic (and tectonic) activity (Part II). This elevated dried out layer cake is then slowly cut through with channels carved by new streams. If we walk along the stream channel, we can look into the layers of the cake (the strata).

A fossil hunter will walk along such channels looking for signs of fossils eroding out of the layers that correspond to the time period in which she (or he) is interested. If she finds something of interest, she then digs into the cake at that point, looking for the rest of the skeleton

---

*   Each layer in this "cake" is referred to as a *stratum*—plural strata. If we carefully dig up these layers millions of years later, our studies are called *stratigraphy*.

†   Actually, we do a normalized measurement for maximal accuracy. We also measure the amount of the radioactive precursor of the argon (potassium). The ratio of radiogenic argon to the parental potassium gives us the age we seek.

and other surrounding materials that might be helpful. Such useful items might include stone tools or the remains of other plants and animals that lived at the same time, carrying information about what the local environment was like.

We might imagine that our pushed up lake bottom layer cake with volcanic dust icing layers requires a complicated, unlikely series of events. In fact, such layer cakes are made fairly frequently in many parts of the world. For example, the Glen Iris Gorge in Fletchworth Park in New York is a dramatic example of one of these events. A high-resolution image of this particular layer cake can be found in online Endnote 18 to this chapter. The Grand Canyon in the western US is another particularly large, spectacular example and there are many others around the world.*

Fortunately for us, areas in East Africa that were home to the important events in human evolution have been volcanically active throughout the period we care about. These regions have produced many such datable strata sets. We have fossils of some of our ancestors in very nice layer cakes providing us with a good deal of information about when and where our immediate ancestors first evolved and lived.

That is the good news. There is also bad news.

First, fossils form relatively rarely. Most bones are destroyed by scavengers or fall into the wrong kinds of locations so that they are chemically dissolved or trampled to dust before they can fossilize.

Second, most fossils that do form come to be buried too deeply for us to find. For obvious reasons, we cannot systematically dig up the entire surface of the Earth. We need to find some indication of where we should dig (like bones eroding out of a stream cut) if our work is to be cost-effective. At best, we will have in hand only a relatively small number of fossils.

Finally, there is one last problem. We will be concerned with human ancestors and their relationships to their tools and to the animals they may have hunted or scavenged. We can tentatively identify the associated artifacts or prey animals because they or their fossils show up in the same layers nearby ancestral human fossils. This proximity can give us useful information, but with a caveat. We cannot be sure that the fossil human ancestor actually made the tools or hunted the fossil animal discovered in the same layers or strata.

Each of these problems will limit us somewhat. However, by virtue of the hard life's work of many gifted, dedicated people, we have a significant number of different fossil finds of human ancestors. These finds

---

*   A Google image search of "Grand Canyon" is illuminating.

give us enough contextualized fossils that we can be fairly confident of the interpretations of the record that will be most important to us here.

Investigators who specialize in this kind of analysis of the fossil record are referred to as *paleontologists*. Those who specialize specifically in the study of fossil human ancestors are referred to as *paleoanthropologists*. As we said previously, paleoanthropologists use the generic term *hominids* for bipedal apes. We now have the context and tools we need to explore the rich details of our origins as humans around two million years ago.

Before and after – an introduction to the stories of Lucy and the Nariokotome boy*

A good way to begin the detailed exploration of our origins is with a useful pair-wise comparison. We can contrast the first relatively complete skeleton of one of the earliest African humans, living around 1.7 million years ago, with the skeleton of a pre-human australopith who lived around 3.5 million years ago. These fossil skeletons (and others we will discuss later) were all discovered in stratigraphic contexts allowing them to be dated. The comparison of these two skeletons will allow us to ask the first crucial questions required by our theory. Could the earliest clear humans (as assessed by brain expansion) already throw with elite human skill? Equally importantly, was this elite skill unique to humans—that is, were our pre-human australopithecine ancestors, living long before the first humans, *unable* to throw with elite human skill? If the answer to either of these question is no, our theory is incorrect.

The early complete *human* skeleton we will use was found by our Stony Brook colleague Richard Leakey and his collaborators Allan Walker, Kimoya Kimeu, and their associates while digging along Lake Turkana in East Africa. It is often referred to as *Nariokotome boy*, after the region where this young male skeleton was found (see Figures 7.2 and 7.3).

The first australopith we will focus on is the famous Lucy fossil (Figures 7.2 and 7.3) discovered by Donald Johanson and his colleagues.[19]

---

* See Leakey and Lewin (1992), Walker and Shipman (1996), and Johanson and Edgar (1996) for discussions of these fossils and their history.

Some important differences between Lucy and the Nariokotome boy are illustrated in Figures 7.3. We will refer to details of this figure throughout our discussion. For now, recall what we have said about diet and anatomy. Lucy and all the australopiths were pot-bellied as contemporary chimps are, indicating a large digestive tract. This, in turn, reflects a diet containing substantial amounts of high-volume, low-nutritional-value foods (leaves, for example). Thus, Lucy apparently did not eat a diet that was dramatically different than an individually-foraging contemporary chimp.

In striking contrast, Nariokotome and all members of early *Homo* show the sleek, narrow waist characteristic of healthy humans. This narrow waist reflects a smaller digestive tract of a carnivore, of course. The larger body size of Nariokotome than Lucy also suggests this richer diet.[20]

Further, notice in Figure 7.4 that Lucy (like later australopiths) has an ape-sized brain.[21] In contrast, Nariokotome already has a brain dramatically larger than any ape or australopith. As we saw in Chapter 6, human brain enlargement reflects life history redesign. The brain enlargement in Nariokotome almost certainly represents the initial stages of this profound adaptive change. This life history redesign apparently requires the social support of the initial form of the human village (Chapter 6). The evolution of the human village, in turn, requires access to inexpensive coercion (Chapter 5). Moreover, remote killing through elite human throwing is apparently the only way our ancestors could have first gained access to inexpensive coercion.

Thus, we now have in hand what we need to use Lucy and the Nariokotome boy to test a very strong prediction of our theory. Nariokotome should be able to throw with elite human skill and Lucy should not.

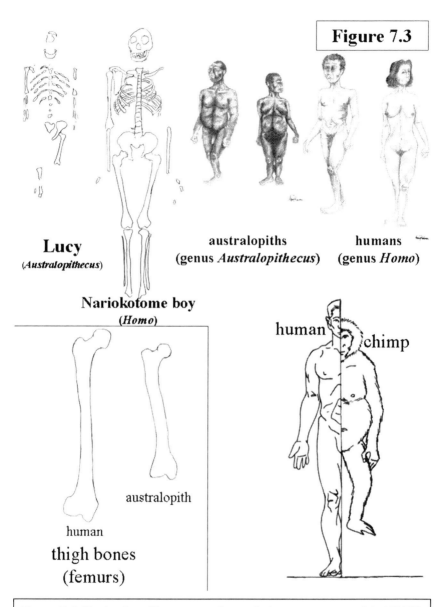

Lucy
(Australopithecus)

australopiths
(genus *Australopithecus*)

humans
(genus *Homo*)

Nariokotome boy
(Homo)

human   chimp

australopith

human

thigh bones
(femurs)

Figure 7.3

Figure 7.3: The bodies of humans and our relatives are compared. At LOWER RIGHT, humans and our closest living relatives, chimps, are compared. Notice the smaller body, shorter legs, and pot-bellied body shape of the chimp. These chimp traits are all shared with australopiths and, presumably, with the common ancestor of humans, chimps, and australopiths (see phylogenetic tree of these relationships in Figure 7.2). At UPPER RIGHT is an artist's conception of the smaller body, shorter legs, and pot-bellied body shape of australopiths in comparison to modern humans. Notice also the smaller brain size (see Figure 7.4 below for brain size information).

At UPPER LEFT are the fossil skeletons of australopith Lucy and the early human Nariokotome boy (text). Lucy lived around 3.5 million years ago and Nariokotome boy around 1.6 to 1.7 million years ago. At LOWER LEFT is a comparison of the human and australopith thigh bone (femur). The larger, heavier human femur is indicative of our larger body size and longer legs (text). *Drawings of Lucy, Nariokotome and femurs by P. Bingham based on several sources including images in Johansson and Edgar (1996) All other drawings by Amy Radican.*

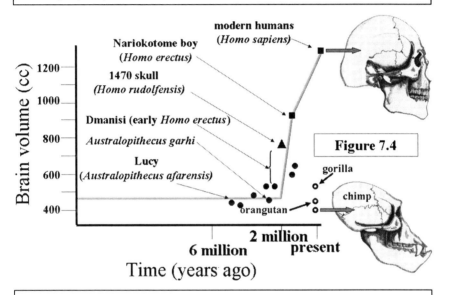

Figure 7.4: The grey line shows changes in brain size (cranial volume in cubic centimeters) in various ape and hominid species over the last approximately sixteen million years. The line is horizontal from about sixteen million to about two million years ago, indicating several things. First, in view of the brain sizes in surviving non-human apes (see chimp, gorilla and orangutan brain sizes at lower right), it is likely that ape brain size underwent little or no significant change during this period. Second, fossil australopiths (genus *Australopithecus*), potentially ancestral to humans (Figure 7.2), had brain sizes in this small ape range.

In contrast to the australopiths and living apes, humans (genus *Homo*) show dramatic brain enlargement. Notice several things. First, this enlargement apparently occurs rapidly, over a period of one million years or less. Moreover, the earliest signs of enlargement emerge relatively abruptly, over a period perhaps as short as one hundred thousand years or less. Thus, human brain expansion was an explosive event on an evolutionary time scale. Second, the explosive growth of human brain size apparently began around two million years ago. This brain expansion apparently dates the origin of the dramatic, uniquely human alternations in life history—the birth of the human village (text and Chapter 6). [Note that the volume units, cc's, are approximately equal to the weight units, gms, used in some other figures.]

Controlled violence – the elite human throwing motion

Consider what is involved in the elite human throwing motion. Paul grew up in rural North America playing baseball for all the years of his late childhood and adolescence. Thus, he can throw with skills approximating (at least crudely) those of an ancestral human who threw "for a living" (Figure 7.5).

If he were to stand fifteen meters (about twelve yards) away and throw baseball-sized stones at you with maximal violence you would be immediately disabled. In all likelihood, you would ultimately die from the consequences of the first few throws—compression fractures of the skull, a shattered sternum, or something similar—in the absence of modern medical care. This lethal capacity is the reason professional American baseball players wear batting helmets when facing a major league pitcher from about twenty meters distance and are still occasionally injured seriously.

To achieve the spectacular momentum of a well-thrown stone (or baseball), we use our entire bodies—literally, from toe to head—in an extremely rapid, highly coordinated and violent motion. Moreover, to achieve the phenomenal accuracy and control that make these high-momentum projectiles so effective, we execute this violent motion sequence under exquisite control.

To get a more complete understanding of these properties, you might want to put your own body through the throwing motion while paying careful attention. You will find your body doing the following things (Figure 7.5). (We describe the right-handed sequence. The left-handed sequence is the mirror image.)

You will slightly lift your left foot while rotating your shoulders into throwing position (right shoulder back). You then drive violently forward off the inside edge of your planted back (right) foot. Your left foot lands about a meter in front of your back foot. You have already started to accelerate your throwing hand forward.

On the platform of your planted feet you begin an extremely rapid sequence of torquing (violent rotational) movements. Your ankles and knees move to allow violent rotation of your hips and upper body.

This sequence is initiated when your hips torque *relative to your planted feet*. Immediately after the beginning of this hip rotation, your chest and shoulders (upper torso) begins a comparably violent rotation *relative to your hips.* This sequence adds acceleration onto acceleration (shoulders onto hips) producing extremely rapid further acceleration of your throwing hand forward.

This building of speed on speed from the bottom up continues. The arms are cocked in the throwing motion and are carried along for a few instants by the pelvis and shoulder rotations. After the shoulder rotation begins, the muscles of the arms contract to reinforce the shoulder rotation and to accelerate the right forearm and throwing hand directly—more acceleration built on acceleration. Near the end of this process, the muscles of the hand and fingers carry out one last acceleration before releasing the stone (or baseball).

During the second half of the throwing motion, a wave of counter-torquing (deceleration) follows the earlier wave of acceleration up the body in the same sequence. As a result, the rapidly moving arm and hand whip against the decelerated body during the release of the projectile. The body finishes the throwing motion under control and is able to reload and throw again almost immediately.

The net effect of all this acceleration on acceleration on acceleration is the release of a relatively massive projectile from the hand at speeds in excess of 140 kilometers per hour (90 mph) with sufficient momentum to fracture a human skull catastrophically. Notice the blurring of the hands in Figure 7.5. This reflects the extremely high speeds that hands achieve as a result of this orchestrated whole-body sequence.

When a human throws this way it is analogous to a cheetah running or a dolphin swimming. We are executing an elite behavior. It is a behavior we are "born" to do. Just as the cheetah's and dolphin's bodies have been redesigned for their elite specializations, so has ours.

That we are, in fact, designed to throw with elite skill is not in doubt. Rather, our challenge is to recognize which features of our skeletons most fully reflect this specialized redesign. This question will continue to benefit from more expert attention in the future. However, we will argue here that we can already recognize some features that are very likely to be a part of our redesign imparting throwing virtuosity. We can also recognize these skeletal adaptations in the fossil skeletons of our ancestors.

In the remaining sections of this chapter, we will review some of the most telling examples of skeletal redesign that are likely to have evolved to support elite throwing.

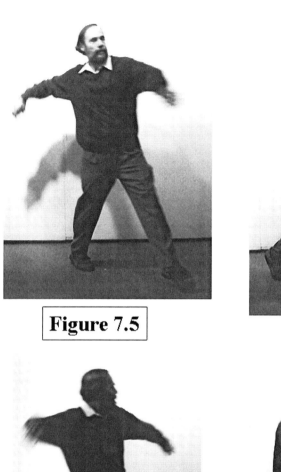

Figure 7.5

Figure 7.5: Paul shows the human throwing motion from two different views (text). Notice the planting of the back foot on the side (LOWER LEFT FRAME) and the subsequent rotation of the back foot (other frames) as the front foot is driven forward and plants. From the platform of the planted feet,

the legs, pelvis, and upper body torque (rotate violently) in rapid succession adding acceleration onto acceleration. Finally the shoulder, arm, and hand muscles complete the process releasing the projectile. This sequence of movements produces extreme whip-like final movement of the hand (note blurring of the hands in the images due to high speed of movement) resulting in velocities in excess of 90 mph for projectiles weighing several hundred grams. This projectile speed and mass corresponds to amounts of momentum sufficient to fracture human skulls, rib cage, or long bones— sufficient to kill adult humans (text)

Bodies and brains, apes and humans

It has long been recognized that the first humans are called human because of a complex suite of distinctive anatomical features they share. Sudden expansion of brain size is one of these (Figure 7.4). We will argue that many or all of the *remaining* diagnostic features of *Homo* are redesigns for elite throwing.

These new human features include their large, sleek bodies.[22] The sleekness of human bodies is a clue not only about the human diet (above), it also is suggestive about throwing. Visualize again the human throwing motion. The violent rotation of the torso relative to the hips would be difficult or impossible for a pot-bellied australopith (as it is not possible for contemporary pot-bellied chimps; Figure 7.3).

Nariokotome boy has the sleek human body, as we said. Lucy and the other australopiths do not (Figure 7.4).

This is part of a more general first impression. When we look at the skeleton of Nariokotome boy, it looks very human, very much like ours. Lucy's skeleton does not look nearly as much like ours. This is what we expect if Nariokotome was redesigned to throw with the same elite skill as contemporary humans, but Lucy and her kind were not.

We can examine this first impression in much more detail as follows.

Aircraft and skeletons – engineering by design

An aircraft designer supposedly once famously remarked that an aircraft represents "a series of compromises flying in close formation." He was referring to the fact that different parts of the aircraft can have disjoint properties for different reasons. For example, one set of design changes might increase the carrying capacity of the plane while reducing its aerodynamic efficiency. Another set might increase the fuel load the plane can carry while decreasing the amount of freight it can transport. Still another might increase the maneuverability of the aircraft while reducing its range, and so on and so on.

The crucial point here is that every design improvement degrades the aircraft from some other point of view. A well-designed ground support fighter would generally be a lousy strategic bomber and vice versa.

Precisely the same logic of trade-offs also applies to the design engineering achieved by natural selection. There is no free lunch. Every new capability an animal acquires will come at the cost of loss or degradation of some other capability. Moreover, animals need many different capabilities to negotiate their lives. Thus, almost every part of the animal must be a multipurpose trade-off.

Think for a moment about your hands, for example. You use them for many crucial adaptive activities such as eating, working, mating, childcare, and many others. The device designed to fulfill all these diverse functions can do many things well; but always at a cost. For example, the delicacy of our fingers permits enormous dexterity, but leaves our hands very vulnerable to damage when we are forced to use them as shock weapons.

These features of evolutionary design are crucial to us now. As we look at skeletal structures through the rest of the chapter, we will be asking *not* whether a particular bone looks as if it is designed to throw. *Rather*, we will ask whether the trade-offs in design of the bone have been shifted toward elite throwing, perhaps at the cost of some other function.

For example, we will explore the changes in the hands and feet of the first humans. We will argue that they look like what we expect of redesign to throw. However, there is reason to believe that australopith hands and feet made them elite tree climbers—perhaps to escape big cats among other things.[23] The changes in human hands and feet that make us elite throwers make us less competent as tree climbers. This is a trade-off that must have made sense for natural selection to permit the changes in early humans.

Axis of power – throwing and the human pelvis[24]

Mary Marzke and her colleagues have made crucial contributions to the problem of throwing and clubbing in the fossil record.[25] We will make extensive use of this work in the broader context of hominid evolutionary anatomy.[26]

Several details of the redesign of the human pelvis relative to the australopith pelvis give us a strong test of our theory. As we said, we do not expect any skeletal structure, including the human pelvis, to be redesigned for the exclusive purpose of elite throwing. Rather, we expect the pelvis to be redesigned in a way that allows all its diverse functions, but also that it is "tuned" to allow it to produce the elite human throwing motion while also doing all the other things it must do. Equivalently, when we find a muscle that appears to be redesigned to permit elite throwing, we anticipate that it will also have other functions.*

This multifunctionality of muscles can be a problem. We might worry whether we have identified the right function of a particular redesign. However, if we see redesign of an entire suite of different structures in just the way that is required to permit the elite human throwing motion and if these changes all occur for the first time together *as a package*, we can be fairly certain that we have identified the evolutionary origin of elite human throwing.

We will argue that the diverse large muscles attaching to the pelvis comprise just such an elite throwing package. Moreover, the sections that follow will argue that other parts of the body are also part of this throwing package.

The elite human throwing motion requires the violent rotation of the most massive parts of the trunk or body axis to impart very high velocity to a thrown projectile as we saw above. This violent rotation is produced by the controlled contraction of a series of very large muscles, including the ones shown in Figure 7.6.[27] These muscles are attached to the hip bone or pelvis as shown in Figures 7.6 and 7.8.

It is useful to begin with one of these muscles, the *gluteus maximus*. This is actually a bilaterally symmetric pair of muscles, one on each side of your body. This muscle makes your rear end, the cheeks of your butt in popular slang. Your gluts (for short) are the largest single muscles in your body. Moreover, their massive size is unique to humans among the living apes. (To get a feeling for this, contrast your own human rear end to that of our closest living relative, the chimp, in Figure 7.7.)

So, very large gluts are unique to the sole *surviving* hominid.[28] Are enlarged gluts unique to humans among *all* the hominids?

Consider where the gluteus maximus is attached on the human skeleton and, therefore, how its contraction will move the skeleton. Panel A of Figure 7.8 shows the differences between an australopith pelvis and two human pelves—a modern one and the one belonging

---

* For example, see subsection **Born to throw (and to run?)** below.

to the ancient human Nariokotome boy.[*,29] For the moment, focus on the modern human and australopith pelves.

You are looking at these pelves from the top (a "liver's eye view"). The pelves are aligned at the spinal column (marked by an "x" in panel A of Figure 7.8). Notice that the top margin (the *crest*) of the large flat curved blade-like portions of the pelvis (the iliac blades) projects laterally away from the spine in an arc. This arc extends a little further back from the spine in the human pelvis than in the australopith pelvis.

This has the effect of improving the angle for any muscle attached to the backmost part of this crest and to a structure nearer the body axis allowing the muscle to exert more torquing force.[†] This backward-extending part of the crest is where the gluteus maximus muscle attaches to the pelvis (panel B, Figure 7.8). Thus, we have the largest muscle in the human body (the gluts) given extra pulling power by having its pelvic attachment moved backward in humans compared to australopiths.[30] What is this new power good for, exactly?

Return to Figure 7.6. The gluts attach to the underside of the backmost part of the iliac crest. They then wrap outward and laterally *around* the hip and connect to a large ligament (the iliotibial tract) which connects just below the knee. As a result, the gluts torque the pelvis relative to the knee (and the planted feet) during the throwing motion.[31] Actually, contraction of the glut on one side accelerates the pelvis at the beginning of the torquing motion, and contraction of the glut on the opposite side decelerates the pelvis later in the throwing motion.

It is striking that these enormous muscles are actually relatively poorly designed to help strongly with any other major motion (such as swinging the leg from front to back during walking or running). The gluts appear to be designed as we would expect if throwing were a major function among the diverse set of compromises they were designed to fulfill. Thus, much of the power of this new human muscle is readily interpreted as *power for throwing*.

---

*   The Nariokotome pelvis has been distorted a little during its approximately 1.8 million years in the ground.

†   More formally, we can say that this repositioning of the glut attachment site improves the equivalent of the *moment arm* for rotational or torsional movement of the pelvis with respect to the planted feet and lower legs.

Return again to 7.6 and notice the smaller muscle also attached to the Y-shaped iliotibial tract ligament—the tensor fasciae lata. This muscle torques the pelvis in the opposite direction from the glut on the same side. We might expect these bilaterally paired muscles also to be involved in throwing, analogously to the paired gluts. If so, and if throwing were unique to humans, we might expect the pelvic attachment of the tensor fasciae lata to reflect this. Apparently, it does.

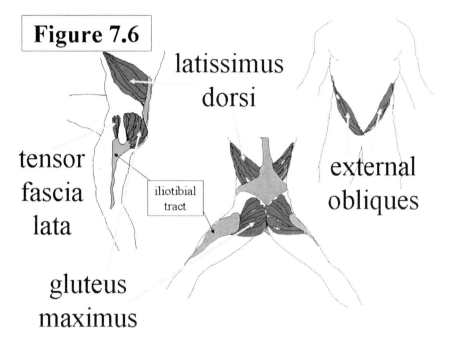

Figure 7.6

latissimus dorsi

tensor fascia lata

iliotibial tract

external obliques

gluteus maximus

Figure 7.6: Shown are four of the major muscles that torque (violently rotate) segments of the torso (text; Figure 7.5). The gluteus maximus and tensor fascia lata muscles attach to the pelvis (Figure 7.8) and run around the hip and downward (through the iliotibial tract ligament) to attach below the knee. These two paired muscles contract on one side and then the other to accelerate (and then decelerate) the hips relative to the planted feet during the throwing motion. The human enlargement of these muscles is illustrated in Figure 7.7.

The latissimus dorsi muscles attach to the back of the pelvis (through a ligament) and run around the back to attach to the base of the arms. The abdominal oblique muscles attach to the rib cage and run down and across the stomach to attach to the pelvis at the sides and front (see Figure 7.8). These two muscles apparently contract on one side and then the other to accelerate (and then decelerate) the shoulders and arms relative to the pelvis during the throwing motion.

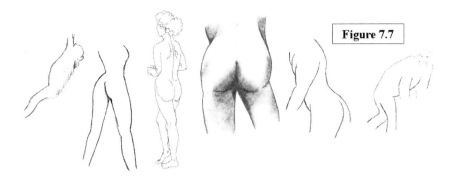

Figure 7.7: Contrast the large, conspicuous rear ends of humans (central four images) with the 'under-developed' chimp bottoms (leftmost and rightmost images). This reflects the dramatic difference in size of the ape and human gluteus maximus muscles (Figure 7.6). These large 'gluts' are unique to humans and apparently represent one of the many examples of how our body has been redesigned to permit human elite aimed throwing. *Central pair of drawings by Amy Radican, others by P Bingham.*

Notice that the crest of the iliac blade of the pelvis continues its long arc away from the spine in both australopiths and humans (panel A, Figure 7.8). However, in humans, this arc extends much further forward. Each tensor fasciae lata attaches (underneath) to the front of this long curved crest (panel B, Figure 7.8). This repositioned attachment gives the muscle a better angle for torquing the pelvis relative to the axis of the body (relative to the planted feet).[*] Again, the human pelvis differs from the australopith pelvis as we expect if australopiths were not yet redesigned to execute the elite human throwing motion as humans are.

After the pelvis is torqued, relative to the planted feet during the elite human throwing motion, the shoulders and arms need to be torqued relative to the pelvis adding acceleration to acceleration. Another set of large muscles extends *upward* from the pelvis, whose contractions probably help drive this violent motion.[32]

First are the abdominal oblique muscles in front (Figure 7.6). As their name suggests, these muscles run obliquely from the rib cage diagonally across the abdomen down to the pelvis. The contraction of the obliques on one side is expected to accelerate the shoulders at the beginning of the torquing motion associated with throwing and contraction of the

---

[*]   As above, this repositioning is expected to improve the equivalent of the moment arm for torsional or rotational movement of the pelvis relative to the planted feet and lower legs.

other set on the opposite side should decelerate the shoulders near the end of the throw.

We might expect the attachments of these muscles to be altered to improve their capacity to support the elite human throwing motion, and our skeletons indicate that they have been. Specifically, parts of the external obliques attach to the same much altered forward-most part of the crest of the iliac blade as the tensor fasciae lata does (only on top of the bone instead of below; panel B, Figure 7.8). Thus, the repositioning of this structure forward (Figure 7.8) should improve the torque of the external oblique. Other fibers of the obliques attach to the very front of the pelvis to what is called the pubic arch (the bone you can feel right behind and slightly above your genitals). Notice that this bone is also much further forward of the body axis in humans than in australopiths, as expected if australopiths had not yet evolved the elite human throwing motion.

When the paired obliques are apparently accelerating and decelerating the shoulder rotations from the front, the paired latissimus dorsi muscles are probably doing the same thing from the back (Figure 7.6). These muscles (*lats* for short) attach to the base of the arm and run obliquely across and down the back to attach to the pelvis (through the large ligament indicated in light gray in panel B of Figure 7.6) at the backmost part of the iliac crest. Notice that this pelvic attachment is precisely the repositioned backmost part of the iliac crest to which the gluteus maximus is also attached (from below). Thus, the attachments for the lats are also apparently redesigned as expected if elite throwing were an adaptive goal.* Again, this redesign occurs in humans, but not in australopiths.

So, we can identify changes in the human pelvis that support the elite human throwing motion (in addition to whatever other purposes they may have). Moreover, we do NOT see those changes in australopiths. Our theory absolutely requires that these "throwing changes" be present in the very first humans (the first clear members of our genus *Homo*). Apparently they are. Notice that Nariokotome boy's pelvis has all the human changes supporting elite throwing—the backmost part of the iliac crest projects further back from the body axis, the forward-most part of the crest projects further forward, and the pubic arch projects further forward. Moreover, no clearly australopith pelves show these throwing related changes. (We will discuss possible *transitional* animals in a moment.)

---

*   The positioning of these various powerful muscle sets in opposition across their pelvic attachments (lats from above, gluts from below, for example) effectively allows them to pull directly against one another. Good engineering.

Thus, the fossil pelves of these two hominids support our theory. As predicted, the earliest clearly human fossil we currently have already possessed a pelvis apparently redesigned to support elite throwing. In contrast, our pre-human australopith ancestors apparently did not. Equivalently, we fail to break the theory on the facts of the hominid pelvis.

Before leaving this subject, one last issue is interesting. The massive redesign of the human pelvis relative to the australopith pelvis has long been recognized. Many explanations have been proposed including improved walking/running or easier childbirth. Our theory and analysis suggests that most of this massive redesign is not primarily for these things, but rather for elite throwing.

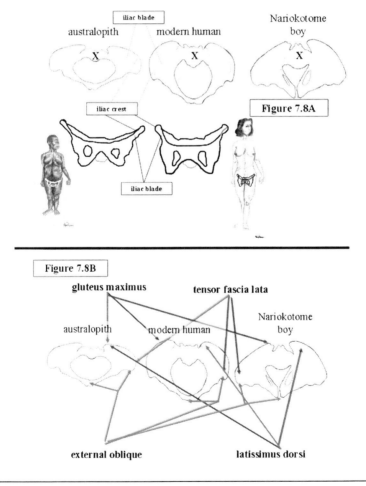

Figure 7.8: PANEL A shows the structures of the human and australopith hip bones (pelves; pelvis, singular). At the top of the panel, pelves are seen from the top—a 'liver's eye view'. Notice how differently shaped the modern

human and australopith pelves are. The iliac crest of the australopith pelvis flares out to accommodate the large gut of the animal, producing a pot-bellied body shape. The iliac crest is also extended further back and further forward in the modern human relative to the australopith. Finally, the front or pubic arch of the pelvis in much further forward in the modern human.

PANEL B shows how the repositioned portions of the modern human pelvis are the attachment points for the major muscles torquing segments of the torso during the elite human throwing motion (Figure 7.6 and 7.7). Notice that Nariokotome boy's pelvis has these attachment site modifications, indicating this very early human could probably throw with elite human skill (text). *Drawings of human australopith individuals by Amy Radican, others by P. Bingham.*

Arms, legs and throwing

The arms and legs of humans have some additional information for us. Look back again at Figure 7.3. Notice that humans have long legs (compared to arm length, for example). This probably has multiple implications—multiple compromises walking (running, throwing) in close formation. Not all of these are of interest to us here, but two of them are.

First, the shorter chimp-like legs of australopiths are probably an element of a continuing adaptation to some level of elite tree climbing. There is other evidence for this ability in the skeleton.[33] Moreover, one likely reason for their retaining this ability is the commitment to *arboreal retreat* which is blindingly rapid climbing up a tree to escape from dangers like the big cat predators.*

Thus, our long legs suggest that the first humans might have had a different strategy for dealing with the big cats. Intimidation of the cats by a group of elite throwing humans is a credible alternative.† The trade-off of elite arboreal retreat for elite throwing would apparently have been worth it to natural selection.

---

*   Arboreal retreat remains an important approach to predator avoidance in contemporary chimps and savanna baboons, for example.

†   It might seem a little implausible at first glance that humans armed only with throwing stones could intimidate a big cat. However, notice that there are no cat hospitals in the wild. A fractured jaw or forepaw is a death sentence (by starvation) for these predators. If ancestral humans traveled in packs and were careful to avoid predators in large numbers, they might well have been able to train the local feline hunters to avoid them. Notice also that these ancestral humans would still have been able to climb trees, as we do today, just not with the elite speed of chimps or australopiths. Thus, simply forcing the big cats to hang back and hesitate would probably have been sufficient.

Second, our longer legs might be related to throwing directly. Our legs play such an important role in throwing, their redesign might be relevant. We do not currently know enough to say more on this particular matter, but longer legs are at least suggestive and we will use them again later. In particular, our large, robust thigh bone (femur, technically) will become important. (Compare australopith and human femurs in Figure 7.3.)

Hands for tool use and hands for throwing[*]

The human hand is different than the hand of any other living ape.[34] Traditionally this has been thought to be connected to our elite use of tools.[35] We will argue here that this traditional view is wrong – or, at least, very incomplete. Specifically, we propose that the unique new features of human hand design relative to australopiths are (mostly or entirely) for elite throwing.

We face two limitations in understanding fossil hands. First, the hand is an extreme example of a "series of compromises working in close formation." Primate hands do so many different things that their design in constrained in many different ways. Second, we do not have nearly as many fossil hominid hand bones from the early human period (around 1.8 million years ago) as we would like.

However, in spite of these limitations, heroic effort has been invested in understanding the evolution of the human hand. There are several insights from this work that are useful to us.[36]

The australopith ancestors of *Homo* were bipeds as we saw. Their hands are no longer needed for locomotion as in other apes. Australopith hands were thus free to be redesigned and they were. Fingers became shorter in relation to the thumb, a trait we share with them (Figure 7.9, middle panel).[37] Our question then is whether we can recognize hand adaptations to elite throwing and ask when they emerged. The evidence is not adequate to give us an unambiguous answer. However, available fossils are nonetheless highly suggestive and contribute to the larger picture emerging from looking at the entire body as a suite of throwing adaptations.

The two best pieces of evidence can be visualized by manipulating a baseball (or a throwing stone) with your dominant (throwing) hand. First, grip the baseball as if you were going to throw it (Figure 7.9, top panel) but as delicately as possible without dropping it. Now wrap the thumb

---

[*]   See Napier and Tuttle (1993) for a review of the tool-user view of human hands by one of its greatest exponents.

and forefinger of your other, non-dominant hand around the dominant hand thumb from the outside and pinch the fleshy part between the bases of the thumb and index finger of your dominant hand. These actions will let you feel the level of contraction of the muscles in this part of the throwing hand. Now grip the ball (or stone) firmly as if you are about to throw it. The group of muscles you are feeling will contract powerfully. This ability to grip a projectile tightly during the violent throwing motion is certainly critical to elite human throwing.

Fortunately, we know a lot about the evolution of some of these muscles. The most important one for us is called the first dorsal interosseous muscle (DI-1, for short). This is the muscle that produces the hard bulge on the outside surface of the portion of your throwing hand you are feeling when you grip the ball tightly (see arrow in top panel of Figure 7.9). This muscle is massively redesigned in humans relative to chimps.[38] Its human-specific attachments along the inner surface of the basal thumb bone (called the first metacarpal) can be seen on the bone itself, including in fossil bones.[*] Australopiths do not have the attachments reflecting this DI-1 redesign; however, the earliest members of *Homo* from approximately 1.8 million years ago apparently do.

Other members of this muscle group have substantially different mechanical properties in the human hand relative to the chimp hand.[†, 39] It is more difficult to assess the properties of these muscles from fossil hominid hand bones. However, these muscles pull hard on the basal thumb bone (first metacarpal) similarly to DI-1 and this bone is substantially more robust in *Homo* than in australopiths (Figure 7.9, middle panel). This enlargement suggests that the overt redesign of DI-1 in *Homo* may be symptomatic of a more general, global redesign of the hand for throwing.

A second challenge when throwing is to release the tightly held projectile at just the right instant from a hand moving at 90 to 100 miles per hour (145 to 160 kilometers per hour). This takes stupendous dexterity—part of our elite throwing repertoire. Look again at your throwing hand gripping the baseball. The fleshy pads on the inner surface (opposite your fingernails) of your thumb and first two fingers contact the ball firmly (Figure 7.9, top panel). These pads both hold and sense the ball. They must be exquisitely effective and sensitive.

_____

[*]   This presents as a rough area on the bone surface not visible in most photographs.

[†]   Most important of these muscles is the opponens pollicis. Agur, et al. (1999) is a human anatomy textbook with images of these muscles.

Put down the ball and look at these fleshy pads. Move them with the tips of other fingers. Notice that they move extensively and that you can feel this movement. Now test movement at the very tip of your finger. Here movement is much more limited. This is because the very tip of your finger's flesh is attached to the flared tip (called an *apical tuft*) of the underlying finger bone (Figure 7.9, bottom panel). Thus, any movement in the fleshy pad contacting the ball produces stretching of skin and underlying tissue relative to this immobilized neighboring tip bit. The extensive innervation of the ends of your fingers easily detects this stretching, giving you the delicate control you need to "throw like a human."

These enlarged apical tufts can be seen in the fossil record. As with the DI-1 attachments, this enlargement appears to arise with the first members of *Homo*.[40]

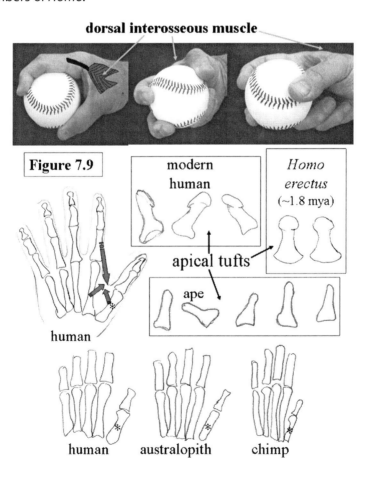

Figure 7.9: At TOP are three views of a modern human hand grasping a baseball as a projectile in preparation for aimed, high-momentum throwing. Diagrammed on the leftmost hand are the two flared portions of first dorsal interosseous (DI-1) muscle (text). One tract of this muscle attaches to the first (thumb) metacarpal and the other to the second (index finger) metacarpal. Both tracts then connect to the lower index finger bone (proximal phalanx) through a tendon. Contraction of this muscle supports the power throwing grip shown. The broad attachment of the DI-1 muscle to the first metacarpal providing power to this grip is apparently unique to members of *Homo*. Also notice the role of the fleshy tips of the thumb and index and second fingers in gripping the ball.

At CENTER LEFT are the major modern human hand bones with the first metacarpal indicated by the asterisk. The forces generated by contraction of the DI-1 muscle are indicated by the arrows.

At BOTTOM are the central hand bones (metacarpals and first phalanges) from a human, australopith (reconstructed from multiple sources), and chimp hand. The first metacarpal is marked with an asterisk. Note that the thumb is longer in both australopiths and humans; however, the thumb (first) metacarpal is substantially more robust in humans, arguably to provide support for the powerful contraction of the DI-1 and related muscles (text).

At CENTER RIGHT are the apical tufts of the finger tip bones (distal phalanges) from modern humans, Dmanisi *Homo erectus* (1.8 million years ago) and from modern apes (chimps and gorillas). The human enlargement of the tip of this bone is not shown by australopiths and arguably represents an adaptation to hyper-sensitive manipulation, including accurate release of a high-speed thrown projectile (text). *Drawings by P. Bingham (see text for sources).*

Hands and tools, again

Before leaving the hand, it is important to return to the classic suggestion that human hands are redesigned for elite tool use.[41] Our theory suggests an alternative view. The human hand is arguably redesigned for elite throwing. This throwing hand is then used to make tools, but is not primarily designed for this purpose.

Many animals use tools and do so with hands, paws, mouths, or beaks designed for other purposes. There is not any good reason to believe that a chimp or australopith hand could not have formed the anatomical foundation of a uniquely elite tool culture just as the human hand ultimately did. Some of the details of this culture would no doubt have been different (reflecting the differences in its implementing hand), but elite it could still have been. The fact that a piano sonata is simpler than it would have been if humans had twelve fingers does not disqualify Mozart as elite music.

On the basis of these considerations, it is circular to argue that human hands are designed for tool use because they are transformed around the same time as the first new human-like tools show up. The hypothesis that the human hand was primarily redesigned for elite throwing is precisely as capable of accounting for the hand evidence as the tool use hypothesis and much better at accounting for the other parts of the transformed human anatomy.

We will see in later chapters that there are very sound reasons for believing that the uniquely elite human tool use is an *effect* of human cooperation, *not a cause* (Chapter 10). Thus, we anticipate that the human hand was redesigned first for elite throwing, allowing uniquely human cooperation, and this throwing hand was then recruited for elite tool use supported by uniquely human cooperation.

At the feet of a throwing animal

Feet are not the first thing we think of when we think of throwing. But they actually have a crucial role to play. We drive violently off a firmly planted back foot when we throw. Moreover, this planted back foot rotates sharply during the throwing motion. More precisely, we plant on the side of the back foot behind the big toe, then the foot extends and rotates into a more toes-back configuration by the end of the throw. Compare the feet in the different frames of Figure 7.5 and observe your own back foot during a full-out throw.

This elite throwing motion would be expected to require a redesign of the foot. As with the hand, we do not yet have enough fossil evidence to be completely confident. However, the pieces we do have are consistent with the larger picture presented by the body *in toto*. We will focus here on the simplest and, arguably, most clear case—the changes producing the human *big toe* complex.

To appreciate these changes we must first understand the platform on which they are built. Figure 7.10 compares the foot bones of humans, chimps, and a reconstruction of an australopith foot.[42] Notice first that the chimp big toe actually projects out away from the rest of the foot. This is because chimp feet are a lot like hands. They can grasp things, especially tree branches during high-speed navigation of their rainforest habitats. In sharp contrast, *both* australopiths and humans have a big toe which lies more parallel to the other toes and forms a unit with them rather than acting like the chimp's "foot-thumb."

This foot bone structure is one of the diverse pieces of evidence that Lucy and her kind walked along the ground bipedally like we do, rather than on all fours like chimps and gorillas prefer to do (Figure 7.2).[43]

Our question then becomes whether we can see indications either that Lucy's foot was already designed to throw with elite human skill (falsifying our theory) or, conversely, whether changes in the foot allowing elite human throwing failed to occur until the emergence of the first humans (or their *immediate* ancestors) as our theory requires. The modest information we currently have is consistent with the predictions of our theory. To understand this, return once again to the back foot in elite human throwing (Figure 7.5).

At the beginning of the motion, we push violently off the side of the foot along a surface including the big toe and the ball-like bulge just behind the big toe. This large, heavy projection is probably essential to permit the elite throwing motion to be executed day after day throughout the life of a professionally throwing ancestral human. This projection is produced by the large human first metatarsal bone (Figure 7.10) and the heavy ligaments and soft tissue parts attached to that bone. This bone is substantially enlarged in humans relative to Lucy and her kind (Figure 7.10).

This enlarged human first metatarsal is important for a second, related reason. When we drive forward off the side of the ball of the foot behind the big toe, the foot tends to collapse under the strain. This must be prevented by muscles that resist this undesired twisting. One of the muscles likely to participate in this control is the peroneus longus (Figure 7.10). This same muscle is probably also involved in rotating the foot more forward later in the throwing motion (Figure 7.5). This muscle is anchored to the smaller calf bone (fibula). It runs down the leg along the inner back edge, connects to a tendon that runs under the arch of the foot and attaches to the very same first metatarsal bone we push off of when we throw (Figure 7.10).

Thus, the first metatarsal not only supports the enlarged structure we push off of to throw, it also provides the connection for one of the muscles that probably stabilizes and torques the foot during the throwing motion. The enlargement of the first metacarpal in humans but not in Lucy and her kind is certainly consistent with the hypothesis that Lucy could not throw with elite human skill.

Unfortunately, among the many pieces of information we lack about ancestral human feet are good examples of the first metatarsal bone from early clear humans. These bones were not recovered with the Nariokotome fossil, for example. So we cannot be certain whether the enlargement occurred in the very first humans, as our theory requires. Nevertheless, the foot bones we do have are consistent with our theory.[44] For example, the very early ancestral finds at Dmanisi (below) also include enlarged human-like first metatarsals.

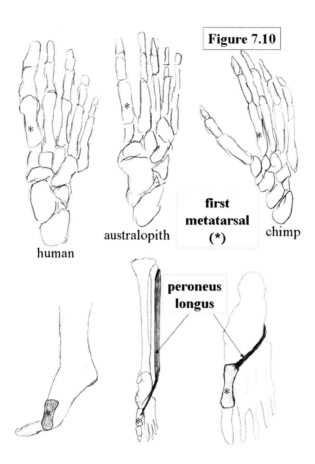

Figure 7.10

first
metatarsal
(*)

australopith   chimp

human

peroneus
longus

Figure 7-10: Chimp, human, and (reconstructed) australopith foot bones are shown at TOP. Notice that the hominid (human and australopith) big toes have been redesigned for bipedal locomotion while the chimp big toe retains the capacity to grip for tree navigation. The first metatarsal bone is indicated by an asterisk throughout the figure. This bone is highly robust in humans, including early *Homo erectus* (Dmanisi; text).

This enlarged metatarsal forms the foundation for the robust ball behind the big toe that humans drive off of when they initiate the throwing motion (Figure 7.5). This large metatarsal is also the connection point in the foot for the peroneus longus muscle (BOTTOM). This muscle controls and torques the back foot during the violent human throwing motion. The two images at the BOTTOM RIGHT illustrate right feet viewed from the bottom. Note the wrapping of the peroneus longus connecting tendon around the outside of the ankle to attach to the first metatarsal bone..

> The modern human foot has apparently been redesigned to support the elite human throwing motion. The fossil evidence indicates that this redesign occurs with the first humans, apparently not in their australopith ancestors. *Drawings by P. Bingham (see text for sources).*

Finally, notice that the evidence gives us a similar view of the human foot as we have already had of the hand and torso. Specifically, the human foot is often thought of as having been redesigned relative to the australopith foot for some improved version of bipedal locomotion, perhaps including elite running.[45] However, the hypothesis that the human foot is redesigned to support elite throwing is (at least) equally successful in accounting for the foot bone evidence.

Head and shoulders above the rest – the human head is still while the shoulders throw

Focus on a target, relax, and throw in slow motion. Now focus on your head and your shoulders and throw again. You will find that your head and eyes remain locked on the target (like an elite fire-control system) while your shoulders and arms move and rotate violently. This requires a substantial independence of movement of the head and shoulders relative to one another.

Now impersonate an old-fashioned robot whose entire upper body (head and chest) rotate as a unit. This movement is unnatural for humans. However, the other great apes move their upper bodies somewhat like the robot. This "stiff necked" movement results from powerful connections between the shoulder complex (pectoral girdle) and the head in apes. The heavy muscles are repositioned in humans to free the head from the shoulders.[46]

We cannot assess the connections of these muscles to fossil skulls very well. However, the shoulder complex in australopiths looks more like chimps than like our own whereas the earliest known members of *Homo* have shoulder complexes much like ours.[47] This strongly suggests that head freedom for throwing arose with the first members of *Homo*.

Predicting Dmanisi – forecasting new observations

Paul first published the basics of the theory we are exploring in 1999 in the professional scientific literature.[48] Thus, enough time has passed that we should expect new observations that were forecast by the original paper. This kind of true prediction is a strong test of a theory. One particularly good example of such a successful forecast is provided by the recent discoveries at Dmanisi in the Central Asian nation of Georgia, as briefly mentioned above.

Dmanisi is currently being excavated by a international team of paleoanthropologists, geologists, and archaeologists.[*,49] The team includes our SUNY (Binghamton) colleague, friend, and one of the world's experts on early *Homo*, Philip Rightmire. The dig's other leaders are David Lordipanidze (Georgian State Museum) and Reid Ferring (North Texas State).

Dmanisi gives us a suite of vital observations.

First, the context is well dated. We can be reasonably confident that the hominids found at the site lived 1.8 million years before the present. Thus, they are probably the right age to be among the very first human ancestors to leave Africa. (Apparently, australopiths never left their birth continent.) These fossils also carry another useful possible implication. The Dmanisi hominids probably already had some new strategy that allowed them to expand their range in a way their ancestors never did. It is very attractive to propose that this new strategy was a new scale of social cooperation with all its diverse adaptive consequences.

Second, the Dmanisi dig has yielded five relatively complete skulls to date. These all show cranial volumes well below Nariokotome (Figure 7.4),[50] ranging down to near australopith/ape in size. In spite of this small size, however, the details of facial anatomy indicate that these animals are members of early *Homo*. Thus, the Dmanisi hominids look like representatives of animals that have just begun the evolutionary redesign to the human life history that will become so much more extreme in their later descendents and relatives (including us).

If our theory is correct in detail, these animals were probably hunters/ power-scavengers. They should already be elite human-like throwers. The remainder of the Dmanisi evidence strongly suggests that these hominids are *precisely* what our theory predicts.

Third, the Dmanisi hominids were hunting/power-scavenging as indicated by fossil prey bones found intermixed with the hominid fossils. Moreover, the reconstructed environment of the Dmanisi fossils would have represented an ideal environment for such a lifestyle.

---

*   See the cover story in the April 2005 issue of National Geographic for an excellent summary of this work.

Fourth, though we do not yet have a fossil pelvis, we have other skeletal elements that indicate that these animals were probably redesigned for elite human throwing. A Damanisi femur is large and robust—like *Homo*, unlike australopiths (Figure 7.3). Damanisi shoulder bones look human-like, not australopith-like. Though we have not discussed shoulder bones in detail, they are almost certainly among the structures redesigned for elite human throwing. Finally, a robust first metatarsal indicates humanlike throwing foot platform and peroneus longus attachment as probably required for elite throwing (Figure 7.10).

Fifth, the Dmanisi hominid skeletons were deposited up on a plateau formed by wind-blown soil. Thus, this site has no indigenous rocks. All the stone tools found with these animals were transported to the site by them. Remarkably, the Dmanisi dig has generated far more throwing-size manuports that traditional stone tools (like sharp flakes for cutting).[*,51] This is precisely what our theory predicts and requires.

Born to throw (and to run?)

We have argued that from their very beginning, *Homo* (the first humans) were already redesigned to throw with elite skill. Humans are apparently "born to throw." Since our original publication of this proposal,[†] Dennis Bramble and Dan Lieberman (2004) proposed instead that humans are "born to run".[‡,52] It is clear that humans have acquired an elite distance running capability that the other great apes do not share. Indeed, several of the evolutionary adaptations to throwing (above and others we have not reviewed here) probably also facilitate our ability to "run marathons".[53]

We propose that both suggestions are very likely to be correct. Humans are a "series of compromises, throwing and running, in close formation."

---

* This same pattern of a few sharp tools and many manuports probably also exists in the African record of early *Homo*. However, African sites have large amounts of indigenous stone so that unworked manuports often fade into the background noise of naturally occurring stones. There are a few specific African sites that have been suggested to show the expected elevated levels of imported stones sized properly for human throwing just as Dmanisi clearly does (Potts, 1988).

† Bingham, (1999) and (2000).

‡ Also, see Heinrich (2002) for a wonderfully readable account of this idea.

However, two things are crucial for us to recognize about this possibility. First, elite throwing and running are very likely to be important elements of a new hunting/power scavenging adaptation. Running down prey and keeping up with the big cats impose demanding requirements. We look like animals adapted to these requirements.

Second, however, elite distance running has evolved repeatedly in mammalian carnivores.[54] Think, for example, of pack hunters like wolves and African hunting dogs. Thus, elite running (and hunting and scavenging) is *not sufficient*—or necessarily even relevant—to produce a new kind of ecologically dominant animal. Moreover, there is not any theoretical reason we know of to believe that running would have any such effect.

Thus, of these two elite human capabilities, we argue that elite throwing is, by far, the more important one if we wish to understand our unique evolutionary trajectory.

Where are we and where are we going?

We have argued in this chapter that we can now interpret the entire fossil record of human origins with new power and clarity. The available evidence supports the strong, simple picture required by our theory. The first animal in the history of life on Earth to be able to kill adult conspecifics from a distance arose between about two and a half and two million years ago. This was probably the original elite throwing hominid power scavenger/hunter. Very rapidly thereafter, the descendents of this animal evolved a massive new scale of kinship-independent social cooperation (supporting brain expansion through life history redesign).

The economy of this reinterpretation of the hominid fossil record is compelling. However, we have just begun. If our theory is right, it should give us comparably simple, direct interpretations of every other feature of the human story. We will argue in the chapters that follow that it does.

෧෨

First Interlude
Sex and the power of counter-intuitive predictions

Many of us grew up in a human village where almost everyone had a mother and a father. In spite of the capacity for discreet infidelity (Chapter 4), most parents are committed to their children, the products of their nuclear family. However, there is very good reason to believe that our comfortable contemporary world is **not** typical of much of the human past.

Several theories of the origins of human sexual/mating behavior have been popular among biologists and anthropologists in the late 20th Century.* All of these assume, explicitly or implicitly, that simple kin-selected behavior (Chapter 3) is central to the evolution of human sexuality. For example, it has sometimes been argued that the unique life history of human infants and children evolved as a result of newly increased support and investment from fathers.† This new ancestral-male parental investment is claimed to have come with *pair-bonding* and, thus, increased certainty of paternity. In other words, the male

---

* See Fisher (1983), Hrdy (1999), Angier (1999), and Buss (2003) for recent reviews and discussions of these theories.

† In our closest living relatives, chimps and gorillas, the post-mating contribution of fathers is mostly restricted to indirect protection of the young from violence by other males, in other words, to prevent infanticide. Females carry essentially all the day-to-day burden of childcare and feeding.

invests his time and resources because he has new confidence in his genetic relatedness to the offspring.*

Our theory makes a fundamentally different prediction about the origins of human pair bonding, as we will see in Chapter 8. We do not ignore the power of kin-selection; however, human sexual/breeding behavior apparently emerges under the dominant guidance of our unique adaptive strategy, the management of **non-kin** conflicts of interest. Humans have come to the approximation of monogamy that characterizes many contemporary human societies by a circuitous and surprising route, indeed.

Our theory predicts that our ancestors' sexual behaviors probably ranged across a spectrum between two alternating extreme patterns. The first pattern, somewhat like our own, consisted of mothers, fathers, and children in a "family" with a tolerable approximation to sexually exclusive monogamous mating.

In contrast, the second ancestral mating pattern was very different than many of us know. Though adults probably formed social pair bonds, both women *and* men mated promiscuously, sometimes even publicly, in groups and by *consensus* of both sexes. No one knew for sure who a child's biological father was under these mating conditions. The very nature of family was substantially different than in many contemporary cultures.

Ancestral infants and children resulting from such promiscuous mating systems would still have had the benefit of increased paternal contributions to their care from the cooperative human village. But this behavior had a more subtle origin than simple kin-selection. In fact, this characteristically human paternal contribution was, instead, also an effect of our unique management of non-kin conflicts of interest.

Our theory is unprecedented in predicting these two alternating extremes to *both* be characteristic human adaptations. The empirical evidence strongly supports these predictions, as we will see. In other words, humans who are promiscuous, even to the point of engaging in public sexual orgies, can be behaving just as "naturally" as humans who fall in love and live for years in faithful, sexually monogamous relationships. It all depends on the immediate details of their circumstances and the specific adaptive challenges they face.

Our sexual anatomy and psychology are expected to have been built by design information that did well under *both* promiscuous and monogamous mating systems, depending on the context. How we look

---

* Of course, this model begs the question of why humans might have begun to be-have this way while virtually no other mammals do.

at sex and at one another is deeply affected by this complex evolutionary history.

As radically different as the ancestral *promiscuous* human village may seem from our own, it will prove **not** to be different in the ways that truly matter. Children were still loved, taught, and protected. Adults still cooperated, still often respected their neighbors, and still thought of themselves as members of a common community, the human village. The compelling ethical foundations of human behavior were still central to everyday life. The only things that were different were sexual/mating behaviors and the redesigned family structures these behaviors produce.

The notion that humans might form loving, effective families just as well when mating promiscuously as when mating monogamously is a little unexpected for most of us. We can look around our world and traditional nuclear families appear to be everywhere, in Africa, Asia, Australia, Europe, and the Americas. Surely, this means that humans are designed to live exclusively this way. In fact, some investigators of human sexual behavior have drawn exactly this conclusion. It has been suggested that finding similarities between, for example, late 19th Century Japan and mid-20th Century Brazil indicated that a behavior was *cross-culturally universal* and, thus, inherently human.

This inference of the universality of monogamy as the natural human mating strategy is deeply flawed for quite straightforward reasons. When we look at Japan and Brazil or almost any other recent human society, we are looking at a relatively new human social system, the *state*. As we will see, our theory predicts the state and its properties (Chapters 13-15). However, properties we take for granted, like *monogamy* (or mild polygyny), are quite idiosyncratic to specific local social conditions, including the presence and protection of the state. They are **not** universal.

It is important to note that when we talk about human ancestral promiscuity, we mean multiple partner mating under sexual democracy, not the polygyny of recent hierarchical states dominated by small groups of elite males (Chapter 13).[*]

In the following chapter, we see that, for the uniquely human mammal, promiscuity is the best adaptive choice under certain

---

[*]   We often think of societies that are violent, male-dominated, and more or less polygynous (based on the outcome of male status competition) as characteristic of the ancestral human condition. We will find in Chapter 13 that this is almost certainly **not** the case. Such male-dominated societies are probably a quite recent (and ephemeral) product of the early forms of the state (archaic states) over the last few thousand years, **not** characteristic of most of our two-million-year evolutionary history as a species.

conditions and monogamy under others. We will provide extensive evidence that we all carry the evolved potential for both patterns of sexual behavior.

These counter-intuitive predictions about our core sexual behaviors and feelings are particularly valuable as scientific evidence. We would probably not have come up with them from the vantage point of our home base, state cultures. But our new theory *does* predict them. Thus, if we find what we predict, this empirical discovery is an especially compelling piece of evidence that our theory might be correct. On to Chapter 8, and what we will argue is precisely this insight from the evidence.

☙

Chapter 8
Promiscuity and monogamy – sex and family life in the uniquely human village

A troop of accidental apes rests for the evening. They enjoy many bene-fits from their new adaptive trick—kinship-independent social cooperation (Chapters 5 and 7). But their lives remain risky. Each year one or two adults will die from injuries delivered by large dangerous prey or competing preda-tors. Disease is also a constant, but unpredictable companion. The animals "know" that anyone of their number could be gone tomorrow, no matter what each of them might do.

Relying on a single mate for reproduction in such a risky environment makes no sense for these animals. That mate might die in the blink of an eye. Males and females must "plan" for this eventuality so that their offspring have the best chance of survival. Each sex "buys life insurance" by having multiple mates.*

These accidental apes have lived with this relentless logic for tens of thousands of years. Their design information is now highly refined to build bodies and behaviors that work to achieve the best strategic outcome for that design information in this uniquely cooperative, but high-risk environ-ment.

Today the troop is well fed after a successful round of power scav-enging. The other local predators are now off hunting elsewhere, so the troop settles down to an unusually restful evening. Several adult female

---

* Each sex is buying insurance for the life of his/her design information, of course. We will develop the adaptive logic of this strategy in detail below.

troop members have recently weaned their youngsters. These children are now being sustained by a rich subset of the foods hunted and gathered by various members of the troop including the males who might have fathered these youngsters or who are being coerced to contribute by potential fathers.* Relief from the severe calorie drain of lactation has allowed these females to gain some weight and, much more importantly, they have begun menstruating (advertising cycles of ovulation) again.

This condition of the females has been very big news among the males in the troop. They have been looking forward to this new opportunity to mate. A male exchanges a flirtatious glance with one of these females. By unspoken mutual consent, the pair slips behind a bush at the periphery of the troop's night camp. They are both intensely aroused; the male is rigidly erect and the female wet and dilated.

The pair begins a boisterous, joyful mating. As the female approaches orgasm she begins an ever-escalating vocalization of ecstatic pleasure. The other members of the troop are drawn to the pair. (This recruitment of others is probably the unconscious "purpose" of the female's orgasm "song," in fact.) The other males and fertile females alike find themselves irresistibly excited by the sight of the mating pair. As the pair reaches orgasm, other males approach the first female and one immediately begins to mate with her. Other males solicit the remaining fertile females and begin mating.

There is an excess of males because a number of females are pregnant or still nursing. Each male waits his turn. Each female finds that her latest orgasm with one male simply increases her desire for the next. Each male finds that the site of other mating couples brings him back to full arousal within seconds after his previous orgasm. For the next hour, each female and each male mate repeatedly until every female has mated with every male and their sexual gathering slowly winds down.

The males and females lie together in warm afterglow. They share a transcendental sense of satisfaction, joy and a common future.

Thousands of years later, the descendents of this troop find themselves migrating into a new territory where their kind has never been. Perhaps they have entered Asia from their ancestral African home for the first time. This new territory is spectacularly rich in resources, especially in the absence of competition with a large number of other accidental ape troops (villages) with their comparably elite skills.

Moreover, this new territory "feels" different. The climate is not quite like what they have known. Though they do not know why, this is a very good omen. It means that some of the worst diseases and parasites of their ancestral homeland have been left behind.

---

*   See Chapter 6 for the very specific meaning *children* and *childhood* has for us.

These migrant accidental apes bring their lives with them, mating patterns and all. However, these mating patterns slowly begin to change. Individuals who survive the rigors and dangers of youth to reach reproductive age now have new prospects. They have a very good chance of surviving for many years—even into middle or old age. Adults are much safer in this new environment than in the one their ancestors recently left.

Though individual apes are not consciously aware of this, the logic of life insurance has now changed and natural selection will produce new adaptations to this change. Because both parents are much more likely to survive long enough to see their mutual offspring through to independence, it makes sense for both parents to produce and invest in youngsters they know to be their own genetic offspring. Relentlessly, over the next several thousand generations, subtle alterations in behavior reflecting this new adaptive logic will be rewarded through better than average reproductive success. Inevitably, the mating behaviors and attitudes of the accidental ape are redesigned.

These new-model accidental apes will find themselves bonding with one (or a few) individuals of the opposite sex, eventually coming to consider "falling in love" to be an inevitable, even desirable, part of their lives. Promiscuous mating will remain, but will be much rarer and, in some local troops, mating with extra partners will even be driven entirely underground.

This new system is not exclusively monogamous. Each sex has an incentive to cheat secretly on her/his "monogamous spouse." Thus, this new system is really a mixture of ancestral promiscuous mating and purely monogamous mating.

Over the coming millennia, the remote descendents of these now redesigned accidental apes will find themselves once again in hard times with death stalking the daily footsteps of even healthy young adults. They will respond as their ancestors did with promiscuous purchasing of much-needed life insurance. As need arises over the ensuing two million years of their evolutionary history, the members of this lineage will alternate between its promiscuous and monogamous mating strategies, depending on adult mortality rates. Their design information is thereby given the best chance for replication under whatever contingencies these humans meet.

*As a result of this complex evolutionary history, this accidental ape's descendents (us) will be not be "naturally" promiscuous or "naturally" monogamous. They will be **either** depending on local actuarial circumstances. Promiscuity or monogamy will not be the characteristic reproductive system of these humans—rather the **village** will be that characteristic system. Individuals in the human village will love and raise their children and will cooperate to defend and enhance their shared interests. And they* *will do so when adult members are mating promiscuously and when they are more or less monogamous.*

One hundred thousand generations later, the descendents of these first human villages (now calling themselves women and men) will walk on the Moon and dream of roaming amongst the stars. But they will carry with them bodies and minds that are the legacy of adjusting mating behavior to the demands and risks of their environment from relatively extensive promiscuity to a passable approximation of monogamy and back again. They will remain devoted to falling in love, but some will also sometimes long for other mates, different lovers, new opportunities.

Before we begin

In this chapter we will explore the deepest roots of behaviors that are near the core of each of us—our sexual and parenting behaviors. Those of us old enough have fallen in love have known both the terrible pain of rejection and the transcendent joy of mutual affection. Those of us who are fortunate have loved our children and been loved by our parents. We have been nurtured in the protecting bosom of our villages. We the lucky cannot really imagine life without love, family, and community.

Our sexuality is the arena of many of the most intense experiences of our lives. We seek these experiences—love, lust, commitment, connection—with every bit as much intensity as we seek food and water. Most of us cannot live without them. Our challenge in this chapter is to understand how these universal, but (in part) uniquely human, longings and joys came to be.

As we proceed, we will find some of our long-held beliefs about these intensely private, utterly compelling parts of our lives challenged. We will take a quite different view of the evolutionary origins of human sexuality and parenting than may be currently popular. Some have claimed that monogamy is the natural human condition. Others that promiscuity is this natural human condition. We will challenge both those polar-opposite claims

As we explore human sexuality, it is important for us to query candidly our private feelings, experiences, and observations as sexual beings. The discovery of new insight by each of us into her/his sexual life history is a strong new test of whether our new theoretical perspective is likely to be fundamentally correct. At the end of this piece of our journey, we can find something of value—humane self-knowledge and mutual understanding.

One reason we sometimes misconstrue our inherent sexuality

Some human cultures were changed by a relatively recent development. Under novel conditions that first arose about 5500 years ago, small groups of elite males began to exert highly disproportionate coercive control over local societies (Chapter 13). As with all humans everywhere, these males exploited their control of decisive coercive means in pursuit of self-interest (sometimes unconsciously). These pursuits included reconfiguring some of the public rules for human sexual behavior including promotion of multiple wives (polygyny) for elite males.

These new male-dominated cultures were ancestral to many contemporary societies.* Moreover, these hierarchical ancestral polities invented written language so that their dominant members wrote much of what we think of as our ancient history. These (unconsciously) self-interested writers depicted the world to be *inherently* as they wished it to be, of course.

As a result of these features of our very recent past, we are sometimes tricked into thinking of some aspects of human sexual behavior as being preferentially connected to male domination and male interests. This is a profoundly misleading perspective. Most of our two-million-year human history probably consisted of relatively democratized cultures, societies where women and men were all relatively powerful as individuals (Chapters 7, 11 and 12). Men and women are both quite well adapted to such environments.

We will shortly see some surprising things about our evolved sexual capabilities and proclivities. In examining this evidence, we will sometimes see humans mating promiscuously, including females mating with multiple males, for example. The crucial goal throughout our exploration of evolved human sexuality is to remember the long-term ancestral context. Do not visualize the quirky recent "unnatural" male domination of the public rules for human sexual behavior. Think, instead, of each co-equal woman and man making individual choices in pursuit of her/his own interests in the context of the cooperative ancestral village. This egalitarian setting probably reflects how our ancestors lived throughout almost all of our two-million-year history. This democratized

---

* Indeed, some level of male domination of the public sphere and of the nominal legalisms of sexuality remains a property of some of the contemporary descendents of these cultures.

environment is the only context in which the evolution of these behaviors can be properly understood, we argue.

Sexual minds and bodies

The fictional tale of the accidental apes at the beginning of this chapter synopsizes the evolution of our sexual and child-rearing behaviors as our theory apparently predicts them. Much more theoretical and empirical work is needed if we are to understand fully how all the details of our sexuality evolved within the uniquely human ancestral village. In spite of the limitations of our current knowledge, the scenario we explore here is likely to contain fundamentally accurate insights. The remainder of this chapter will examine the evidence that something like this story is our authentic past.

Our theory makes some precise, important predictions. We expect that ancestral humans employed inexpensive conjoint coercive threat in pursuit of self-interest (Chapter 5). This adaptive approach will include individual acts having the effect of enforcing sexual behaviors that serve the interests of the individuals enforcing them.*

When the individual self-interests of the members of a coercively dominant subgroup (usually a majority) of the ancestral village are served by promiscuous mating, that will be the behavior pattern that emerges and promiscuity will be enforced.† Alternatively, when sexual monogamy serves the interests of most of the individual members of a coercively dominant consensus, some approximation of monogamy will be enforced.

Adult mortality risks appear to be the most important factor in determining how humans mate. When adults have a relatively high probability of premature death, monogamous mating is a bad idea

---

* This enforcement would have been largely unconscious most of the time in our evolutionary past—particularly, early in that past. As we will discuss in more detail in Chapter 10, this unconscious enforcement will generally take several forms. It will sometimes be enforcement of behavior patterns that seem (consciously or unconsciously) to further the interests of the enforcers directly. Alternatively, this enforcement will also often take the form of policing conformity to behavior patterns that seem to prevent other non-kin individuals from exploiting the resources of the village for self-interested gain at the expense of coercing individuals (Chapter 10).

† The technical term for promiscuous mating systems is *polygynandry*. This combines the term for individual females mating with multiple males, *polyandry*, and individual males mating with multiple females, *polygyny*.

for both parents of any individual youngster. The reasons for this are straightforward.

When the likelihood of premature death is high, the individual nuclear family is extremely vulnerable. By dispersing their reproductive efforts beyond the nuclear family, adults are more likely to leave offspring that survive to independent adulthood should they or their nominal mate die prematurely.

Conversely, when the probability of premature death is low, it can make adaptive sense for each adult member of a pair to focus his/her efforts on the pair's common offspring and to observe a high level of sexual fidelity.

We will discuss this logic in more detail in the following section. For now, it is sufficient to understand the fundamental argument. Our theory predicts that human reproductive behavior is unique in only one way. We are Earth's first **kinship-independent social breeders** (Chapter 6). We are completely committed to this element of our breeding strategy and to *no other* part of our sexual and child-rearing repertoire. Everything else we do in this arena of our lives, including who and how many we mate with, is *contingent, variable*.

Women and men – the logic of monogamy and promiscuity

Even though the human village is characterized by kinship-independent social cooperation, this cooperation does not displace kin cooperation. Rather, kinship-independent cooperation is "added on top of" kin cooperation, so to speak. In other words, kin cooperation persists in humans similarly (though not identically) to non-human animals (Chapter 4). In humans, multiple copies of the resulting kin-selected social units (families) are knitted together into the uniquely human cooperative village.* Understanding how these non-kin and kin-specific modes of social cooperation articulate with one another is crucial in understanding our evolved mating strategies.

Human youngsters are aided in their growth and development by the larger village, **but** they remain uniquely dependent on the extra help they receive from close kin. These close kin always include *mother*. However, who *else* gives extra kin help to the growing youngster is

---

* We will talk below about human *families* and *spouses* even in environments where promiscuous mating is common. This is not merely a literary convenience. Humans are clearly adapted to pair bonding even when they are actually mating promiscuously.

variable, sometimes including large direct contributions from a specific individual *father*, sometimes less so.

This help contributed to the growing youngster by close kin is referred to technically as *parental investment* whether the help comes from the biological mother (maternal investment) and father (paternal investment) of the child or not. Ancestral adults were highly adapted to understanding their likely genetic relatedness to youngsters in the human village and to partition their help accordingly, in part.*

Consider the options for parental investment by ancestral adults in an environment of high adult mortality risks. If a couple mates monoga-mously, each of them can expect extensive parental investment from the other and their children can do well *if* they both survive long enough. However, the couple runs the terrible risk that one or both of them will expire before their children have reached independence.

How, precisely, can two mates *buy life insurance* against a very real threat of premature death?[1] Each can do so by mating with others, by mating promiscuously, as we suggested above. Here is how this appar-ently works in detail.

For the female, her promiscuous mating has created paternity uncer-tainty. In addition to her spouse, other males in the village will be aware (intuitively, not necessarily consciously) that they may have fathered one or another of her offspring. The female can call on these males in the future to contribute extra parental investment for these youngsters. They are more likely to respond to this call when those youngsters might be their close kin (genetic offspring).† Thus, because of her promiscuous mating, the female has a more diversified set of assets, claims she can make if her mate dies prematurely.

Moreover, by creating paternity uncertainty, the female increases the incentive for all these possible fathers to cooperate with one another to provision her children. Each such genetically self-interested male will have an incentive to coerce other males when he can (even those who cannot be a genetic father) to cooperate to provision his potential children. These behaviors all reinforce the fundamental human trick of

---

*   The phrase "in part" refers to the other requirements for contributing to larger non-kin cooperative enterprises of the village. In view of the coercive power of the larg-er village, these additional obligations can sometimes trump help to close kin depend-ing, on the sex of the adult and local circumstances.

†   Remember that natural selection "plays the odds." It produces proximate psy-chological devices that "recognize" the probability of the presence of specific pieces of personal design information in another. It thereby takes into account the probability of paternity just as surely as it factored in segregation of genes during gametogenesis (Chapter 3).

kinship-independent social cooperation, secondarily helping to create the cooperative human village (Chapters 5 and 6).

This network of possible fathers is the female's life insurance. It is very useful to think through the implications of this adaptive approach. If her mate dies prematurely, she may seek a new mate. However, in the harsh ancestral environment, any potential new mate will not want to contribute parental investment to her existing offspring by another male. Indeed, he may even kill these youngsters or insist that she kill them as a *sine qua non* to becoming her mate.* However, if she has been sexually active in the past with a new prospective mate, her offspring may actually be his. He will have a reduced incentive to see them killed and is much more likely to tolerate them.† Again, paternity uncertainty serves the female's design information well.[2]

Thus, our ancestral mothers purchased vital life insurance for their young—for our ancestors—by mating promiscuously.

What about our ancestral fathers? They face an analogous problem. What if they or their spouse dies prematurely? The male's premature death is partially "insured" by his spouse's promiscuity, for the reasons we just discussed. This is one reason that males are *relatively* agreeable to their spouse's promiscuity under high mortality risk conditions. However, this particular source of insurance comes at the risk that his spouse's offspring (thus insured) are not his to begin with.

Thus, he also needs other sources of insurance to offset this cost of his wife's purchase of insurance. By mating with females who are the spouses of other males, he acquires this additional insurance and, of course, these other females are agreeable to this for reasons of their own. Any offspring of other females that he fathers are born into other families. Perhaps both spouses of these other families will survive to rear these offspring to maturity.

Several details of this situation in these ancestral humans are highly enlightening. Adults of both sexes agree openly that promiscuous mating is a good idea. However, each individual (male or female) would prefer to have the option of mating promiscuously while requiring her/his spouse to be strictly (or mostly) monogamous—having one's cake

---

* Recall our infanticidal male lions from Chapter 3. The logic is precisely the same here.

† It is hard for us not to be ethically harsh in judging these infanticidal behaviors in our ancestors. However, we should remember the terrible struggles and mortal threats they faced. As we sit here, snug in the bosom of the modern state, we would do well to reserve judgment. Our ancestors were often strong and brave in ways we can only dimly comprehend. We would not be here without their strength and courage.

and eating it, too (for the specific reasons we will explore in detail in a moment). This creates potential conflicts of interest between spouses and beyond.

Of course, management of conflicts of interest (including ones like these) is apparently the fundamental human trick (Chapter 5). Self-interested adults will act coercively in concert with other adults sharing their interests with the effect that these conjoint interests are enforced.

In the discussion above, we spoke of "buying" life insurance. Each sex pays something in order to purchase this insurance. These costs create the conflict of interest problem that must be solved. These details will also help us in a moment to see why human villages tend to convert to more monogamous mating systems when adult mortality risks decline.

First, consider the costs of insurance to a typical ancestral male. He is mating with multiple females. To a casual contemporary observer, this might seem like a benefit to him, not a cost. But this view is naïve. Promiscuous mating imposes a significant cost on him—the cost of paternity uncertainty. The same females who are mating with him are also mating with other males. Indeed, while mating with many females, he may not have fathered any offspring anywhere in the ancestral village.

Even if he *has* fathered offspring somewhere in the village, he does not know which ones are his. Thus, he must distribute his parental help (paternal investment) among multiple possible offspring, rather than focusing them exclusively on offspring he is reasonably sure are his.

Again, this uncertainty and its consequences are the price each individual male pays to purchase life insurance in a high adult mortality risk environment.

Second, consider the costs to typical ancestral female. By mating promiscuously, she gains a call on the resources of multiple males, as we saw. However, these males will make complex self-interested *strategic* use of these resources. They will contribute some to parental investment in their possible offspring, as the female would wish them to do. However, recall that the males are in a competitive mating situation. They are mating with females who are also choosing to mate with other males. Males will use some of their resources *not* for parental investment, but for recruitment of additional mating opportunities. More bluntly, males will use some of these resources to pay to mate with the "sexiest" available females.*

---

* It is purely a matter of semantics, of course, as to whether we would call this pattern of ancestral behavior prostitution.

Who are these sexiest females? Pregnant or nursing females who cannot become pregnant?* No! Males will generally choose to purchase sex from females who are neither pregnant nor undergoing nutritionally stressful nursing. They will look for non-pregnant, but robustly well-fed females (as assessed, in part, by sexually selected fat deposits in female breasts and rear ends, for example). In other words, males will give a fraction of their resources to females who need them *least*—healthy, non-pregnant, non-nursing—rather than to females most in need of those resources.

This *sub-optimal* distribution of male resources is the price each individual female pays to purchase life insurance in a high adult mortality risk environment.

With these costs and interests in mind, consider the adults in a promiscuous mating environment. First, a female—what is her optimal strategy? To mate with other males in addition to her nominal spouse. But, at the same time, she would prefer that her own spouse mate only with her and focus all his resources on her offspring. If she were to insist on this arrangement, she would be a free rider on her spouse and on the larger promiscuous mating system, accepting the mating system's benefits while evading its costs. Why can she not get away with this behavior? Of course, her spouse will object and will probably mate promiscuously despite her wishes. Moreover, all the other females will object. She is "stealing" resources from their spouses while not allowing her own spouse to reciprocate. Other females will use their uniquely human access to inexpensive coercive threat to prevent her from pursuing this strategy.†

Now consider a male in an ancestral promiscuous mating environment. He would prefer to mate with other females, but to have his spouse mate only rarely with other males (to purchase life insurance), thereby increasing his paternity certainty above that of other males. Again, if he were to insist on such an arrangement he would be free

---

* Recall that nursing females in the nutritionally stressed ancestral environment probably did not ovulate. Thus, sex with nursing females was probably a relatively poor reproductive opportunity.

† Here is the cost/benefit logic of this coercion. The free riding female can be ostracized or killed if she does not comply with the rules of the promiscuous mating system. The costs of the necessary coercive threat are low for ancestral human females (Chapter 5). The benefit or payoff is the generation of a new unattached male who can contribute his resources to them and their future offspring. Coalitions of females exerting the necessary coercive threat will generally also be in the position to exclude any by-standing females who do not contribute to coercion from reaping the benefits of their enforcement actions.

riding on the promiscuous mating system. Of course, his spouse will object and will probably mate more promiscuously despite his wishes. Moreover, the other males will object. He is mating with their spouses while not allowing them to mate as often with his. They will use their uniquely human access to inexpensive coercive threat to prevent him from pursuing this strategy.*

Thus, we expect the relatively promiscuous ancestral human mating system to be sustained by self-interested, uniquely human, inexpensive coercion (Chapter 5).

Now we are prepared to understand the circumstances under which ancestral humans might have opted for a (relatively) monogamous mating system. What will happen to a promiscuous mating system if adult mortality risks become very low? Consider the logic for a typical ancestral female—again, this time in an environment with low adult mortality risks. If she and her spouse mated monogamously, he would be much more willing to invest all (or, at least, most) of his resources in her and their joint offspring   Moreover, he would be more likely to invest these resources at the times when she and her offspring needed them most—when she is pregnant or nursing. Further, he is likely to live long enough to continue to provide this optimal stream of resources until their children to grow to independence.

Under these new conditions, the cost/benefit ratio of promiscuous purchasing of life insurance is no longer good for our female and she "wishes" to cease to purchase it. (The process of change in mating system is evolutionary here. No individual conscious awareness of the logic is necessary, of course.) Other females will agree and will collaborate to ostracize (or kill) promiscuous females who threaten this mating strategy.

Conversely, her mate will also have an incentive to mate monogamously and focus his resources on his children. Notice that he is giving up most opportunities to mate with other females. Thus, he will have few offspring by other females. However, he can also be relatively confident that his spouse's offspring are his own. In fact, these two effects precisely offset one another on average. The number of extra offspring he fathers by his spouse will offset the reduced number of offspring he fathers by other females.[3]

---

*   The cost/benefit logic is precisely analogous to the female case. Ostracizing the free riding male will be inexpensive to the other males. This modest expense is offset by the resulting access to new mating opportunities with the former spouse of the ostracized (or dead) free riding male. By-standing males who do not contribute to this ostracism can generally be coercively excluded from its benefits.

However, the male still obtains a net gain. He now knows better where to focus his parental (paternal) investment. He can lavish his resources more exclusively at the periods of peak need by his spouse and his children, as we discussed above. As well, he will be able to adjust his efforts so that they are focused more exclusively on his known children rather than dispersing them among the much larger number of children he *might* have fathered under promiscuous mating, many of which will not be his biological offspring. Finally, again, both he and his wife have a good chance of surviving long enough for their better-provisioned children to grow to independence.

Again, promiscuity-based life insurance has become a bad investment for our male in a low adult mortality risk environment and he will "wish" to cease to purchase it. Most males will agree and will collaborate to ostracize (or kill) males who actively seduce (or rape) their mates, free riding on the monogamous mating systems.

Now we can assemble the larger picture of our evolved sexual desires and behaviors. We expect two million years of human history to have been characterized by many alternative periods of life in uncrowded, benevolent environments (low risk) and crowded, hostile environments (high risk). We all have inherited the machineries **both** for monogamy **and** for promiscuity with our strategy being dictated by our immediate circumstances.

Notice also that the preceding discussions of social coercive logic referred to overt behaviors known to spouses and others. Free riding that is discreet and undetected remains a potentially successful strategy for each sex. Correspondingly, each sex has evolved exquisitely sensitive psychological mechanisms of sexual jealousy in defense against this threat.

Thus, overall, we expect ourselves to be highly adapted to pursuing either monogamy or promiscuity as our publicly acknowledged mating strategies—again, depending on circumstances. Moreover, we expect individuals of each sex to be highly adapted to occasional strategic, discreet free riding in publicly monogamous mating systems.

This picture of the origins of human sexuality makes a number of readily testable predictions about our evolved sexual anatomy, physiology, and psychology. In the rest of this chapter, we will explore how well these predictions are fulfilled. Moreover, if we can restrain our tendency to react with ethical disapproval to people who behave differently than we do, we can grow to have a much deeper, more humane understanding of ourselves, of the opposite sex, and, indeed, of all other people.

Promiscuity and modernity

We are suggesting that humans are innately *both* monogamous and promiscuous, contingent on their immediate circumstances. One prediction of this hypothesis is that a few humans will seek to be actively promiscuous even in a monogamous environment. This prediction arises from the fact that there is inevitably individual variability, both genetic variation and variation in experience.[*]

Thus, we would expect a certain fraction of people to gravitate to promiscuous mating systems even in contemporary low-adult-mortality-risk environments like North America. There is evidence that this is, in fact, the case.[†]

However, aggressively promiscuous North Americans are a partially hidden minority. It is difficult for us to evaluate the significance of their behaviors. We can learn more by looking at cultures in which promiscuity is the publicly acknowledge majority mating system as in the following section.

Many fathers – the ethnography of promiscuity

One thing becomes immediately obvious when we move into the empirical record of the wider pan-human world beyond our modern states. Relatively faithful monogamy is clearly not the only human strategy.

The diverse aboriginal cultures of the Amazonian basin in South America are a magnificent place to start. Many of these cultures escaped substantive contact with modern states well into the 20th Century. Thus, they potentially approximate some elements of the ancient human condition. Moreover, they have retained these properties into an era when our state level science has partially outgrown its extreme sexual prudery in observing and reporting human behavior.[‡] These societies have given us a priceless gift of insight into our ancestral past.

---

[*]   Likewise, some individuals will choose monogamy even in a high-risk environment where promiscuity is the better risk management strategy.

[†]   See Gould's 2000 book *The Lifestyle* for a journalist's account of one such subculture. See Laumann, et al. (1994) for a vast technical survey of contemporary sexual behavior.

[‡]   The extreme public pretense of sexual prudery was one of the social tools of male domination of human sexuality in recent hierarchical societies (Chapter 13).

These Amazonian cultures practice reproductive behavior, which includes what ethnographers call *partible paternity*—the belief that an individual child can have multiple fathers consisting of all the men who mated with the mother around the time she became pregnant with that child. These breeding systems are generally highly promiscuous.

The numerous local cultures practicing partible paternity include examples speaking highly divergent languages and practicing very different subsistence strategies (hunting/gathering or swidden, slash-and-burn horticulture, for example). These details tell us something profoundly important. These partible paternity breeding systems are almost certainly ancient, existing for sufficiently long times that the slow, gradual processes of change in language and subsistence practice have produced dramatic cumulative consequences. The language divergence data indicate that some of these cultures have been separated from one another for (at least) thousands of years. This strongly suggests, in turn, that partible paternity breeding systems characterized one or more (possibly all) of the ancestral human populations that originally moved into this area as humans first spread into the Western Hemisphere (twelve thousand to twenty-five thousand years ago).[4]

Moreover, at the time of state-level contact with these peoples, they were extremely warlike, including raiding for females resulting in frequent death for adults of both sexes.[5] Thus, these cultures displayed the high adult mortality risks our theoretical picture predicts to be important.

The essential features of the Amazonian partible paternity breeding systems are extremely enlightening. The belief in partible paternity derives from the fact that many (often most) children are conceived by women who are actively mating with more than one man. These women usually have a husband and generally (though not always) mate with him more often than with other men. Thus, a woman's offspring are often fathered by her husband, but not always.*

Under these conditions, it is useful for us to introduce the term *social* in a particular way. When a couple is formally bonded in some local equivalent of marriage, we will say that they are *social spouses*. This term has no *a priori* implications regarding the couple's sexual behavior. They may be socially monogamous, but sexually promiscuous, for example. Likewise, when we say a male is a *social father* of a child, we are

---

* Pursuit of a mate in an otherwise promiscuous environment is probably part of individual risk management. Thus, social pair-bonding (text), rather than exclusively faithful monogamous pair-bonding, might represent the strategic "goal" of our intense "falling in love" and "attachment" behaviors.

not specifying whether he is or is not the *biological* (genetic) father of that offspring.

As paternity is always in doubt in a mammal like us, the distinction between social and biological *fathers* is always important.* Conversely, as maternity is never in doubt, the distinction between social and biological *mothers* becomes relevant only very rarely.

How should individual adults behave in a promiscuous mating environment?[†] First, as we discussed above, each adult male should mate with multiple females to increase the probability that he has offspring somewhere by his social spouse or otherwise. Subject to other limitations and tactical issues, he should mate with as many women as he can, as often as he can. Of course, the level of paternal investment required by the offspring of these other females may dictate a more limited level of promiscuity.

Second, females should be solicitous of promiscuous males (limited by tactical considerations) again, for the reasons we have already seen. Moreover, females should solicit resources from the males with whom they mate.

This is precisely the pattern we see in the partible paternity cultures of Amazonia. Indeed, the term *partible paternity* refers to the practice of naming co-fathers for most children born. These are fathers in addition to the social spouse of the mother. Co-fathers are generally males who have "contributed" to building the fetus by mating with the mother around the time of conception. Moreover, co-fathers are expected to invest time and resources to these children as they grow.

A number of Amazonian cultures practicing promiscuous mating have been studied by professional ethnographers and anthropologists.[6] The Canela are a well-know case investigated by William and Jean Crocker.[7] The Crockers describe some of Bill Crocker's earlier pioneering observations eloquently.[‡]

> [The] wild Boar day feast [occurring once every few years] was held in a
> hut on a garden clearing…small groups of young women disappear into the
> woods followed by groups of young men. An hour later these groups returned

---

* Prior to the very new invention of DNA sequence-based paternity testing.

† Of course, the rigorously correct question to ask is *What behaviors best served the interests of their design information in ancient ancestral promiscuous environments?* These are the behaviors we expect to be produced by proximate psychological mechanisms built by design information selected in this ancestral environment.

‡ From Crocker and Crocker, 1994, *The Canela: Bonding through Kinship, Ritual and Sex.*

separately…. *Such groups vanished into the bushes and emerged [repeatedly] all afternoon, some individuals joining them more than once….*

[Later] Canela assistants confirmed that the groups in the woods…were, indeed, having sex. Informants added that since each group has more men than women, some women satisfied several men. (p. 145)

On another, annual ceremonial occasion [] adolescent women and adult men assemble 500 yards outside the village along a road at about seven in the morning…. The Red man [a member of one of the moieties or "subgroups" making up the village] asked several Red women in low tones which men they favored. Then he…tapped each chosen Black [another moiety] man. At the same time, the Black messenger tap[ped] certain Red men so that they would know they were a Black woman's choice. These two male messengers…were asking women which men [other than their husbands]…they wanted to go hunting with.

Later, small groups of individuals ambled off into the bushes eventually to pair off with their partners for the day.

Once deep in the small woods scattered in the savanna, the man hunted and the woman waited under a tree. If she was sufficiently pleased with the game he brought her, she might give him sex. (pp. 145-146)

A Ceremonial Chief sing-dance day could occur many times during a year to install a singing chief, a town crier [or other dignitaries].

This installation…takes place in the plaza [in the center of the village], while the tribe spends most of the day performing sing-dances in separate male [groups]….[These groups] invite wives of men of [a different group] to come out from their houses and dance with them, interspersed in the male row, arms around shoulders.

Approximately 50 men collect about 10 women…. Then the men, perspiring and elated, take the [willing] women away for sex. (pp. 146-147).

A similar arrangement is set up on tribal work days, when the elders decide most of the men will perform services [for the village] together…. On such occasions, each…moiety files out swiftly to different sectors of the work area. Three to five women are assigned…to each moiety for the day. They flirt with the men while they work and chat with them while they eat…. Then the women… walk out about 30 yards in different directions from the moiety's central location to prepare concealed, comfortable nests in the low bushes in which to please a number of men sequentially. A man picks a woman he is not related to [and who is not his wife] and walks out to her. He swiftly completes the sex act and stands up. Then the next man, seeing the first one stand, walks out to the woman's nest.

This traditional placing of women so that men can go to them sequentially for sex also occurs at the end of a two-week hunting encampment away from the village. Such hunting trips serve to obtain enough meat so that the whole tribe can eat well during a 10-day festival without leaving the village. Again, the men are

> *divided into two large groups by…moiety, and the half dozen women assigned to each moiety are spouses of men in the other moiety. The women…have sex with the men…after the last sing-dance, during the night before their return to the village.*
>
> *This arrangement for sequential sex also takes place informally on any day the men work in their…moiety units close enough so that they can join each other to race back to the village in the mid-afternoon, each moiety carrying a log. While such races, preceded by group sequential sex, do not occur every day, they are a potential element of the daily cycle of events when the members of the tribe are living in their central village….These events are thus daily in nature, not ceremonial. (pp 147-148)*

> *A frequent cause of fun, when I was moving around with male companions, was the discovery of a sex tryst spot. The discoverer would gleefully summon his friends to the spot, and the group, thoroughly enthralled, debated what each mark in the sand indicated….*
>
> *I once found myself conversing with the men going to and from such an episode. The tryst had been informally arranged; it probably occurred spontaneously. (p. 149)*

Notice that these occasions of promiscuous mating represent a spectrum from relatively rare formal, ceremonial occasions at one extreme to informal, everyday occurrences at the other. According to the Crockers' informants, individual extramarital trysts were also a common occurrence, accepted as long as proper discretion was observed. Virtually every adult member of the Canela village had multiple extramarital lovers in practice.

The Crockers' descriptions are notable for another reason. The Canela report viewing sexual promiscuity as "sharing" and as part of the larger cooperative enterprise that the village represents. This looks like what our theory predicts.

The ethnographic record suggests that mating behavior of this general form was probably quite common—possibly even the rule—in *high mortality rate* pre-state human societies. There is considerable variation in how public the acknowledgement of promiscuity was in these situations. However, the underlying patterns are strikingly similar across every continent.[8]

At first glance, sexual behavior like that of the Canela and other pre-state societies might be somewhat ethically and esthetically shocking from our provincial perspective. However, it is vital to recall that these individuals were making the best choices they could in the interests of their children. High-risk environments impose different challenges than the relatively safe environments many of us occupy.

Promiscuous mating – the male hardware

Our theory apparently predicts that humans should be adapted to both promiscuity and monogamy—alternative strategies in different environments. Many of us are already quite familiar with our properties that look like they might be adaptations to monogamy. For example, we fall in love and experience intense sexual jealousy and parental love.[9]

Thus, it will be more illuminating to examine what is less familiar to many of us, the evidence that we are also adapted to promiscuous mating. A simple place to begin is with the basic anatomical hardware of human mating. In the case of males, this hardware is the penis and testes.

Two details are of primary interest.

First, testes size can tell us something about sperm competition. Larger testes produce more sperm, allowing larger numbers to be deposited in the female at ejaculation. If the female is mating with multiple individuals, those males who deposit larger numbers of sperm may be more likely to win the competition to fertilize the egg. Thus, the larger the testes, the more likely the male's ancestors were adapted to competing in a promiscuous mating environment.

The human testes are larger than those of some other apes, even those with much larger body sizes than ours.[10] Specifically, gorillas and orangutans have testis sizes around 30 grams and 35 grams with body sizes of about 75 kilograms (165 pounds) and 170 kilograms (375 pounds), respectively. Dominant individual males in these two species hold harems of females and, thus, probably do not experience much sperm competition from other males. In contrast, chimps engage in relatively promiscuous mating, with each female mating with multiple males. Chimp testis size is much larger, around 120 grams, even though these males mass only about 44 kilograms (approximately 100 pounds).

Humans are in between. Our body sizes are around the same size as orangutans and substantially smaller than gorillas (healthy human males usually mass around 160 to 190 pounds). Yet our testis size (around 48 grams) is larger.

One simple interpretation of these data is that ancestral males were exposed to some sperm competition, but not always at levels as high as chimps.[11]

Second is the shape of the human penis. Recall the frequent practice of sequential sex among the Canela. This is where females sometimes choose to mate with one male after another in rapid sequence. Consider how a male might gain a reproductive advantage in this intense sperm competition. One way is by removing as much of the sperm previously deposited by other males as he can before depositing his own.[12] The human penis is apparently very well designed to do this. It is shaped as a "squeegee" to remove fluid, including semen, from the vagina during the repeated thrusting of human mating before the male deposits his own at orgasm (Figure 8.1).

A detail of human mating is potentially enlightening here. Contrast our mating behavior with that of an herbivore (like a buffalo or a horse, say). In the herbivore case, the male mounts the female, inserts, ejaculates, and withdraws—all in a few seconds. In contrast, human mating events commonly involve a prolonged sequence of insertions and (partial) withdrawals before ejaculation.[13] We take this pattern for granted; however, from an herbivore's perspective, our mating is peculiar. Of course, if one of the purposes of ancestral human male mating behavior probably sometimes included the removal of sperm recently deposited by another male, this pattern is exactly what we should expect—"squeegee before depositing" (Figure 8.1).

As we said in earlier chapters, one of the pleasures of (and evidence for) scientific theories is when they begin to reveal the prosaic as unexpectedly illuminating. Our view of the evolution of human mating apparently accounts for some otherwise quirky (and deceptively mundane) features of our anatomy and behavior.

We take the odd shape of the human penis and the prolonged act of human mating for granted. We should not. They are probably telling us something important.

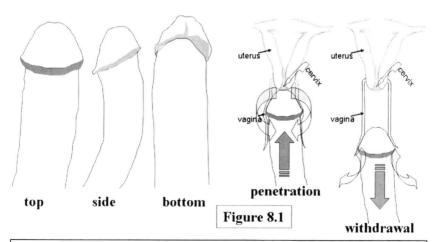

Figure 8.1: At LEFT are views of an erect human penis. Notice that the tip (head, glans) of the penis has an apparently arbitrary shape. However, when the penis is grasped, gasket-like, by the vaginal wall during mating (RIGHT), the in and out motion of mating allows the tip of the penis to systematically

remove fluids (including previously deposited sperm) from the vagina. The strong pressure of the powerful, deep **_penetration_** of the vagina by the erect penis with each fore-stroke forces fluid past the glans as indicated by the convergent curved gray arrows. This fluid is then removed from the vagina by the squeegee-like action of the glans during the **_withdrawal_** at each out stroke as indicated by the divergent curved gray arrows. During protracted intercourse, this process is expected to be very successful at removing sperm recently deposited by another male (text).

Promiscuous mating – the female hardware

It is difficult or impossible for others to tell when a human female is fertile (when she is ovulating). Indeed, most women do not even know themselves when they are ovulating. This pattern is referred to as hidden or *occult ovulation*.

As with penis shape, we take occult ovulation for granted—as natural and unremarkable. Precisely to the contrary, it is highly informative. For example, our closest living relative, the chimpanzee, has a very different pattern of ovulation. Chimp females produce an enormous, bright red swelling around their vaginal opening when they are ovulating. Thus, everyone knows—the opposite of occult.

We do not understand everything about the evolutionary ratio-nale of this chimp system. However, in general, the logic seems to be that a female chimp is mating with members of a set of male chimps who are all (likely to be) closely related to one another. Moreover, she is surrounded by (mostly) unrelated females. Her best option is to publicly mate with multiple males. The conflicts of interest between males that might otherwise cause male competition to explode into violence are moderated to some degree by male kinship. Moreover, by mating this way she advertises to other females that she has acquired a set of possible fathers who are likely to defend her offspring against infanti-cide from unrelated females.

How did humans get to be so utterly different than our closest living relatives? Our much more distant cousins, the monkeys, probably have something to teach us about this. In a classic body of work, Sarah Hrdy showed that langur monkey breeding systems are very different than chimp systems.[14] Indeed, they look much more like the lion systems we discussed in Chapter 3.

A single dominant langur male captures a small troop of females who are close kin to one another. He then attempts to mate with them exclusively, in spite of other males prowling around. He is constantly vigilant, even using violence to coerce his females to stay close to him so that he can monitor their behavior. Of course, just like the lions in Chapter 3, a new male taking over a set of females has an incentive to engage in strategic infanticide, killing youngsters fathered by his (non-kin) predecessor.

Hrdy has shown that langur males do, indeed, commit infanticide as expected. However, they do not do so inevitably as the langur females have developed counter strategies to reduce male infanticide. One of these counter strategies is of particular interest to us here. Even though the male attempts to guard his females, he cannot always do so. When he is not watching, females often slip away and actively solicit matings with surrounding males.

The logic of this behavior is apparently to obscure paternity (without the knowledge of the currently dominant male) so that if one of these surrounding males ultimately takes over the group in the future he will "know" that some of their offspring may be his and may forgo infanticide.

To make this work, langur females must hide when they are ovulating. Otherwise, the dominant male would vigilantly guard a female who happened to be ovulating at any particular moment, foreclosing her options. Ancestral langur females who hid their ovulation would leave more surviving offspring on average and design information producing hidden ovulation would do better.

In fact, the langur female reproductive system looks very much like the human system in many respects. The external female genitalia are simple and do not show any swelling or other advertising of ovulation.

Notice the problem that human females in the promiscuously mating ancestral human village and the langur females share. They are both strategically mating with males who are not close kin to each other—unlike chimp females who are mating with (potentially) **kin** males (brothers, half-brothers, and cousins). Though some details of the ultimate objectives of langur and human mating are different, they share a common requirement. The females have an urgent adaptive need to obscure paternity while not having non-kin male conflicts of interest explode into violence, putting the females and their offspring at risk of becoming collateral damage.

Occult ovulation allows the females to achieve this objective. We just saw this logic in langurs. It allows females to slip away discreetly. In ancestral human females, this same logic would have applied as it did in Canela-like personal trysting. An ancestral human female able to hide ovulation would be in a much better position to purchase life insurance and would leave more surviving offspring on average.

The punch line is that occult human ovulation is extremely suggestive about ancestral sexual practice. Another deceptively mundane detail of our sexuality is what we would predict if ancestral humans were adapted to the challenges of the promiscuously mating ancestral human village. We have inherited these adaptations along with those for monogamy.

The female orgasm – choosing fathers in the ancestral village?

Though there are dissenters to this view,[15] the female orgasm looks like an adaptation.[16] Most importantly from our point of view here, the female orgasm is not merely a subjective event that many women find pleasurable. It also involves a rather well coordinated set of physical events.[*]

These complex physical events apparently include a phenomenon sometimes referred to as *cervical tenting*.[17] This phenomenon has been well documented on videotape using cameras inserted in the vagina during female orgasm. The essential details are as follows. At intense orgasm, the muscles of the cervical region contract in a coordinated way causing the mouth of the cervix to dip repeatedly into the pool of fluids in the vagina including, of course, semen when this is present (see Figure 8.1 for the position of the cervix). Baker and Bellis (1994) plausibly argue that cervical tenting is likely to increase the number of sperm taken up from the vagina into the upper reproductive tract where they have a chance to fertilize an egg.[†]

The question then is what the female orgasm is designed to achieve. Subjective reward encouraging participation in an activity essential to

---

* This concordance between complex, coordinated physical events and subjective events is quite unexpected under the hypothesis that female orgasm is not an adaptation produced by natural selection as has been implausibly suggested by Lloyd (2005). We emphasize specifically that our theory now provides an adaptive rationale for the female orgasm that Lloyd lacked when writing her book.

† Sperm remaining in the vagina are ultimately discharged and have no chance of fertilizing an egg. See the online endnotes from this section for technical discussion of competing views that we consider unlikely to be correct.

reproduction is likely only part of the answer.* However, cervical tenting is something else. Cervical tenting is certainly *not* necessary to achieve fertilization. Non-orgasmic women still become pregnant efficiently, perhaps as a result of the sperm delivered directly to the cervix by the penis during ejaculation.[18] However, if Baker and Bellis are correct that cervical tenting increases the number of sperm taken up, tenting might have another, subtly different function.

Females mating promiscuously in the ancestral environment wished to mate with multiple males to purchase life insurance, as we have seen. However, they also have an incentive (unconsciously) to choose the fittest males to actually father their offspring. These would be males who are healthier, smarter, faster, stronger, etc. The ancestral female can have both these things to the extent that she can bias paternity in favor of a particular male while mating with many. It is certainly plausible that orgasm with cervical tenting would have this effect. She would (unconsciously) time her orgasm to take up larger numbers of sperm from desirable males.†

Again, an inherited property of our reproductive systems is apparently well explained as an adaptation to mating in the promiscuous ancestral village.

The Coolidge effect in men and women

A series of under-appreciated observations strongly supports the picture that mating systems similar to the Canela were an important part of our ancestral past. These are details of the human male (and, perhaps, female) sexual response.

The phenomenon in question is sometimes referred to as the *Coolidge effect*.[19] This name comes from an oft-told story about the 30th President of the United States, Calvin Coolidge. Bermant tells the story as follows and as recounted in Donald Symons' outstanding 1987 book *The Evolution of Human Sexuality*:

---

*   Though almost certainly correct at some level, this answer is ultimately unsatisfying because of our lack of understanding of the underlying logic of terms like *subjective* and *reward* in view of our profound ignorance about the conscious mind.

†   Of course, males would have developed counter-strategies to this female strategy in an attempt to increase their chances of paternity. These counter-strategies might include increasing sperm number directly deposited at the cervix. Active manipulation of female orgasm and monopolization of the female through prolonged lovemaking might also reflect this male goal.

> *One day the President and Mrs. Coolidge were visiting a government farm. Soon after their arrival, they were taken off on separate tours. When Mrs. Coolidge passed the chicken pens she paused to ask the man in charge if the rooster copulates more than once a day. "Dozens of times" was the reply. "Please tell that to the President," Mrs. Coolidge requested. When the President passed the pens and was told about the rooster, he asked, "Same hen every time?" "Oh no, Mr. President, a different one each time." The President nodded slowly, then said, "Tell that to Mrs. Coolidge". (p. 211)*

This funny little story communicates an important message. In species where females commonly mate with multiple males over a short period of time, the male sexual response is adapted accordingly. Such males mate with one receptive female and then become relatively sexually indifferent to that female for a substantial period of time. No further improvement in reproductive success can be achieved by more matings with this female. However, if a new female is presented, these same males immediately become sexually aroused.

What concerns us here is that humans show the Coolidge Effect. Human males generally find it difficult to become sexually aroused to orgasm with the same female partner more than once over a period of an hour or more. However, when the same male is presented with the opportunity to mate with more than one female, he can be repeatedly aroused to orgasm over a period as short as minutes or seconds—generally once with each available female if no other males are present.

The Coolidge Effect in human males has an interesting additional feature. When two males have sex with the same female, each male becomes immediately re-aroused as a result of the orgasm of the other male with the female. Under these conditions, these males can have repeated orgasms with the same female, again over a period as short as minutes. This cycle can be easily repeated multiple times over a period as short as an hour. Apparently, the only limitation on such repeated mating is generalized fatigue on the part of the males or, more commonly, mounting indifference on the part of the female. Again, note how different the outcome is here if one simply removes one of the two males.

Of course, this pattern in the male sexual response would be highly adaptive in a promiscuously mating ancestral village. Males who behaved and felt this way would leave more offspring, on average. We have apparently inherited the design information selected as a result.

Though the phenomenon is less well documented, there is some indication of a partially related psychological feature in human females. At least some women find one orgasm less than fully satisfying. Orgasm with a single male produces the desire for another orgasm. This desire is often difficult to fulfill in the contemporary monogamous environment. It is sometimes easier to fulfill in a mating system like the Canela village.

This adaptation to desire for multiple mates would likely be highly adaptive—purchasing life insurance, again—in ancestral females. Females whose sexual appetites worked this way would leave more surviving offspring, on average. We have apparently inherited the design information selected as a result.[*]

Menopause, menstruation, and human reproduction

Our theory predicts that we are the planet's first kinship-independent social breeder, sometimes mating monogamously, sometimes promiscuously, as indeed, we appear to be.

The next question to ask is whether this theory is potentially complete. Put differently, does the assumption of completeness force us to ignore anything important or to make strained, ad hoc arguments to shoehorn any of the details of human reproductive behavior into our theory? At this writing the answer to both these questions appears to be no.[†] It looks as if our new perspective on the evolution of human reproductive behavior is coherent and stronger than any other available theory. Additional examples give us greater confidence that this is, in fact, the case.[20]

A striking feature of human reproduction is menopause. In most mammals, females continue to ovulate and attempt to reproduce until shortly before death. In contrast, human females commonly cease ovulation while still physically healthy (typically around fifty years of

---

[*]   A significant fraction of women do not experience orgasm during intercourse (reviewed in Lloyd, 2005). This has been taken as evidence against an adaptive reproductive function for female orgasm. However, if our picture of evolved human sexual behavior is correct, contemporary women frequently find themselves mating under conditions (a single male in relative isolation) where orgasm in fact has no adaptive function. It would be of great interest to determine if women not orgasmic with a single male become orgasmic when having sex with more than one male.

[†]   In contrast, all other theories of the evolution of human sexuality that we are aware of appear to be seriously incomplete and/or internally inconsistent.

age). Women generally continue to live productive lives for twenty to thirty years after they cease ovulation.

In a landmark paper, George Williams (1957) pointed out that this uniquely human feature has all the properties of an evolved stage of life, very much like puberty, for example. He suggested that one evolutionary rational for this might be that an older ancestral female might have done better (for her design information) by assisting in the reproduction of close kin (mature children and grandchildren) than by attempting to produce new offspring herself. Kin-selection (Chapter 3) would allow the evolution of menopause, on this view.

The question then becomes why menopause should have evolved so (apparently) uniquely in humans on this hypothesis.[*] This question has been widely debated, with no fully satisfactory answer.[21] Our theory puts us in the position to attempt to do better. First, the persistence of the human village creates an expanded opportunity for older individuals to contribute to the survival and reproduction of kin young. Moreover, because maternity (unlike paternity) is always unambiguous, older adult females are in a unique position to exploit this opportunity. Given the serious dangers of uniquely difficult human childbirth (Chapter 6) and the length of time required to rear a human neonate to independence, shifting parental (maternal) investment from new offspring to grandchildren and nieces/nephews of younger siblings might make good adaptive sense.

Particularly important in this respect is the uniquely large role *culturally transmitted* information plays in the human adaptation. This prominence of cultural information is predicted by our theory (Chapter 10). Grandmothers as repositories and transmitters of this information to close kin might have a large impact.

Our theory also predicts a second adaptive contribution from menopause allowing us to account better for all its properties. Specifically, it is striking that many women remain sexually motivated and active long after inception of menopause. This is not accounted for on earlier interpretations. Our theory of the human village and its strategic promiscuity suggests a simple interpretation of this otherwise mysterious detail. In the ancestral environment, post-menopausal women may have mated with males for purposes of gaining resources that could be distributed to the female's kin.

---

[*] Some anthropologists have proposed what is called the *grandmother hypothesis* for both menopause and human uniqueness (see online Endnote 21). These earlier proposals are interesting, but they leave the most important question unanswered. Why do only humans (apparently) have non-reproductive grandmothers? Our theory suggests an answer to this question.

In effect, on this view, menopause allows manipulation of males. Post-menopausal sexually active women offer a deceptive reproductive opportunity to males who, in turn, will offer resources generated by uniquely human cooperation in return for that opportunity. But these menopausal women are able to do this without incurring the severe risks of late-life pregnancy and childbirth.

This *sexual hypothesis* for menopause might have some other interesting implications. For example, human menstrual flow is rather heavier than in the other apes. There are many possible interpretations of this property. However, if post-menopausal females are offering (deceptive) reproductive opportunities, it is in the individual interest of fertile females to advertise competitively the *authentic* reproductive opportunity each of them represents. Females who do this would probably leave more surviving offspring on average. In other words, expanding the volume of menstrual flow is a straightforward way in which natural selection might address the competitive, adaptive problem presented by sexy post-menopausal women.

Likewise, males will adapt to this deceptive manipulation.[*] For example, males are expected to become much more sensitive to signs of aging, with a strong preference for unambiguously young females. This ostensibly evolved taste in mates appears to characterize many contemporary males.

In summary, this arms race between individually self-interested females in the promiscuously mating ancestral human village might account well for some previously mysterious features of human reproductive physiology.

Finally, in the uniquely cooperative human village, post-menopausal females might cooperate to provide valuable sexual experience to one another's younger sons and grandsons without attendant risks of pregnancy.[†] In view of the complex strategies of the human village (above), such experience might be highly useful with younger females later. Older females appear to play this role in the Canella, for example.[22] This service represents something of value to be exchanged in the public

---

[*]   The frequency of pregnancy in the nutritionally challenging ancestral environment might have been sufficiently low that unambiguous detection of menopausal females by their inability to become pregnant might have required years of observation.

[†]   Young, inexperienced ancestral males were probably relatively poor sources of short-term paternal investment. Their economic skills (hunting, farming, and "manufacturing") were still relatively undeveloped. Thus, they would have been relatively undesirable as co-fathers and are expected to have had difficulty in finding mating opportunities. Thus, sexual training by older women at this early time in their development might have been a significant asset in a promiscuous ancestral village.

"market" of the ancestral village. Non-human animals would not engage in this kind of cooperation because it involves a conflict of interest—"you teach sex to my grandson, but I ignore yours." This capacity opened by menopause might also represent one of its adaptive applications.

The women and men of the modern state in the contemporary pan-global village – where we stand

Our theory makes a number of new predictions about the evolutionary origins of our sexual bodies and minds. This novel perspective is quite different than other currently popular theories of the evolution of human sexuality. Moreover, we argue that our approach does a more complete job of unifying what we know, as a correct theory should. It will be of great interest to see if future investigations can test the implications of our theory more extensively.[23]

As we said at the outset of this chapter, one of the best tests of our theory is how well it accounts for our own individual sexual feelings and behaviors. The intense power and complex ambivalence of our sexuality should be better explained by this approach than any other if this approach is, in fact, fundamentally correct. Each of us becomes a scientist again at this juncture. We can recall and review a lifetime of sexual feelings, experiences, and observations. Are they well explained?

This personal exploration of the evidence will be easier for those readers who have grown up in the mainstream of pluralistic democratic societies. Those of us who grow up in hierarchical, male-dominated communities will sometimes have artificial, externally imposed beliefs about natural human sexuality standing between us and insight. But all of us share a common humanity, and all of us can ultimately come to understand and accept the true human condition.

Work in the area of human sexual behavior has long been difficult to carry out. Among other things, our narrow perspective and ideological/religious/political objections have systematically interfered with a coherent strategy to fund this work consistently. It would be of great, lasting value to all of humanity for a well-endowed private foundation (relatively immune from short-term political demagoguery) to take this work on over a period of decades under conditions allowing us to achieve the additional insight we need. Solid public health decision-making, as well as greater self-awareness and peace of mind for our descendants would be the fruit.

Finally, a new theoretical perspective on the evolution of human sexuality is an important element of a hopeful and exciting confrontation of the possibilities for the human future for many reasons. A few examples will illustrate this.

First is a more deeply humane understanding of one another, as we have already suggested. For example, even sympathetic sex researchers have traditionally been drawn to the conclusion that many features of our sexual anatomy, physiology, and behavior are products of selection for successful *hostile manipulation* of men by women and of women by men in the ancestral past.* In other words, the most fundamental truth about our sexuality can appear to be that it evolved in an environment of monogamous mating where the most successful women and men were those who best cheated on their mates.

Cheating spouses are certainly one element of the human evolutionary past. However, our new theoretical perspective clearly indicates that many devices (behaviors, anatomies) that can function to support cheating also allowed each sex to purchase life insurance *by mutual consent* in the ancestral village or in some high-risk contemporary environments. Indeed, some devices (like the Coolidge Effect) are probably best interpreted in precisely this way. On this view, our surprisingly complex sexual heritage is the product of a very human ancestral cooperative enterprise (the promiscuous, cooperative, uniquely human village), not merely a reflection of thousands of generations of "desperate housewives" and "caddy husbands."

Second is a practical, public health example. Our theory predicts that humans will react to increased adult mortality risks by increased promiscuity. This behavior was often a rational approach in the ancestral environment, as we saw. However, this pattern becomes a tragically ironic, deadly trap when the source of increased mortality is, itself, a sexually transmitted infection (STI). Life-threatening STIs and our ancestral sexual behavior patterns create a devastating death spiral—a fatal positive feedback loop. We mate more promiscuously in the face of increased adult mortality, but our increased promiscuity drives further increases in adult mortality. For example, this new perspective on human sexual behavior may help us deal better with the AIDS pandemic in areas

---

* See an example in Robin Baker's powerfully evocative 1996 popular book *Sperm Wars*. Though Baker has sometimes been controversial, he also seems to be good spirited. Yet he concludes that our sexual anatomy and behavior is heavily shaped by the evolved goals of illicit manipulation. In spite of this limited perspective, we recommend this book highly. It strongly broadens our perspective on human sexuality.

where it continues to be virulent (as in some regions of sub-Saharan Africa, for example).

Third, an abiding scientific challenge has been to understand how (or even if) some of the varieties of human sexual feelings and practice might be products of our evolutionary past. For example, can homo-sexuality, bisexuality, and trans-sexuality be viewed as alternative repro-ductive strategies? It is beyond our scope here to explore these issues in detail. However, the new perspective on sexuality indicated by our theory might be useful to future investigations of these important issues. Two examples will illustrate what we mean. On the one hand, in a promiscuously mating environment, males (especially younger siblings) might have greater reproductive success by devoting their parental investment to the offspring of their maternal sisters. "Gay uncles" might have been among our more successful male ancestors. On the other hand, homosexual mating might well broaden into bi-sexual mating in a promiscuous sequential mating environment. It would be of great interest to explore the potential reproductive logic of this possibility in detail.

Finally, thinking about our evolved sexuality gives us greater insight into our common humanity. Comprehending our evolved minds, in all their rich complexity, opens a doorway to a new kind of mutual respect, tolerance, and affection. We will argue in later chapters that better understanding of yet other features of evolved minds will open this door wider still.

૭ᴖ9

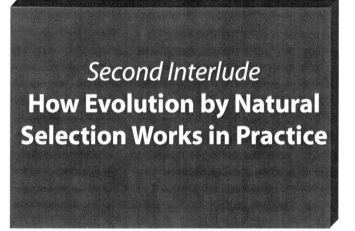

Second Interlude
How Evolution by Natural Selection Works in Practice

What do evolved devices look like?

When we examined the human fossil record (Chapter 7), we saw that devices produced by natural selection ended up being a "series of compromises flying (walking, throwing, etc.) in close formation." This property is unavoidable when natural selection "designs" a "multiuse vehicle" like an animal body.

We can look at the properties of evolved devices from a slightly different perspective. This new outlook will be useful in understanding the uniquely human trick of using language with elite skill (Chapter 9). Think of a bird wing. It has lift and control surfaces that allow these animals to achieve truly breath-taking virtuosity in flight.* Suppose our goal is to understand how a wing works and how it came to be designed. We might begin by looking at its structural elements, like the wing bones in the Figure below. However, if we set out to understand the design of these bones purely on the basis of the requirements of flight, we are fools indeed. The bird's wing bones were produced by evolutionary modification of limb bones originally designed to let the first land animals walk around on all fours. This ancient original limb bone set was subsequently redesigned many times over several hundred million years. It probably passed through modifications for use as forelimbs for

---

* See David Attenborough's magnificent 10-part series *The Life of Birds*. This astonishing videographic piece is available through PBS at http://www.pbs.org/lifeofbirds/.

the bipedal dinosaurs ancestral to birds, then through further redesign for proto-bird gliding and, ultimately, for powered bird flight.

At no point in this process did natural selection (metaphorically) stand back and say, "Back to the drawing board to start from scratch." *Natural selection always operates by going forward from what exists at each moment.* The process has been likened to redesigning a ship at sea that continues to function seamlessly throughout the redesign process. Thus, change will be incremental and "add-on"-dominated, not elegantly designed from the ground up.

In other words, biological devices may function powerfully, but they are always also quirky, even bizarre, historical accidents. For example, we humans have hair on our heads, not feathers. This is certainly *not* because hair is better than feathers, but rather because our ancestors were hairy mammals rather that feathery dinosaurs.

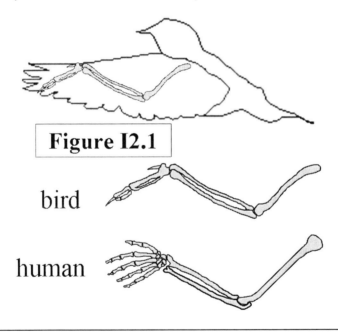

Figure I2.1

bird

human

Figure I2.1: TOP: The major skeletal elements of a bird wing are overlaid in their (approximate) positions.

BOTTOM: The skeletal elements of a bird wing and a human arm (resized to the same scale) illustrate the very extensive similarities (homologies) between these two limb structures—derived by natural selection from the forelimb of a common ground-walking, quadrupedal evolutionary ancestor.

Return to the wing in the figure. By comparing the bones of a bird wing with those of a human arm, we immediately see extensive similarity. By changing overall size, subtly altering proportions and fusing the "finger" bones into a solid wing-tip support, we can easily morph the human arm into the bird wing. This *is not* because elite throwing and playing Mozart have the same inherent requirements as powered flight.* Rather, it is because both the bird wing and the human arm arose through serial modification of the same ancestral starting material retaining many features of these ancient structures just as we have retained hair on our heads.

Once we understand this process, we are no longer hopelessly confused by trying to account for every detail of a bird's wing as a specific adaptation to powered flight. We can make authentic headway. Indeed, we can even learn some unexpected new things about the ancestral animal's forelimb in the process.

Applying this perspective to the evolution of language

As we look at what our theory predicts about the evolution of human language in Chapter 9, remember the bird wing. The detailed properties of human language will be strongly influenced by its communicative function, as 20th Century linguistics assumed. However, in contrast, many of these properties will also be very substantially determined by ancient, pre-human evolutionary history. Ancestral pieces, some serving communication, some not, were (metaphorically) cobbled together and reconfigured by natural selection to produce our communicative capabilities. Once we understand this design history of human language, linguistics becomes even more important to the human knowledge enterprise than we could have anticipated.

In the following chapter, we will start by looking at non-human animals, returning, as we must, to the *conflict of interest* problem and to the *confluent* interests of close kin (Chapter 3). Keeping these elements

---

*   If we had any doubt of this point, we can merely look at the wings of bats, insects, and pterosaurs to see three very different, independently evolved structural solutions to powered flight. Each solves the problem in a wildly different detailed way, though the bats and pterosaurs had the same set of ancestral animal "arm" bones. At this writing http://lpmpjogja.diknas.go.id/kc/b/bat/bat-wing.gif shows an excellent comparison of bird and bat wings. http://www.ucmp.berkeley.edu/vertebrates/flight/pterosaurwing. gif displays pterosaur wing bones very well. Google image searches for all these structures are productive.

at the center of our exploration, we will ask when non-human animals exchange information and when they do not. We will find that conflicts of interest appear to be the sole determining factor. Then we will explore how natural selection would be predicted to redeploy and hone ancestral animal communication skills to produce elite human language. This evolution should occur once the original elite-throwing, pre-human australiopith began to manage the conflict of interest problem in what became the uniquely human way.

We will argue that our approach gives us the first credible, useful theory of the causal processes producing one of our most astonishing properties, elite language. Our bodies and minds have been reshaped to enable the sharing of truly vast quantities of information.

‿

Chapter 9
Voices from the past:
The evolution of
'language'

A great irony of the human knowledge enterprise in the early 21st Century is that we can use our uniquely human language abilities to discuss and analyze Darwinian evolution; yet, we have not had a credible Darwinian theory of language evolution!

One of the many reasons to believe that our theory of human uniqueness is likely to be correct is that it gives us this long missing explanation for the origins of our astonishing ability to exchange information. Our goal in this chapter is to understand how our theory predicts all our uniquely human communication capacities, including not only "speech," literally, but other modes like sign and pragmatic demonstration, as well. Indeed, we will argue that language evolution can only be understood from this broader perspective. This quest will also yield some unexpected additional fruit.

*There is something else about our theory's perspective that will be important at every moment of our exploration. We tend to see language as a qualitative thing. Humans talk, non-humans do not. Indeed, much of professional linguistics is based on this implicit qualitative-difference assumption. In one sense, this extreme picture is right. Our abilities to exchange information are, indeed, very dramatically different than for any non-human animal. However, in a different sense, this extreme picture is a misleading conceit, we will argue. When humans exchange information and speak to one another, they are **not** doing something **qualitatively** different than communication in non-human animals. **Rather,** human communication merely reflects the dramatic **quantitative** improvement* in universal animal capabilities.* This novel refinement evolved to support the adaptive exchange of culturally transmitted information on a vast new scale in humans, as predicted by our theory.

A rough analogy we have used elsewhere will help deepen our grasp of this vital point.† Horses and dolphins are both mammals that can swim. Dolphins are **elite** swimmers, but horses are not. Humans communicate the way dolphins swim, with elite skill. In contrast, non-human animals communicate the way horses swim, with much more modest refinement.

For reasons that will soon emerge, it will serve us well to begin with communication in a "lowly" rodent:

A young "country" rat is born in the pine forests of the Middle East. She grows up feeding on the pine seeds that are the only abundant local food source. These seeds are held in hardened pinecones adapted to await a forest fire to open them and give the resulting seedlings access to sunlight in a newly burned-over spot. For the young rat to get at the seeds, she needs to know how to get into these hardened-target pinecones. She will not survive otherwise. This skill requires precisely executed stripping of the cone, a task requiring substantial amounts of detailed know-how. If we bring the country rat into the laboratory, she will feed quite happily on pinecones. Over generations of country rats, bred in the lab, youngsters grow up "stripping" pine cones.

In contrast, the "city" cousin of the country rat does not normally feed on pinecones and will not strip and eat them in the laboratory. This "non-stripping" pattern is likewise passed on through many lab generations.

If we put adult country and city rats in the same cage with pinecones, the country rats happily feed on them and the city rats never will. Indeed, the city rats must be supplemented with feed to keep them from starving. We humans are mystified. We think, "surely NOT," the city rats will simply copy the country rat feeding strategy. They do not.

Thus, at first glance, pinecone stripping looks like a genetically controlled behavior, with the country and city rat populations having recently evolved different feeding adaptations. However, Joseph Terkel, on whose elegant studies this story is based, went a step further.‡ He asked how the behavior was transmitted when the country and city rats were inter-bred. He found that the behavior was apparently **maternally inherited**. This statement means that if your mother was a pine cone stripper, you would strip and if

---

*   Grasping this perspective also improves our understanding of how much more we have in common with non-human animals than we are accustomed to believe.

†   Bingham (2008).

‡   See Terkel's chapter in Heyes and Galef (1996).

she was not, you would not grow up to strip cones, even if your father was a stripper.

While rare, special maternally-transmitted genes are known, these particular pieces of design information are poor candidates for control of sophisticated behaviors like pinecone stripping. Thus, Terkel asked whether pinecone stripping behavior might be **culturally** *transmitted from mother to offspring.*

He could ask this question because the kin-selected mechanism leading mothers to rear only their own close kin offspring in their natural environment could be subverted by humans in the artificial laboratory environment. The rat mother's system involves learning the unique sights, smells, and sounds of each of her youngsters over the first twenty-four hours or so after birth. If a human intervenes very shortly after the births of two litters, pups can be switched and the mothers will unknowingly raise the new pups as their own. When a mother is fooled in this way into raising non-kin youngsters, we call it **cross fostering***.*

When a country and a city rat mother give birth to a litter at the same time, half the pups from each can be switched. The pups then grow up in the presence of mothers either stripping or ignoring the pinecones in their cages. If pinecone stripping is a culturally transmitted behavior, all the pups raised by the country mom should strip (both those genetically hers and those "adopted") and none of the pups raised by the city mom should strip. This is exactly how the pups behave.

This beautiful result teaches us a crucial lesson. Adult rats—generally not close kin in the wild—have not evolved to learn from one another.[*] Thus, they do not copy each others' potentially useful behaviors as we humans would. In contrast, baby rats are adapted to acquire cultural information quickly and effectively, but only from their mothers.

As we will see, this behavior is a specific example of what appears to be a very general phenomenon among most mammals. Individuals follow a single, simple rule about cultural traditions: "Do not believe anybody but your mother." This fact clearly indicates that the conflict of interest problem is likely to be involved here. The only reliably kin mammals are mother and offspring. [Mammalian paternity is almost always in doubt (Chapters 3, 4, and 8).] Thus, the only individual in a youngster's environment whose interests are reliably (partially) confluent with its own is its mother.

This simple, yet profound lesson about conflicts of interest and communication will be a vital tool in understanding the evolution of human language.

---

[*]   As we will see in Chapter 10, minds are not generalized devices. Under most conditions, they can only do those things they are specifically, explicitly designed to do.

"It's the conflicts of interest, stupid"*

Biologists, linguists, and anthropologists have been flummoxed for generations by the problem of human language evolution. Our theory apparently puts us in a position to do much better than these previous attempts. The simple title above drives the crucial point home. We will make one simple argument. Adaptive communication is limited by only a single factor—our old friend, the conflict of interest problem. Moreover, these conflicts of interest are like all the others. They can be cheaply managed by an animal that can project threat from a distance (Chapter 5).

Thus, once our last pre-human ancestors evolved the capacity for elite aimed throwing (Chapter 7), the evolution of elite communication would have been inevitable and almost certainly very rapid. Human speech is merely one of a family of predictable, straightforward features of our adaptation to this novel opportunity.

We will begin by exploring the fundamental logic of human and non-human communication. If we take care and have patience with this exploration, it will reward our efforts handsomely. Then, at the end of the chapter, we will look at how our ancestors probably used their capacity for inexpensive coercion to police information exchange between potentially hostile non-kin conspecifics. Again, this new management of conflicts of interest was both *necessary* and *sufficient* for the evolution of uniquely human elite communication, including language, on our theory.

Rats, bees, mammals and herds – communication in non-human animals

Language is communication. Do non-human animals communicate? They certainly do. Terkel's pinecone-stripping rats are clearly receiving information from their mothers. Even if we do not know all the details, transmission of cultural information is not in doubt.

Moreover, information of this sort can certainly be highly adaptive. Yet an adult city rat stranded in the country would usually starve in the presence of plenty of pinecones to sustain her. Her own cultural heritage

---

* This section title paraphrases a famous American political slogan coined by James Carville for Bill Clinton's 1992 American presidential campaign: "It's the economy, stupid."

("garbage can lid-flipping," perhaps) would be very adaptive in the city, not so much in the country.

If culturally transmitted information is so valuable, why is it so narrowly transmitted in non-human animals? Baby rats learn well from their mothers, but grow into adults who are utterly "thick". They starve in the presence of other adult rats feeding happily on pine seeds. Many observations indicate that diverse non-human mammals apparently also transmit cultural behavioral information exclusively from mother to offspring while they fail to learn from other adults. These behaviors include hunting skills taught by cheetah mothers, fruit foraging skills by orangutan mothers, salmon fishing skills by grizzly mothers, and many other well known cases.[*,1] As we have already suggested, this odd transmission rule is exactly what we would predict if communication of culturally transmitted information is limited exclusively by a single factor—the conflict of interest problem.

Before exploring this argument more carefully, it is enlightening to evaluate alternatives to our claim. Perhaps there are also important limitations other than conflicts of interest constraining non-human animal transmission of cultural information.

It has been suggested that abstract symbolic capabilities are unique to humans and that the prior evolution of these capabilities was a *sine qua non* for language.[†] This ostensibly unique symbolic ability evolved as a consequence of our brain enlargement, perhaps. We will see in more detail later in the chapter that this hypothesis is extremely unlikely. One of many reasons to doubt the uniqueness of human *abstraction* is the existence of a tiny-brained non-human animal that uses overtly symbolic gesture to communicate information.

We refer to the well-known waggle dance of some species of honey bees.[‡,2] This waggle dance is a form of communication where bees arriving back at the home hive "tell" their sisters how to reach a rich flower patch they have discovered. Foraging bees measure the angle between the patch and the point of a line drawn vertically from the sun downward to the horizon (viewed from the perspective of the hive). Call this the "angle to goodies." Naturally built hives of most bees are vertical. The bees enter the hive and transform horizontal coordinates anchored to the sun into vertical coordinates anchored in the direction of gravitation pull, where up equals the direction of the sun—an overtly

---

*   See Heyes and Galef (1996); Avital and Jablonka (2000); and Whiten, et al. (1999).

†   See Deacon (1997) for a well-known recent exposition of this proposal.

‡   See Gould and Towne (1987) for a recent review.

symbolic transformation. The bee then dances along a line deflected from the vertical, the symbolic sun, by the angle to the goodies. Moreover, other details of the dance apparently contain symbolic information about distance to the flower patch and its quality. If the hive happens to be horizontal, the dance directly indicates the actual direction to the goodies.[3]

Symbolic communication is apparently not unique to humans.* In view of this, could it be that we are looking at human language entirely wrongly? Could non-human animals be communicating just as promiscuously as humans do and we have just missed it? Call this the *Dr. Doolittle hypothesis*. Famously, for example, many animals herd, flock, or school together. Moreover, these herds often behave in substantial synchrony, sometimes moving as if they are a single "super-organism." Clearly, their members, mostly non-kin adults, are exchanging information. However, this exchange is apparently much more limited in magnitude than typical human communication.

Specifically, information exchange here apparently reflects a feature of *selfish herding*.[4] Each member of the herd/flock/school is present for self-interested reasons that have no direct connection to the interests of the others in the herd. For example, each member of the herd will seek to use the other non-kin members as predator shields whenever possible. A well-known case is *stotting* in gazelles, where individuals bounce dramatically across the landscape when predators approach rather than merely running away. They are saying to the predator "Don't bother. I'm too fast and fit. Take the other guy."[†,5]

Of course, in the process of all this self-interested behavior, each member of the herd will generate public information as an unavoidable by-product. For example, darting to avoid a predator is an obvious, if unintended, signal to other conspecifics. In this case, others in the herd are expected to be adapted to pirate this information in their own interests.

Selfish herding apparently teaches us a vital new lesson about communication between non-kin. When an individual produces

---

*   We should not be surprised by this conclusion. Animal minds are symbolic devices. For example, "recognizing" an object does not involve any tiny pictures of the object in the brain, but rather abstract (possibly digital) representations of the object. Moreover, abstract animal minds are manifestly capable of interpreting the behavior of other animals, during mating, say. Thus, it would be astonishing, indeed, if this abstract computational capacity were not harnessed to symbolic exchange of information under the right circumstances.

†   Search "gazelle stotting" on Google Images for some dramatic pictures of this behavior.

information that is not likely to be faked, this information will be attended to by non-kin others if it is adaptively valuable.* Again, the crucial point is that the information is unlikely to be faked. In contrast, information that might or might not be true such as, "this is a good source of food, eat it" is fakable or *contingent*. Such contingent information almost never is exchanged between non-kin non-human adults.† In contrast, non-contingent, non-fakable information is exchanged. It is actually "plucked from the ether" in pursuit of self-interest all the time.

Thus, the problem with non-human animal information exchange is apparently *not* the inherent ability of animals to understand the implications of the behavior of other animals. Rather, it is the conflict of interest problem. Contingent, fakable information can be used to manipulate and mislead. It can be used for *hostile manipulation*. For example, recommending a food source that is actually toxic is an excellent way to eliminate a non-kin conspecific competitor.

Hostile manipulation of non-kin competitors would be a useful adaptive competitive strategy. Thus, we expect target animals to have evolved to be highly resistant to such manipulation. Non-human animals will systematically ignore potentially hostile (contingent) information. Remember our thick adult rats. Of course, if conspecifics are ignoring possible hostile manipulation, there will be little selection to produce false signals. Thus, conspecific silence—where non-kin individuals mostly ignore one another—is expected and observed.

This picture accounts remarkably well for what we see in non-human animal communication. Mom is the only reliably kin individual in your environment if you are a mammal. Therefore, learning your mother's cultural traditions is likely to be adaptive and is expected. Likewise, the honeybee waggle dance is expected. The waggle dancing bees in a hive consist of relatively close kin (Chapter 3). Any residual conflicts of interest are suppressed by worker bee policing (Chapter 5).

One last example is illuminating. When sexual adults mate, the process generally involves dramatically choreographed cooperation between two non-kin conspecific adults. This cooperation is so extensive and important that the animals are completely redesigned (geni-

---

* It is also required that this kind of information be produced often enough that animals are shaped by natural selection to recognize it. Also, notice that stotting involves a lot of energy and at least a modest risk of injury (fatal leg fractures from stepping in an animal hole, for example). Thus, stotting is unlikely to be undertaken purely to manipulate other conspecifics. In other words, stotting is unlikely to be a "fake" signal.

† Contingent information is exchanged between *kin* animals with *confluent* interests as we have already seen in young mammals above.

talia, minds, etc.) to support it. Indeed, elaborate coded informational signals are generally exchanged between potential mates. Some of this information is non-fakable, non-contingent information; however, some of it is contingent, like the production of mating pheromones by female moths to attract males, for example.[6]

This pattern of information exchange is precisely what our picture predicts. Once each member of a pair has decided to mate—perhaps after exchange of some non-fakable, "quality-indicating" information— there are generally no remaining conflicts of interest, for the moment. Each individual either mates and successfully reproduces or no mating occurs and neither reproduces. There is no free rider strategy available at this instant. Thus, this example is non-kin cooperation relatively unobstructed by intra-species conflicts of interest.

In overview, we argue that the large body of empirical evidence from the study of non-human animal communication has a single simple, but absolutely vital punch line. The *sole* limitation on the evolution of exchange of information is the conflict of interest problem. It follows that an animal that evolves the capacity to manage the conflict of interest problem should evolve expanded information exchange between non-kin.

It is vital to be clear. Some individually adaptive strategy *must* exist for suppressing hostile manipulation *before* the systematic exchange of contingent information between non-kin conspecifics can evolve. One way to achieve this result would be to evolve the ability to cost-effectively ostracize those using information for hostile manipulation. Otherwise, non-kin exchange of contingent information can never evolve—not anywhere, not ever.

However, given the dramatic *potential* individual advantages of exchange of contingent information, a new adaptation to elite communication should emerge explosively on an evolutionary time scale *when the appropriate conditions exist*.

This is the simplest statement of our theory's account of the evolution of elite human information exchange, including language. Throughout the rest of the chapter we will explore and expand on this fundamental insight. What we will be doing is looking at human language and its evolution through the lens of our theory. We will claim that this view of language explains its properties, context, and evolution very comfortably and naturally.

In contrast, traditional language-first or language-is-one-of-a-kind hypotheses can account for only a few pieces of empirical evidence and do so only by making tortured *ad hoc* and/or *sui generis* assumptions.

A good theory must account simply, yet broadly for all features of the elite, uniquely human exchange of information, as ours apparently does.

In other words, what follows is an argument from parsimony (Introduction). It is not an analytical demonstration that our theory *must* be correct. Rather, we claim that our theory is very likely to be correct (much better than its current competitors) because of its simplicity and broad explanatory power. As we have said, this is the nature of most scientific arguments (Preface, Introduction).

Do as I do – and as I say and sign

The first essential point to grasp about elite human communication is that *speech*, per se, (literally talking) is just the tip of a much larger iceberg. Consider how we train neurosurgeons, for example. Do we take them into a classroom and say, "Close your eyes and *listen* very carefully to what I *say*"? Of course, we do no such thing. Rather, we begin with classroom and textbook work making extensive use of detailed diagrams, pictures, animations, films, and energetically gesticulating lecturers. Then we take them into an operating "theater" to observe a surgeon who may also speak to them, but not necessarily. We next let them assist at the operating table beside a skilled surgeon who will sometimes use speech to spice up or disambiguate the lesson, but whose movements and choices are usually the most important element of learning.*

Notice that we could teach these skills to a congenitally deaf, non-speaking human. Could we train a non-speaking chimp to do neurosurgery? Of course, we could not, even though these animals have significant abilities to transmit their own more modest cultural traditions. This, again, is our first crucial insight. Humans are not merely speakers. Rather, we are *elite communicators* with speech being merely one of several overlapping, partially redundant, sometimes synergistic modalities of communication. This is precisely what our theory predicts.†

---

* Notice that you can substitute almost any other profession here and get the same answer from professional athlete to experimental scientist, carpenter, pilot, plumber, etc. Even highly abstract professions, like mathematics, are taught with extensive use of demonstration calculation, graphic support, analogical gesturing, and so on.

† It has been proposed that the evolution of language might have been the first step in the evolution of human uniqueness—the *language-first hypothesis*. This hypothesis is deeply flawed for several reasons, including ignoring or miscomprehending the conflict of interest problem. The observation that humans have equally elite communicative skill in multiple other modalities, of course, argues empirically against such speech-first models.

This example carries even more insight. Suppose we *did* set out to teach a congenitally deaf non-speaker to do surgery. Demonstration would remain a major part of the process. In principle, demonstration, alone, could be sufficient for the entire task. However, demonstration alone would be inefficient. We would have to construct enough different teaching cases that the student could abstract the "rules" from her own guided personal experience. One of the important teaching roles of speech is to communicate abstractions like rules, best practices, principles, natural laws, and so on. In other words, speech can also relate indirect experiences; experiences individuals do not necessarily have for themselves.

How would our non-speaking student surgeon deal with this issue? She would most likely use manual sign language, probably American Sign Language (ASL) in the United States, for example. Humans use sign language with the same virtuosity as they use speech. This fact was not clear in the early 20th Century. However, the work in the linguistics of sign language pioneered by William Stokoe at Gallaudet University revealed what we now know to be true. Human sign language is just as sophisticated as our speech. Indeed, the properties of sign language are essentially identical to those of spoken language (below).[*,7] This remarkable insight is also crucial to us.

If you are unfamiliar with the power of symbolic sign language, you can quickly grasp how uniquely good humans are at it. The lover's quarrel scene in the film version of the play *Children of a Lesser God* is a powerful lesson.[†] The actors are Marlie Matlin and William Hurt. What makes this fictionalized event scientifically useful is that Matlin does not merely *play* a congenitally deaf signer, she is one in real life. Thus, when we watch her sign we learn several amazing things. First, we see that Matlin can transmit information, even abstract conceptual information, through sign as rapidly and vividly as a hearing person could through speech. Second, and even more important, Matlin plays the emotionally charged, evocative scene just as compellingly and powerfully as an equally gifted speaking actor would. In other words, the fact that Matlin has never ever heard a spoken word in her life does not prevent her from being fully *human* in every possible sense of that word.

Humans are not fundamentally "talking" creatures. Rather, we are fundamentally *elite communicating* creatures in many modes. We are

---

*   See Maher's outstanding 1996 biography of Stokoe.

†   This scene runs from about 1hr 25min to 1hr 34min into the commercially available version of the film.

capable of maturing into fully human adults through the use of any or all of a diverse set of channels of communication.[8]

Now compare Matlin's performance to the sign language skills of non-human animals. The famous gorilla signer, Koko, speaks to us on a video tape commercially available from PBS and free in online video clips.[*] Two things are remarkable here. First, Koko can apparently use abstract symbolic sign to communicate.[†,9] This capability does not surprise us, but it is important to have this confirmation. Second, however, she does this with remarkably less skill and complexity than do humans. Koko might sign, "give nut." She will never sign, "Give me a nut or I'll call the SPCA and have their lawyers file a cruelty-to-animals suit against you."

Thus, humans and non-humans alike are capable of abstract symbolic communication through manual signing. But humans display an elite level of performance at signing that non-human animals lack.

One last remarkable piece of evidence argues that humans have been adaptively redesigned for elite exchange of information in many modalities—not just speech. Think of how we use our eyes to communicate. Our gaze can convey different types of information. In the correct context, a person's gaze can speak volumes about intent or desire. In fact, our eyes have apparently been modified to make this information channel more clear and salient.

Look in the mirror or into the eyes of a friend. Notice the stark contrast between the central pupils of your eyes and the surrounding "whites." The whites of human eyes are a novel design feature.[10] The eyes of most individuals among the great apes have black "whites."[‡] Again, looking in the mirror, imagine the whites of your eye turning black. As a result of this ape property, the difference between the iris/pupil and the surrounding eyeball is slight. The direction of gaze of a typical ape individual is hidden, occult. In contrast, the human gaze is overt and extremely conspicuous to others. An animal conveying information to

---

*   A video entitled *A Conversation with Koko* (available at http://www.pbs.org/wnet/nature/koko/ at this writing) illustrates the capacity, but distinctly "below-human" skill level, of gorillas for abstract symbolic sign language. A YouTube video of Koko and Fred Rogers (http://youtube.com/watch?v=Q5RrORtDZuQ) is also illuminating.

†   Koko's willingness and ability to communicate with her human handlers probably reflects an artificial transfer of kin-selected gorilla skills to an adaptively novel situation. This would be somewhat analogous to the cross-fostering result in the pine cone-feeding rat experiments.

‡   A Google image search of "ape eyes" will produce some excellent examples.

others as part of cooperative enterprise might find such a prominent gaze beneficial.

In summary, let us be perfectly clear about this vital part of our argument. As long recognized (though not always universally acknowledged) non-human animal communication reveals all the essential features inherent in human communication. What has never happened before humans is the evolutionary modification and redeployment of these universal animal features to enhance their performance dramatically, allowing communication to become a truly elite capability.

The challenge, then, is to recognize the circumstances under which selection for this particular elite redesign occurred for the first and only time in the history of our planet. Our theory predicts that this elite redesign would occur, *inevitably and rapidly*, in the first animal that could control conflicts of interest (Chapters 5 and 7). ***No other assumptions or processes are apparently necessary for natural selection to produce speech or, indeed, the entire panoply of elite human communicative abilities.***

If our larger theory is right, the resulting picture of elite human communication can now allow us to make rapid new progress. We can ask whether human language looks (and evolved) precisely as this picture predicts.

Why does human speech "look" as it does?

Though the theory of language evolution we are developing is simple and, we believe, compelling, a few linguists would object to some of our arguments—violently in some cases. We counter-argue that many of these objections derive from an ideological commitment to the one-of-a-kind status of human language, not from any weakness in our empirical evidence or theoretical arguments.[11] We can get a clearer picture of why human speech has the specific detailed properties it does and confront some of these specialist objections at the same time.

First, non-human apes can sign, but they apparently never generate a complex speech stream; they cannot literally "talk." Is this confirmation that the human capacity to generate a complex sound stream is unique? No, it is not. Some non-human animals can be trained to generate speech-like sound streams, though as with Koko's signing, not with human virtuosity. One famous example is the grey parrot, with Irene

Pepperberg's studies being the gold standard.[*,12] Likewise, dogs can also be taught to utter words and short phrases.[†]

Of course, these animals may or may not have much cognitive access to what they are saying, though Pepperberg's work is suggestive that sometimes they do. However, this is not the issue at the moment. What matters is that the capacity to generate passably useful speech sounds is certainly not uniquely human.

Second, speech has a highly combinatorial, hierarchically nested structure. Brief meaningless sounds (called *phonemes* by linguists) are combined to make words (*morphemes*, roughly) which generally have some specific symbolic meaning. Words are then combined into phrases, clauses, and sentences (by the rules of grammar) to create immediate, contingent, contextualized meaning and to express complex ideas beyond the power of individual words.[‡,13]

These three nested levels of the organization of language are given the technical names *phonology*, *morphology*, and *syntax*, respectively. The connection of these nested levels to meaning is called *semantics*. This use of combinatoriality to give the system its rich complexity (generatively) is sometimes referred to as *recursion* for esoteric historical reasons.

This hierarchical, combinatorial structure and its connection to meaning and intent have long been considered unique to human language by some professional linguists. We argue that this traditional view is merely a working assumption, not a manifest fact about the world. Moreover, this working assumption is increasingly untenable in light of our knowledge about minds, both human and non-human. Investi-

---

* At this writing, you can see Irene and Alex the grey parrot in brief clips at http://www.youtube.com/watch?v=XcLLk-r1aSs and
http://youtube.com/watch?v=WGiARReTwBw&feature=related. You can also see a grey parrot named Einstein speaking and mimicking sounds at http://www.youtube.com/watch?v=K7ht0a2-OnA&mode=related&search=. See Pepperberg's excellent 1999 book on her work with Alex.

† Search "talking dog" at http://www.youtube.com/. You will find many interesting examples.

‡ See Pinker (1994) and Jackendoff (2002). Notice that language comprehension and production are not simple *bottom-up* processes. When we articulate a vowel sound, for example, it is significantly different in almost every context. We understand phonemes, in part, because of their context—that is, as part of a larger whole. This element of parsing is sometimes called *top-down*. To see that your mind does top-down processing think about what it means that you are able to read the following sentence: "Evary larje wird in thes santence is speled incarectly."

gating a few of these properties of minds and language is extremely illuminating.

Begin with the fact that human sign language has an equivalent hierarchical, combinatorial structure to speech (see below). Moreover, sign uses these structural properties to achieve near-infinite *generatively*, again, just like speech. Thus, all the important *structural* features of language are not properties of speech alone, but are generic to human communication in general—possibly in all modes. This is important because it says that the use of physical and manual gesture (body movement) to convey meaning has the same underlying structure and cognitive requirements as speech.

As we have already seen, non-human animals can apparently use gesture symbolically. Moreover, there is very good reason to think that non-human animals understand (*parse* in linguist's vernacular) gesture in the same hierarchically nested way that humans do. One reason to believe this is that the waggle dance is inherently combinatorial, as we have seen. Direction can be varied independently of distance indicators, for example.

However, there is another, more powerful argument. Begin with the insight that humans do not merely interpret speech and symbolic sign in a hierarchically nested, combinatorial fashion. We apparently understand the entire world this way. This insight, so relevant to language, was developed in detail by Irving Biederman in his classic 1987 paper.[14] Figure 9.1 helps us grasp Biederman's message. In the top 1/3 of the figure we see the elementary shapes of a cylinder, a crescent, and a cube. These elementary shapes Biederman calls *geons*. The first part of the argument is that geons are to parsing the perceptual world, what phonemes are to parsing speech/sign. They have little or no inherent meaning, but can be assembled into composite structures that do have meaning.

To see this, consider the middle 1/3 of Figure 9.1. The elementary geon shapes are combined. But now we do *not* see "crescent stuck to cylinder," for example. Instead we see "cup" or "mug." We unconsciously combine geons into meaningful objects. This level of geon combinatoriality is simply the perceptual equivalent of morphology in language.

Finally, look at the bottom 1/3 of Figure 9.1. Here the geons at the top, with necessary stretching and shrinking of individual dimensions, are combined to make simple objects which are combined, in turn, to create an "scene," perhaps of a coffee cup and a brief case on a table. This is the perceptual equivalent of syntax, combining simple objects (morphemes/simple meaningful objects) into complex, contingent local meaning (a sentence/a scene).

This clearly suggests, as Beiderman points out, that hierarchical combinatorial organization is just the structure of perception; that is, *all* perception, not merely language comprehension.

Of course, language requires both perception (comprehension) and generation. There is equally good reason to believe that human body movements, including not only symbolic gesture and speech articulation by the vocal apparatus, but also throwing, walking, and all the rest, are organized hierarchically.[15] Thus, the hierarchical structure of language *generation* likely also simply reflects the inherent structure of all actions or behaviors, not merely communicative ones.

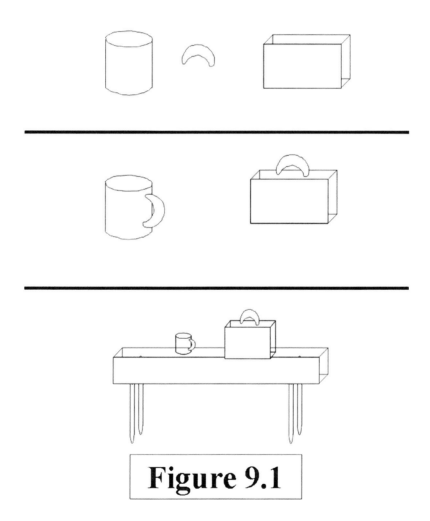

Figure 9.1

Moreover, there is no reason to believe that human movement and perception are qualitatively different from the equivalent non-human processes. Thus, human and non-human animals apparently generate gestures (vocal or manual) that have inherently hierarchical, combinatorial structure.[16] Moreover, non-human perceptual systems are probably designed to parse these gestures (and everything else about the world) in the same way we do (below).

On this view, the structure of language is merely the structure of the mind—all minds, including non-human animal minds. In humans, these universal features are merely refined for quantitatively improved information exchange. This refinement, once again, is the result of an evolutionary process we predict to begin swiftly once the conflict of interest problem is managed.

Could we still attempt to rescue the "language is one-of-a-kind" picture in view of this insight? In the attempt, we might argue that the entire human mind has been radically reorganized under the influence of evolved language and that our assumptions about how non-human animal minds perceive and control movement are wrong. Biederman's observations cannot exclude this possibility. However, there are several good reasons to reject this view.

One reason is parsimony. It would be startling, indeed, if things as fundamental as perception and motor control were completely redesigned in humans rather than being carried over and conserved from our pre-human ancestors.

A second reason to reject the view that human minds are fundamentally unlike non-human minds at the most basic level is more direct. We humans can *generalize* perception. Suppose you were to step off an airplane in a part of the world you had never seen before—the Australian Outback, for example. You stand before a species of tree you have never seen before. You do not say, "Oh my, what is that?" Instead, you immediately recognize the object as a tree, even if it is a very exotic tree.

This capacity to generalize is almost certainly due to our unconscious decomposition of objects and scenes into geons and then reconstructing, parsing, what we are seeing. Our ability to react evocatively to cartoon characters, consisting of nothing more than a few elementary geons, clearly also supports this view of our perception, for example. Thus, a tree or a human being consists of a predictable set of cylinders (torso, limbs, digits, trunk, branches) and spheres (head, fruit) arrayed in particular relationships and with particular proportions.

If this picture of perceptual generalization is correct, we can use it to ask if non-human animals perceive as humans do. In fact, even pigeons—not the "rocket scientists" of the bird world—can generalize across perceptual categories with ease.[17] This strongly suggests that human and animal minds do these things similarly. All minds perceive by parsing hierarchically and combinatorially.

All of these insights have a single, simple underlying message. Human language looks just as it should on our theory, by all criteria that we know how to apply. Human language looks like it results from the combination and refinement of previously existing non-human animal capabilities—processes produced by natural selection in ancestral humans for one purpose. That purpose is more efficient exchange of larger amounts of information. *All that would be required to produce such an evolved refinement is a change in selection so that sharing of more information between many more individuals became adaptively valuable to individuals (to their design information).*

As we have said, our theory predicts that precisely this circumstance occurred with the initial evolution of *Homo*. The evolution of remote killing capability allowed the subsequent evolution of control of conflicts of interest including the coercive suppression of hostile manipulation. This new capacity, in turn, eliminated the *sole* impediment to the adaptive sharing of contingent information between non-kin conspecifics. The rest was simply conventional evolution by natural selection.

Talking minds

Now that we have a credible theory of language evolution, we can approach the analysis of language with the same confidence as we might undertake in the analysis of a novel animal forelimb, for example. Just as with our bird wing in the Second Interlude preceding this chapter, we can eventually expect to sort out the huge contributions of historical accident (ancestry) and the probably modest additions of new design to our communicative abilities.

Again, we are arguing that the complex structure of language revealed by linguistics is merely the structure of the ancestral animal mind, redeployed to allow enhanced information exchange in a new animal who can manage the conflict of interest problem.[20]

This perspective is also very important from another point of view. We do not understand what kind of a device a mind is. No one does.* We still cannot draw a formal functional diagram of the mind as a material device—not even a "toy" simplification containing the essential attributes of such a device. We do not know whether it is

---

* Nevertheless, our theory gives us some important new insights to some areas of the mind problem, as we will see in Chapter 10.

fish or fowl or how much of it might be a computer-like executor of algorithms or a *Darwin machine*, for example.[18] We do not even know how to define words like *conscious*, *meaning*, or *intent* in a literal, mechanistic sense. They are useful operational and subjective terms only.

Thus, the proposal that the structure of language might be extensively determined, or even completely dominated by the ancient structure of the universal animal mind [just as the bones of the bird wing reflect the structure of the ancient ancestral land animal leg (Second Interlude)] would mean that language could be a direct pipeline into the heart of the mind problem. The assembly of meaningless small units (phonemes) into meaningful larger units (morphemes and sentences) almost certainly holds rich clues to how a meaningless meat machine can be assembled to create a "meaning-full" mind. This is a much bigger payoff than merely understanding the nature of language as if it were a novel evolutionary widget added on to a mind of otherwise unknown functional organization. Correctly understanding language as an evolved device can pull linguistics center-stage in the future of the human knowledge enterprise.

For readers who are comfortable with the picture we have built so far, skip to the last two sections of the chapter for a discussion of the evidence that language evolved in the human lineage over the last two million years as our theory predicts. These sections also contain useful new details about the evolutionary logic of this process. The two sections immediately below flesh out our discussion of several issues for the benefit of curious readers and any still-skeptical professional linguists.

Animal minds

Think again about our Middle Eastern pine cone-feeding rats. The newborns are designed to acquire their feeding skills from their mothers, but not as adults from other, non-kin adults. Though we do not know exactly how these rats learn this skill, we know that many other mammalian youngsters actively mimic or imitate their mothers as part of the same evolved capacity to acquire information.

Laboratory animals are notoriously bad at human-designed imitation tasks.[19] This has been taken to mean that imitation is hard. We suggest otherwise. Rather, non-human animals may not imitate well in the highly artificial social situations human experimenters produce because of their evolved indifference to most exogenous input.[20] This behavior is a consequence of the conflict of interest problem, as we said.

In fact, under the right conditions, animals are able to learn by imitation quite well. For example, mammalian calves will sometimes mimic their mothers with remarkable enthusiasm.[21]

Even if this perspective is correct, we might nevertheless think that animal imitation is unrelated to human language. Modeling (by mom) and imitation (by the youngster) does not seem much like a speech exchange at first glance. Imitation might be *iconic* and direct while language is abstract and *symbolic*, for example.* We argue precisely to the contrary. Both communicative acts are fundamentally identical, differing only in second-order details.

Consider what imitation involves. The animal is not seeing itself in a 360° mirror. It is watching another animal move its body, then transforming that information into a form that allows it to control its own body similarly, including the parts of its body it cannot see. In fact, the imitator almost always needs to understand the intent or purpose of a behavior to imitate it reliably.[22] So imitation probably involves the communication of intent.

Still, is not this communication of intent non-symbolic, distinct from human language? We argue that, on the contrary, it *is* symbolic. Specifically, we suggest that the imitating animal is decomposing the demonstrating animal's body into components (geons), monitoring the relative movements of key components, then inferring intent. Moreover, this inference is combinatorial, just as language is.

A pet dog makes a good example. When you pick up his favorite fetching stick, he might run immediately to the door, inferring your intent and preparing for a romp in the yard or the park. Now suppose, instead, that you pick up a rolled up newspaper for purposes of whacking the dog across the rump, perhaps in training him not to urinate on the carpet. He might run and hide under the bed rather than scampering to the front door.

If you visualize these two scenes, in each case you are part of the scene. Neither the fetching stick nor the newspaper will normally be salient to the dog unless it is in your hand. However, the significance of one cylindrical object (the stick) is very different than the second (the rolled up paper) in the context of your hand. This is just combinatoriality again. The perception of the cylinder is combined with the perception of your hand.

This combinatoriality is also hierarchically nested. Your picking up the fetching stick outside in the open communicates a different intent than when you are in the house. Likewise, the rolled up newspaper in your hand might provoke a different response outside where the

---

*   This is the view taken by many traditional linguists, in fact.

dog knows he can easily out run you. The perception of the cylinder is combined with the perception of your hand, which is nested within either the house environment or the outdoor environment.

The punch line is simple. We argue that animals are quite capable of inferring intent from the symbolic interpretation of the movements (gestures) of other animals. Moreover, both the generation and interpretation of these gestures apparently involves hierarchically nested levels of combinatoriality just like elite human communication, including speech.

As before, this is not a detailed analytical argument. Rather, it is a case for plausibility. We suggest that this view of animal communicative abilities is at least as reasonable and likely as the older views that insist on seeing human capabilities as qualitatively novel. Once again, it follows from this approach that human language is merely the enlargement and improvement of the ancestral capacity of non-human animals to interpret the behaviors of other animals as symbolic (abstract) indications of intent. The idea that non-human animal gesture might be the antecedent of human language is not new to us and we will have more to say about its development in a moment.

This view of the foundation of language in non-human behavior clarifies how elite communication (speech and/or sign) might evolve very rapidly under new selection in *any* animal. In fact, a new body of evidence suggests directly that this picture is realistic. There is a non-human animal that has also recently come under selection for improved communication.

This story begins with investigators in an animal behavior laboratory who wanted to find out if they could give away the location of food hidden in one of several boxes to a chimp by gazing or pointing at the correct box. While humans get the point immediately, the chimps do not.[23] This result is probably not because chimps are stupid. Rather, they evolved in a social situation in which such cues from other non-kin adults would never have been reliable and are, thus, simply ignored.[24]

This surprising discovery led to a series of related studies. The most interesting insight from our point of view is that one experimental animal *did* get the pointing and eye-gaze cues. The animal is the domesticated dog. This is salient, because wolves, the wild ancestors of dogs are less competent at this task.[25] Dogs evolved from wolves under intense selection by and around humans very recently (perhaps beginning as little as ten thousand years ago).[26] Among the skills that are adaptive for these human camp followers and collaborators is the ability to communicate with humans. If elite communication evolves rapidly from non-human capacities under the appropriate selection, dogs should show abilities just like these.[27]

Dogs share one last informative capacity with humans. We are able to acquire enormous numbers of vocabulary words as we learn a new language. One of the ways we do this is by what specialists call *fast mapping*. Here is an example. We are shown five objects. Four are familiar (cup, ball, plate, toothbrush, for example) while the fifth object is something we have never seen before (an exotic religious icon from a foreign culture, for example). The group of objects is indicated collectively and we are then asked to retrieve the "glink" (a non-word to us). We immediately, unconsciously assume that the unknown object is a glink without explicitly being instructed. Indeed, months later we will be able to associate the word glink with this object based on this single exposure.

Dogs can learn human words. They sometimes comprehend hundreds of different words. They can easily retrieve a ball on verbal command from our group of five objects. Much more interesting is that if you ask a dog with the right vocabulary to bring the glink he (or she) will draw the same inference that humans do. Moreover, the dog will remember what glink "means" for many weeks after this initial exposure. Thus, dogs also acquire new vocabulary words by *fast mapping*.[28] Yet again, the capacities supporting human language are apparently universal to non-human animals. Also, dogs may have rapidly improved their performance at fast mapping under brief human selection.

Remember that dogs also have the capacity to generate a modest speech stream (above). Perhaps, if we had the patience to carry out the right pattern of selection on dogs over the next several thousand years, our descendents and theirs might become conversational companions. It would be most interesting, indeed, to hear all about how the world looks to a dog.

Why do we talk? – Keeping secrets from the enemy and teaching our friends

One of the most conspicuous features of speech is *arbitrariness of sign*[29.] This concept simply is that a word for an object like "book" in English *means* the same thing as "kniha" in Czech, for example. Words have no *a priori* relationship to the objects they designate or represent. Equally striking is that radically different arbitrary signs are used in every different human language.

This aspect of language has other telling properties. Human languages diverge rapidly when two speaking populations are isolated from one another. Similar ancestral arbitrary signs are gradually changed into very different arbitrary signs.[30] Moreover, one contributing source of this divergence is active redesign by users to hide meaning from "outsiders." The common development of near-incomprehensible local dialects by marginalized populations (*out-groups*) within modern states is well known, for example.[*,31]

Collectively, these empirical observations strongly suggest that arbitrariness of sign is an adaptive strategy for encryption of meaning among members of a local social entity. It is not an *a priori* or essential property of language in its purely communicative role.

It is illuminating that our theory predicts precisely this property. Specifically, as we will see in later chapters, the scale of human social cooperation—how big our kinship-independent cooperative social units are—is highly contingent, in readily intelligible ways (Chapters 11-16).

At the moment, the relevant implication of this contingency is that ancestral humans would have evolved throughout our entire approximately two-million-year history in a social world consisting of a highly cooperative *in-group* surrounded by dangerous, potentially threatening *out-groups*. Our ancestors would have been adapted to communicate actively with their in-group while attempting to disguise their communication from out-group individuals. Members of each group would anticipate and fear hostile manipulation by the other.

Another way of grasping this vital point is that human kinship-independent social coalitions evolve to the maximal adaptive size under local circumstances (Chapters 11-16). At that point, each such group would behave toward other such groups on the landscape analogously to non-kin conspecific individuals in non-human animals—that is, as if different groups have unmanageable conflicts of interest with one another (Chapter 3).

Ancient ancestral human cooperative coalitions would almost always have been sufficiently small that the interfaces between potentially hostile coalitions would have been frequently encountered by most individuals many times over a lifetime. Thus, our language faculty is expected to evolve to allow us rapidly to develop "unbreakable" codes for communication with in-group members in the presence of hostile out-group members. This, of course, is just arbitrariness of sign.

In fact, we wish to go one step further. Specifically, we propose that speech originally evolved (at least in part) as one of several alternative modes with the *specific, dedicated* purpose of carrying on *encrypted*

---

*   North American readers will be aware of *Ebonics*, the African American dialect of English, for example. Google "ebonics." See http://www.cal.org/topics/dialects/aae.html at this writing.

communication.[32] Sound, of course, lends itself to encryption because of the absence of obvious connection to physical body actions and objects—unlike sign. On this view, speech evolved not because it is "better" than sign for communication, but because sound signals are more readily, naturally encrypted than manual gestural signals. On this view, two million years later we are all "code talkers."[33]

Speech almost certainly has other properties—beyond encryption—to recommend it over sign. A likely candidate is facilitating teaching by allowing hands to be fully engaged in "showing" while the voice does the "talking," for example. A conservative model might be that encryption was one of several adaptive goals producing speech.

Fossil voices and minds – the paleoanthropological record of human communication

As we have seen, our theory makes strong, specific predictions about the evolution of elite human communication, including language. We expect these elite refinements to begin evolving very rapidly after the evolution of elite human throwing allowing the solution of the conflict of interest problem (Chapters 5 and 7). This is a very different evolutionary history than predicted by many anthropologists and linguists. Moreover, it represents a clear and continuing opportunity to falsify our theory—to break it on the facts of the empirical record.

The fossil record is a rich arena for possible falsification. There are three ways in which we can readily use the fossil record in this pursuit. First, we can focus narrowly on the modality of speech itself. Strictly speaking, our theory does not require that uniquely human speech (as opposed to sign and other modalities) begins evolving *immediately* at the birth of *Homo* around 1.8 million years ago; however, we strongly suspect that this will prove to be the case. Moreover, our theory *unambiguously* predicts that elite, fully human speech evolved long before the last approximately hundred thousand years—contrary to one currently popular view.[34]

To evaluate these predictions, we look at the human vocal tract. This part of our anatomy carries overt modifications dedicated specifically to speech. Figure 9.2 illustrates the structure of the human vocal tract and how it differs from the other contemporary great apes and, by inference, from our pre-human ancestors.[35] The human throat has been radically altered to modify the vibrating air columns between the vocal chords

and the lips allowing the complex, subtly modulated human vocalizations that are so different than the chimp's pant-hoot.*

Unfortunately, most of this apparatus consists of soft tissues that
essentially never fossilize. However, one small bone, *the hyoid*, resides
right in the middle of all this redesign (Figure 9.2C). The hyoid forms
attachments for a number of the small muscles whose masterful manipulation is required to recite the Gettysburg address. We certainly expect
this bone to be modified in an elite speaking animal compared to a
chimp and it appears to be (Figure 9.2C).

This is an evolutionary change we can hope to score in the fossil
record. However, the hyoid is small and delicate, a bit like the wishbone
of a chicken. It often fails to be preserved in fossil hominids. For example,
hyoids were not recovered with the Lucy or the Nariokotome erectus
fossil hominid individuals (Chapter 7). Nevertheless, we do have a few
hyoids from other sources and they are most informative.

Neandertals are a local human population that occupied much of
southern Eurasia from about three hundred thousand years ago until
they were apparently replaced by our immediate ancestors as these
behaviorally modern humans exploded out of Africa around forty
to sixty thousand years ago (Chapter 11). Judging from sequences of
mitochondrial DNA extracted from fossil Neandertals, compared to our
DNA sequences, we last had a common ancestor with Neandertals around
500,000 years ago. In spite of this long divergence time, *Neandertal hyoids
are essentially identical to ours* and very different from a chimp hyoid (Figure
9.2C). This clearly suggests that Neandertals had fully articulate speech in
our view,[36] though this remains controversial for some.[37] Fortunately, there
are many additional pieces of fossil evidence giving us a broader picture.

The Neandertal evidence gives us one likely time point. Humans probably spoke by at least 500,000 years ago. Our theory also requires that
speech does *not* evolve before the first humans roughly two million years
ago (Chapter 7). On our theory, speech could not evolve *before* skeletal
redesign for elite human throwing and, thus, advanced control of conflicts
of interest. A new australopith fossil, the *Dikika child*, supports this prediction. This fossil is about 3.3 million years old and has an ape-like hyoid.[38]

Thus, as our theory predicts, hominids apparently did not have articulate speech *until* the emergence of animals that could control conflicts
of interest, early *Homo*. But then they evolved speech rapidly so that
five-hundred-thousand-year-old hominids talked.

---

* Notice that this redesign famously makes humans much more vulnerable to choking on their food than chimps are. This clearly suggests that the redesign to speak was
adaptive—one of a "series of compromises (talking) in close formation."

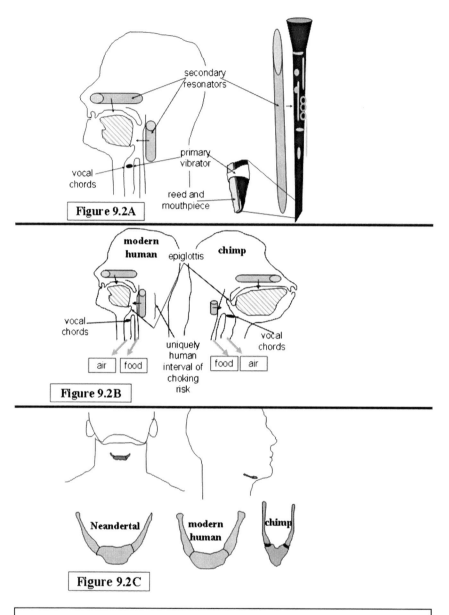

Figure 9.2A

Figure 9.2B

Figure 9.2C

Figure 9.2: Key features of the human speech apparatus are illustrated.

PANEL A: The physics of sound generation by the human speech apparatus is illustrated by analogy with the clarinet. A primary vibrator (vocal chords or mouthpiece with reed) emits a set of vibrations whose frequency (higher or lower pitches) can be partially controlled. These vibrations stimulate secondary vibrations in the connected air columns (secondary resonators; air columns in the throat and mouth or the barrel of the clarinet). These vibrations in the secondary resonators are the primary sounds we hear. These secondary vibrations can be extensively controlled by manipulating the shape and length of the vibrating air columns (with throat musculature or the keys on the clarinet). Humans modulate this speech apparatus with unconscious virtuosity.

PANEL B: The human and chimp vocal tracts are compared. Notice that the epiglottis (the flap of tissue that covers the opening to the lungs when food is swallowed) is at the back of the mouth in chimps but has migrated substantially down the neck in humans (to enlarge one of the secondary resonators for speech). This creates an extended path segment in the throat where swallowed food can block the airways in humans. This redesign, thus, imposes a risk of choking (a cost) on humans that must be offset by a corresponding benefit, presumably speech.

PANEL C: The shape of the modern human hyoid bone is shown at bottom center. The position of this bone in the throat is illustrated at the top of the panel. The chimp and human hyoids are very different. The Neandertal hyoid strongly resembles the modern human bone and is not at all like the chimp hyoid. The black disc segments of the chimp hyoid are cartilage that often fails to calcify. In contrast, the human and Neandertal hyoids are a single fused bone.

Let us be absolutely clear. Our theory predicts that we will see hyoid redesign for speech beginning very soon after the emergence of *Homo*. Future hyoids finds from *Homo* individuals living between 0.5 to 1.8 million years ago will be of the very greatest interest.

Second, we can focus more generally on elite communication in all modalities rather than simply speech. This allows us to approach this problem in two other ways, each leading to a very similar conclusion. One of these approaches is the brain system that apparently recognizes intent from movement (including gesture). Using measurements of electrical activity, we can ask when individual neurons in the brain are active and when they are quiet. By doing extensive measurements of many individual neurons in the brains of macaques (monkeys) who can be trained to undertake complex, voluntary movement, it is possible to ask which neurons let the animal execute a "purposeful" movement—picking up a jelly bean, for example.

Many neurons participate in different parts of this action—*motor neurons* that control muscle action directly, for example. However, there is a subclass of neurons that are of special interest. These are called *mirror neurons* and they fire up *both* when the monkey executes the movement *and* when the monkey observes a human experimenter executing that same movement. These neurons look as if they are "recognizing" purposeful movement no matter who executes that movement. They can "understand" purpose in others.[39]

Further supporting this view are the following remarkable properties of these neurons. They remain *inactive* when the human experimenter pantomimes picking up the jellybean, but with no jellybean target for the behavior. Even more remarkably, the jellybean can be placed in the monkey's field of view, then obscured by a small screen. When the human experimenter reaches behind the screen to grasp the (invisible) jellybean the mirror neurons still fire. However, if the experimenter puts the screen in front of a vacant place (no jellybean) and reaches behind the card, the mirror neurons do not respond. Notice that both acts of reaching behind the card look exactly the same to the monkey. The only thing that matters for the firing of the mirror neuron is the monkey's prior knowledge that the jellybean is present in one case and not in the other.

Thus, again, this mirror system seems to "understand" not merely actions in view, but also the purpose or goal of those actions. Thereby they probably allow the animal to understand the intent of another animal from that other animal's goal-directed movements.

We are still a very long way from understanding how this knowledge might translate into our subjective experience of comprehension of communicative intent. However, these results are useful to us because mirror neurons having these properties are concentrated in specific regions of the monkey's brain, including the area homologous to one of the prominent language areas of the human brain, Broca's area (Figure 9.3).[40] Moreover, Broca's area in our closest relative, the chimp, is selectively activated when chimps engage in communicative gestures.[41]

Thus, it has been suggested that elite human communication arose first through modifying the mirror neuron system to enhance exchange of abstract symbolic gesture (including sign language). This machinery was then further modified to allow vocal gestures to be used for communication as well as the ostensibly original manual gestures.[*,42]

Beautifully consistent with this hypothesis is the observation that Broca's language area is involved in human sign language as well as in speech, apparently in very analogous ways in the two communication systems.[43]

Now we are in the position to return to the fossil record with fresh eyes. Let us begin with the good news. Broca's area is conspicuously enlarged, relative to the rest of the brain in humans. This enlargement almost certainly reflects, at least in part, the evolution of elite human communication.[†] Moreover, this enlargement is strongly asymmetric.

---

*   See Corballis (1991 and 2002) for excellent general discussions of these issues. See Hewes (1992) for a review of the early work.

†   Notice that the mirror system is probably also involved in elite human "showing," learning by demonstration and the like. Thus, Broca's area enlargement may well correlate with *all* modalities of elite human information exchange.

The left hemisphere region (Broca's area *sensu stricto*) is much larger than the corresponding right hemisphere region in most people.

As well, the brain pushes against the inside of the skull throughout development and life, leaving a detailed imprint on the inner surface of the skull. When a skull fossilizes, this imprint is often preserved and can be examined by coating the inner surface of the skull with liquid rubber-like materials, allowing the rubber to polymerize and retrieving this *endocast* by collapsing and reinflating the resulting *balloon.*[*]

Now we have to confront the bad news. While the surface of the brain contains enormous amounts of useful detail, this surface is covered with a tough leather-like coat (the *dura*) which tends to fuzz over this detail in endocasts limiting what we can see in fossil skulls.

Fortunately, Broca's area protrudes so conspicuously that we can recognize it in endocasts even through the dura. Thus, we can ask the following question, *When does the hominid brain show enlargement of Broca's area?* This question is probably the same as asking when human ancestors first began to evolve elite human communication. This answer is very clear. Australopiths *do not* show enlarged human-like Broca's areas, but *every Homo* skull examined to date *does* show this feature.[44] These include skulls from near the origin of *Homo* around 1.8 million years ago.

This is precisely what our theory predicts and requires. Elite communication apparently evolves very rapidly in the first animals that have acquired the capacity to control conflicts of interest. Along with the evidence of the hyoid, this represents additional strong support for our theory's decisive predictions about language.

Third and finally, we can now ask about the relationship between elite human communication and the evolution of our cognitive virtuosity. Our theory makes very clear and specific predictions about this relationship. We expect that elite communication (facilitating access to enlarged, cooperatively generated information streams) and brain expansion (facilitating use and management of that information) must evolve together. This is part of our answer to the *why did* question from Chapter 6, again. Thus, brain expansion itself is probably symptomatic of elite communication and it shows the expected chronology (Chapter 7). The strength of this argument will become clearer as we explore the evolution of our minds in Chapter 10.

---

[*]   See page 81 in Johanson and Edgar (1996) for some breathtaking images of endocasts. Google Image search of "endocast" produces several useful images including a high-resolution image at http://www.skullsunlimited.com/graphics/endocast-lg.jpg at this writing.

Before we move on to Chapter 10, it will serve us well to stop and reconsider elite communication and some of the adaptive problems it both solves and creates.

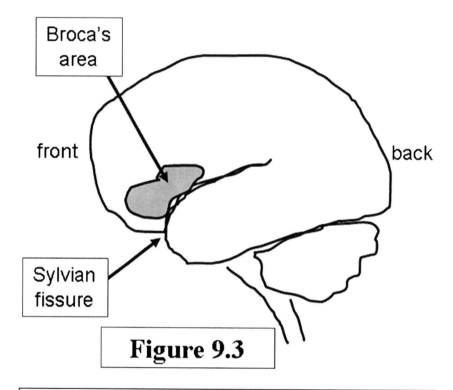

Broca's area

front

back

Sylvian fissure

Figure 9.3

Figure 9.3: Broca's area in the human brain is indicated in gray on the left hemisphere of the human brain. This area is a bulge flanked by a deep grove or fissure in the brain (the Sylvian fissure). As a result, the enlargement of Broca's area in the left hemisphere is easily recognized by its imprint on the inner surface of the skull, including in fossil skulls (text).

Liars and the lazy – the everyday problems with elite communication

Suppose natural selection produces a new kind of animal who can cost-effectively exclude cheaters and free riders from the conjoint social enterprises we call *cooperation* (Chapters 5 and 7). This new adaptive strategy will include control of the hostile manipulation problem, we have suggested. It is helpful to visualize the ancestral human village, consisting of multiple non-kin adults.[45] Throughout the following discussion of strategic logic, remember that we are talking about *ulti-*

mate causation. Proximate psychologies of individuals (how they feel or think) will cause them to behave *as if* they understood this ultimate, strategic logic though their conscious subjective experiences will usually be something very different (Chapters 3 and 10). For example, they might feel *moral outrage* at betrayal in the course of ostracizing free riders, but not consciously understand the cost-benefit structure of the coercive behaviors that result.

In Chapter 5 we discussed free riding using the example of a cooperative hunt where the free rider attempts to take meat from a carcass he did not help obtain. By recovering the free rider's ill-gotten gains, ostracizing cooperators are directly compensated for enforcing cooperation. In this chapter we are concerned with communication. Here the fruits of cooperation are new information, not something material.

Suppose an ancestral individual chooses to free ride on the process of acquiring this new information. In other words he/she decides to sit back and let others take the risks and spend the time/effort to accumulate useful facts and insights about the world. Further, suppose, for simplicity, that bringing in new useful information is the most important contribution each member of a local cooperative coalition performs.* Thus, the future usefulness of this "informational free rider" to others in the coalition is relatively low. Therefore, the free rider will consume future production without making a sufficient contribution to generating that product. Ostracism of this individual now would be a net gain to each remaining individual in the immediate future. In a remote-killing animal this ostracism could well be cost-effective (Chapter 5). Box 9.1 describes more completely how we can apply the logic of self-interested coercion to the problem of enforcing reliable exchange of contingent information.

Box 9.1: The specifics of enforcement of honest, productive communication

Ostracizing informational free riders will not evolve merely on the basis that excluding miscreants is cheap and, therefore, potentially cost effective. Even under these conditions, each individual will still have an incentive to let others ostracize the free rider while consuming the

---

* For example, a scouting party comes back with accumulated information about good sources of food. The actual hunt might be easy and productive once the village is "well-informed."

long-term future benefits of enforcing honest communication.*,[46] However, there is one other issue that makes ostracizing this free rider individually adaptive. Think about an adult living in an ancestral human village approximately 1.8 million years ago. This individual will own things. There are many examples of such property. For example, each adult might have tools, hides, food caches, and many other personal items of adaptive value. Further, adults have mates who can be of value as new mates to others. Finally, adults of either sex might have adolescent offspring who are sources of household labor. In other words, our free rider is a property owner with assets that can be seized. †

Now when the informational free rider is ostracized, his/her assets can be seized immediately, but only by those who actively participate in ostracizism. Given the relatively low costs of coercive violence in remote-killing early humans, this undertaking can be highly profitable (in the adaptive sense) in its own right. The future net returns of ostracizing this non-contributing free rider are gravy for everyone else in the coalition, whether they contributed to the ostracism or not.

Does this make non-contributors to the ostracism free riders still? No. Rather, those who DO contribute become *compensated law enforcement providers*. Remember, again, that they achieve a net immediate, individual gain from the act of ostracism itself. Of course, compensated professional law enforcement remains a central, vital part of all human social systems through the present moment.

Being lazy is not the only parasitic strategy. It is also possible to provide false or highly selected information designed to get the non-kin

---

*   Letting others do the ostracizing is, itself, actually a form of free riding. The original cheater is formally called a *first order* free rider. One who fails to ostracize is then called a *second order* free rider. Of course, second order free riders could also be ostracized, but one who fails to contribute to this undertaking becomes a *third order free rider* and so on. Second order free riders, third order free riders, and so on are called, collectively, *higher order free riders*. Because of the infinite regress of possible higher order free riding strategies, it is likely that ostracism to enforce participation in ostracism is rarely, if ever, a viable strategy.

†   Again, it is vital to grasp that this calculation of cost/benefit sounds a little inhuman and mercenary to us. That is because any such decisions to ostracize (ultimate causation) are actually mediated by an extensive and delicate functioning of our evolved ethical sense (proximate causation). A free rider is ostracized only after she/he has "crossed the line" and acted in ways that are clearly outrageous. Moreover, expulsion (or worse) is only undertaken on the basis of a broad consensus (generally democratic in the ancestral village).

targets of the information to behave in ways that selectively benefit the provider of that information (or her/his close kin). For simplicity, we can call such social actors *liars*. Of course, liars can be ostracized with the same self-interested logic as the lazy slackers above (Box 9.1).

Thus, in the ancestral village of remote-killing proto-humans, each individual faces many other individuals empowered and incentivized to ostracize liars and the lazy. Members of the ancestral human village would have been required to contribute their share of adaptive information if they were to remain active members of the local coalition.* Who will prosper in this environment? *Individuals who can convey information and who can comprehend and absorb that information will do better— much better.*

It is crucial to grasp this point. *Everything* we observe about elite human communication, including speech and *much* more, apparently emerges from it. What happens when an individual brings a valuable new piece of information back to the ancestral village? This new information might be a new insight into how the world works. Or it might be a hot piece of perishable tactical intelligence like the migratory passage of a prey herd, say. If she is successful in conveying this information, the adaptive cooperative performance of the village is increased and she and her close kin benefit. Moreover, she has paid her what-have-you-done-for-us-lately membership dues, another individual benefit. The better she is at generating a communicative stream (pantomime, sound, pictures drawn in the dirt, etc.), the better her adaptive prospects.

Now consider the others in the village. The better they are at comprehending her informational stream, the better they and their close kin will be served by the ongoing cooperative performance of the human village.

These are apparently the simplest first fundamentals. This pattern of selection will redesign ancestral animal capabilities to produce all of elite human communication (above). But there is more. As individuals bring information back to the village, different possible courses of cooperative action will be suggested by each. Moreover, those different courses of action will tend to serve the interests of different members of the village (and their close kin) somewhat differently.

This creates strong additional selection for communicative ability. An individual who is most skilled at conveying new proposals for action to the other members of the coalition, has the best chance of

---

* Of course, an individual might specialize in some other service to the coalition of sufficient value that others would be willing to specialize in scouting and science for the village, in return. However, this does not change the fundamental logic here.

influencing the coalition to pursue the cooperative course that best serves this individual's adaptive interests. Being articulate pays off.

Of course, under these conditions, it is in the interests of other members of the coalition to assimilate and evaluate all incoming information and proposals to best protect the adaptive interests of themselves and their close kin. This creates strong selection for rapid, accurate comprehension. Moreover, harnessing the mutual interests of others to assess accurately the implications of new information would also be adaptively wise. Evaluating information this way is just *public doubting*. This process remains the central functional part of the human knowledge enterprise to this day (Preface, Chapter 1, and 10). Being "smart" and publicly "skeptical" pays off.

Thus, the cooperative animal created and sustained by inexpensive coercion will impose and experience relentless selection for improved generation and comprehension of informational streams. This is just a specific case of natural selection, of course, nothing more. This pattern of selection would have shaped the capacities of the first humans just as surely and strongly as any other case of selection.

What is our picture in detail? A pre-human australopith is redesigned to throw with elite skill thus acquiring access to inexpensive coercive threat for the first time in Earth's history (Chapters 5 and 7). This animal can now afford to police non-kin information exchange. The innate pre-human communicative abilities of this animal will inevitably be refined and reconfigured rapidly over time by the intense new selection to generate and comprehend informational streams. This, and only this, is the story of the evolution of language on our theory.

Let us be perfectly clear. The reconfiguration and redeployment of ancestral animal capabilities to produce elite human speech or elite cheetah speed hunting are NOT different in any important, qualitative way. Each case merely reflects the adaptation of an animal to an idiosyncratic, species-typical way of life—elite exchange of culturally transmitted information or elite hunting of speedy medium-sized prey, respectively. If we fail to grasp this and, instead, insist on seeing human speech as qualitatively new, we are doomed to ineffectual phantom chasing, on our theory.

༄

Chapter 10
Wisdom of Crowds – implications of uniquely human information sharing

*Is it not reasonable to anticipate that our understanding
of the human mind would be aided greatly by knowing the purpose
for which it was designed?*

George C. Williams*

George William's characteristically no-nonsense sentence above is not merely one of the most trenchant statements ever made about human biology. It also turns out to be one of the most important sentences ever uttered for humanists and social scientists and about the ethics and politics of the human future for all of us. In this chapter we will begin to confront why this is so.

*Human minds **appear** to be qualitatively different and, vastly more powerful, than non-human animal minds. We write symphonies, build massive passenger aircraft, and leave footprints on the Moon. Non-human animals apparently have trouble accurately counting their fingers and toes. Though non-human animals do some amazing things in the narrow domains of their species-typical adaptations, they cannot compete with*

---

*   This quote is from p. 16 of George Williams' epoch-making 1966 book, *Adaptation and Natural Selection*. All subsequent thought about evolutionary biology, and about human evolution in particular (including our own), was profoundly influenced by this book. It is still a richly rewarding read.

*the worldwide dominance and spectacularly variegated expertise of the human species**.*

No theory of human uniqueness can be credible if it does not provide a front-of-the-tee-shirt explanation for our spectacular minds. Our theory gives us this essential, parsimonious explanation of human cognitive virtuosity, as required. The simplicity and transparency of this explanation is another reason to believe that our theory is likely to be fundamentally correct.

Before we begin there is some bad news. No one, including us, knows very much about precisely what kind of physical device a mind is. We have myriad juicy clues at scales ranging from molecules and cells to perceptions and behaviors, but no real global insight or overall grasp of mind.[1]

However, there is also some good news. We do not need to be aware of how a mind works as a proximate device to learn a great deal about it. We need only know what natural selection has shaped the mind to do.[2] *Minds do what they are adapted to do—nothing more and nothing less. They can be extremely powerful devices. Think again of the bird wing from the Second Interlude. Its exquisite deployment of lift and control surfaces allow a small hawk to pursue quarry through a dense forest at blinding speeds. By understanding what this wing is designed to do, we can predict many of its properties. The same is true of the hawk's mind. Understanding its adaptive purpose allows us to predict many of the mind's pragmatic features—the behaviors it will produce and the information it will perceive and process (somehow). The hawk's wing and body are deployed in species-typical high-speed aerobatics by its mind.*

Our theory makes a very specific prediction about what human minds are designed to do. They are, first and foremost, devices for navigating the unprecedented human social world. The world created by our unique deployment of coercion resulting in management of the conflict of interest problem (Chapters 5-9). Because ancestral humans could kill from a distance, they ceased to see conflicts of interest as mere facts about the world to be submitted to as non-human animals mostly do. Instead, ancestral humans "saw" conflicts of interest as opportunities to be engaged. These were challenges that were now manageable and adaptively malleable.[†]

---

*   See Gazzaniga (2008) for an excellent, readable review of our current understanding of the proximate devices underlying human minds. See Dennett (2005) and Metzinger (2009) for discussions of the subtle problem of the material basis of subjective awareness, an issue that will not concern us directly here.

†   We are speaking metaphorically, as we often do for economy, of course. Humans evolved to manage the conflict of interest problem unconsciously, not to consciously understand it.

This uniquely human perspective has profound implications for how the universe appears to our evolved minds. We see the world as we must if we are to monitor and control conflicts of interest, moment-to-moment, day-to-day. For example, every action by other people either serves our confluent interests with other members of our cooperative coalitions (good) or it interferes with those interests (evil). We display spectacular (generally unconscious) virtuosity at calculating these interests and responding coercively or cooperatively to them. We see the entire human world through this ethical filter—sometimes even projecting this perspective onto the larger extra-human universe.[*,†,3]

This characteristically uniquely human moralistic approach to the world is intimately connected to our elite human "smarts," our cognitive virtuosity, in a simple, direct way. Specifically, uniquely large cooperative human social coalitions are sustained by the coercive behaviors directed by our proximate ethical sense. Moreover, these enlarged social aggregates hold vastly greater amounts of culturally transmitted information than any non-human animal ever had access to. It is this culturally transmitted information that makes us smarter than other animals. Again, we possess all this information as a direct, transparent consequence of our unprecedented control of the conflict of interest problem (implemented by our ethical psychology).

Two million years of conventional Darwinian selection on the genetic repertoire of the human lineage has shaped our minds to have a rich, powerful connection to this uniquely human cultural informational heritage—utilizing it, managing it, adding to it, and transmitting it to those with confluent interests. This ancient, enormous cultural informational stream empowers us to "understand". As a result, we can manipulate the physical universe on an entirely new scale and with unprecedented command, as no non-human animal can.

---

*   This unique ethical slant of human minds has long been recognized empirically. See, for example, Waddington's 1960 book *The Ethical Animal,* Alexander's 1987 book *The Biology of Moral Systems,* Wright's 1994 book *The Moral Animal,* and Ridley's 1996 book *Origins of Virtue.* Our theory allows us to predict and understand this evolved property of our minds for the first time.

†   Both Thomas Huxley and George Williams famously argued that natural selection is an amoral (or actively evil) force against which we much struggle (Paradis and Williams, 1989). Our theory allows us to update this profoundly important argument very substantially. We will claim in this chapter that the human ethical sense we would use to struggle against the consequences of natural selection is, itself, a product of natural selection. However, we now understand the circumstances under which the evolved human ethical sense might be entrained to serve the interests of all humans (Chapter 17) arguably allowing "good" to triumph over "evil" in the Huxley/Williams sense of those terms. Also see Fourth and Fifth Interludes.

The bottom line – and some essential fine points

It is important to be very clear from the beginning. The species-typical properties of human minds consist of two domains on our theory. First, our non-human ancestral ethical and political psychology has been extensively refined, enlarged, and redirected to support our unprecedented, self-interested management of the conflict of interest problem.

Second, *as a result of our management of conflicts of interest*, we have access to stupendously larger amounts of culturally transmitted information than non-human animals. Various features of our ancestral animal minds have evolved to permit individually self-interested humans to capitalize on this informational heritage. Elite language capabilities (Chapter 9) and brain expansion (Chapter 6) are among the genetically supported adaptations to this vast river of cultural information.

It will also be vital that we think clearly about the differences (and similarities) between genetic design information and culturally transmitted design information. If we are to understand our theory's account of human history (below; Chapters 11-16), we must grasp the adaptive purposes that culturally transmitted information serves. There are (at least) three distinct functional subclasses within the uniquely enlarged human stream of culturally transmitted information.

First, culturally transmitted *know-how* lets us put footprints on the Moon and uncounted millions of other remarkable things. We will call this portion of our cultural heritage *pragmatic* information. Our ecological dominance and cognitive virtuosity results primarily from our remarkably facile acquisition, exchange, and deployment of pragmatic information.

Second, other portions of the cultural information stream will be *the rules of the game* for our uniquely human social cooperation. We will call this *social contract* information. It defines those behaviors that will provoke coercive ostracism and those that will elicit enforceable collaboration. Our ethical/political psychology is highly adapted to negotiating, enforcing, and responding to the details of our social contracts.[*]

---

[*]   Human genetic design information has certainly also become adapted to life under our unprecedented social contracts. Not surprisingly, the pragmatic, functional relationship between genetic and cultural design information in this domain appears to resemble the analogous relationship in the domain of human language (Hauser, 2006).

Third, for reasons that will emerge in later chapters, the *scale* of uniquely human social cooperation has been sharply limited throughout almost all of our two-million-year history. Our ancestors always lived within highly cooperative units (human villages) sustained by the self-interested management of the conflict of interest problem *among* their members. However, these relatively small ancient units had continuous, *unmanageable* conflicts of interest with members of neighboring units. We are also highly adapted to dealing with these chronic external conflicts.*

This adaptation to cooperating-to-compete creates cultural information streams dedicated to managing our interactions with individuals external to our cooperative coalitions. This portion of our culturally transmitted heritage includes *identifier* information. Identifier information is adapted to label members of our cooperating local group reliably, perforce also labeling "outsiders" who will be hostile. Identifier information will grow more prominent as human social units grow larger and we will elaborate on its properties in later chapters (Chapters 11-16).

One final reminder will serve us well. A new, clear understanding of our evolved orientation to the huge human cultural patrimony is vital if we are to have a science of human history (Chapters 11-16). Here we stand, two million years into this history, looking out onto our universe with a stupendously deep comprehension of its workings and the irresistible impulse to employ that comprehension for what each individual considers to be the "good." This evolved perspective forms the only foundation we can build on; but it is a vantage point offering realistic hopes of a democratized, humane, peaceful, fulfilling, pan-human future (Chapter 17). For now, we need to begin by understanding our minds.

Human genius – simple, transparent origins

A mind can only do the amazing things it does because it is a complex, orderly device. Moreover, minds do not get that way out of thin air. If we put a few hundred electronic components in a box and shake it, we do not get a functioning device. For example, our computers require that their components be arrayed in just exactly the right way and that the

---

* Note again that conflict *between* human coalitions is, itself, a cooperative endeavor fraught with conflicts of interest *among* the members of each competing group. Thus, adaptation to uniquely human scales of group conflict (war) must be an *effect* of the human management of conflicts of interest, not a *cause* of the evolution of our cooperation as sometimes erroneously proposed.

software controlling their moment-to-moment behavior be just right (bug-free). This takes reliable, tested design information—lots of it.

A mind is a device facing all these same engineering problems. Its capacity to manage the execution of complex "purposeful" tasks requires the input of huge amounts of reliable, tested information. Where does all this design information come from? For all animals (including humans) it apparently has three distinct sources—genetic information, individual experience, and culturally transmitted information. Each of these information sources has its own peculiar properties and limitations. Understanding our theory's predictions about each information source will bring the secret to uniquely human intelligence into sharp focus, we argue.

First, we have already discussed how genetic design information works (Chapters 1-3). The question then is whether natural selection acting on human genetic information could possibly be the *sole* or direct source of our cognitive virtuosity, our unique genius. Are human brains somehow genetically rewired so that they inherently work better than non-human animal brains? This *genetic hypothesis* for human intelligence is still widely accepted; however, we argue that it is naïve, inadequate, and unnecessary.

Human brains certainly have been genetically rewired. For example, they are enlarged (Chapter 6) and they can "talk" (Chapter 9). However, though our brains are uniquely large among the apes, some other animals have brains nearly as large as ours in proportion to their bodies.[*,4] Size alone is unlikely to be *sufficient* to produce cognitive virtuosity. Likewise, our elite communication simply gives us the capacity to exchange the information we already have. This capability is useful only as long as the information we possess is *reliable*. A fool might be just as articulate as a person of great skill and wisdom. Put differently, a larger brain, in isolation, speaking or not, is just a bigger box of disconnected components. It has to have additional *reliable* information to achieve genius.

We argue that qualitative changes in the genetically built human mind are relatively unimportant. They are analogous to turning a robin's mind into a hawk's mind. These changes alone simply produce a species-typical redeployment of prosaic, finite mental assets, not qualitative changes in fundamental kind or function. We are more interested in where we get all the *extra* reliable information needed to create our super minds.

There is another reason to think that genetic redesign is unlikely to be the central causal process behind human intelligence. Genetic design

---

[*]   Dolphins have brains proportionately nearly as large as ours, for example.

information has been under selection to produce the smartest possible individual animals for at least five hundred million years. In practice, the upper limit on this *purely genetic* approach to intelligence was almost certainly reached long before the first human evolved.[5] As expected on this view, the mouse and human genomes are actually strikingly similar.[6] We humans appear to have an off-the-shelf mammalian genome with only the usual, mundane species-typical evolutionary tinkering.[7]

The punch line is that human cognitive virtuosity is *supported* by some changes in genetic infrastructure, but almost certainly does *not derive* fundamentally from a new infusion of large amounts or new forms of genetic information. This is a profoundly important insight. We humans may rule the world *not* primarily because of our biological, genetic endowment, but as a result of some other source of information.

Second, all genetically built brains are apparently designed to learn from experience. However, learning in this way is risky and time-intensive. We can be maimed or killed while acquiring experience, as some of us are, particularly as children. Learning also involves *opportunity costs*—time away from foraging, feeding, sleeping, mating, etc. There are only twenty-four hours in a human day, just as in the day of a chimp or a forest hawk.

The nurturing safety of the human village probably allows human youngsters to survive a little more adventuresome exploration than non-human youngsters can. However, taking its high, irreducible costs into account, individual experience is probably not where most of the vast *extra* information comes from to turn our large brains (big boxes of disconnected components) into elite minds.

Third and finally is culturally transmitted design information. We know unambiguously that this kind of information is one source of the adaptive behavioral sophistication in many non-human animals. There is extensive observational evidence that our closest relatives, the other great apes, have culturally transmitted adaptive traditions.[8] As well, there is beautifully rigorous experimental evidence that even rodents have culturally transmitted adaptive traditions as we saw in Chapter 9.[9]

We argue that our elite *human* minds emerge from the massive expansion of this third stream of information. A vast enlargement of access to reliable, culturally transmitted information is the fundamental source of human cognitive virtuosity. This enlargement is transparently predictable on our theory.

A moment spent to bring this crucial prediction into sharp focus is very well invested. Culturally transmitted design information (cultural information, for short) contains some portion of the sum of all the individual experience of *all* the other animals, past and present, who exchange *reliable* information with us, directly and indirectly. Suppose you and we *each* go out and invest a little time and risk acquiring different new insights into the world. Then we exchange our insights perhaps using our elite human skills at information sharing (Chapter 9). We each "make a profit" by getting three insights for the price of one. Now multiply this trick by the hundreds of members of an ancestral human cooperative village—huge profit margins, indeed!

If this massive use of cultural information is such a powerful trick we must then ask why only humans employ it on a large scale. Indeed, this turns out to be the sixty-four thousand dollar question. The answer lies in the conflict of interest problem and its uniquely human solution.

To see the logic of this answer imagine we are members of an ancestral information-sharing human village. A potentially wonderful alternative strategy is open to us. We can laze under a tree while everyone else works away at gathering information. Then we consume their information without contributing any. We might even actively disseminate "false" information to manipulate the village to do what is in our interest, but not in theirs. Either way we are cheaters, free riders (Chapter 5).* Of course, if this parasitic strategy truly were available, those who used it would profit preferentially. As a result, the information-sharing village would rapidly dissolve, collapse; actually, it would never evolve in the first place.

To make the information-sharing village work or even for it to evolve at all, we must be able to ostracize liars, fools and the lazy, and do it inexpensively (cost-effectively) (Chapters 5 and 9). Thus, remote-killing ancestral humans are expected to be the first animals in Earth's history to be able to build a cultural information-rich social world on a vast new scale (Figure 10.1; Box 9.1; Chapter 9).†

For a non-human mammal, the rule is "don't believe anybody but your mother". She is the only reliably close kin individual in its world (Chapters 3 and 9). The non-human mammal's cultural matrimony passes mostly through the tiny channel of a single mind at each generation— mother to offspring. In stunning contrast, humans are the *pedagogical animal*. We learn from huge numbers of non-kin every moment of every day. We are the owners of a staggeringly massive cultural heritage.

---

*   Actually, most of us have difficulty clearly imagining this approach. Our evolved human ethical psychology makes it unattractive.

†   See BOX 9.1 in Chapter 9 for the detailed logic of humans enforcing "honest information sharing" by ostracizing "liars" and the lazy.

Thus, our unprecedented brilliance is apparently the result of a fundamentally new *social* process. A process supported by and helping to drive species-typical genetic redesign to be sure, but a social phenomenon, nonetheless.

It is hard for us to be objective about how profound and pervasive this human property is. A large part of our cultural heritage is absorbed pre-consciously, the way we learn our first language as very young toddlers. Thus, we "feel" that we are so much more than the cultural heritage we are consciously aware of acquiring (the ability to read or to drive a car, for example). We are, indeed, much more than these things. But this "more" is mostly the product of the portion of our cultural heritage we acquire before we can remember.[10]

One obvious implication of this picture is that the cooperative human village does not merely allow the redesign of our life history to permit uniquely human brain expansion (Chapter 6).* The village is *also* the source of all the extra cultural information that allows this enlarged brain to perform with such uniquely human virtuosity.†

Wise crowds, limited individuals

As we said, our massive cultural informational repertoire is apparently the fundamental source of our cognitive virtuosity. We also possess a related, powerful additional social capability. The capacity to aggregate opinion as a tool to insight.‡ When the members of a large collection of humans are all incentivized to find the best answer to a question, we can do amazingly well by pooling our perceptions. The key is to combine our insights in the right way (averaging them, for example, if we are guessing the number of jellybeans in a jar).

Well-aggregated human opinion can even out-perform experts in many cases. This uniquely human process requires control of the

---

*   This is the answer to our *why did* question about human brain enlargement in Chapter 6.

†   An enlarged brain, by itself, is useless, as we said. It is just a gas-guzzling (energy intensive) extravagance. In an adult, the brain is about 2 percent of body mass, but consumes about 20 percent of the calories we burn. In a young human child still growing its brain, this energy expense may approach 50 percent of calories burned. Only by providing extra adaptive benefit can such an energy-hogging device become a good evolutionary investment.

‡   Surowiecki's 2004 book *The Wisdom of Crowds* is an excellent, accessible introduction to work in this area.

conflict of interest problem, of course, or no such pooling would ever occur. It must be the case that the best available individually self-interested strategy for each member of the aggregate is to actively strive for the best possible *shared* answer—not to shirk or actively mislead. Once more, free riders must be expelled through the cost-effective deployment of coercive threat. The uniquely human capacity to organize and police this kind of information sharing is decisive to the reliability of our cultural information (Box 9.1; Chapter 9).

There is another important detail of this process of *aggregating insight*. Though voting for the number of jellybeans in a jar can be an effective way to get a very good answer, we can do even better. The route to doing better is public, cooperative *doubting*. Suppose you know our first averaged guess at the number of jellybeans is too low. Perhaps, your vantage point allows you to see that there are equal numbers of five colors of beans. Further, you can also count the beans of one color in a small part of the jar whose contribution to the total volume you can estimate. You review your calculation for the members of our group—you publicly doubt our estimate. Now everyone in the group evaluates your doubt and, if they find it cogent, they revise their estimates accordingly, doing their own counts and volume estimates. Our new *average estimate* might be significantly improved.

This little toy example illustrates the central role of public doubting in our unique trick of generating massive amounts of *reliable* cultural information. A bigger, multi-dimensional, long-term version of this process is where all of science, scholarship, and our cooperative knowing come from.

Public doubting is central to the uniquely human intellectual adaptation. It allows us not merely to share the insights we might come across alone, but to "wire together" our individual minds into a much more powerful *cooperative mind*. We will call this human trick the *human knowledge enterprise* henceforth. We are all engaged in this enterprise—as both proposers and doubters—almost every moment of our public lives.*

We said at the beginning of the book that the most fundamental tool of science is not technical hardware or even individual scientific minds. Rather, we said that the most fundamental tool was doubt (Intro

---

* Indeed, the institutions in which most of us work perform better or worse according to how well they actively enable this process.

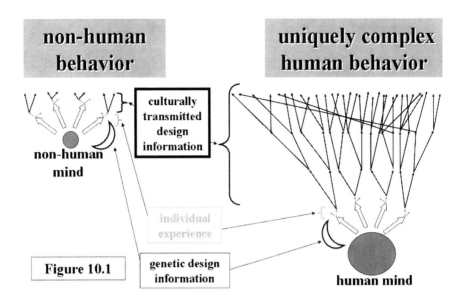

non-human behavior

uniquely complex human behavior

culturally transmitted design information

non-human mind

individual experience

Figure 10.1

genetic design information

human mind

Figure 10.1: Individual human minds are powerful because of the uniquely massive heritage of culturally transmitted design information made available to them through the human control of non-kin conflicts of interest (text and Chapter 9).

The grey circles represent **newborn** brains (one human, one non-human) and the arrows represent informational processes turning these into **adult** brains with species-typical behaviors. The source of the design information supporting these developmental processes can be genetic (box arrows), cultural (black arrows), or individual experience (grey arrows). To understand the power of the human adaptation to information sharing we must be aware of two things.

First, non-human mammals acquire their cultural heritages almost exclusively from their mothers, the one individual with whom they have reliably confluent interests (Chapter 3). Call the cross-sectional area of their cultural information stream *one unit*.

Second, an ancestral human acquired her/his cultural heritage from each of the hundreds of individuals making up the ancestral village, as a result of the capacity of the individuals making up the village to control the conflicts of interest between non-kin (text and Chapters 5 and 9). Moreover, each of these hundreds of individuals, in turn, learned from hundreds of individuals in the previous generation and so on. Thus, the cross-sectional area of the ancestral human cultural information stream is *hundreds times hundreds of units*, many orders of magnitude larger than for a non-human animal. This information stream in the contemporary human world is probably millions of times larger than for a non-human animal.

duction). Now we can understand more clearly why this is true. Doubt is the central tool of *all* human knowledge and insight in any area, from agricultural practice to economic policy, from the rules of baseball to global climate change or universal health care.

Human genius, wealth and power – consider a concert pianist

An elite young pianist is playing a Chopin concerto. Her hands show a breathtaking dexterity and command. Where did she get this skill? She had many teachers, each of whom had many teachers (possibly including Emanuel Axe), each of whom had many teachers and so on and so on back through Franz Liszt and even further back through Wolfie Mozart and beyond.

Her contemporary skill is simply the result of the uniquely human trick of sharing a massive body of cultural information held by huge, multigenerational non-kin coalitions. But our pianist has an interesting additional lesson to teach us about the source of the power of the human mind. Is our pianist naked? Probably not. Did she make the clothes she wears? Probably not. Did she grow the food she eats, did she build the house she lives in, could she service the taxi that brought her to Avery Fisher or the Royal Albert Hall this night, and so on and so on? The answers, of course, are almost always, 'No'.

Yet, our pianist, most likely arrived at her concert on time, well fed, well rested, and properly attired. We all take this process for granted, but we should not. Herein lays the second major contribution of cultural information to human genius. We do not merely acquire massive amounts of cultural information personally, individually. We also get to use the fruits of information we do not ourselves possess.

It is convenient to say that we have *direct access* to the cultural information (skills, insights, etc.) that we hold within our own individual minds. We will call this *embodied expertise*. For example, our pianist has direct access to information about music scores and the piano keyboard that make her an elite performer.

It is also convenient to say that we have *indirect access* to the cultural information that others hold, but whose benefits we enjoy. Thus, our pianist has indirect access to the culturally transmitted information held by the tailors, farmers, auto mechanics, and so on whose goods and services she consumes.

Each of us produces something using the cultural information we possess directly. Then we exchange some of what we produce for some of what others produce gaining indirect access to their personal information stores. We often execute this exchange through the convenient intermediate record keeping device of *currency* (an interesting human invention), but that feature need not concern us at the moment. We think of all these exchanges in the contemporary world as constituting our *public economic lives*. This economic sharing of information is crucial to our unique dominance of the planet. Does our theory account for the fact that only humans do this? Yes, it does.

The outstanding 20th Century economist Jack Hirshleifer once pointedly commented, "what can be exchanged can be stolen." He was succinctly describing the fundamental problem with public economic behavior. It is, of course, just the conflict of interest problem once again. If we exchange what each of us knows how to make among ourselves, there is a competing strategy that can work even better: Make nothing, steal what we need from others.

Unchecked, those who play the *thief strategy* will become ever more numerous until the cooperative enterprise is destroyed, as we all know intuitively (Chapter 1). Thus, if a means is at hand for cooperators to inexpensively recover whatever has been stolen and, thereby, to cost-effectively preempt thieves, this system of cooperative exchange can be extremely successful. It will provide indirect access to vast amounts of culturally transmitted information to its members.[11] Of course, humans have apparently had access to the necessary inexpensive coercive threat for around two million years (Chapters 5 and 7). Thus, we have evolved to be the *economic animal* as well as the *pedagogical animal*.

It is vital to understand this elementary bit of economics in order to have a more complete understanding of why humans appear to be so much smarter than other animals. This effect results from our unique access to inexpensive coercion for *two* reasons. First we are individually smarter because of the massive amount of cultural information to which we gain direct access. Second, however, we can do so much more than other animals, mostly because of our indirect access to additional culturally transmitted information held by many others.

Thus, our theory not only gives us a uniquely simple answer to the *human cognitive virtuosity* problem; it also gives us a crucial insight into the human condition. We are all rather individually conceited about how intelligent and capable we are. But this conceit is fundamentally misleading. In a very real sense, all the intelligence and capability we have is a product of our social cooperation—of our common humanity. Without the culturally transmitted information we possess, both directly and indirectly, our lives would truly be "solitary, poor, nasty, (and) brutish." More importantly, we would be no more "mindful" of our origins and our status in the universe than any other of Earth's animals.

This uniquely rich, knowledgeable, and powerful status is quite a return, indeed, on our access to inexpensive coercion. This vital, central insight will also serve us very well when we turn in later chapters to understanding our history.

BOX 10.1: Behavior in the face of adaptive novelty

As powerful as minds are, they are also limited, idiosyncratic devices. As amazing as our hawk's wings are as flying hardware, the hawk cannot play a C major scale on the piano with their tips. So it is with our uniquely human minds. They do what they are designed to do with exquisite virtuosity, but fail utterly at what they are not designed to do. For example, we can recognize an elegant logical deduction or a human social faux paux, but we cannot pick a small, camouflaged prey bird out of the forest foliage.

One other vital point. Think of our hawk again. When snared in a human net trap, it may flap its wings violently in a vain attempt to escape, injuring, even killing itself. Again, so it is with our human minds. We can find ourselves in situations with which our minds are not designed to contend. When this happens, we misperceive our options and generally behave non-productively or even self-destructively.

We can add other useful details to this picture. As we said earlier, natural selection produces animals who "anticipate the past." This is fundamentally true of all evolved devices including hawk wings and human minds. When the present looks like the past, as it often does, this strategy works very well. When the present is fundamentally different than the past, anticipating the past can be a recipe for disaster.

These cautionary notes notwithstanding, our minds tend to *recreate* the past more or less faithfully in the process of *anticipating* this past. This feature is especially true of the uniquely human social world. Thus, very often, our minds are in the same position as the hawk's in forest pursuit of prey—performing with virtuoso brilliance in a world where they were "born to fly"—at least until that world presents a novel situation.

BOX 10.2: The pedagogical economic animal. Something wonderful and new in the universe

We now have a much clearer picture of the evolution of our unprecedented human minds. They are the products of our unique management of the conflict of interest problem. This unique species-typical strategy opens our access to a vast universe of culturally transmitted information. This access, in turn, utterly transforms us. We can comprehend the universe (and ourselves) as no non-human animal could ever have remotely imagined. We can also use this comprehension to empower our interactions with the world, building both vast knowledge enterprises and equally vast physical/technical tools. These abilities are not really the product of our genetic redesign (though that is crucially enabling), but of our social collaboration. And this collaboration, in turn, once again, is only possible because of our ability to manage the conflict of interest problem cost effectively.

We are something truly new and wonderful in the universe. Moreover, our subjective experience of that universe is likewise profoundly, utterly new. Through this unique relationship to information can come true self-understanding and ever greater and universal humanity. Herein lays our promise and our destiny, quite an astonishing payoff for the simple capacity to project death from a distance.

Phantoms in the mind – conscious and unconscious, proximate, and ultimate

We have been looking at how our minds came to be so powerful. This is a different question than how we subjectively experience our minds, what it "feels like" to be or to "live in" such a mind. This difference between the objective and the subjective is a very common source of confusion. We can inoculate ourselves against such perennial futility by being clear about what we can and cannot know from subjective introspection.

First, our subjectively conscious minds are a thin skin over our vastly more massive unconscious minds. Psychologists have recognized this property for at least several centuries. Recent studies clearly corroborate this picture.[12] Thus, when we ask someone to testify about "why they did something" we are generally asking a highly artificial question.

To get a useful, substantive understanding of the origins of our behavior we need much more than introspection and conscious reports. Indeed, introspection is often actively misleading. Though most of us know about this limitation in our awareness, we still have difficulty in letting go of the intuition that we consciously understand the origins of our behavior. We will work toward surrendering this self-defeating illusion as the chapter progresses.

Second, even as we begin to understand our *unconscious* motivations, we still remain profoundly ignorant. All our motivations, *conscious or unconscious*, are merely outputs of *proximate* psychological mechanisms of the sort we began to explore in Chapters 3 and 4. Remember the fundamental logic of such devices. Subjectively, we might eat because we "feel hungry." However, the feeling of hunger is the product of a proximate psychological device designed by natural selection to cause us to behave *as if* we understood the Second Law of Thermodynamics. Specifically, our minds let us act *as if* we knew that, as non-equilibrium thermodynamic systems, we must take in high-energy chemical compounds to avoid running to equilibrium (starving to death). Our conscious reports about our *purpose* in eating are much more likely to involve feeling hungry than evolved adaptations to the Second Law.

It is obvious to us by now that elementary adaptive behaviors like feeding and sexuality have this two-level origin. They are the direct products of proximate psychological devices whose outputs we have some subjective awareness of—hunger, lust, and so on. However, again, these proximate devices result from selection of design information, which builds us to behave *as if* we understood the larger ultimate logic behind our behaviors, though we actually have no *a priori* awareness or understanding of this ultimate logic.*

---

*   Until the human knowledge enterprise *very recently* began to give us this insight.

In fact, *all* of our behaviors have this same two-level causation including our ethical, political, religious, scientific, and economic behaviors. For example, we may "feel" that we espouse a particular public policy position because of our belief in something larger than ourselves, something independent of our individual adaptive self-interest. However, the psychological device causing us to have this feeling has been designed to serve our *individual* biological self-interest in the ancestral human environment.* Again, *note well* that our subjective feelings about the motivations of our behaviors in these areas have no more connection to the ultimate, adaptive purposes of these behaviors than the feeling of hunger has to the Second Law.

If we truly wish to understand ourselves—if we wish to grasp fully what it means to be human—we must hold in doubt all the naïve beliefs we have about why we and others behave as we do. We must seek answers armed with a coherent theory of human behavior and knowledge of what the ultimate purposes of our various behaviors are.

This task is far more difficult than we might, at first, imagine. All of us (authors included) hold subjective beliefs about our motivations that we are loath to subject to systematic doubt and analysis. As our questioning brings us closer to these cherished beliefs, our emotional hot-buttons are pushed. We are offended and threatened. We run the risk of becoming resolutely impervious to detached self-analysis.

We say again, we *must* strive to overcome these emotional barriers if authentic self-understanding and grasp of the human condition are our goals. We must be willing to entertain the possibility that the human social world is not at all what we subjectively imagined. We can learn, with work and patience, to do this. In the end, it is worth the investment. It gives us a newly rich and ultimately humane comprehension of one another and of our world.

Joie de vie, righteous rage and mortal fear – the fundamentals of the human ethical sense

Armed with a good theory of the evolutionary origins of our proximate psychologies, our subjective experience can become a useful analytical tool rather than throwing up confusing misdirection. Think again about human eating. Feeling hungry tells us nothing about

---

* Often, these proximate devices will also be serving our biological self-interest in the contemporary world; but, of course, this will not always be the case if our current situation involves fundamentally new properties or features not adapted to by ancestral humans (Box 10.1).

the Second Law. However, *what* we desire, *subjectively*, to eat gives us rich data against which to test our theories of the ultimate causation of human eating behavior. In fact, the Second Law (and a few esoteric details of physiology and cultural and evolutionary history) can predict our tastes in food very well.

Analogously, our theory tells us what kind of a social animal we are. We can use these insights to predict some of our proximate "tastes." Comparing these predictions to what we actual do and feel as social creatures then becomes another useful test of our theory. We have already discussed our private reproductive and family lives (Chapters 6 and 8). Here, we will focus on our public lives. What kind of *public* social animal are we? Of course, we are asking about the ultimate or most fundamental answer to this question. Indeed, this ultimate answer might even "feel wrong" subjectively at first hearing.

Our theory is quite clear about the nature of this ultimate description of the public human animal. We are highly adapted to the self-interested, individually profitable projection of coercive threat against conspecifics. In most circumstances, the only way for us to pursue this strategy profitably is to project threat conjointly with others.* Thus, we can usually employ this fundamental human behavior only when pursuing that subset of our self-interests that is confluent with the self-interests of others. When we do this, we can realistically (if somewhat metaphorically) describe our species-typical pattern of public social behavior as the *enforcement of cooperation* (Chapter 5).

Another way to grasp this ultimate logic of our social behavior is to look at humans the way an extraterrestrial biologist might. Collections of animals with the above properties (us) should come together to carry out adaptive tasks—to get things done. The members of these collections will, under the appropriate circumstances, coordinate very effectively, each individual a cog in a well-oiled social machine.

However, these cooperating animals are often non-kin and have conflicts of interest. So our extraterrestrial scientist observes that humans also monitor one another with intense vigilance. If one of them should try to steal systematically from the cooperative project, these animals would display threat aggressively and, if necessary, explode into potentially lethal violence directed at the miscreant. Given this capacity for extremely dangerous conjoint violence, each member of

---

*   Only conjoint projection of threat provides the extreme reductions in individual cost that, in turn, creates a net individual adaptive profit opportunity for pre-emptive ostracism (Chapter 5).

the group would also normally be scrupulously careful not to provoke this reaction—busily working away at the common goals of the enterprise.*

Still thinking as our extraterrestrial biologist, how might we make scientific use of the fact that this particular animal claims to be able to actively report some features of its internal subjective mental states? We might collect such self-reports as a part of our investigation. Because the unique social adaptation of this animal is its most important property, we might expect the mental states managing its public social behavior to be among the animal's most vivid, salient experiences. Indeed, these intensely evocative states should contribute to these animals behaving *as if* they actually, consciously understood the ultimate, strategic logic of their public social behavior.

We have been studying these animals for several hundred thousand years now (we are *long-lived* extraterrestrials). We have accumulated quite a large set of interviews. These interviews span the time interval when humans evolved ever increasing scales of their unique social enterprise from the tiny *local* ancestral village to the contemporary *global* village. In the dense, terse, detached language of the professional observer, we might summarize the punch lines from all these subject interviews something like the following:[†,13]

(1) Humans report a strong positive feeling (they variously call it "satis-faction," "fulfillment," "making a contribution," "making a difference") when they are participating in a well-coordinated, efficient, adaptively productive conjoint social enterprise. This positive feeling is expressed in two different contexts.

(1a) Humans get a positive feeling when their contributions to the conjoint public enterprise provide resources for themselves, their mates, their offspring, and other close relatives. It is convenient to call this positive feeling **family satisfaction** *because of its extensive and direct connection to successful reproduction in this animal. The adaptive logic of behaviors producing family-satisfaction is obvious. Recently, their science has matured*

---

* Individuals will often even be well advised to actively advertise and defend their actions as being for the "good" according to the local *social contract*.

† There is a vast literature on the human ethical sense (see linked online endnote). Michael McCullough's 2008 book *Beyond Revenge* for an excellent, accessible recent review of the personal domain of human subjective public ethical psychology. As a psychologist, McCullough is afflicted with 20th Century evolutionary theory that is woefully outdated and inaccurate (in the authors' view). However, his intimate mastery of the relevant observational and experimental psychology makes this book an illuminating read, nevertheless.

to the point that humans are able to understand and acknowledge the ulti-mate logic behind these subjective feelings.

(1b) Humans also get a positive feeling when they believe that their participation in the conjoint enterprise contributes to something they subjectively refer to as **the greater good** or a higher purpose. It is conve-nient to call this positive feeling **higher purpose**. The subjective self-report from these animals is that this higher-purpose has little or nothing to do with themselves as individuals. Humans are extremely prone to misunder-stand the ultimate logic of this subjective feeling. Moreover, at this writing, their science of themselves has yet to advance to the point that they have a clear analytical understanding of their evolution.

The ancestral adaptive logic of higher-purpose appears to have two elements. The first is that the welfare of the individual and her/his close kin was indirectly dependent on the performance of their collaborative social units, sustained by self-interested coercion. Thus, individuals who contrib-uted to the successful functioning of those units would have done better, and design information building proximate mechanisms producing this outcome would have been selected. Call this **social selection 1**.[*]

Social selection 1 is weak in any event and extremely scale sensitive. Its adaptive rationale is quickly lost as social units grow in size. In even modestly large social units the adaptive "echo" (direct benefit to personal design information) of the individual's contribution to the greater good becomes negligible.[†]

The second element of the adaptive logic of the higher-purpose mech-anism is much more important and more general. Specifically, the coercive environment formed by the ancestral human social unit was overwhelm-ingly powerful and consisted of self-interested individuals (Chapter 5, the authors). This coercive power was capable of having profound effects on the survival and reproduction of individuals subjected to it. Tens of thou-sands of generations of ancestral humans were shaped by this intense selective pressure. Thus, humans domesticated themselves, just as they later domesticated other animals. Somewhat ironically, obtaining psychological reward from "altruistic" behaviors that feed the interests of the surrounding coercive coalition was actually highly adaptive **to the individual** as a result. Call this **social selection 2.**

---

[*]   It is crucial to never forget *that social selection 1* can *only* operate in the presence of the self-interested (that is, cost-effective; Chapter 5) projection of coercive threat.

[†]   This particular piece of the puzzle has been understood by some humans. See especially Mancur Olsons' classic 1965 book *The Logic of Collective Action.*

Social selection 2 is also scale sensitive, but in the opposite direction from social selection 1. As social units become larger, their coercive power over the individual generally becomes greater.

Most humans have great difficulty in understanding this adaptive logic. Often, they persist in stopping at the subjective, proximate level. People who contribute as dictated by the higher-purpose psychology are "good" and, they believe, this is just a "fact about the world" (natural or divinely created). When they do try to think more seriously about this part of their evolved psychology, they are extremely prone to invoke social selection 1 (even where it is so weak as to be irrelevant). They are almost completely unaware of social selection 2 (see Box 10.3 for our discussion of this vital issue; the authors).

This subjective distortion contaminates their social sciences badly. At this writing there still exists a massive, worldwide "scientific culture" built around the theory that social selection 1 is the sole human evolutionary legacy. This propensity most often expresses itself in one of the most characteristic (and self-defeating) delusion of the evolved human mind, the group selection fallacy.

(2) Humans report the most extreme arousal and complete clarity of action when they detect particularly egregious or inveterate cheaters and free riders. Such individuals are described as worthless parasites, predators, or criminals when their actions subvert conjoint behaviors supporting the family-satisfaction feeling. Even stronger are the feelings against individuals who subvert activities producing the higher-purpose feeling. Such individuals are often referred to with near-hysterical venom as traitors, heretics, satanic, truly evil, and so on. This stronger reaction to higher-purpose violations reflects the greater potential to mobilize conjoint coercive power here than in the family-satisfaction domain.

*Individual humans classify behaviors that support their social enterprises (interests) as moral or ethical. So it is convenient to call this strong negative subjective reaction **moral outrage**.*

The evolved function of moral outrage seems clear. It is the subjective component of the proximate psychological mechanism producing self-interested conjoint coercion—the foundational adaptation of this animal. If the threatening individuals are from outside the cooperative coalition, the violent response is often called war by humans. If the threatening individuals are from inside the enterprise, the response is generally called justice.

As expected from its central role in this animal's adaptation, the most extravagant conspecific violence humans are capable of occurs under moral outrage. For example, millions of individuals have been stoned (or shot with projectiles) to death over the last hundred thousand years and moral outrage is the most common subjective state of the individual killers. Further, for example, humans were killed in batches of fifty thousand to one hundred thousand at a stroke from the air in a mid-20th Century planetary-scale "war" of "good" nations against "evil" nations. The planners of these raids also frequently expressed moral outrage at the "evil" target nations.

*(3) In view of the scale and ferocity of conjoint violence stimulated by moral outrage, humans are highly adapted to avoid becoming individual targets of this violence. They feel an acutely intense negative feeling (a fear) when they believe (consciously or unconsciously) that they are in danger of provoking moral outrage in powerful aggregates of conspecifics. It is convenient to call the subjective component of this proximate mechanism by the same name humans themselves give it, **guilt**.*

(4) Moral outrage is a two-layered proximate device apparently serving two interconnected strategic needs. When individual humans find them-selves direct victims of a parasitic, predatory act they do not merely suffer the adaptive loss from that act. The predatory act also potentially carries the message that its target is powerless to resist. This is equivalent to identifying the target *individual as a social outcast, perhaps a chronic free rider. The target must not submit to this re-labeling, but must resist it at all costs—usually through expressing his/her access to coercive power by mobilizing a violent response (or powerful, overt threat).*

*These strategic requirements have shaped a powerful additional proxi-mate psychological element to moral outrage. We can borrow human terms for the subjective output of this element and call it **dignity feeling**. (Humans also use many other words in addition to dignity for this subjec-tive output, including pride, honor, status, gravitas, and the like.)*

*Satisfying dignity feeling is one of the strongest impulses humans have. The adaptive necessity of having access to coercive power to defend self-interest is very well illustrated by this. The behaviors generated by dignity feeling, in turn, create a second layer to the human guilt response. To avoid the violent aggression from offending dignity feeling, miscreants actively engage in **forgiveness seeking**. If this pursuit of forgiveness is genuine—explicitly acknowledging the dignity of the target, that is, submitting to the target's access to coercive threat—the target's proximate psychology is designed to **forgive**. In forgiving, the coercive power of the target is rein-forced and enhanced. **Magnanimity** is a strong form of forgiveness, gener-ally a signal of access to great coercive power.*

Finally, in view of the central role of access to conjoint coercive threat, and its enabling proximate dignity feeling, humans have one last proximate device in this context. They constantly evaluate each individual's dignity (power) and they are intensely hostile (contemptuous) of those lacking dignity. It is convenient to call the subjective output of this device **power-lessness contempt**. *This last proximate psychological device produced highly toxic effects in the recent novel environments of hierarchical distribution of coercive power* (below; Chapter 13; authors' note).

(5) Humans subjectively insist on seeing the entire universe in terms of "good" and "evil" and as having a "higher purpose". As usual, at this early point in the biological emergence of a self-aware organism, this subjective perspective borders on the perverse in its ferocious intensity. As yet, humans have not achieved a level of self-understanding that would allow them to see that their "good" and "evil" and their "purposes" are actually human inventions coming into their solar system with their unique biological evolution. Presumably, they will eventually come to see this unique characteristic of themselves if they survive long enough and manage to deploy their capabilities for knowledge generation sufficiently effectively. They would thereby come to the key decision node that all self-aware species throughout the universe inevitably reach. It will be of the greatest interest to see their particular accommodation to this insight when the time comes.

(6) These animals have created some new variations on their ancestral pattern very recently (the last few hundred generations or about 0.4 percent of their unique evolutionary existence). First, they have established new enormous social units (Chapters 11-16, the authors). *These units are one million to ten million times larger than their ancestral villages! Moreover, these new large units first arose in the context of an unprecedented concentration of the vital human social asset—coercive power—in the hands of small subsets of members of these units (what humans call elites).*
These very recent developments have thrust humans into situations with important elements of adaptive novelty. They sometimes find themselves in a world that is simultaneously *self-created, yet hostile and alien. How they ultimately cope with this profound new evolutionary challenge will determine* **everything** *about their future on this planet. Time is short for them and their destiny hangs in the balance.*

The point of this detached, clinical, extraterrestrial description of our public ethical psychology is to help us begin to see it for what it apparently is. Our public ethical feelings are intense and emotionally evocative. This tends to hide their adaptive purpose from us. Think about the symptom list we just read. Is it accurate and relatively complete? Is it what our theory predicts and requires? We suggest that the answer to both questions is, "Yes."

Finally, we might immediately feel that we have just devalued human life. If our sense of a *higher purpose* is just the subjective output of a proximate mechanism designed to serve self-interest, how can we still feel that our transient, individual lives have larger meaning? We will have much more to say about this later (Chapter 17 and Postscript).

Importantly, for now, we should recognize that knowing the evolutionary origin of a subjective experience need not constrain its worth to us in the future as Earth's first self-aware animal. We are perfectly free, for example, to agree that a peaceful, transparent, just, wealthy future for all Earth's children is a higher purpose on which we will insist. We can realistically implement this purpose if we achieve sufficient self-awareness. Our theory apparently predicts that this utopian-sounding suggestion can become realistic *under the appropriate conditions* (Chapter 17).

Box 10.3: Human self-domestication and the group selection fallacy

Our imaginary extra-terrestrial anthropologist above introduced a vital insight. We are self-domesticated creatures. It will enrich our grasp of our common humanity to understand more about what this means.

Consider a non-human domesticated animal. Dog breeds selected for their abilities to retrieve game or herd sheep are useful cases. What is the adaptive logic of these behaviors? Ancestral abilities (from the wolf) have been magnified and refined to improve the performance of some specific dog-human collaborative enterprise. Humans also make a reciprocal contribution to this enterprise and both dog and humans can "win" adaptively. *However*, if we interpret this last (true) statement as a *sufficient* description of the adaptive *origins* of the retriever and sheep dog breeds, we are profoundly mistaken.

Rather, the adaptive logic is as follows. Ancestral dogs that could retrieve or herd were selectively fed, protected, and allowed to breed by humans. Humans exploited their nearly complete coercive power over dogs to achieve this outcome. The resulting selective reproductive success of specific dog individuals is just natural selection imposed by human choice (often called *artificial selection*). The fact that selected dog behaviors redound to the benefit of dog-human projects is a product of human selection, **not** directly of the adaptive benefit of the collaborative behavior itself.

Now, suppose we could speak to the retriever or the sheep dog and ask them how they feel when they are "doing their thing."* Retrievers and sheep dogs often display great vitality and apparent joie de vie while going about their duties. They seem to enjoy executing these breed-typical tasks. This joy in execution seems to be particularly strong *in the presence of humans*.

How might a dog imagine the evolution of its abilities and feelings here? This is surely an artificial question, of course. Contemporary dogs have no science and, thus, no access to ultimate causation. However, if dogs started developing their own science in the future, how would their evolved psychology complicate answering this question? They are selected to please us humans. Might this psychology lead them to emphasize the contribution of their evolved behaviors to the dog-human relationship? Almost certainly it would. Their initial "hypothesis" for the ultimate origins of their retrieving and herding behaviors might, therefore, confer causal power on the fruits of dog-human cooperation.†

As we have seen, this hypothesis is fundamentally wrong. The adaptive benefits of dog-human cooperation are a secondary effect whose primary cause is the self-interested application of coercive power by humans. Once again, the mutual benefits of these collaborative activities are not *sufficient* to account for the evolution of these dog behaviors.

When our imaginary early science-developing dogs make the mistake of attributing sufficient ultimate causal power to the fruits of dog-human collaboration, they are making the intellectual error of (unconscious) *coercive audience pandering*. This tendency to pander to coercive power is a fundamental aberration in their ability to understand themselves.

---

*   Talking to a dog probably is not quite as far-fetched as it might sound (Chapter 9).

†   Different kinds of dog-human collaborative behaviors could have evolved on the direct basis of mutual benefit between members of these two different species. This can happen when the cooperative activities occur in an arena where members of different species have little conflict of interest (like the ants and aphids in Chapter 5, for example). However, these are not the cases of interest here.

We are just like these early science-developing dogs. Our public social behaviors are shaped by the irresistible demands of the potent
coercive human coalitions in which we have been embedded for the
last two million years. We are human-domesticated animals just as
surely as a retriever is. We derive subjective (proximate) psychological
pleasure from doing "public-spirited" things. Moreover, these public-
spirited actions are among the things we are most anxious to pub-
licize, (unconsciously) satisfying the intrusive demands of our highly
coercive coalitions.* The subjective satisfaction we derive from others
recognizing our public-spirited acts is analogous to our subjective
pleasure when members of the opposite sex find us attractive. In each
case our pleasure indentifies potential individual adaptive payoffs including evasion of fatal ostracism and support of our future reproduction, in these cases.

Thus, we are just as prone to coercive audience pandering as dogs
are. This psychological aberration leads us directly to one of the most
profoundly pervasive delusions in contemporary human science, the
group selection fallacy (Chapter 3). When we commit this fallacy we are
generally confusing the *agent* of selection (the human group or village) with the *unit* of selection. We are also pandering to our coercive
audience by (unconsciously) claiming loyalty to the unit unalloyed
with exclusive self-interest. We are being *good domesticated animals*.

It is not that we do not genuinely feel public-spirited. All of us who
are psychologically normal often do, us included. Rather, the point is
that when we say we feel this way because we are committed to the
interests of the group (institution, country, cause, etc.) this is an unconscious pretense, not a legitimate depiction of our actual (ultimate)
motivations. If we wish to discover our capacity for being *authentically*
public-spirited, we need a deeper level of self-awareness—and a good
theory of humanness.

We will argue that the illusions and fallacies from our (mostly unconscious) pandering to the coercive human audience currently haunt
and defeat our authentically humane goals at every level from our
individual lives to the corridors of national and international political

---

* Recall that even (or especially) iconically altruistic figures like Gandhi, Mother Te-
resa, and Martin Luther King, were extremely attentive to publicity and to their public
images.

power. We can only discuss the human future with wisdom (and real-ism) if we take the time to understand and subdue the delusion that uniquely human cooperation is produced by group selection.

As we first come to fully grasp our status as self-domesticated animals, we can feel diminished and dehumanized. However, if we do not let this tem-porary revulsion repel and defeat us, we come through this phase to a new level of humane insight. The evolved psychology our ancestors selected in one another—and that we inherit—can be consciously deployed by transparent, democratized consent to the self-aware construction of a vastly more humane and fulfilling future (Chapter 17 and Postscript).*

Right makes might and might makes right – conformity, coercive power and the human ethical sense

This picture we are building of our uniquely human public psychology makes several important predictions about our political behavior. These predictions will be vital to us in later chapters as we look at how our two-million-year history has unfolded.

First, we need to understand how ancestral humans made public ethical and policy judgments. As we said, they would have done this through a proximate psychological mechanism that likely had no conscious awareness of the ultimate logic, the game theory of social cooperation (Chapter 5). Again, this psychological mechanism's subjec-tive outputs would have been to the logic of their social cooperation as hunger is to the Second Law.

Public doubting would normally have played a central role in such decision-making. Each individual might have expressed an opinion (self-interested, unconsciously or consciously) about a course of action. Exchange of such opinions by all allows individuals to begin to iden-tify and agree on a course of action that best maps on to the confluent interests of a substantial majority. This dialog is a *negotiation* in the most literal sense. The overwhelming coercive power of a large majority then forces everyone who wishes to remain with the coalition to get on board with the consensus. This is simply *democratic governance*, of course.

---

* Full awareness of our self-domesticated (and self-domesticating) heritage will also be vital to dealing with the fundamental source of all pathology in human cultures—the elite concentration and resulting narrowly interested deployment of coercive power and threat (below; Fourth Interlude; Chapters 13 and 17).

Under these conditions, the ancestral human public political sense is expected to have a very important property. This proximate mechanism would tend to perceive overwhelming coercive threat from a large majority consensus as representing the "right." Also, finding oneself in danger of being the target of such concentrations of coercive power would tend to provoke an aversive feeling of being "in the wrong" (guilt).

Very often both these reactions to concentrations of power would have been perfectly appropriate and ethical from the self-referential point of view of the local cooperative human coalition. Again, the coercively powerful aggregates we have been discussing would have identified and consolidated their confluent interests during transparent public doubting and would reflect some local approximation of the public good.

Of course, this logic of social consensus is realistic rather than Pollyannaish only when access to decisive local coercive threat is broadly (democratically) distributed, as it probably was in ancestral, elite-throwing humans (Chapters 5 and 7). Under these conditions, individuals arrive at a workable approximation of the common good as the best *available* individually self-interested option (Chapter 5).

However, human history includes the continuous invention of new coercive means and some of these new weapons have lent themselves to concentration of overwhelming coercive power in the hands of the few. We will have much more to say about the details of such concentrations of coercive threat later (Fourth Interlude; Chapters 13-17). For now, our focus is on how the evolved human ethical sense is expected to respond to such coercive asymmetries.

Our theory seems very clear about this evolved response. Humans will tend to perceive concentrations of coercive power, even in the hands of the few (we will call such concentrations *elite* power), as the *right*. This will be the case for those controlling such power. They will entertain little doubt that they deserve the power they hold.* They will think of themselves as ethically superior, perhaps even ordained by a higher power. Conversely, those who are the targets of such overwhelming power will tend to submit and can even sometimes be made to entertain the possibility that their subservient condition is earned or, at least, ordained.†

---

*   Acton's empirical dictum that "absolute power corrupts, absolutely," thus, also has robust theoretical support.

†   Of course, the relentless pursuit of self-interest will make it perennially difficult for elite, concentrated power to restrain those it manipulates. Much more about this issue in later chapters.

As we turn later to how our history has unfolded, this *ethical power response* by our evolved, uniquely human psychology will be vital to recall. We spoke about how a hawk's flying behavior can be both its essence (when it is pursuing its adapted behavior) and its downfall (when it is trapped in a net, an evolutionarily novel situation; BOX 10.1). The human ethical power response has often put our ancestors in a similar position, sometimes serving their common needs, sometimes ensnaring them in brutal egocentrism and subjugation.

For now, the good news is that improved knowledge of the rules of our social world can, *under the right circumstances*, put us in a position to act to control and prevent repetition of the most egregious inhumanities of the human past (Fifth Interlude; Chapter 17).

Finally, think for a minute about how ancestral humans might have recognized cheaters and free riders. Again, this mechanism would have caused them to behave *as if* they understood the game theory of social cooperation without actually understanding it. Important work remains to be done in this area, but we can make one very useful statement. Public doubting would have been crucial and central. But it could not be practically applied to every individual behavior at every moment. Some routine, shorthand method of evaluating possible social cheating would sometimes be needed. A simple approach is not to ask whether all the many things others might be doing are *cooperative*. Instead, we might simply insist that others limit what they do and that they do just what we are also doing (*conform*), thereby simplifying the monitoring process.

As this approach predicts, ancestral humans apparently evolved to be extremely *conformist*. We have inherited this psychological mechanism.* We do not attempt to decide whether every behavior by every other member of our cooperative coalitions might actually be an act of a social parasite. Monitoring at this level of detail is an impossible task. Instead, we merely ask whether other people are doing what we do. If we are all embedded in a mutually self-interested joint venture we should all be behaving similarly.

Of course, this simplification of the monitoring problem comes at a cost. Stifling free riding also sometimes inadvertently stifles

---

* We sometimes entertain the conceit that we are non-conformist or that we are tolerant of non-conformity. But the range of non-conformity in question is actually minute, narrow. We insist on all kinds of speech, dress, and behavioral conformity allowing tiny amounts of strictly limited *non-conformity*. If you doubt this fact, go to the mall with your pants on your head or start addressing people with the wrong gender pronouns.

innovation. Economic systems grow (or fail to grow) depending, in part, on how this particular difficulty is managed, as we will see in later chapters. The suppression of innovation by elite concentrations of coercive power will prove especially important—and toxic (Chapters 13-17).

Cultural information and Darwinian evolution

Biologists and sophisticated anthropologists have long recognized that natural selection or Darwinian processes can act on any kind of design information, as long as this information has the necessary properties.[14] This most emphatically includes culturally transmitted information. A clear understanding of the implications of Darwinian evolution of cultural information in the uniquely human context will be one of the most profoundly important research areas of the 21st Century. Our limited goal in this section is to provide a few crucial first insights useful to this massive project.

The fact that any kind of information can be subject to natural selection (under the right circumstances) is often misunderstood, sometimes in a naïve, self-defeating attempts to be more humane while actually engaging in (unconscious) coercive audience pandering (Box 10.3). The following quote is one notorious example of this misconception from the famous science writer Stephen J. Gould.[15]

> *"Cultural…change manifestly operates on the radically different substrate of Lamarckian inheritance, or the passage of acquired characters to subsequent generations."*

Of course, the word "manifestly" in Gould's sentence is a red flag. He wishes us to accept the rest of the sentence without question (without doubting). Gould's use of manifestly is a categorical statement, not the culmination of a careful argument. This makes the sentence a non-scientific statement of belief and, thus, not particularly likely to be true. However, Gould's statement is useful to us as data about how we feel intuitively about cultural information. The notion that all our insights and beliefs might result from the blind, purposeless process of Darwinian evolution is a little shocking. As we will see later, we can eventually gain active, conscious control over cultural change in ways our ancestors never did. But first we must understand how this process has worked in our past.

A logical error in Gould's sentence above is a useful starting point. He imagines that Darwinian processes are not shaping culturally transmitted information because it is not being transmitted in the same intergenerational fashion as genetic information is. This view is false. In fact, all that is necessary for information to be modified by Darwinian processes is that the information show variation (change over time) and that this information affects the behavior of the vehicle possessing that information in ways that affect the probability that this information is transmitted to other vehicles (replicated). How, when, where, and how fast this information transmission occurs is largely irrelevant.

This argument may seem a bit complicated and abstract. To see it more clearly, think of a single piece of culturally transmitted information that can be replicated by being somehow transmitted from one mind to another. Thus, it can come to be present in multiple copies in its inevitably Malthusian world. Copies of this piece of information will sometimes vary a little from one another (the Second Law again; Chapter 2). Now suppose that a specific variant of this piece of information causes a behavior in the individual possessing it that leads that individual to live longer or be more conspicuous to other individuals around her (or him). Finally, suppose that this possessing agent transmits new copies of this piece of information to the minds of those around her for as long as she lives and more efficiently if they are paying extra attention to her. That is it. This piece of information will spread through the population of vehicles with greater efficiency than other pieces. The versions of it that spread the best will persist longest over many generations in the population.

Indeed, an individual vehicle may acquire and retransmit different versions of a piece of culturally transmitted information repeatedly through her/his individual lifetime. For example, both authors helped teach our children computer keyboard skills. However, we are old enough that we learned mechanical typewriter skills as children and electronic typewriter skills as young adults, before acquiring the present computer version of these skills. Moreover, we acquired each new version of these skills by observing others, acquiring (replicating) pieces of their culturally transmitted repertoire.

Now add one last element. Suppose some pieces of design information evolve to shape individuals (vehicles) so that they pay most attention to other vehicles that appear to be healthiest, longest lived, most powerful, most reproductively successful (sexiest), and so on.* The effect will be that the population will appear to adapt very efficiently and will pass this adaptation on by teaching others. This is what Gould mistakes for a Lamarckian process. It is not. It is Darwinian.† The only salient differences between genetic and cultural design information are the details of transmission. Indeed, *Lamarckian* evolution is arguably a figment, an oxymoron. We know of no cases of evolution that are better explained as Lamarckian rather than Darwinian.

The crucial point from this elementary introduction is that *both* genetic and cultural information will be sources of exquisitely adaptive design. Moreover, the uniquely massive scale of culturally transmitted information in humans means that the cultural design information stream will be important to understanding our behaviors.

Another implication of the similarity of the logic of Darwinian selection on genetic and cultural design information is enlightening. Our personal repertoire of culturally transmitted information will be adapted to cause us to behave as if we have conflicts of interest with distinctively different personal repertoires of cultural design information. The logic producing this *ethnocentric bias* is extensively analogous to the logic of *kin-selection* on genetic design information (Chapter 3).

Yet another vital consideration is that cultural design information is just as likely to deploy coercive threat in service of its "interests" as is genetic design information.‡ The logic of this deployment by genetic design information is apparently clear (Chapter 5). The same fundamentals will certainly apply to cultural information's use of coercion. The challenge remains to map fully the second-order details of this process and the interaction between the two streams of design information in controlling social coercion.§

---

*   The pieces of information in question are expected to include *both* genetic and cultural design information.

†   We are not discussing more advanced issues here. For example, how do the "interests" of genetic and culturally transmitted design information interact? However, these issues need not divert us at the moment.

‡   We use the same metaphorical shorthand in this paragraph as elsewhere. More precisely, culturally transmitted information will evolve to cause its vehicles to behave *as if* they are deploying coercive threat in service of the interests of culturally transmitted information.

§   Note that the genetic and cultural information streams will generally collaborate intimately in the control of coercive threat, just as they do in controlling human

Finally, Darwinian processes acting on cultural information will often produce behaviors that are well adapted to the "success" of their human agents (the vehicles of the information). Much of our culturally defined behavior is adaptive for this reason.[16] However, remember that Darwinian processes are just long chains of historically accidental, short-term solutions to adaptive problems in the past. Thus, whether produced by genetic or cultural information, behavioral solutions to adaptive problems can be weirdly quirky at the same time they are adaptively sophisticated (see Second Interlude).

Science and democratic society – true believers and the human cultural information stream

Most of us feel subjectively as if we believe in something and that this belief is vital and central to the meaning of our lives. What are these beliefs on our theory? We think we know enough to begin to construct a usefully detailed, but still partial, answer to this question. *Beliefs* originally included the details of the *social contract* under which ancestral human cooperative enterprises were carried out, as we said previously. These are encoded in a form that can be communicated and thus, both negotiated (through public doubting) and transmitted.

Among the many implications of this picture is that public statements of belief also have the essential, ultimate function of acting as *recruiting speeches* designed to get others to join the conjoint enterprises we favor under the terms of the social contract we propose.* These contracts and undertakings are produced by proximate mechanisms that served the self-interests of our ancestors. Again, we are generally not conscious that they have this property. Public statements of belief by *large sets* of individuals are even more effective recruiting devices because of the coercive power of large groups. Such large groups are (unconsciously) saying "here are the contractual conditions for joining a going concern."

---

language (Chapter 9; Pinker 1994) or our subjective moral psychology (above; Hauser, 2006).

* Remember that only *conjoint* enterprises can survive and prosper in a world dominated by an animal with access to coercion through projection of threat from a distance (Chapter 5).

Notice also that *competing belief systems* will vie for members.*
This is just the inevitable consequence of cultural design information
being selected for successful replication in the very Malthusian world
of the human village. Of course, this form of competition often involves
survival and transmission during public doubting. Public doubting this
way imposes competitive, Darwinian selection on beliefs. As a result of
the central role of this process, we are expected to be highly adapted
both to promulgating beliefs *and* to responding to statements of belief
by others when we feel (often unconsciously) that these beliefs may
either further or subvert our individual interests.

This process of competition among belief systems is crucial to
understand. Of course, *science* is merely the functioning of this process
with the most prosaic of purposes. It is *not* the negotiation of the high-
er-order details of our social contracts, but just the working out of the
mechanical/technical details of the world in which our cooperative
contracts will be executed. By employing democratized social doubting
to belief to get the most complete, clear picture of our environment,
we give our mutual social enterprise the best chance of success. Again,
we are expected to be adapted to do this well—humans are naturally
scientists. Science is how we develop pragmatic information.

However, pursuit of self-interest can corrupt what is ostensibly
science, just as thoroughly as it can corrupt our political and economic
dialogs. For example, a "science" of gender properties by elite males in
a male-dominated society is certain to be toxic to everyone else and
self-serving to elite males. Likewise, for example, a "science" of individual human differences by benefactors of a race-based slavocracy is a
preposterous notion.

This problem for science is more pervasive than is sometimes real-
ized even by professional scientists. In the context of formal institutions
(universities and institutes, for example), members of scientific subcultures and disciplines run the constant risk of becoming self-interested
priests of arcane knowledge cults. Such scientists ostensibly pursue
insight, while actually (often unconsciously) merely defending and
sustaining their privileged access to adaptively useful assets. At best,
such academic subcultures live in a sterile intellectual echo chamber.

---

*   Again, actually, beliefs (as information) will be selected to cause their agents (humans) to behave as if the beliefs are competing for disciples. However, as with genetic
design information (Chapters 2 and 3), we can use the shorthand of speaking as if beliefs
themselves had goals and behaviors.

At worst, they actively suppress real progress in their ostensible areas of expertise.*

Fortunately, the solution to this perennial threat to the scientific endeavor is the same as the answer in any other human domain, democratized public doubting. In the face of a democratic wise crowd, pseudoscience designed to protect narrow interests is exposed and discredited. This essential doubting can be (and is) implemented in two important ways.

First, professional scientists MUST also write for all members of the pan-global human wise crowd. Only when our specialist insights survive pan-human doubting can they legitimately be called knowledge.†

Second, when the knowledge enterprise is working effectively, students are much more important than certain contemporary views of education acknowledge. They are not merely engaged in the passive reception of insight, they are also central to the essential, continuously ongoing, active public doubting enterprise (BOX 10.4).

---

* These toxic outcomes ensue when collaborative social control of institutional assets trumps generation of new insight as the (often unconscious) local adaptive goal.

† A new academic specialty calling itself *science studies* has grown up over the last several decades. One of its ostensible goals is improving the social value and outcomes of science by better understanding its process (see, for example, Collins, 2009, for a brief discussion). This is a worthy goal, in general. In practice, science studies has been myopic in its understanding of how science works as an historical process. More importantly, science studies advocates have apparently sought to ameliorate the toxic effects of one set of pretentious experts by creating another equally pretentious set.

BOX 10.4: "Education" and knowledge in the pedagogical animal

Education is sometimes taken to mean the passing of knowledge from an authoritative voice (teacher) to a passive acolyte (student). We argue that is a profoundly pathological view of the two-million-year-old role of education in the uniquely human adaptation. Instead, we believe that students are the cutting edge of the public doubting enterprise. Teachers initially propose knowledge, merely acting as informed advocates of a position they believe based on past experiences and social contracts. Students act as skeptical consumers, revisers, and editors of these propositions and social contracts. Propositions surviving and refined by this doubting become (always provisional) *knowledge*.

Under these conditions, education is not merely a necessary adjunct to the transmission of knowledge, it is also a central active, functional part of the ever-unfolding knowledge enterprise. In every "classroom" on the planet, every day for the last two million years, coalitions of individuals have been subjecting pieces of what we think we know to new doubt. Students continuously vet and re-vet potential knowledge for continuing membership in the cultural stream, our informational heritage.

In our experience as university teachers, this approach to education is richly fulfilling for both students and teachers. Everyone shares information, everyone is enriched in new learning, and everyone is contributing to the negotiation of future social contracts. The knowledge confronted by students and teachers collaboratively is relentlessly polished and sharpened, while the inevitable dreck, noise, inefficiency, and inelegance are culled from the information stream.

This crucial creative role of students also allows them to take personal ownership of knowledge in the only way we humans have evolved to do—by having that knowledge survive their own personal doubting, supported by public disputation.

The voice of authority embodied in "experts" is merely an artifact of the distortion of the knowledge enterprise by relatively recent elite concentrations of coercive power (Chapter 13). It has no role to play in a properly functioning human knowledge enterprise, in our view.

Might, right, bizarre belief and history

These various features of *belief* have merely brought us back, full-circle to the uniquely human adaptation to using large, non-kin coalitions to generate huge amounts of reliable culturally transmitted information. This enterprise works well to serve its revolutionary adaptive function only when large numbers of individuals are free to express public doubt and, thereby, contribute and gain access to reliable information.

It is also vital to recognize that universal public doubt implies universal (democratic) access to coercive threat. This is an inescapable implication of our fundamental theory (Chapter 5). Only those with the power to contribute to ostracizing proponents of beliefs hostile to their own interests can have confidence in the usefulness of the products of the wise crowd.

Such universal access to coercion was probably the human condition throughout most of our evolutionary history until some conspicuous, very recent episodes beginning around fifty-five hundred years ago (Chapter 13). However, to understand our history over the last fifty-five hundred years, it will be necessary to understand the effects on belief when universal access to coercive power breaks down.

Consider how the process of belief negotiation works if small elites find themselves in possession of disproportionate coercive power in contrast to the ancestral condition of broad access to effective threat. Such elites face many special adaptive problems (Chapter 13); however, in the context of the human knowledge enterprise and belief, they face a specific difficulty. Elites rarely have absolute power (outside of the occasional heavily militarized slavocracy) and, thus, elite interests are best served if non-elites can be convinced to sign on to (or at least submit to) the elite-dominated social contract. Beliefs justifying elite domination as a moral and/or pragmatic necessity are especially useful.

Before we discuss how these self-interested elite goals affect the knowledge/belief enterprise, recall that human proximate psychologies will tend to see decisive coercive power as a sign of being in the right (above). Thus, both elites and underclass members will have a perverse tendency to see elite domination as "natural" simply because of the decisive power elites hold, by definition, the quintessence of circular political reasoning.

Elites will propose belief systems in which they (kings, nobles, pashas, raghs, mullahs, popes, etc.) are ordained to control the social enterprise—or to *rule*, to *interpret the law*, *to shepherd their flocks*, and so on. Again, obviously, authoritarian belief systems of this form originate from the human knowledge enterprise entrained to the interests of elite coercive power. Less obviously, racist, sexist, fundamentalist, and ethnocentric belief systems are apparently also of this form. They are designed to serve elite interests (Fourth Interlude; Chapters 13-16).

At one level, these insights are not new. Machiavelli, Hobbes, Aristotle, and many others grasped many of the empirical features of such systems. However, our theory gives us more robust confidence that we know where elite-serving belief systems of all types come from.*

However, this application of the theory to belief yields another important insight. Recall that Darwinian processes, whether acting on genetic or cultural design information, are fundamentally "purposeless" in and of themselves. They are near-term mechanical processes blind to their long-term consequences. Information that wins the Darwinian competition today is here tomorrow and all other information is lost—*no matter the consequences for tomorrow.*

While this can lead to highly adapted, efficient design, it can also produce bizarre outcomes. Everything depends on the instantaneous logic of this foresight-less process. A non-human biological example is illuminating. If longer male tail feathers are a signal of a healthier, genetically superior father for their offspring, females birds may become adapted to select for the longest tail feathers when they choose mates. However, as tail feathers get longer under this selection, females will still continue to mate with the males with the longest *available* tail feathers, even to the point that male plumage becomes so extravagantly elaborated and enlarged that the males' life expectancy is reduced.† This process is referred to as *sexual selection* by evolutionary specialists, but it also illustrates the more general and fundamental process of *runaway selection*.

Now consider human elites promulgating self-serving belief systems. How will this process play out? First, notice that elite domination of a local social system creates a prize for self-interested others to steal from the current elite. If a new set of individuals can seize coercive dominance, they reap great adaptive rewards.

Such takeovers will impose tactical/military requirements (Chapter 13). Also necessary will be a new, competing social contract. However,

---

*   Our theory also suggests avenues to sweep such elite belief systems from our social world, as we will see in later chapters.

†   Hence, the famous African widow bird (most of the males die young as a consequence of their female-selected tail feathers) and the well-known peacock among, many other examples.

this new belief system is NOT the product of a change in the local physical environment, a new insight, or new rules for social cooperation. Instead, the actual situation is just "meet the new boss, same as the old boss".* So the new competing belief system will generally distinguish itself from the original (something it MUST do for recruiting purposes) in arbitrary ways such as developing artificially distinct identifier beliefs, often piled on top of earlier artificial, elite-justifying identifier beliefs already in place.

This detail becomes very interesting when we consider that elite-dominated social units will undergo competitive elite takeovers, again and again. This is a recipe for *runaway selection* for arbitrary beliefs. Elite belief systems are predicted to become ever more bizarre and disconnected from the physical world.† However, they will also often be believed when they are part of a successful elite social take over. Under this condition these beliefs are sustained by the single criterion human evolved proximate psychologies find most salient: promulgation by decisive coercive power.

Finally, we note one last feature of the public negotiation of elite-justifying belief systems. When a seizure of elite power is being attempted, the promotion of new identifier beliefs and the questioning (even ridiculing) of entrenched beliefs is inevitably part of the recruiting process. Thus, concentrations of elite power must always interpret public doubting of their beliefs, even bizarre, arbitrary elements, as part of a nascent *hostile take-over*. In fact, it almost always has such a purpose (consciously or not).

This is true, of course, because underclass members of the social project have strong adaptive incentives to seize the enterprise in elite-dominated polities. Thus, elites must suppress all public doubting with *maximum prejudice*. Burning of "heretics" at the stake, publicly sacrificing non-believers, and so on will be an *inevitable* element of elite concentrations of coercive power. These violent, conspicuous, *unavoidable* symptoms of elite-dominated human social units will be most useful to us in later chapters (Box 17.4).

We propose that this framework is a useful new theory of the origins of one of the most quirky features of human uniqueness, bizarre religious beliefs including both traditional *theistic* belief systems, as well as more recent *secular* religions like fascism, communism, and radically

---

*   Lyrics form "Won't Get Fooled Again" Pete Townsend and The Who.

†   Recall that the adaptive function of identifier information is purely self-referential, connected *exclusively* to the internal workings of human coalitions. Its real-world verisimilitude is irrelevant.

doctrinaire capitalism. Again, this theory of the evolution of elite beliefs appears to follow from our theory of human uniqueness.[*] Extensive further investigation will be required to test and refine this picture.[†] However, we note several elements of empirical support.

First, all the large traditional institutionalized theistic religious belief systems currently in existence first arose over a very brief historical period sometimes referred to as the *Axial Age*.[17] This brief interval corresponds to the rise of what historians call early *archaic states*. As we will argue in Chapter 13, archaic states are social units predicted by our theory to arise from new technical developments that allow unprecedented elite control of decisive coercive threat. All of these elite-serving religious belief systems come to prominence, initially, as tactical tools of elite seizure of archaic states as a matter of historical fact (Chapter 13). They were then apparently progressively elaborated and enforced as still other interest groups struggled to seize or retain control of these elite-dominated social units, as expected.

Second, bizarre, doctrinaire, racist, or ethnocentric beliefs are often promoted as an agent of ostensibly democratic seizure of the local human social enterprise. However, the surreal elements of these beliefs apparently persist (and become exaggerated) only when these initial *ostensibly* democratized coalitions actually come to posses *elite* power. Examples include Christianity in the Roman Empire, Islam in the original Islamic Empires, and Bolshevism or Nazism in early 20th Century Eurasia.[18] Such belief systems were and are, of course, also part of the contemporary recruiting packages of the likes of some politically active American Christian fundamentalists, Ayatollah Khomeini, and Osama bin Laden. As predicted by our theory of elite identifier beliefs, such efforts generally represent attempted elite takeovers, *not* eventual democratization.[19]

Of course, again, theistic belief systems are not the only culprits here. An evangelist of a secular or scientific doctrine who claims that expert specialists should decide public policy rather than allowing the normal functioning of transparent, democratized doubting, is just as suspect as a theistic evangelist. Moreover, the "science" of such a secular evangelist's position is just as suspect as the Vatican's pre-Galilean astronomy.

---

[*]   It is possible that there are other sources of runaway selection on human beliefs. It will be of great interest to investigate this issue in detail in the future.

[†]   J. Souza, A. Shah, P. Bingham, manuscript in preparation.

"History is the lie made up by the winner"*

A better understanding of our evolved minds throws strong new light on *history*. These insights will serve us well as we build a theory of history in the following chapters. First, we are well advised to ask how reliable the written historical record is as a product of the human mind. Of course, the written historical record is produced by self-interested individuals, everywhere and always. Among our most important public political acts are predicted to be recruitment and ethical self-justification. When we argue for a public policy, for example, our self-interest is the primary criterion we employ (usually unconsciously) in choosing the policies we support and seek to preserve as *history*. Each of us tends to object to this claim on subjective grounds. However, it is almost certainly correct. This element of our behavior is mostly hidden from us, unconscious.

Most sound evolutionary theories, including ours, predict that psychologically normal people will usually be incapable of passionately arguing a position that does not potentially serve their ultimate individual interests.[†] These individual self-interests may also coincide with the interests of others. This possible concordance of interest is of special importance to us because of our adaptation to conjoint projection of threat and to cooperative solutions in pursuit of those interests (Chapter 5). This prediction seems to be very well supported by all the evidence we have.[‡,20] There are many obvious examples. Rich people arguing for

---

*   This quote is variously attributed, including to Napoleon. In the same vein, Voltaire is reputed to have said that "history is the lie commonly agreed upon," while Henry Ford is reputed to have said that "history is bunk."

†   Two crucial points: First, recall that self-interest can involve satisfying the demands of coercive coalitions (Box 10.2). Here self-interest can sometimes be entrained to the larger good *under the right circumstances* (Chapter 17). Second, of course, in general, we always run the risk of being mistaken when we (unconsciously) identify our ultimate interests through proximate devices. In other words, human perceptions are sometimes mistaken, even perceptions of self-interest.

‡   There is much confusion on this issue in some subcultures of contemporary *behavioral economics*. It is sometimes argued that people often do not pursue self-interest. However, most of the strategies for making this assessment fail utterly to recognize that *effective* self-interest must also *recognize the surrounding coercive consensus*. Most of the *self-interested* behaviors that are avoided by subjects in these analyses are precisely the behaviors that would provoke a response from surrounding people to *free rider* (Chapter 5).

flat or regressive income tax rates and poor people arguing for progressive tax rates are simple, easy cases.

However, even when we feel we are arguing a high-minded or even self-sacrificial policy (high-income people arguing for progressive taxation, for example), actually we generally are not. High-income people arguing for a more progressive tax rate are essentially always individuals who believe they can influence the way that governmental money is eventually spent, including in ways that serve the ultimate interests of those high-income individuals.*

This brief synopsis of the basics of human political psychology is vital if we are to understand *history* (also see Fourth Interlude). Everyone writing history is a self-interested individual, just as surely as we and you are. Moreover, the sole criterion we apply in asking if a self-serving policy position is *worth* promoting publicly is whether enough others are likely to agree with us because the policy also serves their interests. *There are apparently no other criteria.* In fact, we are not even concerned with the *numbers* of other people who agree with us, only with the *coercive power* they possess. This criteria will be important in Chapter 13 when we examine the emergence of elite coercive power. Yet again, we are not conscious of this feature of our behavior most of the time.

Fortunately, the contemporary disciplines of history and archaeology are well aware of these defects in the historical record. Many scholars have set about decoding written history to begin to give us a more unvarnished picture of the past. As we set out to test our theory's predictions about human history, we are fortunate, indeed, to be the beneficiaries of this massive, beautiful body of work.†

---

* We will argue later (Chapter 17) that is it possible for psychologically normal people to argue for policies that serve the common good of the global human population, but only under very special circumstances. For now, when we feel our political positions reflect the "national interest" or the "universal interest" we are mostly naïve and self-deceptive. Also see Fourth Interlude.

† Of course, professional scholars, entrenched in long-lived university bureaucracies, run the constant risk of becoming enmeshed in academic subcultures primarily designed to compete for institutional resources. When this happens these individuals become (unconsciously) members of narrowly self-interested groups rather than detached scholars serving the public coalition of the whole. They cease to be honest participants in the universal wise crowd (Chapter 17).

Last refuge of a scoundrel and the better angels of our nature – public human debate

Samuel Johnson is reputed to have remarked that "patriotism is the last refuge of a scoundrel." We all understand this comment intuitively. Ostensibly public spirited declarations are often self-serving. The world's clay-footed professional politicians are entirely too obvious for us to miss their real objectives.

But theories of evolved political psychology predict that we can substitute any public-spirited declaration for "patriotism" and that all of us are "scoundrels" (to some degree) all the time. *All* the public pronouncements of belief by *any* of us—no matter how well intentioned we believe them to be—are produced by proximate psychological devices designed to serve our own self-interests. We have no direct conscious access to the strategic logic of our public social behavior. Usually, we believe we are acting *in good faith* no matter our ultimate (and unconscious) goals.

Does this mean that authentically public spirited action is beyond us? We will have much more to say about this later, but the short answer is, "No." What is required for our public actions to become more authentically what they purport to be is sound theoretical self-awareness and *transparency*. As long as we are in a position to require of one another that all beliefs and claims be subjected to universal, democratized doubt and explicit justification, we are all forced to argue for the pursuit of self-interest in ways that serve the larger common interest as well.

This may seem like a faint, pale imitation of socially responsible behavior, but it is not. We suggest that this is simply the ancient human adaptive trick, evolved over two million years of relatively democratic control over self-interested coercion. In this environment, the better angels of our natures are liberated. We are capable of spectacular generosity, understanding, and humanity *within* democratized cooperative social units. Authentic, honestly doubting, transparent negotiation of social contracts (beliefs) on a pan-global scale is not merely achievable (*under the right circumstances*), it may even already be being born (Chapters 15-17).

Where to now?

With this chapter we have completed our overview of how our theory predicts and accounts for the features of individual humans, including our minds.[21] The last several pages have touched on how our individual properties might produce and interact with the sometimes enormous social units we have come to live in. To understand the emergence and functioning of these larger societies is to understand our history.

As we will see in a moment, our theory of human uniqueness also yields a theory of history of unprecedented scope, power, and economy (Third Interlude). This extensive predictive power, in turn, yields many new opportunities to falsify the theory against the evidence. There is much opportunity for the global wise crowd to engage in productive public doubting of our claims.

At the same time, a strong new theory of history changes profoundly our view of how our contemporary world came into existence and currently functions. This, in turn, will alter the way we conceive of ourselves, our societies, and our common future.

The remainder of the book will be a tour of these richly illuminating fruits of the new theory.

෬

Third Interlude
What does it mean to have a theory of history? Simplicity beneath the illusion of "booming, buzzing complexity"

Human history is often seen as "one damn thing after another" or a series of historical accidents with no underlying law-like properties. In contrast, a few scholars have attempted to develop theories of history assuming that history does show recurrent features resulting from underlying general laws.[*] To date none of these earlier efforts have been successful. However, we will argue in the remaining chapters of this book that we are now, for the first time, in the position to build a simple, coherent, powerful theory of *all* of human history.[†]

Specifically, the new theory of human uniqueness developed in the preceding chapters inescapably implies a strong, specific new theory of history. Moreover, this theory is so transparent, direct, and general that it is susceptible to many independent, clear, potentially falsifying tests against the vast body of empirical evidence, in contrast to all prior theories of history of which we are aware.

This testing will be an exciting, but remarkably challenging part of our journey. We will all be forced to hold in doubt many of our most cherished

---

[*] Examples of sometimes heroic early attempts at coherent theories of history are diverse. These include Aristotle 2500 years ago, Marx in the 19th Century and Jared Diamond (1997, *Guns, Germs and Steel*) very recently, among many others.

[†] A few earlier theories have claimed to be successful in accounting for narrow, highly local, causal sequences. However, the value of these claims, even if true, is extremely limited because of the failure of these theories to generalize to all of history. Our theory does generalize in this required way.

beliefs about the larger course of human events and the origins of human purpose. Of course, this process is just exactly how science always works. We must hold everything we think we know perpetually in doubt.

Nonetheless, this testing process becomes extremely difficult when the beliefs we are holding in doubt impinge on our personal conceptions of the very meaning of human existence or of an individual human life. Scientific doubting in the domain of human history will take time, effort…and some psychological pain. However, we believe the trip—an illuminating emotional and intellectual roller coaster ride—is well worth this ticket price. Among other things, the journey will, arguably, give us what we *really* need to take wise control of our common human future.

Front of the tee shirt, again – What is our theory of history?

It will serve us well, before we push our roller coaster car over the first precipice in the next chapter, to understand the course of the ride we will take.

First, the human story of the last two million years is well described as a series of *adaptive revolutions*. These revolutions are relatively brief time intervals of spectacular increase in *adaptive sophistication.*[*] These revolutions are separated by periods of *adaptive stasis* (intervals, sometimes very long in duration, characterized by little or no *net* change in adaptive sophistication).

In other words, a graphic plot of human adaptive sophistication over historical time will look like a set of steps—the *adaptive staircase*. It will NOT look like a long, gently up-sloping line (Figure 13.1). Some specialists will object to this empirical description of history, but we will argue that they should not. The notion that the two million years of human history is a relatively uniform long, slow, continuous climb is an unverified assumption or an observational preconception. It is not a rigorously tested hypothesis

---

[*]   As we will see in later chapters, adaptive sophistication can be measured in many ways, all giving us very similar answers. For example, we can examine the sophistication of our physical tools. A massive passenger aircraft is a fundamentally more sophisticated tool than a stone hand axe, for example. Further, we can measure human *carrying capacity* (how many people per square kilometer a culture supports). Two hundred thousand years ago, there were probably no more than one million humans on the entire Earth. Today more people than this live in one of the five boroughs of New York City.

in any cases of which we are aware.* We will defend our interpretation of the archaeological/historical record in detail in the next six chapters.

Second, in view of our clear theory of culturally transmitted information and behavioral/technical sophistication (Chapter 10), we can predict the basis for these adaptive revolutions. They must reflect substantial increases in the amount of culturally transmitted information people have access to, including both direct and indirect access (Chapter 10).

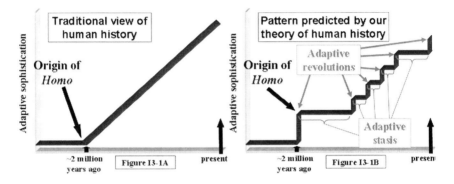

Figure I3-1: What does history look like? We argue that it can best be described as an adaptive staircase (text).

PANEL A: Traditionally, human historical progress has often been visualized as gradual. Small local innovations occur from time to time, unpredictably, producing tiny incremental increases in adaptive sophistication (text). Historical change is the cumulative effect of these many tiny events over long periods of time.

PANEL B: Our theory predicts that human history will consist of relatively brief periods of vast, pervasive innovation (adaptive revolutions), separated by periods (sometimes long) characterized by little or no *net* increase in adaptive sophistication (periods of adaptive stasis). Moreover, our theory predicts precisely why adaptive revolutions and periods of stasis occur (text). These predictions can be tested against the evidence (Chapters 11-17). (Note that the magnitudes of elapsed time (horizontal axis) and adaptive sophistication (vertical axis) are not to scale in this version of the adaptive staircase.)

---

* Reports of *cultural flowerings* are well known in traditional histories. Examples from the Mediterranean basin include the Athenian Greeks, the Imperial Romans, and the early Islamic Empires. However, these are *not* adaptive revolutions, but rather reflect relatively modest political-economic cycles superimposed on the larger adaptive staircase. Their status as "major cultural transitions" is mostly attributable to elite propaganda (Chapter 10; Fourth Interlude), not to rigorous, detached scientific analysis (Chapter 13).

Third, in turn, we can predict with high confidence that the only possible source of increased amounts of culturally transmitted information is an increased scale of human social cooperation. More information can only be developed, stored, transmitted, and used adaptively if more people contribute to the social enterprise.* The details of this effect of social scale are straightforward and we will explore them more thoroughly as our analysis unfolds. Notice also that this assumption will be tested along with the testing of the larger theory itself.

Fourth, conflicts of interest are universal and independent of the scale of non-kin social cooperation. This fundamental principle is to human history what gravity is to cosmology. It is simply not possible to understand anything important about history without fully grasping this law, we argue. For example, nation-states have conflicts of interest precisely analogously to the non-kin individuals we discussed in Chapters 3 and 5. Cooperation between nation-states requires coercive management of these conflicts of interest just as surely and totally as cooperation between non-kin individuals requires it.

Fifth, in contrast to the scale-independence of conflicts of interest, means of coercion that would allow inexpensive, self-interested management of these conflicts of interest are *not* scale-independent. For example, thrown stones are not cost-effective weapons for suppressing conflicts of interest on the scale of nation-states. Thus, each new scale of human social cooperation can *only* emerge under a single, predictable circumstance. *Required is a new coercive technology—one permitting relatively inexpensive and, thus, self-interested, coercion* **on that new scale**.

Sixth, such a new coercive technology is both a *sine qua non* and an irresistible impetus. The new technology does not merely *permit* adaptive revolutions, it *drives* them, inevitably, on our theory. Put differently, when humans are presented with a new weapon, we immediately deploy it in pursuit of self-interest. We are directed in this pursuit by our evolved proximate psychologies, minds highly adapted to precisely this behavior. When we behave this way, we thereby "recreate the ancestral past". However, when the new weapon system allows inexpensive coercion on a new scale, the resulting social system will be the ancestral human social adaptation *writ larger*—a new adaptive revolution.

To be emphatically clear, our theory predicts and requires that the law-like patterns of human history are **dominated by a single recur-**

---

* This apparently remains true even in the very recent past where non-biological forms of information storage (formal written language, electronic data storage, etc.) became available. This scale-dependence of useful access to culturally transmitted information seems unambiguously true before the invention of written language a mere fifty-five hundred years ago.

ring process. A new weapon system of enlarged scale is "invented." This invention is generally serendipitous and initially serves some narrow, local adaptive application—a new hunting tool, for example. Nonetheless, humans immediately deploy this weapon in pursuit of social self-interest, *inadvertently* suppressing conflicts of interest on the new scale permitted by the novel weapon. A new scale of social cooperation emerges, producing an adaptive revolution.

As we will argue, this process has recurred, at ever-increasing scales, repeatedly throughout our two-million-year history *through the present moment*. Again, *all* the salient transitions of the human story apparently arise as a result of this single process.

When did we become "fully human"?

An old saying goes, "When you have a hammer, everything looks like a nail." Our understanding of evolution of genetic design information has grown explosively since Darwin. But this knowledge sometimes becomes a "hammer" and we look to turn all human adaptive change into a "nail." Thus, we are sorely tempted, and often simply presuppose, that most of the changes we see in the record of human behavior over the last two million years are driven primarily by genetic change.

Our theory of history clearly cautions us *not* to make this assumption. While genetic redesign was obviously conspicuous early in our two-million-year history, it is an untested, questionable assumption that it has continued to play a significant role, thereafter.[*] Indeed, our favored working hypothesis, pending new evidence, is that people living five hundred thousand years ago, and possibly much earlier, were fully as human as all of us are today.

Human behavioral sophistication is, first and foremost, a function of the scale of our social cooperation. Thus, *genetically*, all of our ancestors were probably fully human long, long ago. *Socially*, all of us can still become "more human" and more humane. We will have much more to say about this in Chapters 11 and 17.

---

[*]   Indeed, even the genetic changes early in human evolution (approximately one to two million years ago) are only secondarily causal on our theory. They are primarily adaptive responses to the new pattern of human social behavior (Chapters 5, 9, and 10). They do not *drive* human evolution on this view, *they merely support it*.

"Accidents of history" – what limits our ability to predict history?

Newton's Laws predict/generate Kepler's Laws of planetary motion. In contrast, Newton cannot predict that Jupiter is orange and Neptune is blue. The motions of the planets are law-like. In contrast, the chemical composition of the planets is accidental, *contingent* from Newton's vantage (outside the domain of his theory). However, these variable planetary compositions are also irrelevant to Newtonian mechanics. Thus, they do not limit the theory *within* its domain.

To understand what our theory of history can and cannot do, we must comprehend what it explains and what it takes to be contingent (outside its domain). We must also understand how such contingencies limit our ability to use or test the theory.

First, conflicts of interest are managed by humans, never eradicated. Thus, each new scale of human social cooperation will also create new opportunities for self-interested capitalization on the larger social enterprise. No matter its scale, each human social unit will encompass relentless, perennial arms races between the coalition of the whole and sub-populations who will inevitably attempt to exploit the social system from within in pursuit of self-interest. Moreover, the details of these processes will be substantially affected by contingency—accidents of birth, death, assassination, "luck" in the markets, and so on.

These relentless conflicts generate much of the sound and fury of our day-to-day history. However, during periods of adaptive stasis, this noise signifies nothing. It produces superficial cosmetic alterations that have little or no substantive consequence. The resulting *simmering equilibrium* (a steady state) will be predominantly characterized by many important *universal* and *predictable* properties.

In contrast, during periods of adaptive revolution, the pursuit of individual self-interest by members of competing sub-populations helps drive the directional change associated with the revolution. However, even here, the diverse, unpredictable superficial beliefs (Chapter 10) of these competing subgroups are mostly irrelevant and the essential features of the ultimate outcomes are, again, highly predictable.

Second, adaptive sophistication is limited by access to reliable, culturally transmitted information *and* by local circumstance. For example, we cannot currently grow maize on the polar ice caps. These local variables are outside the domain of our theory but still effect the details of historical change. However, this kind of location variable does not defeat us. We will be primarily concerned with human historical change as a function of sequential events, not location.

Third, inventions of new weapons are locally serendipitous, at least in part and on the time scale we care about. Even here, though, we retain some predictive power. For example, as social coalitions get larger, they generate more culturally transmitted insight/information. Their inventive power increases and new weapons come faster at each new scale as a result. Human historical adaptive change is expected to be autocatalytic, its pace increasing over time, as it appears to be (Chapters 11-17).

Fourth, what else about history is *contingent* (not law-like, unpredictable)? Some professional scholars will now be shouting, "Everything that matters!!!" By this objection, they will often mean that local human cultural beliefs are as spectacularly varied as the appearance of different flowers or the sexually selected tail feathers of different bird species— and equally unpredictable. About this fact, they are, of course, perfectly correct.

However, advocates of this view often also mean that most of these locally multifarious beliefs are the *causes* of adaptive differences between cultures and of historical change. Thus, history itself is just a massive string of local historical accidents. In making this claim, proponents of this view are almost perfectly and totally *wrong*, we suggest.

The picture we will defend is simple. To paraphrase the old joke, "Local cultural beliefs are often utterly unpredictable in many details and cultural beliefs affect history. Unfortunately, the details of cultural beliefs that are unpredictable do not affect history and those that affect history are not unpredictable."

More specifically, we mean that cultural beliefs are important to history only as the proximate (as opposed to ultimate) agents of *strategic* human behavior. As wildly unpredictable as local beliefs are superficially, their adaptively—and historically—relevant behavioral consequences are generally very predictable. The ultimate logic of our adaptive behaviors—and the behaviors themselves—are, we will argue, mostly *universal, simple,* and *comprehensible*.

Hubris, disrespect, and evil intent – the ethics of a theory of history

Astonishingly, it has become morally *de rigueur* in some professional academic circles to regard history as manifestly contingent, precisely contrary to the argument we will make. This view sometimes claims to respect the human capacity to determine our own history—to represent an ethical respect for humanity. It is no such thing.

It is scientifically bankrupt—defining a simple, general analytical under-standing of history as both unattainable and immoral to attempt.[*]

In fact, we argue that this view, most emphatically, does *not* respect our common humanity. The fundamental qualities each of us and every one of our ancestors share, led all people everywhere at all times to behave in "human" ways. These ways are, we will argue, most predict-able, indeed. In coming to understand these quintessentially human ways, we do not merely gain the insight we need to explain history. We also empower the better angels of our universally human nature for a realistic pursuit of a more humane future—an endeavor belonging to each of us and on behalf of all of us.

---

[*]   This view can also reflect our unconscious conceit (our belief that the history of an organism as complex as us surely must not be subject to simple underlying laws).

Chapter 11
The human mastery of Earth begins – the behaviorally modern human revolution and the first information economy

The first humans arose around 1.8 million years ago (Chapter 7). There-after, people continued to live in ways that changed very little for next 1.7 million years! The ancient life ways of these people continued, millennium after millennium, in spite of the fact that they had evolved our characteristic breeding system (Chapters 6 and 8), language (Chapter 9), and enlarged brains (Chapter 10) early on.

Through all this time the oceans and polar ice caps penned us into Africa and southern Eurasia. This period was so vastly long that humans differentiated into two subpopulations that looked rather different from one another.

*One of these distinct subpopulations, the **Neandertals**, occupied large chunks of southwestern Eurasia, from present day London to the Balkans, from Germany to the Middle East. The second human subpopulation is called "modern," meaning they looked just like us. **Moderns** occupied Africa and parts of the Middle East.*[*,1]

Neandertals were somewhat shorter and more heavily built, conventional adaptations to the colder Eurasian climates. Moderns were taller and more slender, an adaptation to warmer equatorial climates. The faces of moderns and Neandertals also looked rather different, with Neandertals having more protuberant brow ridges, bigger projecting noses and a receding chin in comparison to moderns.

---

*   Stringer and Gamble (1993); Shreeve (1995); Klein and Edgar (2002).

The boundary between the modern and Neandertal populations in the Middle East oscillated a little north and south repeatedly over several hundred thousand years, but remained essentially stable. In spite of their morphological differences, moderns and Neandertals made the same simple kinds of tools and lived the same hunter-gatherer lives. Their residential sites were mostly small and transient. Large or permanent towns were certainly not part of their experiences.

Even more remarkable than this 1.7 million years of adaptive stasis, was the fact that it ended abruptly, not gradually. About fifty thousand years ago, something changed radically and profoundly. People started making new things, going new places, acting in remarkably new ways.* This sudden change was initiated by a small population of modern humans that started increasing in number rapidly and spilled out of Africa. First they moved into Eurasia, sweeping Neandertals before them.† Ultimately, they spread to all the places of the world. Many of these new homes, including Australia, North America, and South America had never before seen a hominid (Figure 11.1).

*Consider the implications of these properties of our history. Our evolved social psychologies cause each of us to think of local neighbors who look and act just like us as "special" and "unique." The truth is precisely to the contrary. All of us alive today are the recent descendents of a small group of African **modern** people. Moreover, the many local populations and cultures of the contemporary human population is something never seen before in our planet's history. We are a massive global population of **billions** of individuals very recently emerged from a tiny local population.*

We tend to think of ourselves as having African, Asian, European, Native American, Aboriginal, or Polynesian ancestry. But this identity is really like being from London or Chicago, it is recent and superficial. All humans alive today are brothers and sisters in a very real sense. We are very, very much like one another. And we share one more vital legacy. Our tiny group of

---

*   There has been a recent debate among academic specialists about whether this transition was more or less abrupt. Our theory accounts for the evidence producing this confusion and controversy, as we will discuss in a moment.

†   Neandertals were either driven to extinction or numerically overwhelmed and assimilated by these expansionist moderns. As a result, all humans alive today have either **only** modern ancestors or **mostly** modern ancestors with a modest admixture of Neandertal and/ or other *non-modern* ancestry. Attempts to sequence the Neandertal genome from DNA retrieved from fossils are currently underway. These may give us a better understanding of this piece of our family tree. However, notice as this story unfolds that understanding whether we do or do not have some Neandertal ancestry will have little impact on what it means to be a *modern* human.

common modern ancestors was evolutionary revolutionaries. They aban-doned 1.7 million years of history and forged a new world, the world we now all take for granted.

Understanding how our theory accounts for and predicts the revolution these ancestors made is our goal in this chapter. Their invasion of Australia and the Western Hemisphere required a new capacity to contend with the open ocean and with arctic conditions. As they arrived on new continents, they manifested an extraordinary command of their new local environments. They redesigned ecosystems wholesale to serve human needs, perhaps even contributing to the extinction of many large animals that had long survived before human arrival.

By around thirty-five thousand years ago, these moderns were creating sophisticated art—something humans had never done before.* Moreover, the "manufactured" objects of these moderns were unlike anything made by their ancestors. Their tools were more sophisticated and complex, their ornaments newly abstract and esthetically evocative.

These moderns began producing the first ever large "public works" including enormous earthen mounds bigger than a dozen football fields and complex fishing traps extending over thousands of square meters.

The totality of this new scope and sophistication was the first of the adaptive revolutions making up the human story since our origins approximately 1.8 million years ago (Third Interlude). This adaptive revolution changed the human condition in a fundamental way and ultimately produced all of us alive today.

Does our new theory of history predict this revolution? We will argue in this chapter that it does—directly, simply, and powerfully. This insight, in turn, will teach us something very valuable about what it means to be human.

---

* A Google image search with "Paleolithic art" and co-terms like Chauvet, Lascaux, or Venus will produce a trove of powerful images. All the stunningly beautiful things you will see began to be produced suddenly by the first behaviorally modern humans around thirty-five thousand years ago, after nearly 1.8 million years of nothing even remotely approaching this level of artistic sophistication.

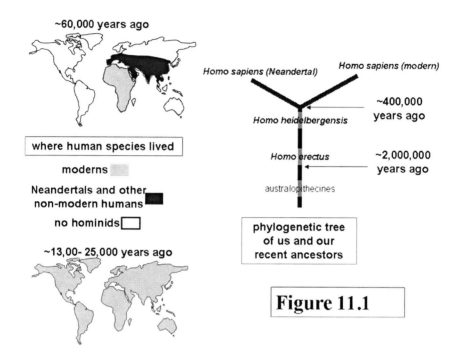

~60,000 years ago

where human species lived

moderns

Neandertals and other
non-modern humans

no hominids

~13,00- 25,000 years ago

Homo sapiens (Neandertal)   *Homo sapiens (modern)*

Homo heidelbergensis   ~400,000 years ago

Homo erectus   ~2,000,000 years ago

australopithecines

phylogenetic tree
of us and our
recent ancestors

Figure 11.1

The origins of the new globetrotting sophisticates – artists, merchants and eco-monopolists

Our modern ancestors could not only do things no earlier human could do, they also had vast new opportunities their ancestors did not have. They quite literally took over the world in a way that no large creature had ever done in the entire history of the Earth. They began the course we still, to this day, pursue—adapting the entire planet to serve human interests.

These people showed new behaviors we still display today, yet they were anatomically indistinguishable from their immediate ancestors. Thus, these *moderns* are often referred to as *behaviorally modern humans*. Their emergence is called the *behaviorally modern human revolution*.

A long-standing, widely popular hypothesis for the origin of this revolution is that some genetic change in the African pre-modern population created modern behavior. Perhaps a new genetic variant filled in some last missing piece of the suite of human mental faculties,[2] or enabled the final emergence of complete language.[3]

On the contrary, we will argue here that such explanations are unlikely—even unintentionally racist. We suggest that Neandertals had just as great a claim to being human as we do. African moderns evolved a new scale of social cooperation, not a new genetic capability. More specifically, their social adaptation was transformed as a consequence of the "accidental" local invention of a new weapon, which, in turn, allowed the uniquely human management of the non-kin conflict of interest problem to be prosecuted on a much larger scale than ever before.

Put differently, a new weapon allowed *law enforcement* to encompass much larger groups of people, yielding the fruits of information sharing on a much larger scale (Chapter 10; Third Interlude). This new scale of social cooperation gave them the same kind of advantage over Neandertals as European colonialists had over Native Americans just four hundred years ago—again, *larger social scale not genetic superiority.*[4]

Everything characterizing modern humans unfolds as a consequence of this single original causal event, the local invention of a new, more effective weapon. Indeed, our theory predicts that, had this new weapon been invented in Eurasia among the Neandertals instead, one of their descendents might be writing this book, trying to understand why and how "archaic Africans" were driven to extinction by "modern," "fully human" Neandertal-kind.

We will argue that the substantially enlarged scale of social cooperation produced by this new weapon engendered the observed *behaviorally modern* up-tick in adaptive sophistication. Included will be some innovative modifications/expansions of the ancient ancestral human trick of cooperative *economic exchange* (Chapter 10) such as the expansion of the domain of *currency* and *capital*. We will further argue that this expanded domain of economic cooperation provided new opportunities for individual specialization, producing improved sophistication of tools and techniques as well as idiosyncratic new skills like art.

In other words, all the features of the behaviorally modern human adaptation have a single, simple origin, an expanded scale of social cooperation. This expanded social cooperation, in turn, was enabled by the application of a new weapon, representing a new means of self-interested coercive suppression of conflicts of interest sufficient to this larger scale. Though the scale is new, the underlying logic of this characteristically human management of conflicts of interest remains unchanged (Chapter 5 and below).

This picture is a very precise. If our theory is right, it *must* be accurate. If our theory is wrong, failure of these highly specific claims should be obvious.

Before testing our simple theory against the empirical evidence, it is useful to return to the competing genetic theories of modernity. As we mentioned in Chapter 9, the hypothesis that language evolves late in moderns fails against the fossil evidence, probably decisively. Behavioral modernity as a product of new genetic language-related endowments has no empirical support.[*]

Another alternative is the *new mind* version of the genetic hypothesis. This hypothesis suggests that moderns were genetically endowed with some unprecedented new capacity for "abstract thought" or the like. This hypothesis has the dubious virtue that it is currently impossible to test and falsify. That alone is *prima facie* reason to be suspicious. But we can do better. We can show that this hypothesis is perfectly unnecessary and almost certainly naïve, as follows.

Consider the pilots, scientists, and engineers who collaborated to put footprints on the Moon, sequence the human genome, or invent quantum mechanics. Only fifteen to twenty generations ago, most of their ancestors were illiterate.[†] They were Medieval peasants living a borderline subsistence on the equivalent of about $2 a day, generation after generation with monotonous predictability. Just 250 generations before that, their ancestors were hunter-gatherers who had never even conceived of a permanent home or a domesticated crop.

Are we compelled to believe that we contemporary humans are the products of an "agricultural" genetic change in our ancestors 270 generations ago, followed by "Moon-walking" genetic changes in our Medieval ancestors after that? Of course, we are most emphatically *not*! Well then, where did this capacity to walk on the Moon while listening to Beethoven or Mick Jagger on an iPod come from?

The details of our most recent new capabilities are the subjects of later chapters, but the general principle is simple. Human genetic change had nothing to do with any of this. We suggest that genetic change is no more likely to account for the behaviorally modern human revolution than it is to account for the "Moon walking revolution."

Again, the prediction of our theory is clear. The behaviorally modern human revolution was a purely social revolution just as surely as the European conquest of North America or the internet revolution. The behaviorally modern revolution had demographic (and, thus, genetic)

---

[*]   Recall that such a theory would be logically incoherent in any event (Chapter 9). Talking does not help us unless we can assure that the information we exchange is reliable. Reliability can only be assured *within* a cooperative coalition and the size of that coalition determines how much information we have (Chapter 10).

[†]   As were ours and yours, as well.

consequences and aftershocks. Some populations prospered, others were lost or were assimilated as small minorities. But, these outcomes were *irrelevant by-products* of *social* processes, *not* the *drivers* of those processes, on our theory.

How revolutionary was the revolution?

Professional scholars specializing in the behaviorally modern human revolution have been concerned with how rapid this historical transition actually was. At first, it appeared to be instantaneous on an evolutionarily timescale, occurring at a point around thirty-five thousand years ago. This perception arose because the revolution was originally recognized in only one idiosyncratic local neighborhood, Western Europe, which was hit by the wave front of the modern population as it spread out of Africa in full throat. It is now clear that this revolution occurred in Africa over a significant period of time. The primary scholarly debate at this writing is whether it occurred relatively rapidly over approximately five to ten thousand years or more slowly, perhaps over fifty thousand years.[5]

Two points about this debate will matter to us here. First, even if this process took fifty thousand years, this is still a rapid revolution on the time scale of the entire 1.8-million-year human story. Second, these contrasting views of the speed of the revolution are based on different weighting of various pieces of the empirical, archaeological record. Our theory puts this diverse evidence in a novel context and allows us to suggest a new resolution of the timing question.

In brief, our theory predicts that the archaeology reflects two phases. The first phase involved the invention of a new weapon in its initial, primitive, poorly performing form, beginning around seventy thousand to one hundred thousand years ago. The social effects of this modest improvement in weaponry, if any, were limited and sporadic. The residues of these weapons themselves represent much of the archaeological evidence for something new at this early time. There is little *other* evidence for any *behavioral modernity*.

The second phase was associated with a relatively dramatic refinement of this new weapon. This era also involved a large increase in the scale of social cooperation engendered by this newly improved weapon, on our theory. This enlarged scale of social cooperation, in turn, produced dramatic, pervasive social consequences, leaving a newly rich archaeological footprint. This second phase began around fifty thousand years ago and was very rapid, taking less than ten thousand years.

Congruence in space and time – a new weapon and a social revolution

Any new scale of social cooperation *must* be preceded by a novel technology that allows self-interested application of coercive threat, resulting, in turn, in the emergence of an expanded scope of enforced cooperation, on our theory (Chapter 5; Third Interlude).

As we will see in more detail below, one of the key features of a new coercive technology is its range. Thus, we predict that a new scale of human social cooperation will be preceded *very briefly* (on an evolutionary time scale) by the invention of a weapon system with substantially increased range.[*]

The behaviorally modern human revolution is our first opportunity to test this prediction. We can attempt to falsify the theory on the empirical evidence with robust confidence. The theory apparently survives this first test remarkably well.

Specifically, for approximately the first 1.7 million years of human history, we have no evidence for any weapons with substantially enhanced range over that of a thrown stone. Then, around fifty thousand years ago, a weapon with a profoundly longer range was developed. This weapon is the spear-thrower—more commonly known by its Aztec name, the *atlatl* (Figure 11.2). This weapon amplifies the normal human throwing motion, allowing a bolt or dart to be lofted to relatively great distances—certainly fifty to seventy meters in the authors' non-expert hands and, in expert hands, perhaps out to over one hundred meters.

To understand the revolution this new weapon produced, it is essential to take a detailed look at the properties of thrown stones (and hand thrown spears) in contrast to the atlatl bolt. It is fruitful to visualize these different weapons at two extremes—up close and very far away. Up close, ten to fifteen meters, thrown stones are effective, even against individuals carrying the kinds of protective or warding gear that would have been available in the Paleolithic—wood or hide shields, for example. This effectiveness is partially because it is impractical to shield the entire body this way, and any exposed bit—like the feet or the head—can easily be hit by an expert thrower, bringing down the target individual. Also a hail of high-velocity stones from nearby will tend to knock the shielding from the hands.

However, as we throw stones from very far away, we must loft or lob them. We are not able to impart the extreme velocity to a thrown

---

[*]   Properties of a new weapon system other than range can also matter as we will see in later chapters, but for now only range need concern us.

stone that would let us throw "on a line" over long distances in the face of gravitational falling. When under long-range fire from lobbed stones, it becomes very practical for a target individual to shelter beneath a hand-held shield as these high-angle, lower velocity projectiles come in. Stones are not very effective at extreme range.

Contrast the thrown stone with the atlatl bolt—a very high-velocity, aerodynamic, sharp, penetrating projectile (Figure 11.2). At close ranges these bolts will be at least as effective as thrown stones. They also have the additional capability to penetrate shielding unless this shield is very heavy which, in turn, can injure the target or make his/her shielding too unwieldy to continue to use (pin-cushioned with bolts).

However, in contrast to thrown stones, the velocity and penetrating power of the sharp-pointed atlatl bolt means that it can still be deadly when rained down on a target from a relatively great distance. This increased range allows much larger numbers of people to engage in self-interested ostracism, supporting an increased scale of social cooperation (see below for details).

Thus, the atlatl has the properties of a weapon sufficient to produce an adaptive revolution on our theory. In order to test this prediction, we need to be able to determine when and where this weapon comes into use.[6]

The *throwing stick* and *bolt shaft* of the atlatl are normally mostly wood and, thus, highly perishable (see Figure 11.2 for details of this weapon). Therefore, these wood elements are not reliable evidence for invention of the atlatl in the archaeological record. However, sharp stone bolt *points* are virtually imperishable and are extremely reliable indicators of the emergence of this weapon. Specifically, atlatl bolt points must be symmetric, aerodynamic and of the proper size (relatively small) (see Figure 11.2C). These characteristics are in sharp contrast to hand-held thrusting spears, where a larger, heavier point is optimal (Figure 11.2C).

One of the world's experts on these weapons is our Stony Brook colleague, John Shea. John has followed our development of the theory since its original publication in 1999 and has recently undertaken the vital work of understanding the history of this weapon. His masterwork on this subject was published in 2006 in The Journal of Archaeological Science.[7]

This work has important implications for us here. Calibrating bolt tip size and shape based on cultures that still used the weapon until very recently, John finds that the atlatl, in its sophisticated modern form, was first developed around fifty thousand years ago—precisely at the beginning of the explosive phase of the behaviorally modern human revolution. Moreover, this invention occurs in exactly the East African/ Middle Eastern population that founded the behaviorally modern human populations.

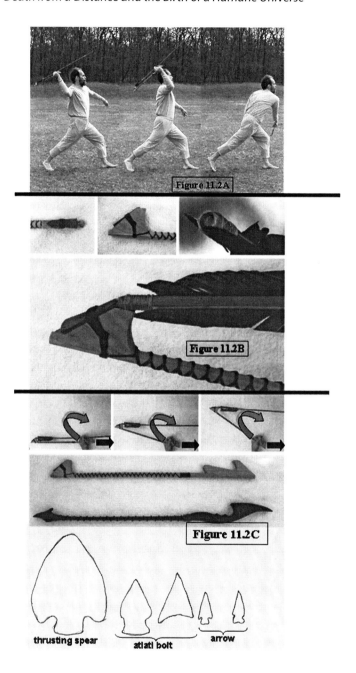

Figure 11-2: Illustrated are key features and properties of the atlatl or spear-thrower.

PANEL A: The throwing motion with the atlatl is illustrated. Notice that the body and arms execute a conventional human throwing motion. The atlatl or throwing stick extends the effective length of the hand, dramatically increasing the velocity of the ejected bolt compared to a spear or stone thrown directly from the hand.

PANEL B: The top right image is an enlarged view of the base of the atlatl bolt, the projectile. The top left two images show the *hook*, the part of the atlatl inserted into the depression in the base of the bolt and imparting the momentum from the throwing motion to the bolt. The image at the bottom of this panel shows the base of the bolt seated in the hook assembly.

PANEL C: The long central image shows two examples of contemporary atlatls. At the right end of each atlatl is a grip region designed to be held tightly by the throwing hand. At the left end is the hook assembly (see Panel B). The sequence of three images at the top of the panel illustrates the acceleration of the bolt by the hand as amplified by the throwing stick. The forward movement of the hand (black arrow) begins the final stage of acceleration of the bolt as the contributions of the trunk to acceleration are completed (see Figure 7.5 for the human throwing motion). Near the end of the throwing motion the rotation of the wrist (gray curved arrow) imparts a strong additional acceleration to the bolt as a result of the mechanical advantage provided by the long atlatl. At the bottom of the panel are tracings of a stone point from a thrusting or stabbing spear. Also shown are two examples of points from atlatl bolts. Notice the much smaller size of the bolt points compared to the thrusting spear point. As well, the bolt points are much thinner in cross-section, producing a highly aerodynamic profile in flight. As a result of these properties, advanced atlatl bolt points can be readily distinguished from earlier thrusting spear points in the archaeological record. Notice also that atlatl bolt points are significantly larger than the points on arrows projected by a bow (Chapter 12).

Thus, the first important prediction of our theory is remarkably and precisely fulfilled. A new weapon with substantially increased range is invented immediately before the ensuing behaviorally modern human revolution in exactly the predicted location.

There is an additional technical wrinkle to this story. Beginning between roughly seventy and one hundred thousand years ago, symmetric points clearly intended for hafting onto some kind of smaller spear or knife handle become wide-spread in Africa.

By John's ethnographic criteria, these points are too large to be "modern" long-range atlatl bolt points. However, their shape clearly suggests that they are designed to be aerodynamic. Some physical anthropologists believe them to be projectile points.[8] Adding fuel to this fire, bone points are also found, again in Africa, for the first time around seventy thousand years ago. Bone points are less dramatically sharp than a stone point, and it is thought that they are only effective with a delivery system like the atlatl (but only at short range).

Our theory suggests a simple organization of this archaeological evidence. Specifically, we propose that the early projectile points (about 50-100,000 years ago) were used with a relatively primitive version of the atlatl, designed to project moderately heavy bolts over short ranges. All ancestral human weapons are invented for some other purpose—hunting, for example—and then are co-opted for social coercion. The early, heavy atlatl might have allowed lower risk hunting of more dangerous large prey or of raft-based hunting of near-shore sea mammals, like seals, for instance. To the extent that this weapon also extended the range of coercive threat (modestly) it would be expected to produce, at most, a very limited increase in the scale of social cooperation. These short-range weapons and their small social effects, if any, are the likely origin of African evidence suggestive of modernity at seventy thousand to one hundred thousand years ago, in our view.*

At some point around fifty thousand years ago, the key additional technical innovations occurred to produce the fully modern, long-range atlatl (Figure 11.2) as John's analysis indicates. This new weapon would have been invented to solve a local adaptive problem—perhaps to allow hunting of even larger terrestrial or marine prey. However, our theory predicts that this weapon would rapidly produce a substantial increase in the scale of social cooperation when co-opted and redeployed for self-interested social coercion. A dramatic adaptive revolution is predicted to ensue, as the evidence indicates that it does.

This foundational insight opens many other doors, allowing us to account for most or all of the diverse features of the behaviorally modern human revolution with an economy and power no earlier theory provides (below).

"Big business" – the crucial details of increasing the scale of human social cooperation

Some important work remains to be done to establish various details of the relationship between the many properties of a dominant coercive weapon and the scale of social cooperation within a society. However, we can provide a very useful initial solution at the level of the first behaviorally modern societies.[9]

As we have seen, human adaptive sophistication is limited by available culturally transmitted information (Chapter 10). Moreover, the

---

* Indeed, the primary criteria for early African modernity are weapons-associated artifacts themselves. Thus, this early phase of ostensibly modern behavior may not actually reflect a significant increase in adaptive sophistication.

amount of culturally transmitted information is certainly limited by the number of people who cooperate to generate, store, and transmit this information—that is, by the scale of human social cooperation. At the moment what matters is how many people come together for cooperative exchanges in one place and at one time at a *local gathering.*

Our theory lets us estimate maximal adaptive local gathering size as follows. We will use the *rational actor fiction* again as we did in Chapter 5. That is, we will pretend that ancestral humans were aware of the ultimate strategic logic of their social actions and behaved rationally on the basis of that knowledge. Of course, it does not actually work that way, as we have seen. Rather, Darwinian processes—acting on genetic and culturally transmitted information—shaped ancestral human proximate mechanisms that caused people to behave *as if* they understood this logic in the narrow specific context where they found themselves.

The definitive property of this context will always be the logic of coercive management of conflicts of interest. Use Figure 11.3 to help understand the following crucial description of how coercion works here.

Suppose a local population only has access to a coercive weapon with an effective lethal range of about fifteen meters, the range of a thrown stone. Thus, individuals encircling a small minority of *free riders*, while remaining in range, would form a circle of $2\pi r$ or roughly ninety meters around. If each individual needs a meter of lateral space to throw, this means about ninety individuals could participate in a self-interested coercion event. Any group up to ninety adults would be able to engage in self-interested coercion producing *enforced cooperation* (Chapter 5; Figure 5.1). Other factors, like local availability of food and water, might also put upper limits on the size of local gatherings. However, these groups will tend to be as large as other conditions allow, but only up to about ninety individuals in order to maximize adaptive access to information and to maximize return/cost ratio of self-interested coercion (Chapter 5).

Why do we say that local gatherings will be limited to about ninety individuals in this case? The logic is apparently straightforward. As local gatherings grow larger, there will be increasing competition for profitable coercive acts. Remember that the logic of human coercion is that it must be *profitable* or *compensated* (Figure 11.3B, left hand case). No more than ninety individuals can encircle and effectively threaten a target and thereby gain a share of the confiscated assets of a free rider or cheater (Chapter 5). Thus, we expect that as local gatherings exceed ninety adults, they will tend to fission into two separate local gatherings.

A related effect of modest weapon range will also conspire to limit the size of local gatherings.[10] If these congregations involve exchange of assets, as they certainly did, "what can be exchanged can be stolen."* When about ninety individuals are present, such attempted theft can be managed through conjoint ostracism. However, if local gatherings grow much larger, a new self-interested strategy becomes viable—parasitic cooperation or *organized crime*. Specifically, a sizable subset of individuals can gang up on single individuals to extort or steal what she/he possesses.

This strategy can be thought of as reversing the free rider problem. An internally cooperative subgroup of about ninety individuals within the "over-sized group" becomes an entity "cheating" on the larger cooperative enterprise—stealing from individuals and smaller subgroups of cooperators. Strategically rational approaches to this parasitic goal would probably include specialized raiding of wealthy large gatherings or running a *protection racket* from within periodic gatherings.

In either case, this group of "criminals" would have coercive dominance over individual targets. Moreover, when multiple target individuals teamed up to ostracize these "gang members," they would find themselves in something approaching one-on-one combat (Figure 11.3C, left hand case). As always, this is an unacceptably costly position (Chapter 5) for *both* the "cooperative" and the "criminal" subgroups. For each group, the best available option is to go their separate ways, to fission into groups of about ninety individuals.

The crucial point is that the equilibrium outcome will be group sizes consistent with everyone participating in coercion and not any larger given the weapons available.[†]

However, now consider what happens when the modern atlatl comes into this world, with its lethal range about ten times greater than a thrown stone. The maximal size of the local gathering increases to somewhere in the range of nine hundred individuals. The limitations we just mentioned still occur, but now at much larger group size. For example, nine hundred individuals can now *all* derive the various self-interested benefits from participating in coercive ostracism and seizure of free rider assets (Figure 11.3B). Moreover, it is now strategically rational to suppress organized crime on this new scale. An entire "criminal gang" of ninety individuals, say, can be cost-effectively ostracized (Figure 11.3C).

---

*   Reprising economist Jack Hirschleifer's famous quip mentioned in Chapter 10.

†   As we have seen many times before with evolved behaviors, this outcome is "insane" in a larger sense. The resulting smaller cooperative gatherings will be much poorer than a larger gathering that functioned, but there is no strategically viable way to create this larger functional unit *as long as decisive local weaponry is insufficient to support the necessary law enforcement.*

The square law for coercion from a distance (Chapter 5) now operates to the benefit of each member of the entire, much larger local gathering.

The consequence of this coercive deployment of a longer range weapon is that local gathering size will increase substantially, producing a dramatic increase in the opportunities for productive cooperation and exchange of goods, mates, and information. This powerful new, enlarged social unit is the agent of the behaviorally modern human revolution, we propose. A single *cause*, the introduction/invention of a new, more effective weapon, is the *sole* rate-limiting event in any populated local environment. All the other features of the behaviorally modern human revolution are merely diverse effects of this single, simple, original cause.

Open societies or closed? – The limitations on cooperation

We need to ask one last question before examining some illuminating details of the empirical record of the behaviorally modern human revolution. Can different local gatherings of the maximal weapon-determined size have had very different memberships at different times, say in the Spring and the Fall, for example? This question is important.[11] If local gatherings can be partially overlapping in individual membership, they can transmit information throughout a potentially vast interconnected network of local gatherings over time. We must understand the answer this question to see, in turn, what the actual *effective size* of behaviorally modern human societies was.

It appears that an indefinitely large social unit cannot be produced by the coercive processes we are exploring. Rather, smaller sub-gatherings occur at various times, but the largest sustainable local gathering is a discreet closed unit, approximately nine hundred individuals in the behaviorally modern case.

The logic is likely to be as follows. When hypothetical, partially overlapping groups gather, the interests of "excluded" individuals are violated. They miss opportunities for exchange of mates, goods, and information. These individuals, in turn, have an incentive to exclude "cosmopolitan" attendees from other gatherings in deference to individuals who bring *all* their assets to recurring meetings of the same local coalition.

This exclusion has the effect of *closing the units* making up the largest possible local gatherings. It is convenient to refer to this closed social system consisting of the maximum local gathering as a *local coalition*.

Of course, some individuals will cross the resulting boundaries of such a world on a more or less permanent basis—for example, in search of mates—essentially defecting. Some illicit crossing of this sort will be tolerated by group members who are also in search of new mates.

Moreover, local coalition boundaries can be systematically crossed in certain other special, narrowly defined circumstances. These occur when external individuals can enter without creating unacceptable conflicts of interest. For example, an individual could take locally produced goods to an unrelated local gathering, where they are now exotic, and exchange them for different local goods, returning with these, also now locally exotic goods to be exchanged with local members of the first local gathering (Box 11.1).

However, anytime such inter-coalitional trade could potentially enhance the competitive position of one local coalition over another, it will tend to be suppressed by the disadvantaged local coalitions.

What type of benefit does this leave for such cosmopolitan trade? Excellent candidates are goods that are difficult to *reverse engineer* or, for other reasons, are difficult for a trading partner to make instead of continuing to acquire through trade. Preserved foods made from locally available components would certainly work. For example, a coastal coalition might trade dried marine fish for powdered acorn flower produced by an inland forest coalition. *Currency* (below) is also an excellent candidate for such exchange.

In contrast, easily exploited insights, readily cloned technologies and state-of-the-art weaponry will tend be suppressed as items of sanctioned exchange. Think of how happy contemporary corporations and nation-states are to exchange goods (indirect access to unknown information) with one another and how loath they are to exchange industrial and military secrets (direct access to information).

In overview, we expect exchange of culturally valuable information to be restricted to local coalitions and to leak slowly out of them and across the landscape. In contrast, certain subsets of locally exotic products might be much more widely exchanged.[12] Thus, again, we expect local coalitions of approximately ninety adults to determine the level of *pre-modern* adaptive sophistication and of about nine hundred adults to determine the, much higher, *behaviorally modern* level of adaptive sophistication.

It is most illuminating to see how well the various elements of this simple picture serve to explain the behaviorally modern adaptive revolution. We will explore several selected, important elements of this evidence.

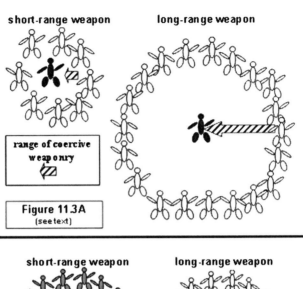

short-range weapon   long-range weapon

range of coercive
weaponry

Figure 11.3A
(see text)

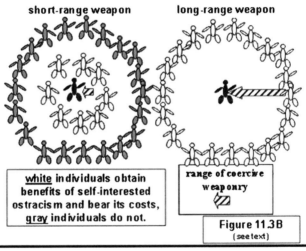

short-range weapon   long-range weapon

white individuals obtain
benefits of self-interested
ostracism and bear its costs,
gray individuals do not.

range of coercive
weaponry

Figure 11.3B
(see text)

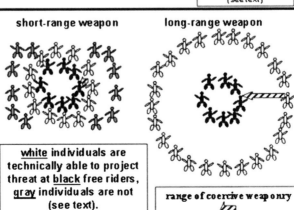

short-range weapon   long-range weapon

white individuals are
technically able to project
threat at black free riders,
gray individuals are not
(see text).

range of coercive weaponry

Figure 11.3C
(see text)

Box 11.1: The adaptive logic of the traveling merchant

When a traveling merchant brings exotic goods into a local gathering (main text above) he (or she) faces potential conflicts of interest with the locals. Why do they not just steal what he has? *Any members of the local gathering who do this (thieves) are also stealing from the other members of the local gathering who may trade with the merchant.*

Thus, for these trade-beneficiaries there are two options in the face of potential thieves. First, they can steal directly from the merchant, themselves, and obtain a one-time benefit. (Victimized merchants will not return.) Second, alternatively, they can protect the merchant by ostracizing local thieves. When they make this second choice, they obtain a net of two benefits *in addition to access to this batch of the merchant's goods.* They *also* obtain the assets of the would-be thieves, seized in compensation for coercive ostracism. Moreover, they *also* gain access to the next batch of the merchant's goods as he returns to the *fair-dealing* local gathering in the future. The best available option for local gathering members benefiting from trade is also to defend the interests of the merchant.

Thus, being an itinerant merchant can be a viable adaptive strategy. Local gatherings become trade outposts for those goods and services that do not contravene the interests of the majority of the local coalitions (main text above).

What evidence can we find?

Our question is how well these theoretical expectations and predictions compare to the way behaviorally modern humans actually lived. We will use two primary kinds of information to explore and understand these people.

First, the archaeological record of the original emergence of modernity in Afroeurasia is extensive and priceless.[*,13] However, we will also make use of the more recent archaeological record from populations

---

*   Stringer and Gamble (1993); Shreeve (1995); Klein and Edgar (2002).

whose decisive coercive weapon remained the atlatl. For example, Native North American societies continued to be atlatl-limited until well into the first millennium—until around 400 to 600 CE (AD). The record of these relatively recent behaviorally modern humans is, thus, extremely fresh by archaeological standards.[*,14]

The second source of information is the records of direct observations of populations that continued to use the atlatl as the decisive coercive weapon system through first contact with European colonialists. Ethnography is the term for such evidence. The most dramatic cases of behaviorally modern ethnography—as we define it—involve the diverse Australian aboriginal populations.[†,15]

Explaining modern behavior – the simple fundamentals

A good way to understand how well and completely our theory accounts for the details of the behaviorally modern revolution is to look at the most conspicuous manifestations of this modernity. One good example are sites used for local gatherings—perhaps seasonally—that persisted long enough (or recurred in the same place often enough) to leave a footprint. Such sites are few and tiny in the pre-modern record. In contrast, with behaviorally modern humans, such sites become very large and conspicuous (see Figure 11.4 and discussion below).[16] Australian aboriginal large local gatherings are another important element of this record, and the maximal sizes of these gatherings—generally one thousand or less—are in line with our expectation of around nine hundred individuals.[17]

Another very conspicuous feature of the archaeological record of behaviorally modern humans is the scope of networks for trading exotic goods. The enlarged scale of local gatherings should increase both the volume of this trade and the distances traversed by merchants.

One well-documented example of this trade involves special stone, raw materials suitable for making advanced tools. The relatively rare and precious types of stone used for this purpose—flints, obsidian (volcanic glass) and the like—turn out to have complex chemistry, which differs detectably from place to place. A dramatic manifestation of this site differentiation is different local outcrops of obsidian, which might be black, brown, or green, for example.

---

*   Dancey and Pacheco (1997); Carr and Case (2005).

†   Altman (1987).

Thus, chemical fingerprinting of stone tools in the archaeological record can tell us where their raw material came from. When archaeologists ask how far a stone moved through trade, the differences between pre-modern and modern human behaviors is very striking. The stones associated with the behaviorally modern humans, traveled much farther, on average.

Of course, equally fundamental is the expected effect of larger cooperative aggregates on technical or adaptive expertise. The *wisdom of crowds* (Chapter 10) becomes much more powerful with larger aggregates incentivized to share information. We expect innovation to be substantially more rapid in this situation. Thus, the observed increase in the sophistication of human tools with modernity is a fully predictable outcome.

Likewise, this increased technical sophistication is expected to enable other new adaptive opportunities. For example, improved watercraft would allow crossing the open ocean and invading Australia and possibly contribute to the invasion of the Western Hemisphere. New technical means for dealing with severe cold might also contribute to the invasion of the New World across the polar ice.

Finally, see Box 11.2 for a discussion of an important feature of how humans achieve technical mastery in larger coalitions.

Box 11.2: Individual specialization and human cooperation

We can be specific about how increased adaptive sophistication might actually be acquired and stored as our cooperative coalitions grow larger. With the increased technical know-how from expanded scales of social cooperation comes increased *economic productivity*. Each individual can now produce more food, shelter, clothing, and/or knowledge than before.

This additional production generates what economists call a *surplus*, that is, more useful stuff per capita beyond the minimal requirements of subsistence. This means that people have more to trade with one another. Essentials can be overproduced by a subset of the population

Also note that if rates of leakage of valuable cultural information across local coalition boundaries (above) is relatively low, as expected, these increased behaviorally modern rates of local innovation will drive much more dramatic differences in local styles of tools or artifacts. In other words, there should be much more extensive variation among local modern cultures than between pre-modern ones. Strikingly, this is precisely what the archaeological record indicates.[18]

Explaining modern behavior – "symbolism"

Our theory accounts for the development of individual specialization and the production of surplus goods. These are both the products and the instantiation of increased abundance of available cultural information. However, this picture also accounts for some more esoteric elements of the record of behaviorally modern humans. The emergence of what are traditionally called *symbolic* items is an illuminating case.

Anthropologists have observed that behaviorally modern humans were the first to make huge numbers of beads, perforated animal teeth, small statuary and other seemingly useless objects. Such symbolic objects were clearly traded across vast distances. These items are sometimes imagined to be involved in *social gesturing*, *identity negotiation*, and many other more or less obscure social acts.[19]

The imaginary humans some cultural anthropologists describe as doing these odd things have an ethereal, otherworldly quality. We argue that this approach is a profound mistake. It is even a subtle kind of racism—thinking of our ancestors as less effective and pragmatic than we are. Early behaviorally modern humans are not extraterrestrial-like science fiction characters. Rather, they are people just like us. They thought as we do, acted as we do, and desired what we desire. The only difference between them and us is the scale of their social enterprise and, thus, the amount of information they had access to, on our theory.

Are we then saying there are no symbolic objects in the mountains of artifacts left behind by early behaviorally modern humans? No, we are not. Rather, we are suggesting that prehistoric symbolic objects and contemporary ones are mostly the same things.

To see what we mean, imagine the President of the United States at a White House press conference. Is he wearing a neck tie? Probably. Is there an American flag somewhere in the background? Probably. Are these symbolic objects? Absolutely. Would these objects be meaty clues if we were extraterrestrial anthropologists studying American politics, economics, and sociology? Absolutely not! "Purely" symbolic objects are not usually very important.

Would the American President have on his person or nearby any other symbolic objects, ones that might be more enlightening about American economics and politics? We are sure he would. For example, he might have a stack of elaborately made, identical little symbolic objects that he could exchange for adaptively useful things—food, clothing, etc. Moreover, copies of these objects are observed to be very widely exchanged throughout the American social system. They are even taxed by the government to pay for compensated, professional social coercion—that is, law enforcement—among many other social functions.

Of course, the symbolic objects in the President's pocket are just *currency*, money. Indeed, we propose that many or most of the symbolic objects from the behaviorally modern human record are precisely the same thing...currency.[*, 20]

It is illuminating to understand how this works. As cooperative coalitions become large, the number of possible adaptively useful exchanges or transactions also becomes correspondingly larger. Imagine how many trades and deals we might do with ninety close colleagues over a few weeks. Now imagine how many we might do with nine hundred members of a professional association gathered each winter at Palm Springs for an Annual Convention. The analogy to the contrast between pre-modern and large behaviorally modern gatherings is obvious. The question is how would we keep accounts in a new environment where exchanges have become much more numerous and multifarious?

Individual memory is inadequate to record everything that has been given and received at local gatherings of nearly one thousand people reliably, particularly in the important cases where these deals are agreements about the future. However, this adaptive problem can be dealt with if exchanges are instantaneously reciprocal so that no complex, long-term record keeping is involved. However, this requirement for instantaneous reciprocity creates a new bottleneck.

---

[*] Some knowledgeable readers will object that this is not likely because these pre-historic symbolic objects were often worn as ornaments—for example, as necklaces, belts and the like. Of course, this is not a good objection. Gold, silver, and diamonds are both currency and ornament in our culture. Indeed, much of the "punch" of their ornamental value derives from their currency value. We note also that some very valuable currency/clothing (tanned animal skins or cloaks of exotic bird feathers, for example) might also have had important currency value while not surviving in the record for us to find.

An example will illustrate the solution to this bottleneck and for simplicity we will speak in the singular. Let us say you specialize in raising chickens, we grow potatoes, and we wish to trade. We may need your chicken today, but you may not want our potatoes until next winter. If there is a marker that we can give you today for your chicken that is a reliable mnemonic for the potatoes we owe you next winter, our instantaneous exchange can still serve remote future needs. A long-term business deal just becomes two instantaneously reciprocal transactions displaced in time.

In the interim you may decide you want a new spear point in August rather than our potatoes in December. You trade our marker to a point maker, who then exchanges the marker with us for potatoes in December. Of course, our marker might be traded many times for many things before being redeemed for potatoes. Indeed, everyone else in our local coalition will have their own set of markers in circulation, as well. Thus, we not only have a marker out for potatoes, we hold markers for tanned animal skins, for example, and other things we received in trade with others. These markers are all redeemable to obtain the many diverse things our other local coalition members have markers out for. Of course, we have a name for such a system. We call it a *cash economy*.*

Each of these many transactions presents a conflict of interest problem. For example, we might decide not to honor the marker for our potatoes. However, our free riding would be easily handled by self-interested coercion. For example, if we renege on business deals our future value to others is low. Thus, seizing our assets by force now, including the potatoes to pay for our marker and other assets to compensate those who take the risk of ostracizing us, satisfies everyone (except us, the cheaters). Everyone wins and the system is sustained.†

The upshot of this logic, applied to every member of a cooperative coalition, is the establishment of a functioning market. Humans are intuitively gifted (possessing an evolved psychology) at using such markets to set exchange values—the chicken/potato equivalence, for example—and, thus, *currency values*.[21]

The functioning of such markets rewards people for producing more and better things. This is just Adam Smith's *invisible hand*, of course.‡

---

*   This picture of the *evolution of money* is based on the authors' unpublished theoretical work in collaboration with Roman Spektor.

†   Currency is a *stable store of value*. Thus, it also allows the gradual accumulation of local resources that could support larger cooperative projects. When currency is used in this way today, we call it *capital*.

‡   Adam Smith (1776) *An inquiry into the nature and causes of the wealth of nations*.

Indeed, our theory predicts that public exchange—that is, market behavior—is just as ancient and uniquely human as our language,[22] both being adaptive products of our control of conflicts of interest. This is just a restatement of our insight from Chapter 10 that we are an *economic animal.*

All that is required to produce this remarkable effect—other than cost-effective coercive enforcement—is that our pieces of currency be so difficult to counterfeit that it is easier to deliver the required real goods/services than to cheat by making our own bogus currency.

It is striking that behaviorally modern human symbolic objects—*putative currency*—have this essential quality. Often, for example, these objects are marine shells that are exotic and easily recognized while at the same time relatively difficult to obtain away from the coast. Similarly, hard-to-obtain and easily recognized animal teeth might be used.* These objects were then subjected to labor-intensive secondary processing such as drilling, polishing, and so on that added further to the difficulty of counterfeiting them.

Ultimately, much later in our history, exotic, rare ("precious") metals and other hard-to-counterfeit media increasingly came to fill this role.[23]

Thus, we expect money to explode into the archaeological record as soon as the scale of human social cooperation is extended by the introduction of a new coercive weapon. This, we propose, is precisely the story of the emergence of a major class of the symbolic artifacts during the behaviorally modern human revolution.

Explaining modern behavior – building really big things

Behaviorally modern humans also began building huge structures that have no obvious mundane adaptive function. These structures are sometimes called *monumental architecture.* A well-know example is the mounds of the Hopewell of the American Midwest built between about 0 to 400 CE (Figure 11.4). Hopewell social cooperation was sustained by the atlatl as the dominant coercion weapon.[24] Thus, we expect this relatively recent culture to, nonetheless, be a good example of the original modern adaptation. Consistent with this picture, the Hopewell engaged in the seasonally large local gatherings and *symbolic exchange* we have already discussed.

---

* Again, many archaeologically perishable items were probably also used for currency.

Because these structures are relatively recent, Hopewell monumental architecture is better preserved and more easily studied than older equivalents. Figure 11.4 shows two examples of these remarkable structures. The archaeology indicates that they were not part of a town or permanent settlement, but rather were places of seasonal congregation of relatively large numbers of people. Apparently, these gatherings represent the local gatherings our theory predicts.

The detailed architecture of these large structures seems to be the quintessential case for *arbitrary social meaning* so much favored by some cultural anthropologists. One favored claim is that they were *ceremonial centers* providing the context for purely ritual activities whose meaning was somehow entirely internal to the Hopewell society.

We argue otherwise. Again, behaviorally modern humans are not "exotics" doing strange things we do not understand, but rather they were just like us.* Therefore, we suggest that these structures merely reflect meeting centers provided in pursuit of local self-interest and being successful only to the extent that they serve the self-interests of their customers, merchants, and vacationers.

On this view, these structures were part of an organized meeting center like the hotel at which our large annual convention above might occur. Services provided would include food, sanitation, sleeping arrangements, and security. This security would be self-interested coercive enforcement of cooperation or fair dealing. Thus, individuals come to local gatherings at the site in pursuit of mates, goods, services, profits, or information, all under the uniquely human coercive umbrella sustained by the atlatl.

This is a strong, specific claim. It should be readily testable. It will be of great interest to see whether professional archaeologists and ethnographers can extend these tests. We have thus far found no disconfirming evidence of this picture. What we currently know is as follows.

First, burials at such sites often contain symbolic objects—*putative currency*.[25] Their presence is consistent with these sites being centers of commercial activity generating this wealth.

Second, these sites should be designed to house local gatherings on the scale of the behaviorally modern maximum of about nine hundred people. The large earthen Hopewell circles in Ohio give us the opportunity

---

*   We might think that contemporary humans have purely social meeting places. However, then we note that churches/synagogues/mosques define social units whose members typically also engage in *preferential* economic/political interactions with one another. To say that these institutions are purely social is to miss their evolved psychology and important pieces of their contemporary function.

to test this prediction (Figure 11.4). These circles are large enough to enclose a campground for this number of people.

Third, several architectural details of this Hopewell circle are provocative. Why enclose at all? If these are local economic enterprises—remember the hotel analogy—it would be necessary to control access. The Hopewell circles generally had one or a small number of openings that looked very much like gated entry points. The modern reconstruction of one of these circles in Figure 11.4 illustrates this feature.

Fourth, coercive law enforcement would also be a perpetual necessity for such a facility. The structure of the Hopewell circles, again, is consistent with this expectation. The circles are enclosed by a bank well raised above the level of the circle floor and easily walked, patrolled (Figure 11.4). Moreover, the entire floor or the meeting/camping area, is within atlatl bolt range of this surrounding bank.

Equally suggestive is the fact that the *borrow pit* from which some of the soil was taken to build the bank is always on the *inside* of the circles. In other words, the defensive moat for the embankment faces *in* rather than *out*. This positioning is expected if the target was internal law enforcement, not external military threat. These features would probably have been chosen by a wise, self-interested local meeting provider with security—coercive enforcement—obligations.

The creation of these structures also represents direct evidence for substantial expansion of the scale of human social cooperation. The structures require formidable amounts of labor to build and maintain. Only relatively large groups could field this much cooperative effort consistently. Moreover, these structures make adaptive sense only when a sufficiently large economic surplus can be generated for trade. Again, the increased adaptive sophistication of enlarged modern coalitions is expected to allow such increases in productivity.

If the earthworks were part of an economic enterprise built by self-interested individuals, we would expect new centers to grow up—cloning and competing with earlier ones. Such site cloning is almost always observed with monumental architecture wherever it is built,[*] and this is certainly true of the Hopewell structures.

Notice that the Hopewell gathering for business at such a commercial site, might also have engaged in social rituals designed to recruit and maintain members (Chapter 10).[26] However, these ritual acts, in isolation, cannot be the *ultimate* or adaptive purpose for such sites, on our theory.

---

[*]   We will have more to say about monumental architecture in the context of later, more complex societies in subsequent chapters.

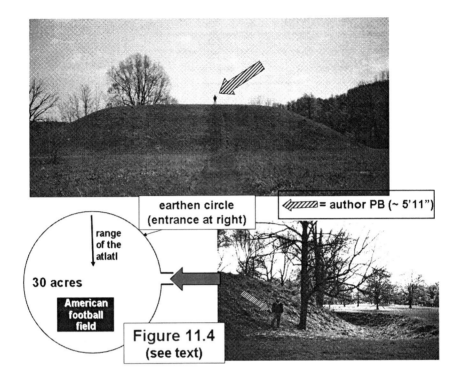

earthen circle
(entrance at right)

= author PB (~ 5'11")

range
of the
atlatl

30 acres

American
football
field

Figure 11.4
(see text)

Explaining modern behavior – "art"

One of the strongest arguments for *arbitrary symbolic behavior* by the first behaviorally modern humans is their production of sophisticated art objects, including statues and paintings. With a tiny number of very modest exceptions, "pure" art is never produced by pre-modern humans throughout their entire approximately 1.7-million-year history. Then suddenly behaviorally modern humans start producing very mature art.[*] This qualitative transition in behavior is utterly remarkable. If our theory is correct it must account simply and directly for all this new art as no earlier theory has plausibly managed to do.

There is an alternative to thinking of art as some arbitrary, self-referential social act without economic or adaptive meaning. In fact, our theory suggests several different, not necessarily mutually exclusive, such explanations.

First, what adaptive function does the use of methamphetamine or free-base cocaine have? Of course, the answer is, as far as we know,

---

[*]   Again, a Google image search with "Paleolithic art" and Chavet, Lascaux, or Venus reminds us of this spectacular art.

none at all. Indeed, under most circumstances, the use of such drugs is maladaptive by any sensible definition. Nevertheless, biological adaptation is directly relevant to understanding them. These drugs apparently artificially stimulate neural mechanisms that give pleasure or fulfillment when stimulated as they are "designed" by natural selection *to be stimulated*. Crystal meth and crack cocaine merely hijack this machinery and generally hyper-stimulate it.

Under these artificial conditions, one set of individuals is willing to give currency in return for drug stimulation, creating a conventional economic market for the drug. The behavior of these individuals, though not adaptive, is generated by "side-ways" functioning of mechanisms produced by natural selection. When a larger market economy exists (above), rational economic actors (drug dealers) then meet the resulting demand.

We suggest that sophisticated art is, in part, like recreational drugs. Good art artificially stimulates perceptual psychological devices in ways that satisfy us. Perhaps they hyper-stimulate these devices in ways that we experience as *beauty*. Of course, contemporary art dealers sell us this beauty fix through a market. There is not the slightest reason to believe that behaviorally modern humans would not have sustained a similar art market.[*]

Sophisticated art also has a second relevant property. When produced by an elite professional, a specialist, art will have the artist's *individual stylistic stamp*. Indeed, this individuality is the mark of a "true artist" as opposed to an accomplished technician. This means that individual works of art become non-counterfeitable. Thus, they are attractive candidates for use as currency—as symbolic stores of value.

The Picasso painting *Boy with a Pipe* sold at auction in 2004 through Sotheby's for $93M *plus commission*!! One might think that purchases like this are made by people who really love Picasso's art. Of course, we know they almost never are. Such purchases are economic transactions, first and foremost—investments in stores of value. Such art objects can also be used as *signals* of various sorts. But the usefulness of art for purposes of *social gestures* is largely contingent on its currency function.

It would be surprising, indeed, if early modern humans did not often use art more or less like we do —which is to say, sometimes as currency.

---

[*]   Pinker (1997) reaches a related conclusion about the adaptive origins of artistic products. Our theory provides the previously lacking explanation for why elite art occurs when and where it does and, of course, why humans have elite capabilities, more generally. Also see Tooby and Cosmides (2001).

Again, notice that all these functions of art require an expanded scale of social cooperation to generate economic surpluses and to enable individual specializations as artist and art dealer. Our theory apparently accounts for the explosive emergence of high art with behaviorally modern humans.

Good theory is powerful

Our theory of human uniqueness implies a theory of history (Third Interlude). Humans have one unique adaptive strategy that we have used throughout the entire course of our history. We project and respond to coercive threat in pursuit of self-interest. When decisive weaponry is of the right form—for example, enabling projection of threat from a distance—it results in our enforcing social cooperation through suppression of conflicts of interest.

When one gives such a "monomaniacal" animal a new weapon, it will be used in an extremely predictable way, resulting in the emergence of a new scale of social cooperation. Moreover, the resulting new scale of social cooperation will produce new levels of adaptive sophistication for the transparent reasons we have discussed.

This theory of history should be general. That is, conflicts of interest should be the universal limitation on human adaptive sophistication and all increases in that sophistication *must require* new coercive means to manage conflicts of interest throughout our history.

This theory inescapably predicts that the behaviorally modern human revolution should be the direct product of a new weapon system and that its properties should all be interpretable as the consequences of an expanded scale of social cooperation.

We have argued here that both these predictions are very strongly supported by the empirical evidence.* The behaviorally modern human revolution was our first opportunity to test our theory of history against the evidence. In Chapter 12 and beyond we will have many more such opportunities. Each test will give us more confidence in our theory, each will also teach us fundamental new lessons about the human condition (Box 11.3).

---

* It will be of the greatest interest to see if continued exploration of the archaeological and ethnographic records of behavioral modernity turns up new observations for which our theory cannot account.

Box 11.3 The biology of human potential

The behaviorally modern human revolution vividly frames a profound lesson for us. It is really the same lesson we have already learned (Box 10.2) but vastly enriched from a new perspective. Our finite evolved *biology* has created and continues to sustain an infinitely expansive adaptive capability, our *social* collaboration. We have already boldly gone where no creature has gone before. But we have just begun.

The human potential to create and manage our world is *not* limited by our individual biology. Individual human physical and mental capabilities are merely the tools of our adaptive behavior, not its limiting ingredient. These tools can be deployed in service of great art, spectacular technology, wonderful science, phenomenal economic cooperation, *and things we have not yet even remotely imagined.*

To see this clearly, return one last time to behaviorally modern humans, to the genetic us. We have apparently had the essential *biological* endowments to create breathtakingly beautiful objects, to walk on the Moon, or to save our children from disease for at least five hundred thousand years or longer. However, these *capabilities* became *actualities* only as the *scale* of our uniquely human *social* cooperation *enabled* them.

Humans are so highly adapted to exploiting kinship-independent social cooperation that richly novel adaptive behaviors *explode* into existence with an expansion of social cooperation, over just a few generations rather than endless millennia. Thus, the only difference between a pre-modern human in a South African rock shelter one hundred thousand years ago and a cosmopolitan citizen of one of Earth's great cities today is *nothing* other than the details of their *social* context and recent *social* history.

This lesson from the behaviorally modern human revolution should be with us always. We have more and much better revolutions yet to make (Chapter 17). Access to the fruits of culturally transmitted information is ultimately what empowers us. Thus far, this access has been almost exclusively limited by the numbers of us who come together in cooperative social entities (Chapter 10). Can our information technology allow us to escape this limitation—to give our descendants access to the fruits of nearly infinite amounts of information? It seems likely that we can ultimately bring this about…and entrain the information to the service of *all* our descendants *under the right circumstances* (Chapter 17).

There is not the slightest reason to imagine that we stand at the end of the process of actualization of human potential here in the early 21st Century. We can realistically hope that all our descendants will look back on us with pity at our ignorance and poverty. What is required for this to transpire is true mastery of the *rules* of the human social adaptation and their wise deployment (Chapter 17). Our potential, as a pan-global, self-aware, adaptively powerful community, to take humane command of our future is enormous.

ﾟ

Chapter 12
Neolithic and "agricultural" revolutions: Humans settle down

The extraterrestrial scouts were impressed by the warmth of the air and the copious flowing water on this planet the natives called Earth. Particularly amazing was the extent to which massive portions of continental landmasses had been harnessed for production of food for the planet's single dominant animal, a species calling itself human. Huge cultivated fields stretched from horizon to horizon. These "amber waves of grain" and "geometric quilts of rice" made it possible for humans to feed the citizens of their enormous nations.

When the time to colonize this newly discovered planet, the extraterrestrial empire had no need to declare war. They merely dispatched their advanced biological weapons to destroy the five major domesticated plant crops. Within a year, the stores of accumulated plant food were depleted and essentially all the six billion humans on Earth had starved to death. The few million who survived ate marine fish, berries, nuts, and wild tubers. They remained widely scattered here and there across their world. Finishing off these stateless hangers-on was trivially easy.

Hanging by a stem –
human-domesticate co-dependence

The point of this science fiction horror vignette is that all of us alive today are utterly dependent on a few domesticated plants. The vast bulk of calories the world's billions of people consume are produced by maize, wheat, rice, soybeans, and potatoes.* These plants have all been so extensively modified genetically by us that they can no longer survive without human care and protection. Likewise, our vast societies cannot survive without them.

The importance of domesticates to human existence has made explaining the evolution of agriculture a central obsession of archaeologists and historians since Darwin. Darwin, himself, explained the biological origin of these peculiar plants themselves. Today, we might paraphrase his insight this way. "Humans apply selection on the genetic design information that builds food plants, reshaping them just as 'natural' selection shapes 'wild' organisms."[†]

Explaining this process of plant domestication turned out to be the easy part. The hard part is discovering *why* massive agricultural revolutions built around domesticated plants happened *when* and *where* they did, rather than one hundred thousand years earlier or a million years later, for example. Many earlier answers to these why, when, and where questions have been proposed, but, we will argue, all have failed. In contrast, our theory predicts all the various diverse *agricultural revolutions* with powerful simplicity.

The emergence of our contemporary ways of feeding ourselves will be the next exciting opportunity to subject our new theory to potential falsification. This exploration turns out to have much more to teach us about ourselves and our societies than merely where our food came from.

How do we know that earlier theories of agriculture failed?

The thoughtful archaeologist, Offer Bar Yosef, recently summed up the state of archaeological knowledge about the birth of agriculture prior to our theory.

> *Investigators are divided on the issue of 'why' human groups became sedentary [settling down into stable communities just before developing agriculture]. Two alternatives are often mentioned. The first suggests that*

---

* Notice that the meat, eggs, and cheese we eat comes from animals almost exclusively fed on the same crop plants we eat. A few other domesticates—rye, barley, millet, taro, sweet potatoes, sugar cane, sugar beets, and others—each also contributes a little of our global diet. See Smith (1995) for the story of our food plants and their chronological origins.

† Of course, humans have also modified most of our food *animals* in this same way.

humans were attracted to spatially restricted and rich resources.... The second explanation proposes that economic and social circumstances enforced sedentism...[1]

This lack of consensus and vague explanation reflects the absence of a generally useful approach to the problem of agricultural revolutions. The two alternatives Bar Yosef mentions are not useful answers. If humans were attracted to concentrated resources, why specifically around twelve thousand years ago in the Middle East rather than at some other time and place? There were other similar geographic areas of concentrated resources in many other places inhabited by humans throughout our vast evolutionary history. Likewise, if some economic or social change led to sedentism, what was that change and, again, why did it happen there and then rather than some other place and time?

One common older answer to these questions invokes climate change. The first symptoms of sedentism and agriculture happened as the Earth was emerging from the last Ice Age. As little as eighteen thousand years ago, massive ice sheets, up to a mile thick, extended from the North Pole to central Eurasia and North America. For example, contemporary Germany and New York State were mostly under massive glaciers.

The Earth's climate ameliorated and the polar ice retreated to its present state between about eighteen thousand and fourteen thousand years ago. In the areas of the world where agriculture would eventually develop, there was a corresponding change in rainfall and temperature. In general, the climate became wetter and warmer.

However, this climate answer to the why, where and when questions is also quite unsatisfactory and incomplete. The last retreat of the ice sheets was simply the latest in a long series of such cycles.* Why did agriculture not develop the last time the climate warmed or the time before that—about one hundred twenty-five thousand or two hundred thirty thousand years ago, respectively? As we have seen, humans of these earlier eras almost certainly already had the same genetic endowments as we do (Chapter 11). Moreover, parts of the world, like Southeast Asia and areas of Africa and Australia, were probably suitable for agriculture even *during* the ice-overs further north. Why not agriculture

---

*   The cyclical changes in global temperature and polar ice are the consequences of subtle wobbling in the pitch of the Earth's axis and in the dimensions of its orbit, altering the amounts of solar radiation that reach various parts of the planet's surface. A Google search of "Milankovitch Cycles" is a useful way to break into the extensive literature on this subject.

in one of these places seventy thousand or one hundred seventy thousand years ago?

We might imagine that agriculture had to await BOTH the behaviorally modern humans we met in Chapter 11 AND climate change. However, as we will see a little later, this answer fails on empirical grounds. The behaviorally modern adaptation was not sufficient to sustain agricultural revolutions. We definitely need a better answer.

The last several decades have produced two surprising and illuminating empirical observations.[2] These discoveries serve to sharpen our understanding of precisely what we need to explain about the rise of agriculture. First, the earliest permanent large villages, traditionally thought to be a symptom of agricultural revolutions, actually arose *before* plant and animal domestication. Second, in a few places, behaviorally modern humans have proven capable of domesticating plants without building permanent villages or showing other symptoms of agricultural civilization.

These discoveries indicate that domestication of plants is not *sufficient* to produce an agricultural revolution. Even more importantly, these observations indicate that highly intensive use of domesticated plants and animals is to agricultural revolutions what Wagner is to opera— loud and conspicuous, but late and derivative.

We can attempt to fix earlier theories to address these objections, but, when we do, our answers raise still other difficult questions, an apparently endless regress. For example, we might imagine that some complicated *combination* of factors must engulf a local behaviorally modern population for an agricultural revolution to ensue. This has been a popular approach in the past, but it has two severe defects. First, if we assume the constellation of factors is complicated enough, the proposal becomes untestable—like blaming the sinking of the Titanic on a complicated interaction between various parts reconstructed from the wreckage while being unaware of the departed iceberg. Second, such theories have very great difficulty in accounting for agricultural revolutions that arise apparently independently in Eurasia and North America, for example. Complicated combinations do not arise repeatedly as an elementary statistical fact.

This unsatisfying lack of simplicity clearly indicates that these combination-of-factors theories are probably just the wrong answers. Earlier work does not sustain traction in accounting for the evidence because it apparently does not have the right theory. We will show that our theory of history contains and implies the first theory of agricultural revolutions that actually works.

We will argue that all the important symptoms of the process leading to agriculture are apparently produced directly and immediately in response to the local introduction of a new coercive weapon, producing, in turn, a new scale of social cooperation. Agriculture *sensu stricto* turns out to be a secondary issue, not the main story line.

To be more accurate and clear in our descriptions, we will call these adaptive revolutions *Neolithic*, borrowing an earlier term meaning New Stone Age. We will find that Neolithic revolutions were diverse. Some produced agriculture while others went in different adaptive directions, depending on the details of local adaptive opportunities. Earlier theories of agriculture are defeated by this local diversity, while our theory predicts it.

Humans are an animal exquisitely adapted to the use of coercion in self-interested ways that generate kinship-independent social cooperation as a by-product (Chapters 5 and 11; Third Interlude). When this human animal acquires a new weapon, the weapon is immediately deployed in this species-typical way. When its properties permit, the weapon's use results in the rapid assembly of a new *scale* of social cooperation. This is just humans doing what humans naturally do. It is our one good trick. Neolithic revolutions are simply the next case of this characteristically human adaptive tactic in the unfolding of history, we argue.

A new weapon – where and when

The inescapable expectation from our theory of history is that *all* large up-ticks in human adaptive sophistication *must* be preceded—*very briefly* on an evolutionary time scale[3]—by the local invention of a fundamentally new coercive technology capable of suppressing conflicts of interest on a new scale (Third Interlude). This is a very precise, testable prediction.

Moreover, there were many different Neolithic revolutions, separated by tens of thousands of miles and thousands of years. Many of these local revolutions were largely or completely independent of one another. Each *must* follow the *local* introduction of a new weapon, on our theory. Thus, the global story of Neolithic change apparently gives us many *different, independent* opportunities to test our theory.

It is fruitful to begin with the large subset of Neolithic revolutions occurring in Eurasia and North America. The new weapon apparently associated with these diverse revolutions is the bow and arrow or just the bow, for short. It was long unclear when and where the bow was invented. Much early confusion arose from ignorance about whether stone points reflected atlatl bolt points (Chapter 11) or arrowheads. However, in the last several decades a great deal of evidence has accumulated that the bow was most likely invented somewhere around the southeastern Mediterranean basin around fourteen thousand years ago or 12,000 BCE (BC). This evidence includes stone points with diagnostic structure as well as the presence of arrow-specific supporting technology, especially grooved stone *shaft strengtheners* which are used for arrows but not for atlatl bolts.[4]

The bow spreads outward from its origin in Southwest Asia up into Europe and across into East Asia. The spread into East Asia probably took about two thousand years.[5] In contrast, the spread of this weapon from Asia across the North Pole was slow. It does not show up in North America until much later, around sixteen hundred to nineteen hundred years ago or 100 to 600 CE (Figure 12.1).[6,7] Alternatively, the bow could have been independently re-invented locally in North America at this time, perhaps in Canada.[8]

From the point of view of our theory, all that matters is that the bow shows up in North America more than ten thousand years *after* its introduction in Eurasia. The bow has some important new features as a coercive technology. Thus, its global pattern of spread creates a remarkably beautiful natural experiment. Our theory predicts that the bow's social effects on these two continents should be displaced by precisely this same ten thousand years. These social effects should *follow* the bow, never precede it, and those follow-on effects should always be very rapid. As will see, this rigorously predicted temporal pattern is, indeed, precisely what the archaeological record indicates.

A study in contrasts – the bow and the atlatl

As the bow is always invented (or introduced) in local human populations that already possess the atlatl, our first question must be whether these two weapons are really different as tools of social coercion. The range of the two weapons is somewhat different. A well-made simple bow has a range of 150 to 200 meters compared to around 100 meters for the atlatl.[9] More sophisticated bows can project further, perhaps out to 250 to 300 meters. These differences in range are relatively modest. Moreover, it is unclear how accurate the bow is at extreme distances exceeding the range of the atlatl. Thus, range differences between the two weapons may be secondary and relatively unimportant.

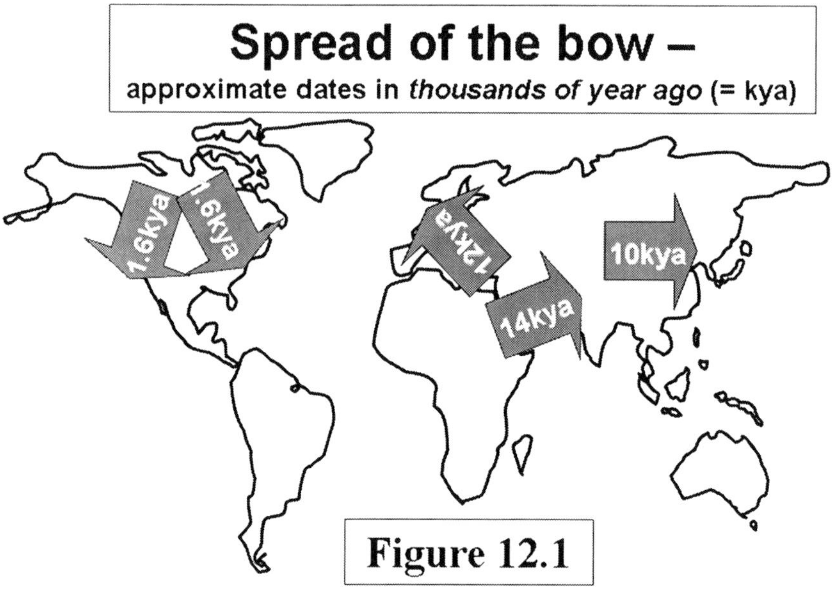

Spread of the bow –
approximate dates in *thousands of year ago* (= kya)

1.6kya

1.6kya

12kya

10kya

14kya

Figure 12.1

However, these two weapons differ very dramatically in another way. The atlatl involves a highly coordinated and violent body movement, including precise control of the long throwing stick as a highly mobile extension of the violently moving hand and arm (Figure 12.2, top panels). Our minds/brains were designed to control release from the hand (Chapter 7), not from a wobbly two and one-half foot extension of the hand. This physical arrangement means that great skill is required to control the trajectory of an atlatl bolt. Acquiring proficiency with this weapon requires extensive experience. Moreover, *maintaining* that proficiency requires consistent, time-consuming practice. These properties of the atlatl are confirmed by our own extensive personal experimentation.

In contrast, the bow requires only minimal, simple, easily mastered movements. Once the bow is drawn, its user aims with tiny, well-controlled movements. The bow is then fired by gentle release of the fingers holding back the base of the arrow notched over the bow string. The rest of the body remains comfortably still throughout the well-aimed release (Figure 12.2, bottom panels).

These features of the bow allow the archer to shoot a "tight group" easily, placing arrow after arrow on the same target with high reliability. Lethal proficiency with this weapon is readily attained and requires only rare, brief practice to maintain. Indeed, the bow is like a modern handgun in this way. A relatively small adult of either sex or even an adolescent child can become a proficient killer within a few hours. As with the atlatl, we have confirmed the properties of this weapon through extensive personal experimentation.

An equivalent way of describing this difference between these two weapons is to say that the *opportunity costs* of proficient use of the bow are much lower than for the atlatl. As we discussed earlier, *opportunity cost* is a formal economist's term referring to the cost of losing the opportunity to do one thing because we are busy doing something else. If atlatl proficiency requires a great deal of drilling, this extensive practice is time away from any other useful activity. Opportunities therefore lost mean high opportunity costs.

The total cost associated with self-interested coercion must include the opportunity costs of the weapon used. Thus, the same coercive episode with the bow is substantially cheaper than with the atlatl. *Law enforcement* now takes less time away from other adaptive opportunities.

Equivalently, new coercive actions that were maladaptively expensive with the atlatl will become cost-effective and adaptive with the bow. Coercive management of conflicts of interest in circumstances that formerly imposed prohibitively high costs can now become cost-effective. We can get an intuitive grasp of this abstract picture by thinking through how certain examples of social cooperation and coercive enforcement would play out with the two weapons and their different opportunity costs.

The large seasonal social gatherings sustained by the atlatl (Chapter 11) are a useful example. One way to deal with the high opportunity cost problem is to invest in intensive practice for a few days before these gatherings and participate in all the cooperative activities of the gathering over the week or two before these polished skills deteriorate unacceptably. As these large gatherings reach a point of diminishing returns, dispersing in small bands across the landscape becomes a better choice than extensive investment in maintaining elite weaponry skills.* In other words, the suppression of conflicts of interest on the large scale allowed by the atlatl is only cost effective in short bursts. The costs of local law enforcement are too high to sustain large local gatherings for long periods.†

---

* Note that enforcement in small bands can be cost effectively achieved with thrown stones and with short-range, low-skill application of the atlatl.

† The alternative of having a few professional police whose individual specialization includes maintaining proficiency with the atlatl is also unacceptable to the large majority of individuals. As always with humans, these "police" would exploit their coercive dominance in pursuit of self-interest, *not* in the interests of most of the members of

Contrast this situation with the bow. Individual participation in coercion with the bow entails little practice. Thus, remaining in large local gatherings for longer periods becomes adaptively sensible. This capability is very important. When large gatherings persist longer they create the opportunity to develop new cooperative strategies that exploit their more extensive manpower. Under some local ecological circumstances it might even become optimal to congregate throughout most or all of the year, to become permanently sedentary.

Second, we can reach a related fundamental conclusion along a different route. Suppose you and your colleagues and family develop the skills and knowledge to produce large amounts of a storable adaptive asset. This might be dried grain or salted fish, for example. These assets become extremely attractive targets for others who did not invest the effort to create them. As we have seen, "what can be exchanged can be stolen."

Anyone who wishes to take these assets from you must simply acquire the coercive means to steal or extort them. Your response must be to sustain the coercive means to repel such theft. However, you cannot spend large amounts of time producing these assets and, *at the same time*, large amounts of time maintaining your weaponry skills—opportunity costs, again. In the context of a high-opportunity-cost weapon your best response to this problem would be not to accumulate extensive assets in the first place. Instead you might hunt and gather what you need as you need it, for example.

In contrast, suppose that the opportunity costs of defending concentrated assets become very low, as the result of local adoption of the bow, for example. Now it can make adaptive (economic) sense to accumulate stored grain, smoked fish, dried meat, mass-produced arrows, and so on.

Combine these two implications of the low opportunity costs of the bow. Large gatherings persist and develop new culturally transmitted practices allowing them to apply economies of scale in labor to generate and protect large amounts of stored assets.

These accumulated assets are often *liquid* (an economist's term meaning that they are easily exchanged for other things). We might trade you a bushel of wheat for a new bow and ten arrows, for example. Thus, long-lived large gatherings sustained by the bow allow their residents to

---

the larger social aggregate. Everyone else would then "vote with their feet," dispersing across the relatively uncrowded pre-Neolithic landscape. This would leave these elite atlatl users without a "town" in which to be "sheriff." It will be of very great interest to see if future evidence might allow us to determine whether such coercive elites played any role in the built-up centers of behaviorally modern seasonal gathering supported by the atlatl (Chapter 11).

apply the tactics of Adam Smith's famous pin factory.* Their members can manufacture more and new liquid assets with much greater efficiency. For example, they might break a process, like harvesting wheat or making arrows, into simple sub-steps that can be assembly-lined to produce a dramatic increase in the productivity of individual effort. We can refer to this scale-dependent increase in productivity as *intensification*.

The combination of accumulation of liquid assets and intensification will produce substantial increases in local adaptive sophistication. Notice that liquid assets for exchange and increased intensification produce substantial increases in the scale of *indirect* access to the fruits of adaptive culturally transmitted information (Chapter 10). This increased adaptive sophistication will occur in longer-lived large gatherings, ultimately supporting permanent towns or equivalent settling down in various patterns across the landscape.[†]

The empirical evidence – the "Fertile Crescent" gets there first

As we discussed, one important theme in earlier speculation about agricultural revolutions was that domestication of plants and animals was the *cause* of subsequent social events such as the first development of permanent, sedentary settlements.[10] Our theory requires precisely the contrary. We predict that the causal sequence is a new coercive technology—the bow—followed by changes in the pattern and scale of social cooperation. Subsequent adaptive changes, including the massive use of domesticated plants and animals, are merely *effects* of this primary cause.[‡,11]

---

*   See Book I, Chapter 1 of Smith's 1776 book *An Inquiry into the Nature and Causes of the Wealth of Nations* (*Wealth of Nations*, for short) for a discussion of this famous example. Smith observed that a chain of individuals, each executing one small step in the manufacture of a pin, could collaborate to produce many, many more pins than the same set of individuals could produce working separately to manufacture pins from beginning to end. This is intuitively obvious to us, of course. We can learn to execute a simple, repetitive motion with spectacular expertise and speed. In contrast, executing a highly complex series of processes cannot be done with nearly as much speed.

†   Do not fall into the group-selection trap of thinking that individuals engage in law enforcement to permit these group-benefiting effects. Rather, coercion is self-interested here as always. It only produces benevolent group effects as a by-product.

‡   Indeed, as we have already hinted, our theory predicts many other effects of this primary cause in addition to agriculture and we will see in moment that these other effects are also seen.

Precisely as our theory predicts, one of the great surprises of late 20th Century archaeology in Southwest Asia/the Middle East was the finding that sedentary hamlets and, later, larger villages *came first* and the domestication of plants and animals *followed*.[12]

Specifically, *within less than one thousand years after the introduction of the bow* in the Middle East, sedentary settlements arise for the first time. There is still occasional controversy about *how* sedentary these first settlements were,[13] but the weight of the evidence is compelling, we argue. For example, new creatures evolved to make a full time living as human household *commensals*—house mice and house sparrows, among others—at this time. These animals required continuous local human occupation to evolve and persist.[14]

The most well-known case of this new sedentism is a local culture in the Levant of the southeastern Mediterranean basin called *Natufian* by archaeologists.*,[15] Beginning about 11,000 BCE, Natufian hamlets contained multiple house structures and employed specialized implements for harvesting substantial amounts of wild grains. These implements included tools that constituted substantial capital assets, like sophisticated sickles/scythes and large mortar and pestle devices for grinding grain. The Natufians were also proficient bow hunters.

Natufian hamlets disappeared transiently during *the little ice age* referred to as the *Younger Dryas* (between about 9000 and 8300 BCE). This climate change would have largely eliminated the dense local wild resources necessary to sustain large sedentary settlements.

However, when the Younger Dryas ended (around 8300 BCE), village life came back with a bang.[16] Villages again grew in size (with the bow in hand) and over the next several thousand years, their residents domesticated many of the plants and animals we still depend upon today, including wheat, rye, barley, goats, sheep, cattle, and pigs.[17]

The near-simultaneous domestication of all these different plants and animals represents spectacular social productivity. This is precisely the property our theory predicts. Creating and allowing a new level of social cooperation will always generate a diverse array of new adaptive consequences just like this one did (Chapter 10).

---

*   Wenke (1996).

Figure 12-2

Spreading out – the bow diffuses across Eurasia and produces other adaptive/agricultural revolutions

The specific case of the *cradle of agriculture* in the Fertile Crescent has received much more archaeological attention than the other later Eurasian agricultural revolutions. Thus, the excavated and analyzed evidence in these other Eurasian areas is less extensive. However, two available sets of observations strongly support the predictions of our theory and none appear to falsify it.

First, the bow apparently spread into Europe producing new cultures, collectively designated *Mesolithic*, in its wake. These cultures were probably more sedentary and ultimately assimilated domesticated plants and animals from Southwest Asia, helping to produce local agricultural revolutions.[18]

Second, though Chinese archaeology has only recently come into its own, the best evidence we have indicates that the bow diffused into East Asia several thousand years after its invention in the Mediterranean basin (ca. 8000 BCE; Figure 12.1). Following rapidly in the wake of the bow are sedentary settlements, succeeded, ultimately, by agriculture.[19]

As with Southwest Asia, permanent settlements apparently followed the bow and preceded new plant domestication, as our theory requires. Moreover, rice is apparently domesticated twice independently and millet once, all at widely separated sites.[20] Thus, plant domestication looks like an *effect*, not a cause and its chronology is precisely consistent with the *cause* being a social revolution produced by the local introduction of the bow.

It will be of the greatest interest to follow the emergence of more archaeological evidence from East Asia. Our theory's very clear predictions should be well tested by these new data over the next several decades.

As robust as the Eurasian evidence is, we do not have to settle. We can add much, much more by moving to a new continent.

The perfect natural laboratory

As powerful as the Eurasian record is, we are limited by the relatively great age of these Neolithic revolutions. They are all eight to eleven thousand years in the past. Time degrades the record. Moreover, the massive upheavals of post-agricultural Eurasia (Chapters 13-16) have had millennia to pave over, plough under, or blow up many of the sites and artifacts we would most like to see.

An outstanding remedy for our ignorance would be a pristine region into which the bow was introduced more recently. Better still, this region should be a wealthy and politically stable modern state so that its archaeology is well developed. Even better yet, this region should be so large and ecologically diverse that we can see all the different local adaptations our theory predicts. For example, agriculture is a relatively efficient dietary adaptation in some ecological settings. However, there are other settings in which agriculture is inefficient, but massively rich sources of wild foods, such as fish or mammal meat, would allow *non-agricultural* Neolithic adaptive revolutions.

Unlike most other theories of agricultural revolutions, ours predicts that such non-agricultural revolutions should *also* be produced by the enhanced social cooperation engendered by introduction of the bow.[*] Notice that these non-agricultural revolutions— occurring in chronological lock step with superficially unrelated nearby agricultural revolutions—are a rich, powerful, and falsifiable prediction of our theory.

A region with all these special properties would be a truly outstanding opportunity to test the theory. But what would make the region absolutely *perfect* would be if it already possessed domesticated plants, perhaps imported from elsewhere, but *without* a social system that would allow them to become a part of a full-blown agricultural revolution.

This last condition would let us test two crucial predictions of the theory. First, domesticated plants are not causes, only effects, on our theory. Thus, their presence, without the bow, should NOT be enough to produce agricultural revolutions. Second, if this first prediction is fulfilled, the presence of these domesticates, already in use in small-scale *horticultural* (family gardening) behaviorally modern adaptations, should allow full-blown agricultural revolutions to happen extremely rapidly *once the bow comes*. There would be no lag in formation of extremely large secondary settlements caused by the (relatively) slow domestication of plants by *artificial selection*. We should simply see a hyper-fast social revolution and its immediate consequences.

This long archaeological/scientific wish list might sound foolishly utopian— surely no such dream region exists. Of course, you have already anticipated that our dream region does, indeed, exist. In fact, it is North America, consisting of present day Canada and the continental United States. This area is one of the richest and most stable regions in the contemporary world. Moreover, the bow was introduced into this region recently, between 100 and 600 CE.

Most of us naively visualize the bow as the quintessentially Native American weapon because it was in wide use when Old World colonialists showed up in force in the 17th Century. However, the bow is actually a recent arrival, making its appearance in North America around the same time as the fall the Roman Empire in Western Europe!

---

[*] Such *non-agricultural revolutions* probably occurred in Eurasia shortly after the coming of the bow, but their recognizable local residues are mostly long since destroyed.

Finally, maize (corn), beans, and squash were domesticated in present-day Central America, especially Mexico and Guatemala.[21] This region is referred to by archaeologists as *Mesoamerica* and these three plants are often called the *Mesoamerican Triad*. For reasons we will explore in Chapter 13, the weapons and social system that produced Mesoamerican agricultural revolutions were apparently not transferable into the different North American ecosystem. However, the food plants of the Mesoamerican Triad thrived in North America in local family gardening horticulture—domesticated plants *without* a Neolithic revolution, as we predict.

We could hardly have been more fortunate. The stage is set. We have an entire pristine continent almost perfectly arranged to subject our theory to powerful test and potential falsification, over and over again.

Time traveling in the perfect natural laboratory

The ages of settlements and artifacts from Eurasian cultures were mostly determined by radiometric dating somewhat analogously to the

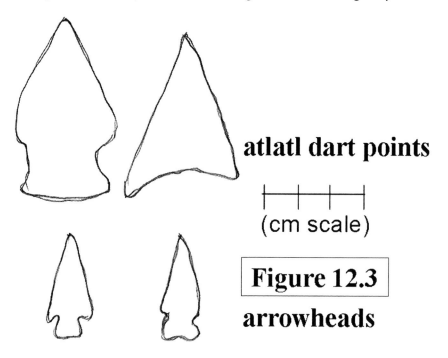

atlatl dart points

(cm scale)

Figure 12.3

arrowheads

way fossils were dated in Chapter 7.[*] This approach is sometimes also used in North America; however, *dendrochronology* or tree ring dating is another useful alternative.

The principle is simple. Each year trees lay down a recognizable ring and the width of the ring is strongly influenced by overall rainfall and average temperature during that year.[†] Rain and temperature averaged over an entire growing season tend to be similar throughout regions of a continent. Thus, all the trees in a large area will form thin rings one dry, cold year and thicker rings another warm, wet year and so on.

Thus, the pattern of rings in any tree from a particular continental area will carry a bar-code-like signature of the interval of years making up its life. If the tree is old enough, say twenty to one hundred years, this bar code will be so extensive as to be completely unique to local trees that grew over the same time period—thick, thin, thin, thick, thick, thick, thin, thick, thin, thin, thin…

If we have a standard of all tree ring thicknesses over thousands of year in an area, we can compare our tree to this standard array and discover that it was "born" in 942 (CE) and cut down in 1027, for example. This is a truly spectacular level of precision. If we have a structure built with several different trees, all cut down in 1027 we can be very confident that this structure was built in or very shortly after 1027. This precision is particularly useful because sedentary humans are very fond of using tree trunks to build things.

This technique helps us tremendously throughout North America where tree ring dating sequences have been painstakingly assembled for the last three thousand years or more. Dendrochronology will turn out be a powerful tool when we look at the American Southwest. Large amounts of wood were used in this region in constructing buildings during the era that concerns us. Moreover, the extreme dryness of the area has preserved tens of thousands of such structural wood elements.[‡,22]

Thus, we can precisely measure the timing of the changes in human social behavior predicted by our theory, like formation of sedentary settlements and initiation of new adaptive technologies like profes-

---

[*]   A short-live isotope of carbon—carbon-14 or "C-14"—is often used to determine ages of artifacts derived from plants, directly or indirectly, with ages of less than twenty thousand years.

[†]   A Google image search with "tree rings" will produce a vast array of vivid examples.

[‡]   See the image in the associated online endnote. Also, Google image searches using "Anasazi" and "stone" or "architecture" will produce useful examples.

sional agriculture,* by the age of their associated wooden building materials.

As well, we can date the coming of the bow with very substantial confidence. In North America, this dating is mostly done using diagnostic stone arrow point types co-buried with datable wood. North American arrowheads are so consistently and dramatically different in size than atlatl dart points that they are easily recognized (Figure 12.3).[†]

Thus, we have all the tools we need. We are ready to time travel around the beautiful natural laboratory of recent prehistoric North America.

The bow comes to Chaco Canyon[‡]

The coming of the bow to the American Southwest is now well dated to 400 to 500 CE.[§,23] Since the coming of humans to North America at least twelve thousand years ago, people in the Southwest never lived in sedentary settlements of any size or for an extended period of time. Just before the coming of the bow to the Southwest, for example, people built simple *pit houses*. These partially subterranean homes were occupied only seasonally and rarely in groups beyond the few an extended family might build.[¶] These recent pre-bow Southwesterners famously

---

* Note that evidence of storage of large amounts of a cultigen like maize in a dated building is extremely strong evidence for "professional" agriculture. Moreover, such observations are generally supported by other kinds of dated evidence – field borders and irrigation canals, for example – creating an essentially unambiguous picture of agriculture's emergence.

† See Justice (1987 and 2002) for wonderful descriptions and images of stone points, including points from most of the prehistoric North American cultures we will discuss in this chapter.

‡ See Stuart (2000) for a delightfully engaging and opinionated description of the Anasazi. Also, see Cordell (1997) for the standard textbook on this remarkable body of work.

§ On current evidence, we can be fairly confident that the bow is in very wide use by around 400 CE. It may have first appeared as early as 100 to 200 CE, but did not come into wide use until around 400 CE.

¶ There is archaeological evidence that humans existed in gradually larger numbers over this twelve-thousand-year period and that their technologies changed over time. However, there is no evidence for any significant increase in overall adaptive sophistication beyond the very modest effects attributable to slow demographic increase.

constructed baskets from woven wild plant fibers.* These baskets often survived in the arid local climate for us to find today. Archaeologists have, thus, long referred to these people as the *Basketmakers*.

Immediately after the arrival of the bow, within less than one hundred years, dramatic changes in the life ways of the Basketmakers ensued.[24] Specifically, the peoples called *Basketmaker II* or *mid Basketmaker* had been growing the cultigens (domesticates) of the Mesoamerican Triad in family garden plots for at least fifteen hundred years. With the arrival of the bow, the *Basketmaker III* or late or advanced Basketmaker phase began. Within the next century, sedentary villages making more intensive use of maize cultivation were established. These new villages announce the development of the culture commonly called *Anasazi* and its various periods are called *Pueblo* periods (below).

These new settlements brought under cultivation extensive new areas of land that had never been used for the limited horticulture/family gardening of the earlier Basketmakers, indicating a new level of adaptive sophistication and technical command.

These sedentary settlements swiftly began accumulating substantial quantities of storable liquid assets, as predicted by our theory. For example, by around 600 CE, hamlets were accumulating enough stored maize to feed themselves for two to four years. Moreover, they were probably generating about two years worth of maize each year. In other words, they were generating a liquid surplus that could be used to fund trade and many other new activities.[25]

These late Basketmakers continued to expand the scale of their social enterprise, building ever-larger settlements, constructing new, more elaborate buildings, intensifying economic activity and agriculture through the *Pueblo I* Anasazi period (ca. 750 to 900 CE). The formally designated Basketmaker-Pueblo transition occurs about three hundred to four hundred years after the coming of the bow. This transition is recognized by a change in architecture within the late Basketmaker permanent settlements, from circular individual structures to room-blocks, ultimately becoming large and apartment-like. This change in architecture reflects increasing adaptation to this new scale of social cooperation and economic intensification (below).

This rapid process continues through the *Pueblo II* Anasazi period and beyond. By approximately 900 CE, massive stone buildings like the famous Pueblo Bonito in Chaco Canyon began to be built.[26] Such structures grew in size so that by around 1100 CE, Pueblo Bonito was five

---

* Uses of these baskets seem to have included gathering and storing wild foods like nuts and dried berries.

stories tall and had approximately eight hundred rooms!*,27 Moreover, Pueblo Bonito was merely one of many similar *Great Houses* built within a few miles of one another in Chaco Canyon. Further, at its peak, the floor of Chaco Canyon contained many hundreds of acres of orderly maize fields systematically irrigated with rainwater channeled through a constructed system from the run-off of the extensive plateau above.[28]

To get a more complete feeling for the remarkable quality of the fit of these phenomena to the predictions of our theory, note that for at least fifteen hundred years *before* the bow, domesticated cultigens were present in this area but they produced little or no increase in social scale or adaptive sophistication. This is the result our theory requires. Conflicts of interest trump *potential* adaptive strategies and new adaptations can *only* arise with new control of conflicts of interest (Chapters 5 and 11; Third Interlude).

As expected, with the coming of the bow and its new capacity to allow control of conflicts of interest, novel scales of social cooperation and unprecedented adaptive sophistication literally *explode* into existence (Figure 12.4). This is a powerful conformity between the predictions of our theory and the empirical observations. After thousands of years of little adaptive change, a new weapon *immediately* precedes a radical restructuring of human social life in the American Southwest.[29]

Our theory predicts many additional details of the adaptive and economic functioning of this revolutionary new society and about its ultimate fate (below). For now it is fruitful to see if other local human populations across the North American continent respond similarly to the coming of the bow.

The bow comes to Cahokia†

The bow first becomes widely distributed in the mid-continental United States slightly later than in the Southwest, around 550 to 650 CE.[30] In contrast to the Southwest, the Midcontinent—the contempo-

---

*   See the associated online endnote for additional images of Pueblo Bonito. Also, a Google image search of "pueblo bonito" with "Chaco" will produce a rich array of images of this Great House. The Chacoan Great Houses were not conventional apartment buildings. They contained rooms used for storage and ceremonial functions as well as possibly residential rooms. Moreover, the pattern of waste disposal at the site suggests that the Great Houses were not necessarily permanently occupied.

†   See Muller (1997) and Milner (1998/2006) for empirically substantive discussions by Mississippian specialists.

rary Midwest and Southeast—is relatively lush and well watered.* Small archaeological sites here are more likely to have been plowed under or paved over by contemporary humans than in the arid, sparsely populated Southwest. Thus, our grasp of the very earliest effects of the bow is a little less detailed here. However, the surviving evidence is quite sufficient to show that the coming of the bow is followed rapidly by powerful social changes, as we predict—again, within less than one hundred years.[31]

In overview, settlement distribution across the landscape changed dramatically shortly after the coming of the bow. Larger settlements and indications of enhanced storage reflecting accumulation of liquid assets characterize many sites from this era.[32] The details are locally variable; however, the area-wide pattern seems clear.[33]

This brief transitional period was followed rapidly by the well-documented Emergent Mississippian (roughly 850 to 1000 CE) and full-blown Mississippian cultures (roughly 1000 to 1400 CE).[34] The Mississippian adaptation involved substantial increases in settlement size and persistence from its inception (roughly 850 CE). The sites also contained evidence of intensification of production. These include storage structures suitable for accumulating large amounts of liquid assets such as maize, for example.

The largest of these settlements included substantial cooperative building projects in the form of huge earthen mounds, which were much larger than anything the earlier behaviorally modern Hopewell built (Chapter 11). These projects include spectacularly large structures like the well-known Monk's Mound at Cahokia near present day St. Louis and many, many other cases.[†,35] These mounds are monumental architecture (Chapter 11) on an entirely new scale. They indicate a substantial degree of coordinated effort by the residents of these permanent settlements.

What is important to notice at the moment is the same dramatic temporal correlation between the coming of the bow and a radical social and adaptive revolution in both the Midwest and the Southeast (Figure 12.4). Again, this correlation is precisely as predicted and required by our theory.

---

* For historical reasons archaeologists refer to this entire portion of the continent as the *Southeast*. The region in question extends, roughly, from present-day Illinois, Ohio, and western Pennsylvania south to the Gulf of Mexico and including the Mississippi River basin. We will use conventional contemporary terms like "Midwest" and "Southeast" to refer to the northern and southern portions of this part of the continent in the text.

† A Google image search of "Monks Mound" and "Cahokia" will produce a rich array of images of this largest of the surviving Mississippian mounds.

The American natural laboratory yields "non-agricultural" Neolithic revolutions - Case 1: The Chumash of the southern California coast[*][36]

One of the most powerful and unique predictions of our theory is that the invention of formal agriculture is an effect, not a cause. Our theory predicts *only* an expanded scale of social cooperation rapidly following the bow. The local *consequences* of this expanded cooperation are predicted to be highly dependent on the available means for adaptive intensification. In other words, in environments with rich wild resources and poor conditions for traditional agricultural intensification, our theory strongly and specifically predicts that the bow will produce non-agricultural revolutions—what we might call *wild food revolutions*.

This prediction is useful and powerful because no other current theory of social complexity predicts wild food revolutions co-causally with agricultural revolutions. Moreover, as we have seen, the North American natural laboratory is the perfect environment to test this crucial prediction.

In fact, wild food revolutions also appear to follow the bow across the continent. The first case we consider are the Chumash who lived along the California coast between present day Los Angeles and Santa Barbara and on the nearby large off-shore islands. These people were fishers, manufacturers, and traders on what is now called the Santa Barbara Channel, a protected marine environment bracketed by the islands and the mainland.

For thousands of years before the coming of the bow, the ancestors of the Chumash lived in this area, harvesting shell fish and other near-shore marine resources as well as engaging in some small-scale trade with inland populations.[37] Local populations may have grown slowly over the several thousand years before the bow, but there were no dramatic changes in ways of life.

The bow arrives in the Santa Barbara Channel around 500 CE.[38] Almost immediately, within less than 150 years, dramatic changes in adaptive behavior ensue.

---

[*]   See Bean and Blackburn (1976), Arnold, (2004), Raab and Jones (2004), and Kennett (2005).

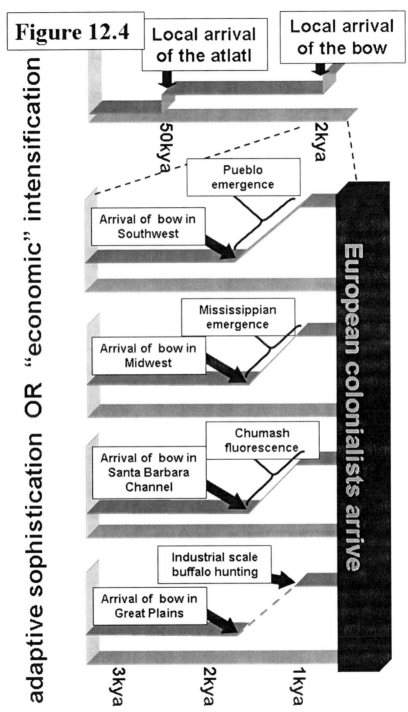

Figure 12.4

For example, more substantial and sedentary villages are built.[39] As well, we see the first development of an expanded trading economy involving marine food from the unprotected *high seas* beyond the channel and intensively produced non-food items exchanged on large scales for resources from inland peoples (marine shell "currency" for acorn flour, for example).[*],[40]

Over the next several hundred years, this new pattern of economic cooperation grows explosively. For example, the famous *tomol*—the ocean-going fishing and trading boat of the Chumash[†]—first begins being made in sufficient numbers to leave an archaeological trace at almost exactly the same time as the first coming of the bow.[41] By around 1050 CE, the Chumash had built a large, vibrant economic system, sustaining several thousand people. This remarkable system thrived until it was ultimately crushed by the oblivious, self-interested elite state power of European colonialists (Chapters 13 and 14).

Indeed, the Chumash system is so dramatic in all its relatively massive scale and abrupt development that it is a favorite of anthropologists and archaeologists trying to understand the abrupt *emergence of complexity*. Into the bargain, this case is also uncomplicated by agriculture.[42] Our theory provides a powerful new answer to these questions of how complexity arises. The bow brings a new scale to law enforcement, supporting expanded, complex social and economic activity.

The American natural laboratory yields "non-agricultural" Neolithic revolutions – Case 2: The buffalo hunters of the Great Plains[‡],[43]

One of the great natural resources of prehistoric North America was the herds of herbivores grazing the grasslands of the Great Plains. These vast populations included at least ten million buffalo or American bison.[44] The herds sometimes covered a local area from horizon to horizon. Since their first arrival in the New World at least twelve thousand years ago, Native American behaviorally modern humans hunted these animals for food at the modest level typical of hunters-gatherers. However, this hunting of buffalo abruptly took on a fundamentally new

---

[*]   See Chapter 11 for a discussion of prehistoric currency.

[†]   A Google image search of "tomol Chumash" will produce numerous images of this famous prehistoric *commercial vessel*.

[‡]   See Chapter 8 in Pringle (1996) for a very readable introduction to this culture.

scale with the rise of what is sometimes called the Late Plains Period—with subintervals having colorful Western names like *Old Woman's Phase* and *One Gun Phase*, after local landmarks around early archaeological digs.

This new Late Plains way of life involved very large sedentary settlements of as many as three thousand individuals.[45] These people were sustained by a substantial economic system. Their participation was based on their ability to produce buffalo products on an industrial scale. They did this by driving large herds of buffalo, hundreds of huge individuals, off high cliffs to plummet several stories to their deaths on the prairie floor below. These *buffalo jumps* were still in use to be documented by the early Europeans. A few famous cases, like the appropriately named *Head-Smashed-In* site in southern Alberta, have been preserved for us to see.[*]

After the amazing sight of hundreds of buffalo tumbling through space and crashing to the prairie floor, large numbers of people would fire up their equipment to produce—Adam Smith pin factory-like—very large amounts of buffalo products.

These "manufactured" buffalo derivatives include many *storable liquid assets.* Among these were hides for clothing and shelter, gut for bow strings, and other body parts for, it has been claimed, everything from fly swatters to children's toys. Perhaps most important of the buffalo products was enormous amounts of *canned food*—pemmican, a combination of dried meat, bone grease, and local berries. This calorie-packed food was dense and easily transported as well as non-spoilable for periods up to several years. Pemmican is thus excellently suited both for local subsistence and for trade, even long-distance trade.[46]

The crucial detail for us at the moment is the chronology of this adaptive buffalo industrial revolution. The bow apparently comes to the upper Great Plains around 100 CE. Over the next several centuries, the Late Plains buffalo economy apparently grows and stabilizes.[47]

The study of this culture is still less complete than for the other North American cases we discussed above. It will be of the greatest interest to see if future work can give us more detailed insight into the sequence of events between the arrival of the bow and the fully operational, large-scale buffalo economy. Our theory's prediction of these details is clear and falsifiable.

There are many other highly suggestive cases of a temporal correlation between the bow and a local adaptive revolution throughout North America.[48] Each of these is a future opportunity to test, and potentially

---

[*]   A Google image search of "Head-Smashed-In" with "Alberta" is illuminating.

falsify, our theory. All of us should follow future work from this vast, fertile natural laboratory with the greatest interest.

Does the Pacific Northwest falsify our theory?[*,49] – Case 3: The Salmon Fishers of the Pacific Northwest

We have yet to discuss one of the most dramatically intensified non-agricultural Neolithic societies in pre-contact North America, the *salmon fishers* of the Pacific Northwest. These people acquired the bow around 200 to 500 CE.[50] Moreover, there is evidence for increased nucleation of larger villages in the period apparently at or shortly after the coming of the bow.[51]

However, the view of many specialists in the Northwest is that intensification occurs much earlier, perhaps a thousand years earlier. If this interpretation is fully correct, our theory is wrong. However, we note that there is another equally attractive hypothesis. The complexity deduced from pre-bow Northwest archaeology may merely reflect behaviorally modern humans living at relatively high population densities and in specific, highly localized areas required by the richest resources on which they depended—including immense annual salmon runs (Chapter 11).[52]

This kind of resource-driven highly localized concentration of population will produce archaeological signatures looking superficially somewhat like those produced by more intensified societies. For example, under such unusual circumstances people may come to *seasonal* gathering spots—a salmon run, for example—in larger numbers and more reliably year-after-year than would commonly be the case for behaviorally modern humans living on dispersed resources. This would generate a larger archaeological footprint at the gathering site, looking somewhat like a sedentary and/or intensified settlement.

Also, as we saw above, our theory predicts that enforced cooperation in large aggregations will be undertaken when the cost/benefit ratio is favorable. Exceptionally predictable and rich resources—again, a salmon run is a good example—will improve the cost/benefit ratio of undertaking atlatl-based coercive management of conflicts of interest over more extended periods. Such exceptional circumstances could lead to somewhat more often and/or longer-duration local aggregation than would otherwise develop in more typical atlatl-based cultures.

---

* See Ames and Maschner (1999) for an empirically authoritative review of the current working picture of these cultures by specialists.

Thus, our theory imposes very specific predictions on what the pre-bow, *behaviorally modern* Northwest should look like. Moreover, our theory also predicts that future work and reinterpretation in the Northwest will show significant changes, especially in intensification, attributable to an increased scale of social cooperation *following* the arrival of the bow. There are strong suggestions in the available record that this might be the case.[53]

These predictions of our theory can be explored and potentially falsified by Northwest specialists. Of course, in view of all the other evidence supporting the theory, we anticipate that such new specialist attention will be very likely ultimately to support the theory.

What more can we expect from a good theory?

We have seen the correlation of the coming of the bow with rapidly ensuing, locally idiosyncratic adaptive revolutions. This is precisely what our theory predicts.

These revolutions are separated in time by nearly ten thousand years, from the Neolithic of the Fertile Crescent to the industrial buffalo hunters of the Great Plains and the maize farmers of Cahokia. The *only* characteristic all of these different areas share, as far as we can tell, is the coming of the bow.

More specifically, each of these different local revolutions was carried out by people who spoke very different languages and came from wildly different cultural backgrounds. Moreover, these people used very different resources in different places—from rice in East Asia, to maize in Chaco Canyon and swordfish in the Santa Barbara Channel.

Similarly, each of these local adaptive revolutions created local architectural and religious traditions that apparently shared nothing in detail with the other cases. In other words, there is nothing in the archaeology of these diverse local revolutions *after* their occurrence to suggest that they shared anything in common *beforehand*, except the recent coming of the bow.

Again, these features of the record are precisely as our theory predicts.

Of course, this excellent empirical support brings us full circle. As the famous early 20th Century British physicist Arthur Eddington once quipped, with tongue only partially in cheek, "No experiment should be believed until first confirmed by theory." He referred, in part, to the synergistic, mutualistic relationship between theory and observation that gives science its power.

Eddington also implied something else. When two theories can ostensibly account for the same evidence, the more cogent and broadly applicable theory is likely to be correct. Individual bits of the empirical evidence we have reviewed can conceivably be interpreted in other ways—narrowly and separately.[54] However, we argue that all the competing interpretations of "agricultural revolutions" of which we are aware, fail to have both the compelling logic and scope of explanatory power of this one.

If this theory really is correct, it should give us a great deal more than just an explanation of the onset of Neolithic revolutions. It should give us many new insights into the properties, organization, and unfolding histories of these societies. This part of the interpretive enterprise could fill many volumes in its own right. Below we will cherry pick a few of the more useful inferences from the theory.

What Neolithic societies teach us about war – conflicts of interest are forever, only the scale of their management changes

The question of what warfare might look like in the ancestral past has recently received a great deal of scholarly attention.[55] We argue that this robust body of work has clearly established empirically what we predict theoretically. War is an ancient and universal human behavior—not some recent aberration of the state, for example. Moreover, our theory's interpretation of the Neolithic record can add substantially to solving the vexatious problem of the causes and logic of war.

The fundamental starting place, as always, must be with conflicts of interest and their management. At the moment, what matters is that any new weapon system has an inherent limitation on the scale at which it can be cost-effectively deployed for the management of these conflicts of interests. For example, when one thousand cooperating archers attack one hundred members of a "criminal gang," each cooperator experiences a one-hundred-fold lower risk that she/he would experience in a one-on-one confrontation (Chapter 5).[*] This can happen because the

---

[*]   Notice, again, that the implication is *not* that one thousand humans ever, or at least very often, actually line up this way against one hundred bad guys. Rather, it is the demonstrable fact that they *could* which makes the threat they deploy necessarily *credible*. The specific calculation of the one-hundred-fold effect mentioned in the test is as fol-

range of the bow allows all the members of a circle of one thousand to be within range of a cluster of one hundred targets, even if the targets group themselves back-to-back (see Figure 12.5 for a graphic illustration of this logic).

Contrast this to a much larger scale where ten thousand cooperating archers face one thousand "criminals" or free riders. Because of the finite range of the bow, now only approximately one thousand cooperators can get into range to shoot at a back-to-back circle of one thousand social "bad guys" (Figure 12.5). Thus, a subset of around one thousand "cooperators" from the larger coalition of ten thousand would find themselves in a very costly one-on-one—thousand-on-thousand—confrontation. This is a dangerous, risky undertaking—usually not an adaptive, self-interested choice.

The implication of this simple arithmetic is that the bow will produce self-interested management of conflicts of interest up to its inherent scale, *but not beyond.* Call a social unit of this scale the *maximum policeable unit* or MPU. The numbers in the example above were chosen because they are roughly realistic. The bow probably sustains local self-interested coercion—*law enforcement*—up to roughly several thousand, but not very far beyond. Let us say the MPU for a bow-based culture is two thousand, for convenience, recognizing that future analysis might revise this number up or down a little bit.* As we have seen, universal, everyday deployment of the bow is cost-effective for all adults.

The implication of this argument, in turn, is absolutely clear. With the coming of the bow, highly cooperative, permanent sedentary social units of roughly two thousand adults will arise. For the reasons we discussed in Chapter 11, these units will be at least partially self-contained or closed.

The cooperative enterprise of each self-contained MPU will have conflicts of interest with other MPUs beyond the capacity of the bow of manage. Thus, these units are predicted to behave toward one another very analogously to non-kin individuals in a non-human animal population. The attitudes and actions of these MPUs toward one another will

---

lows: one thousand (attackers) divided by one hundred (targets) equals ten. Ten squared or ten times ten equals one hundred (Chapter 5).

* This number might be locally variable due to differences in bow technology or the limited ability to deploy large numbers of archers effectively in a heavily forested environment, for example.

range from indifference shading into open, violent hostility according to competitive requirements.

This logic is absolutely vital to understand. The scale of human social cooperation is unavoidably predicted to be strictly and absolutely limited by the scale of effective operation of the weaponry available to coercively police it—no exceptions, anywhere, ever. In other words, members of the MPU will prevent their membership from cheating each other; however, in contrast, there will have no means of self-interested management of conflicts of interest between different MPUs. Thus, cooperation between MPUs will not evolve and confrontation, occasionally violent, will be chronic and inevitable.

This broad, general theoretical picture makes clear, simple predictions about *warfare* during the Neolithic era we have been discussing. As the new weapon, the bow, first comes into a local area, it will be briefly, transiently used by a minority of *early adaptor* self-interested individuals, perhaps even in fundamentally criminal ways—protection rackets, for example. However, very soon a majority of local individuals will acquire the weapon.* This emerging broad access to the new coercive technology will produce a new scale of enforced cooperation throughout the local population, as we have seen. However, this cooperation will now be *more intensive* and *larger in scale*. In other words, the MPU will ultimately be enlarged by the new weapons coercive effects.

This new scale of social cooperation will produce new adaptive capabilities. These new adaptive capabilities, in turn, will increase the number of people who can survive on a particular patch of local habitat. The *local* Malthusian ceiling will be raised. This sustainable number of people is what ecologists call the *carrying capacity* of the local environment.

At first, this increased carrying capacity will provide opportunities for successful reproduction that are accessible with only limited direct competition with others. Thus, different MPUs will tend to adopt an "indifferent" attitude toward one another. Successful social units will grow and fission as their size exceeds the MPU threshold.

---

* The relatively simple know-how to make an effective bow and its arrows would be impossible to monopolize for very long.

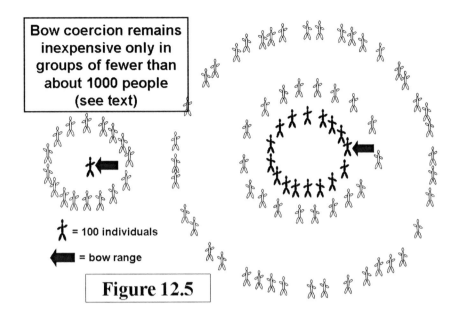

Bow coercion remains inexpensive only in groups of fewer than about 1000 people (see text)

🚶 = 100 individuals

⬅ = bow range

Figure 12.5

Over time, this growth-and-fission process will fill the landscape to the new carrying capacity. In other words, the *Malthusian ceiling* for human population density is social scale-sensitive.[56] The ceiling increases in step with the scale of our social cooperation; however, this ceiling is finite. When it is reached, each social unit now has an "unpoliceable" incentive to pursue its interests through direct competition and conflict with other social units (MPUs) on the landscape.

Humans have been engaging in this kind of *cooperation with conflict* since our origins two million years ago. We are highly adapted to it and Neolithic societies inevitably display it.*

Several of the societies we have discussed here have been well enough studied that we can observe the expected pattern. As the first Chaco Anasazi (ancestral Pueblo), Cahokian Mississippians, or Great Plains buffalo hunters arose, their MPUs replicated themselves across the landscape and create the cultural *fluorescences* we observe in the archaeological record. Ultimately, this process resulted in their hitting the new local Malthusian ceiling and direct competitive acts between MPUs became more common and intense, including acts we would call warfare.

This predicted pattern is general and consistent. For example, during the initial flowering of the Anasazi adaptation, violence was occasionally ferocious, but highly limited.[57] By the later Anasazi period

---

* In spite of this, there is a realistic prospect for a universally peaceful pan-global human future, as we will discuss in Chapter 17.

(after about 1250 CE), after the Great Houses had all been built, there were apparently substantial increases in warfare.[58] Settlements were relocated to defensible positions, like the famous cliff/cave settlements at Mesa Verde and many other places designed to make military access difficult.[*,59]

The famous collapse of the high Anasazi civilization is almost certainly attributable to the playing out of this process.[60] This warfare between MPUs culminated in the building of classic large *Pueblos*, massive multistory room blocks with no windows or doors facing out and which could only be entered with ladders over the outside walls; perfect for pulling up to prevent armed entry while firing arrows and throwing stones down from the roof tops.[†]

Likewise, after the flowering of Cahokia in the Midwest, this great center was first fortified and then abandoned. Heavily fortified smaller settlements, well separated from one another, became the rule.

Finally, something very similar apparently happened on the Great Plains. Indeed, the oft-told story of the *Crow Creek massacre* dates from this era.[61] Sometime in the 1300s a village of about eight hundred people lived in what is now central South Dakota. Their village was typically heavily fortified for this late post-bow era, surrounded by two massive wood palisades separated by a trench approximately twenty feet wide and ten feet deep! Judging from other sites, is it likely that these palisades had crows nests above the top of the walls and protruding outward, allowing defenders to subject attackers to cross fire.

Unfortunately, it was not enough. Crow Creek's defenses were breached and around five hundred people were murdered. Their bodies were later buried in the defensive mote and excavated six hundred years later by 20th Century archaeologists. The dead had been scalped and otherwise mutilated. Men, women, and children were killed, with a conspicuous under-representation of young, reproductive-age females—presumably taken away as a portion of the spoils of a successful attack.[‡]

---

*   See images in associated online endnote. As well, a Google image search of "Mesa Verde Anasazi" or "Hovenweep Anasazi" will produce spectacular images of these defensive structures.

†   A Google image search of "pueblo" will produce vivid examples of this architecture.

‡   A sobering thought, indeed, is that events like this might have been the source of part of the sexual selection on human female attributes, in addition to the prosaic process of mate selection in a peaceful ancestral village.

Though it is beyond our scope here to explore this history further, this pattern of escalating conflict appears to be universal for Neolithic cultures. For example, bow-based New Guinea cultures were also at carrying capacity and extravagantly violent at European contact.[62] It will be of great interest to get a better picture of later Neolithic warfare in the Eurasia than we now have. Such studies will be another opportunity to test the predictions of our theory for the fundamental nature of war (Box 12.1).

One more important detail before we leave Neolithic warfare. A crucial question for us in later chapters will be what conditions might allow multiple Neolithic settlements to become knit together into a larger social unit, analogous to a *state,* perhaps. The descendents of the Pueblo-dwelling late Anasazi were still living in their typically Neolithic, war-like isolated settlements when the first *states* found them, in the form of European colonialists. The interaction of this Pueblo Neolithic culture and the European state will have some very important lessons to teach us (Chapter 13).

Box 12.1 – A general theory of war

It is vital to notice that the preceding section articulates a general theory of war. *War* is merely the consequence of the unmanaged conflicts of interest between human social aggregates, between MPUs. The conflicts of interest that are the fundamental, ultimate cause of warfare can be quantitatively modulated by externals—increases in population or drought causing the Malthusian ceiling to be breached, for example. However, to consider local climate fluctuation as the "cause" of war is a like considering the accelerator pedal to be the cause of an automobile accident. It misses the point. The real issue is the lack of management of conflicts of interest. This theory of war is universally applicable to all scale of human social cooperation, through the present instant (Chapters 13-16).

It follows that invoking ideological or religious motivations for war is to confuse the functioning of proximate psychological mechanisms managing our strategic behavior with the ultimate logic of that behavior (Chapter 10). There is no such thing as a "religious war," though this pretense is sometimes useful propaganda for various self-interested elite individuals (Chapter 13).

The most general lesson is clear. Control of warfare requires cost-effective means of coercive management of conflicts of interest on the scale of that warfare*—always, everywhere, no exceptions. We will also find later that the distribution of access to coercive power *within* a human social unit has an important secondary effect on the frequency and violence of warfare *between* these units (Chapter 16; Box 16.1). These various insights will serve us very well, indeed, as we pursue a more peaceful pan-global human future. (Chapter 17).

The bow and "property"

Our theory yields many anthropological[63] and economic[64] insights. One specific example of an economic implication will serve us well as we move forward. Economists have the following private joke among themselves. Current economic theory merely *stipulates* everything that is actually important and contingent. Economists then "predict" what "should" happen by doing the trivial calculations that ensue from these stipulations. We mostly concur with this self-critical sentiment.

One of the many things that contemporary economics usually stipulate is the existence of property and rights to it.[65] In fact, on our theory, humans have only those rights, including property rights, they can cost-effectively and actively defend, moment to moment. In view of the central, fundamental role of conflicts of interest it can never be otherwise. Moreover, our ability to defend those rights depends *solely* on our access to locally decisive coercive threat, that is, to *real power*.

This blunt, simple picture puts the evolution of property rights into a sharp new focus. Humans have always had personal assets of some sort. For ancient mobile foragers, these would probably have been modest. However, sedentary Neolithic cooperators can generate assets on an entirely new scale, as we have seen. Moreover, the availability of a relatively democratic weapon like the bow means that universal rights to the individual control of one's own assets becomes cost-effectively defensible. These protectable assets, therefore, become *personal property*.

---

*   Actions involved in *law enforcement* at very large social scales can also be construed as war; however, this is *not* the source of the Neolithic warfare we have seen here. Moreover, violence (warfare) associated with *cost-effective* management of large scale conflicts of interest tends to be extremely attenuated after one initial, local demonstration of the credibility of the sustainable coercive threat (see Chapters 13-16).

The logic behind these claims is unexpectedly illuminating. First, anyone stealing liquid assets has now acquired something of value that can be seized back. Moreover, a thief is of little future value as a cooperative colleague. Thus, inexpensive ostracism enforced by joint projection of threat with the bow is an immediately self-interested strategy accessible to any and all (*within* an MPU).

In contrast, those holding property they themselves created in a cooperative enterprise are less vulnerable to cooperative seizure of that property by fellow cooperators because of the future benefits they offer. As well, fellow cooperators will be fellow property owners. Thus, any member of this club attempting to seize the property of another, risks having his own property seized by the coalition of the whole instead— again, a thief offers little future benefits to the coalition of the whole. There are few disincentives to ostracizing a thief.

Notice a detail this logic creates. The *ownership* of property functions as *social* collateral. This property is an asset that can be seized in the event of reliable indications of free riding. In other words, by owning property, each member of the coalition of the whole is funding the coercive umbrella that keeps her/him in line. Property does not merely fund the functioning of the cooperative human social adaptation, *it also funds the coercive umbrella that makes this adaptation possible.*

Thus, property is apparently not merely the thing we usually think of—an asset that can be consumed or traded. It is also something vitally more. Property is what makes large-scale control of conflicts of interest possible and it is our claim to full *citizenship* in the coalition of the whole.

This logic for the *evolution of property* from the development of inexpensive coercion predicts many things we have already seen, such as economic intensification with the coming of the bow. Another prediction is that our evolved minds (Chapter 10) should be inherently suspicious, even contemptuous of the poor, the *collateral-less* non-property holder. We do react this way and we have throughout our history.[*] However, we have not previously understood the evolved psychology of this universal human response.

It will be of great interest to see if our enhanced understanding of the evolution of human economic behavior can support yet further improvements in our *theory of property*.

---

[*]   For example, see Sean Wilentz' (2005) discussion of property requirements for voting in the long, slow emergence of "universal enfranchisement" in the United States. This is merely one of a vast number of documented cases throughout the historical and ethnographic literature.

Ritual – what on Earth is going on?

Our new grasp of the Neolithic gives us novel insight into pragmatic issues like warfare and economics. But it also helps us understand more obscure parts of our behavior, like ritual. One specific class of examples will illustrate these new insights.

In the ancient ancestral past of the Santa Barbara Chumash "Old Woman Momoy" is said to have dosed her grandson with Datura, allowing him to have visions that foretold the future and answered vital questions.

At European contact, spiritual functionaries would deliver carefully calibrated doses of Datura to ritual participants in large public ceremonies. Give too much and the supplicant might suffer massive hallucinations for days with cardiac and other systemic problems, ultimately dying horribly and insane. Give just the right amount and an individual could travel where Old Woman Momoy's grandson went, finding her/his animal spirit guide, seeing the future.[66]

When Europeans reached the Great Plains they found the buffalo hunters gathering in large groups at the summer solstice for the Sun Dance ritual.[67] During this ritual, young men would drive bone pins into their upper bodies deeply enough to allow them to be hauled up a large ceremonial pole on ropes attached to these pins. They would then hang there, for hours or even days, until the pins gradually ripped through their flesh and they fell to the ground.

We can imagine an extraterrestrial anthropologist watching humans for millennia. He (she/it) sees them doing things like this on a truly enormous scale. He throws up his hands (tentacles?) in despair, completely unable to understand this apparently insane, but universal activity; an activity the "natives" call by many names, all meaning "ritual."

Unfortunately, terrestrial anthropologists have been just as flummoxed as their imaginary extraterrestrial colleague. It is absolutely clear that massive, public, idiosyncratic—not to say, bizarre—ritual is a recurring feature of human life. It is equally clear that traditional explanations of this human property have grossly conflated proximate and ultimate causation, producing futility and confusion.

Archaeology of the Neolithic has thrown a powerful new spotlight on ritual behavior. Behaviorally modern humans, perhaps even pre-moderns, probably behaved similarly on a relatively modest scale. However, with the coming of the bow and the Neolithic to Eurasia and North America, these deeply strange, species-typical, large-scale public behaviors grow substantially in scale.[68] The Mississippians and Anasazi apparently begin engaging in these large public rituals at the same time we see all the other signs of their increasing economic complexity—just after the coming of the bow.[69] Likewise the ritual cults surrounding Datura use by the Chumash apparently arose or expanded dramatically with their economic intensification after the coming of the bow.*[,70]

If our theory is correct and largely complete, as we claim, it should explain these superficially arbitrary, nonsensical behaviors. In the following section we argue that our theory does, indeed, explain the *explosion of ritual* with the Neolithic revolutions and, indeed, the very existence of ritual as a characteristic, uniquely human behavior.

Know thy neighbor – uniquely human cooperation and a new theory of ritual

To understand human ritual we must return to the way self-interested humans use culturally transmitted information (Chapter 10). Remember that the crucial human trick is the self-interested deployment of coercion in ways that generate kinship-independent social cooperation.

This uniquely human adaptation means that our relationship to the material and social worlds are very different than for any other animal. Our ability to obtain the material resources we need—food, shelter, water—has probably not been limited very much by other species for more than a million years, a consequence of our unique ecological dominance. Our access to these resources is limited, instead, by only two things—the amount of reliable cultural information we have access to *and* competition from other humans. These two limitations have profound implications for our public behavior.

The first of these limitations is ameliorated by increases in pragmatic information with social scale. Our Malthusian ceilings are raised in larger cooperative polities. Our theory predicts that the sophistication of our adaptive solutions should increase with new coercive means, supporting new scales of this information sharing enterprise (Chapter 10; Third Interlude). The Neolithic adaptive revolutions are further strong support for this picture.

However, the second of these limitations, competition, rather than cooperation with other humans, has some novel effects as our social

---

* The ancestral versions of the Sun Dance did not leave a big enough archaeological footprint for us yet to have noticed its origins. It will be most interesting if we can discover these.

groups get larger. It is here where our theory predicts the social functions of ritual to expand dramatically.

To understand how human competition impinges our social/economic behavior, first recall that non-kin conflicts of interest are a constant part of any social world—everywhere, forever. As we have said, humans have not eliminated conflicts of interest, we merely manage them in a way no other animal does.

Moreover, the way we manage conflicts of interest is through the conjoint, self-interested projection of coercive threat (Chapter 5). Thus, the single most important thing we, as adapted individuals, must do is become a part of the locally decisive coalition of coercive humans. If we succeed, we survive and prosper. If we do *not* succeed, nothing else matters. We are doomed.[*]

As we saw in Chapter 10, we tend to be individually smug about our abilities. As a result, we are oblivious to how dependent upon and vulnerable to our social surroundings we actually are. Being privileged members of a rich, stable modern state, most of us do not have ready, conscious access to this fact about our lives. But stop and think. If you quit your job, moved onto the street, and "foraged" alone for a living, how long would it be before you were dead or imprisoned? Not very long! As a result, people are demoralized and horrified at true feelings of disenfranchisement and social isolation, everywhere and always. Humans are highly adapted to catering to the demands of their coalitions, their MPUs (see Box 10.3).

Our feelings of horror at being isolated are just the subjective consequence of a proximate psychological mechanism that keeps us constantly tuned to the vital coercive coalition of which we are members. What we need to understand now is how this mechanism works in large societies where individuals are often partially or entirely anonymous to one another.

We constantly monitor the information we receive from and provide to others. We are alert to the quality and reliability of that information. We are ready to collaborate in profitable ostracism of liars and we are equally vigilant against provoking others through deceit. This represents the proximate functioning of the mechanisms that support coercively enforced human information sharing (Chapter 10). These behaviors create what we might call a ***mutually informed social aggregate*** or MISA, for short. Of course, a MISA will generally be an MPU (above) or, occasionally, a subset thereof.

---

[*]   Or we must immediately flee to the domain of another coercive human coalition to which we must then become attached or accommodated.

Less obvious, but equally important, is how we recognize and deal with the members of our MPU or MISA. In the ancient ancestral environment, our MISA was small and we knew all its members intimately—and were intimately known, in return. We could conveniently be vigilant that no one was subverting our individual interests. And, as everyone else was doing the same thing, everyone was forced to serve the interests of the aggregate most of the time, at least in public.

However, as our MISA grew in size—as it does in spades with rise of large Neolithic social units—we will be socially connected to too many people to know them all well individually. We cannot monitor what everyone else is doing most of the time. How can our evolved psychology deal with this novel challenge?

Human belief systems are the content of the shared information in our MISA, including the rules or social contract under which our public cooperation operates (Chapter 10). Now, if we bring large numbers of, potentially co-anonymous, people together and require them to declare publicly their beliefs we achieve some surprisingly useful things.

For example, taking the time to acquire all the beliefs of a particular MISA is time-consuming. "Getting educated" is a substantial investment—opportunity costs, again—an investment that would be lost if an individual were subsequently ostracized. Making this investment identifies us as someone who intends to play by the rules, respect the beliefs, of the local MISA.[*]

But there is more, by *publicly* declaring our acceptance of the beliefs of the local MISA, we simultaneously and inevitably identify ourselves as *non-members* to other, potentially competing MISAs. This is profound. We might secretly consider some of the beliefs of our local MISA to be silly; however, if we give any hint of this in public we will be ostracized. Thus, by convincingly acting in public *as if* we share the beliefs of the local MISA we simultaneously contribute to the mutual intimidation of possible "non-believers" in our MISA and cut ourselves off from membership in any competing MISAs. From the point of view of our fellow MISA members we have become "made" people in the jargon of organized crime.

Notice one last thing about the beliefs of our MISA. Those shared pragmatic beliefs that pertain to adaptive activities—foraging, hunting, farming, etc.—will tend to be useful and empirically verifiable/falsifiable

---

[*] Much of the work on theories of religion reviewed here and in other chapters was done by the authors in collaboration with Arnav Shah (manuscript in preparation).

against the evidence of the surrounding material world.* Their structure will predictably reflect the structure of the real world. Likewise, these shared beliefs will include many of our social rules for non-kin cooperation—*thou shalt not kill, thou shalt not steal*, etc. As with *pragmatic* information, these social contract beliefs will likewise be largely predictable and understandably related to the material world.

However, this is not a complete list of the shared beliefs of our MISA. *In contrast* to the pragmatic and social contract beliefs there is a third set. This third set strictly serves a single function—to *identify* us as members of our local MISA and to *distinguish* us from the members of competing MISAs. As individually self-interested members of our MISA, we will insist on this public identification for the reasons above. In so doing we will inflict this public identification on one another.

Notice that this *MISA-identification* information can be *purely self-referential* and need not have any discernible relationship to the material world. Its *sole* adaptive purpose can be in our relationships with fellow MISA members—it need have no other function. Thus, these beliefs have only one relevant criterion for "truth"—they must be shared with the other members of our MISA. They can be completely unpredictable. Indeed, it would be useful if these *identifier beliefs* (Chapter 10) were maximally obscure, esoteric, and counter-intuitive. This would make them harder to learn and to fake, enforcing up-front investment to acquire them and discouraging would-be short-timer free riders.

We have reached the crucial point in the argument. We have a strong new theory of ritual including the bizarre, arbitrary activity it contains. Our uniquely human management of non-kin conflicts of interest creates *two* classes of cultural information. The first is information about the world (pragmatic and social contract information), shared with the non-kin members of our MISA with profound adaptive consequences (Chapter 10). However, the second class of information is purely self-referential (identifier information). It need have no *other* adaptive function than being essential to our reliable identification of the members of our local, uniquely human, MISAs.

Moreover, by making the rituals that demonstrate our possession of these arbitrary, self-referential identifier beliefs both maximally esoteric (costly to learn) and physically risky or demanding (costly to execute), we effectively block anonymous free riders from transiently entering and parasitizing our large MISAs.

---

*   This *realism* applies to those beliefs about the material world that actually affect how we use and interact with it, not necessarily those beliefs that pertain to its remote origins or the agents behind natural processes, of course.

Our evolved psychology makes us exquisitely sensitive to whether others share our beliefs. We have a deep emotional connection to *all* those beliefs, both the real-world ones and the *identifier* ones[71]—we are "true believers" in rituals and all the information that comes with them.

Where to now?

Stepping back to take in the larger picture, we now have in hand an economical, powerfully effective theory of the Neolithic. We arguably understand this vast landscape as we never have before. We are now ready to move further forward in time and ask how the first *states* emerge from this Neolithic foundation. As we travel forward into this era we also encounter the first *written* history. This recorded history will prove to present both a wonderful new opportunity and a difficult new challenge.

௰

Fourth Interlude:
History and the evolved mind of the political animal

Chapters 11 and 12 showed us that our theory of history is apparently highly effective. It accounts simply and economically for the behaviorally modern human revolution and all the many Eurasian and North American Neolithic revolutions. No earlier theory of history has approached this scope and parsimony.

If our theory is, in fact, correct it should continue to account for the rest of history, through the present instant. In Chapters 13 to 17 we will argue that it does. However, as we move forward toward the present, we begin to encounter a new problem, not with our theory but with ourselves. As we have seen, our minds were built by natural selection to have very specific capabilities. An objective comprehension of the underlying ultimate strategic logic of our behaviors is *not* one of these abilities.

More specifically, our evolved psychologies are merely proximate devices (Chapter 10). Their subjective outputs, though a richly rewarding and important part of being human, are far removed from the ultimate evolutionary logic of our behaviors. Our proximate psychologies are designed to cause us to behave AS IF we understood the ultimate logic of our behaviors, NOT to actually give us that ultimate understanding. Our goal here is to understand this ultimate logic.

Before we begin, it is important to recognize that this comprehension of the ultimate logic of our public behavior will not make us less human. In fact, we argue that such an understanding can work beneficially with our proximate psychologies to allow our most humane characteristics to emerge more fully. If we take the time to master ultimate causation in the human world, our rewards can be substantial.

Be forewarned that our proximate ethical and political psychologies will become increasingly, intensely engaged at the expense of our analytical intelligence as we proceed forward in historical time from here. To see an analogy to how this emotional confusion can work, think back to the ultimate logic of biological reproduction from Chapters 2 and 3. We are vehicles built by design information for the "purpose" of replicating that design information. If we look at non-human organisms or at humans far removed from us, we can readily understand the simple logic of natural selection creating such vehicles. However, if we try to look at our individual selves or our mates, siblings, children or parents as *merely* biological vehicles we almost always fail—authors included. Our proximate feelings of love and loyalty for our immediate families overwhelm our analytical intelligence.

Recent history creates the analogous problem. Our feelings about various details of this record trump our analytical intelligence. When this happens, we cease to be scientists and become, instead, the unconscious participants in the very process our science of history is striving to explain.

This is a formidable problem, but not an insurmountable one. With patience, thought, and care we can avoid being defeated. We will begin to work on this new level of historical insight here and will return to it several times in Chapters 13 to 17. This quest is absolutely vital if we are to grasp fully the contemporary human condition.

"History" as evidence

As we have seen, "history is the lie made up by the winner" (Chapter 10). Now we need to look more carefully at some of the implications of this simple, but profound statement.

In the following chapters we come to that part of our history for which we have a written record—roughly the last fifty-five hundred years. Written history will provide us with a phenomenally rich new data source to supplement archaeology. However, this record will be both a blessing and a curse. On the one hand, written history implicitly exaggerates its own importance. If the entire two-million-year human story were telescoped into one year, our written record would cover merely one day, New Year's Eve.* Written history reflects a tiny part of our time on this planet, yet, until the development of modern science, it constituted everything we thought we knew about our past.

On the other hand, this record is—always and everywhere—produced by self-interested advocates. Moreover, this self-driven quality will be just as characteristic of the reportage of many professional historians as it is of the self-advocacy of history's overt political actors.

Equally importantly, none of us can ever be emotionally detached from human political advocacy. Our evolved minds—our proximate psychologies—will not permit it (Chapter 10). We see, mostly unconsciously, each word of recent history as *advocacy* impinging on our vital individual interests, authors included. We react on the basis of those interests. We cannot avoid this reaction any more than we can avoid feeling tired when we have not slept. Moreover, this response is always subjectively ethical. Our proximate psychologies see all political positions as "good" or "evil" to some degree, depending on our interests.

We must *use* the written record while not being defeated by this emotional connection to it. After years of wrestling with this problem we believe this is possible. First, each of us must clearly recognize his/her own interests. Second, we must recognize that the amoral, purposeless universe does not recognize these interests as good or evil. All beliefs and practices, no matter how hideous we may consider them, are still biologically equivalent. Good and evil are self-referential classifications, entirely internal to local human cooperative social adaptations.†

Our proximate psychologies are designed to allow us to pursue self-interest in the public sphere, the domain of uniquely human kinship-independent social cooperation (Chapters 5 and 10). Our best choice is to join coercively powerful local coalitions and use them to enforce our interests while surviving the coercive demands they make on us. Indeed, again, we are *a one note hula-hoop*, a *one trick pony*. This public, cooperative pursuit of individual self-interest in a highly coercive social environment is the ancient, pan-human adaptation (Boxes 10.2 and 10.3). It is what we do and all we do. We are very good at it. Everything about us and about our history emerges from this single adaptive strategy, on our theory.

---

*   The very recent origin of *all* our written history obviously includes the documents making up widely used religious texts, like the Torah, Koran, and Bible, for example.

†   We will argue that a useful *universal* definition of good and evil can emerge from the mutually informed consensus of the entire, democratized pan-global human coalition (Chapter 17). However, we must acquire much more insight before we can discuss this ethical human future. Notice that we are progressively refining our original definitions of good and evil from Chapters 4 and 10.

We emphasize once again that this picture of human social behavior need not be grim or disturbing. We are capable of great humanity and deep fellow-feeling while pursuing this human strategy. Our eventual pragmatic goal is to understand what we are actually doing in our public lives so that our potential for common, pan-global humanity is maximized (Chapter 17).

The *social contract* refers to the publicly negotiated beliefs about how we should behave in our local human non-kin cooperative enterprise (Chapter 10). The social contract constitutes the rules of the game. We generally ostracize those who violate these rules and we expect to be ostracized if we violate them ourselves. The details of the social contract are produced by the processes of the human wise crowd (Chapter 10). The question will be who gets to be a member of the wise crowd defining the social contract.

"Methinks he doth protest too much" – what are we doing when we "explain" our actions?

Psychologists have long recognized an important property of our minds. When we are asked why we did something, we often have an answer. Moreover, most of the time we believe the answer we give is "true." (Those times we consciously lie do not concern us at the moment.) However, in actuality, these truthful explanations are post facto concoctions, called *confabulations*. In other words we behave first, reacting to mostly unconscious mechanisms shaped by ultimate causes, and then consciously interpret our own behavior in retrospect according to our current social environment. The explanations we offer do not reflect the "real" (ultimate) reasons we did the things about which we are being asked.*

This, in turn, presents an interesting conundrum. The question is why we bother to concoct confabulations if they are not true. Such systematic unconscious lying should create selection in others to ignore these

---

*   There are many pieces of evidence supporting these claims. One of the most evocative comes from experiments with *split brain* patients where the two hemispheres of the brain can no longer communicate with one another. These patients can be induced to carry out a behavior by information delivered to the right, non-linguistic brain hemisphere—behavior that is subsequently observed by the uninformed left, linguistic hemisphere. When the patient is then asked why she/he carried out this behavior, the linguistic hemisphere provides a plausible, but completely fictitious explanation, nonetheless consciously believing it to be "true." See Michael Gazzaniga's engaging description of these experiments in his 2005 book *The Ethical Brain*. Also, see Gazzaniga (2000 and 2008).

explanations. We then would evolve to cease to give them. Why have we not ceased?

Our theory strongly suggests an answer to this question. The explanations we offer for our behavior are not intended to be true in any objective sense. Rather, they are offered as elements of our continuous renegotiation of the social contract with the local wise crowd. We are (almost always unconsciously) arguing for a particular cooperative logic for our behavior and, thus, proposing a similar logic to others. Simultaneously, we are also justifying our actions to potentially coercive others. We are trying to avoid being the target of their potentially violent disapproval. We behave this way continuously, in large ways and small, every day of our lives.

This causal picture is vital to grasp. Our explanations of our behavior are a central part of the human cooperative enterprise. They are tools for arriving at and nurturing mutual agreement about how we all should behave, in view of our mutually coercive/cooperative relationships. These explanations are also our social self-justification.[*] Our evolved minds are designed to generate and consume these explanations. This *design feature* is not merely true of our explanations of our individual behaviors, it is also true of all our public policy utterances—economic, political, religious, social, ethical.[†]

Each of us (authors included) has strong emotional connections to cherished beliefs about what is right and what is good. We would not be human without them. But these beliefs are not what we think they are. They are self-interested confabulations—constructions designed to serve and recruit others to our interests.

It will be essential to have these properties of our explanations in mind at every moment as we use, and contend with, the historical record.

"Class" and "interest group"

From Herodotus and Aristotle twenty-five hundred years ago to Hobbes, Locke, Hume, Smith, Ricardo, Marx, and contemporary political writers, our historical politics are always described in a particularly

---

[*]   We are arguing (unconsciously) that the explanation we offer for our behavior *would be* an appropriate, public, prosocial one, *not* that this explanation is accurate in a literal sense.

[†]   See Tilly's 2006 book *Why? What happens when people give reasons…and why* and his 2008 book *Credit and Blame* for a social scientist's recent outstanding reviews of some empirical properties of human public declarations.

striking manner. These authors rarely speak only about individuals. They also write of "interest groups" and "classes" in general and about nobles, peasants, slaves, soldiers, clergy, bureaucrats, bourgeoisie, third estate, and thousands of other more specific subgroups within human societies.

A crucial question for us is what these terms actually describe on our theory. We argue that *interest groups* are real entities and they have both an ancient and a newer set of properties.*

First is the ancient face of interest groups. We each have skills resulting from our acquisition of elements of culturally transmitted information (Chapter 10). *Embodied expertise* is the apt economist's term for such abilities. We are often not conscious of many of the skills and much of the know-how we possess or of their origins. For example, many of us speak and comprehend English with unconscious virtuosity. This skill was culturally transmitted preconsciously, very early in our development.

The abilities we each have that allow us to "hold a job" and be successful cooperative (economic) actors in our local environments also reflect a great deal of comparable embodied expertise. One of the qualities of this kind of know-how is that it represents a tremendous investment. We cannot discard one set of embodied skills and pick up another like we change clothes. Rather, acquiring a new skill set is more like discarding our native language and learning to use a new language exclusively—difficult and a massive new investment.

This investment in embodied expertise has been a property of individual human lives throughout our two-million-year history. We are highly attuned to perceiving, often unconsciously, how our individual interests are served by social contracts that reward the specific embodied expertise we possess. In contrast, we will be threatened by any changes in the local social contract that marginalize our individual skills.

Of course, we will share this self-interest in common with many others possessing similar skills. We will collaborate with them to be adventuresome in endeavors that reward our skill sets and hostile (conservative, reactionary) toward those that threaten to eliminate the needs for our individual skills and knowledge. We will form *interest groups* with those who share our concerns (often sharing our skills) and

---

* We must be very careful not to fall into the group selection fallacy as we use such terminology. Remember that interest groups are made up of individuals all pursuing and protecting individual self-interest in the best way possible given successful management of conflicts of interest within a social aggregate of other self-interested individuals. The "group" exists entirely because of the aggregate consequences of these individual actions and effects.

406   *Death from a Distance and the Birth of a Humane Universe*

cooperate to support or coercively resist social endeavors on this basis. For example, we will try to insist that local economic activity be transacted in our first language.

Second, interest groups existed throughout most of our two-million-year history. The ancestral human trick of projecting coercive threat in pursuit of self-interest probably produced *democratic social systems* throughout almost all of this history. This is the predicted outcome of the weapons that were available (Chapters 5, 7, 11, and 12). Interest group cooperation and competition played out in this broadly distributed access to coercive threat, this "democratized" environment.

However, in contrast, during the last fifty-five hundred years, a new type of human social enterprise arose creating a new set of properties for our interest groups. Societies came into existence in which coercive dominance was held by small subgroups or *elites* (Chapter 13). New technical developments in weaponry made it possible for small elites to acquire coercive dominance over large non-elite groups, undermining earlier democratized systems and creating elite-dominated, hierarchical societies in their wake (Chapter 13).

Under these novel conditions, the ancient evolved human psychology continued to do what it has always done—control and react to projection of coercive threat in pursuit of self-interest. However, the outcome is very different for many of the members of such elite-dominated, hierarchical societies than for individuals in the more democratized cultures of our ancestors and those that sometimes exist today (Chapters 13-17).

At the moment, what concerns us is how different interest groups behave within these intensely hierarchical societies. Most importantly, non-elite, relatively powerless individuals will attempt to defend themselves by recruiting other non-elite individuals with common interests in order to exploit the human trick of conjoint projection of threat (Chapter 5) as effectively as possible in the face of elite coercive threat.

It is the goal of these non-elite interest groups that most concerns us. The technical details of coercive threat in hierarchical societies often preclude the formation of a highly democratized coalition of the whole to unseat elites (Chapter 13). Each individual confronts replacement of the present elite with her/his own newly elite interest group as the *best available alternative* to continued subjugation. Thus, all interest groups, both elite and non-elite, will have this vital function of competing for elite power under these non-democratic conditions.

In such hierarchical cultures, interest groups will be sharply defined and mutually hostile. Of course, existing elites will sometimes foment this mutual antagonism, attempting to divide and conquer potential competitors. These elite actions will further balkanize interest groups and classes. Much of our *written* history originates from these hierarchical states. This fierce, lethal relationship between interest groups will be very common and central to these historical accounts.

What is politics?

We call our characteristically human pursuit of self-interest in the public domain *politics*. All the complex beliefs and feelings we bring into this arena are just the products of our proximate psychologies adapted to uniquely human kinship-independent social cooperation sustained by self-interested coercive threat (Chapters 5 and 10).

It is useful to state this crucial point in the most evocative and illuminating form. All belief systems serve the interests (potentially) of those who espouse them and all are equal as biologically produced cultural devices. This identity of *function* applies equivalently to Nazism, Christianity, Materialism, Islam, Communism, Capitalism, and every other "ism" or belief any of us has ever heard of. They are all copies of the same entities, tools, and products of public negotiation of the human social contract.

This claim of *moral relativism* of beliefs emerges as a prediction of our theory as we will discuss shortly. We will return to how we all, *as ethical creatures*, can deal with this disturbing feature of our lives. It is malleable in pursuit of pan-global humanity when we possess sufficient knowledge (Chapter 17).

For now what matters is that pursuit of access to coercive dominance is a central human evolved need. None of us feels satisfied, or even minimally safe, if we do not possess this access. It is as basic as food, shelter, and sleep. We have this access when two criteria are fulfilled. First, the embodied skills we possess are valued by the coercive coalitions to which we belong. Otherwise, we will be regarded as a free rider and ostracized. Second, the coercive coalitions we join must possess adequate access to coercive threat to defend the mutual interests of their members.

These two criteria are all that matters to our evolved, unconscious minds. Notice that the larger common good of all people is NOT one of these criteria. In the highly democratized ancestral environment (Chapters 5, 7, 11, and 12) the common good and the good of the coercively dominant local coalition were generally nearly the same things. However, in the recent historical era, where coercive technologies support elite domination, the public *social contract* has served the interests of the members of the local elite preferentially. The interests of the local coalition of the whole have been much less relevant. This is merely the inevitable consequence of humans deploying whatever coercive threat they possess in pursuit of individual self-interest as defined by their evolved minds. Armed elites are their own, self-contained wise crowds.

The nature of "good" and "evil"

We are compelled to obey our evolved moral sense. We have no choice. However, we also possess enough empirical evidence to understand that "evil" often perceives itself as "good." Can we make biological sense of what we see? We argue that we can.

First, our ethical sense is designed to support our pursuit of self-interest. Our feelings of ethical outrage at Nazism, for example, are not abstract recognition of some universal truth. Rather, they are careful (though generally unconscious) minding of our individual self-interests. Very few of us, indeed, are likely to be "successful" brownshirts, but almost all of us are likely to be their victims. Our evolved psychologies make us feel this statistical risk.

Second, however, as we come to understand the evolutionary and self-interested origins of our ethical sense, we become empowered to begin to deploy this sense in ways that might come to better approximate the common pan-human interest. We, the authors, choose to define this pan-human global interest as "good," (as, we suspect, will many readers). With this goal in mind, we will briefly frame how our theory explains human political behavior from the ethical perspective we all find so compelling.

Specifically, we believe that it is logically and ethically coherent to argue that good and evil political systems (cooperative human enterprises) are distinguished by how well they serve ALL their members. It could not be simpler. Systems that abuse and exploit substantial numbers of their members are evil; those that do not are good.

Further, we propose that good and evil belief systems are completely distinguished by a single property. That property is whether these beliefs are deployed in the pursuit of self-interest by narrow elites or, in contrast, by broad, pluralistic, democratized coalitions. Moreover, the existence of hierarchical or democratized coalitions is entirely determined by who controls access to coercive means (weaponry). Our claim is that the evidence of Chapters 13 to 17 will very strongly support this empirical position.

We want to be emphatically clear here. We will argue that *all* truly evil outcomes in the public political domain in *all* of human history result from a single underlying cause: The pursuit and exercise of elite coercive dominance by the evolved human mind.*

The problem of elite domination and "evil"

Internally originated democratization of elite-dominated, hierarchical cultures is inherently difficult or even impossible, depending on the local technical coercive environment (Chapters 13-17). When the standing elites are weakened and vulnerable, they are usually merely replaced by another subgroup, recreating hierarchy under the flag of a different elite.

In other words, elementary short-term human social strategies usually perpetuate elite domination (Chapter 13). This logic is all there is. Details of beliefs are merely tools and pawns in the competition to seize elite coercive dominance.

When we possess coercive dominance, whether hierarchically or democratically distributed, we subjectively see our deployment of threat as *law enforcement*. We are *incapable* of recognizing that the targets of our coercion may see it as terrorism or genocide. Likewise, our evolved minds see the beliefs that justify our deployment of the threat we posses as *righteous*. This is what human minds have been designed by natural selection to do—*everywhere* and *always* (Chapters 5 and 10).

A brief historical vignette will illustrate how our evolved human psychologies produce inevitably evil outcomes when coercive power is concentrated in elite hands. When the hierarchical tsarist nobility of early 20th Century Russia was fatally weakened, *proletarian* organizations, including local soviets, successfully seized power (Chapter 15). We mean this literally. These local organizations captured and controlled the technical means of decisive coercion—the weapons.

Though ostensibly democratic, these groups were actually narrowly class conscious. As well, they were locked in a mortal struggle with other equally class conscious interest groups. The armed soviet militias soon replaced the tsarist state coercive apparatus with another equally murderous bureaucracy, the *Stalinist state*. This hierarchical coercive apparatus remains in the hands of Stalin's elite heirs at this writing—after these heirs repulsed

---

* This section begins the authors' own personal declarations of "interests" in history. Of course, in addition to these very general concerns, we also have more narrow interests resulting from our status as professional scholars, scientists, academics, New Yorkers, and Americans. These can also be highly intrusive on our analytical thought. However, our experience is that these narrow interests are easier to fend off than our "cosmic" sense of how history and human societies *should* be or is *destined* to be. It is this cosmic sense of interest that concerns us at the moment.

an attempted takeover by the competing armed goons of Russian business oligarchs/organized criminals.

We might be sympathetic to the plight of the original tsarist Russian peasants and factory workers. We might even feel they were right to believe themselves victimized and, thus, justified in seizing the state. But these feelings and beliefs are, we argue, perfectly irrelevant. All that really matters is what *portion of the whole* holds decisive coercive power. When that power is pluralistically and democratically distributed, the resulting social system can be humane. When that power is concentrated in elite hands—even *proletarian* hands and no matter the supporting beliefs—the outcomes will be brutal and inhuman, always and everywhere.*

Most of us probably recognize this property of human cultures from our own experience. For example, Scientific Materialism, Islam, Communism, Capitalism, or Christianity has each been used, at different times, by coalitions of individuals pursuing either democratic or elitist goals.† The ensuing "good" or "evil" had absolutely *nothing* to do with the details of these belief systems themselves. It had *everything* to do with whose interests these belief systems were serving—small elites or pluralistic democratized coalitions. Interests served, in turn, were determined *exclusively* by who held decisive coercive threat, we will argue (Chapters 13-17).

Why we usually misunderstand the causes of history

> *As theorems and conjectures are to mathematics and quantum mechanics is to physics, so ideas and the principles of government are to politics.*
> Former US Senator Gary Hart (New York Times Book Review, March 22, 2009, p. 14)

---

*   As we will see repeatedly (Chapters 13-17), paternalistic rationales for elite domination are merely self-justifying rubbish, inevitably and universally.

†   Belief systems often include strangely arbitrary elements—conventions like wearing neckties or bones through our noses, for example. As well, belief systems often contain surreal and fantastic elements (angels, spirits, gods, and the like). Extensive exploration of these properties are beyond the scope of our discussion in this book; however, they probably emerge from the internal logic of human kinship-independent social cooperation in a straightforward way (Chapter 10; J. Souza, A. Shah and P. Bingham, manuscript in preparation).

What matters for now is that the *objective truth* of claims about the world is often irrelevant to the political use of such claims. We mean this completely and literally. Factually true claims about the world can be used for evil purposes and factually silly claims about the world can, nonetheless, be part of a good outcome. Notably, however, democratized wise crowds eventually tend to eliminate factually silly beliefs over time, whereas elite wise crowds often defend and nurture them (Chapters 10, 13, and 17).

Not through speeches and majority decisions will the great questions of the day be decided [] but by iron and blood.
Otto von Bismarck to the Budget Committee of the Prussian Landtag on September 29, 1862

We can now see our (mis)understanding of history from another powerful perspective. Throughout most of our two-million-year history, access to coercive threat was apparently relatively universal. We evolved to take this access for granted, much as we unconsciously presuppose the air we breathe (Chapter 10). Under these ancestral democratic conditions, our public debates and decisions about ethics and public policy were decisive. Coercive policing of these decisions followed automatically, unconsciously.

However, with the coming of elite concentrations of coercive power beginning around fifty-five hundred years ago (Chapter 13), our evolved psychologies produced a fundamentally different result. Coercive policing now mostly served narrow elite interests. Public policy debates were replaced by narrowly self-interested false facsimiles. Elites intimidated and indoctrinated while pretending to advocate. Non-elites subverted in search of elite power while pretending to propose.

Our two-million-year-old evolved proximate psychologies are not designed to comprehend this novel coercive situation. Our minds often insist on seeing these faux public ethical and policy debates as somehow real and causal. We find it difficult not to seek the causes of history in this blizzard of disingenuous verbiage.

For example, pick up a book on the history of the English Civil War in the early 17th Century. We will almost inevitably read about the publicly articulated *beliefs* of Royalists, Levellers, Covenanters, Puritans, and other interest groups. We will probably be told that the "disagreements" between these interest groups were the causes of the English Civil War.

Our theory proposes very different causes. The English Civil War resulted from the loss of coercive dominance by a long-standing elite coalition, nominally headed by the ill-fated king, Charles I (Chapter 15). The *public debates* during this episode reflected the processes of recruitment to and consolidation of competing coalitions seeking to seize decisive elite coercive power. Again, we *literally* mean the seizure of weapons and the putting of armed, self-interested actors into the field. The outcome of the Civil War was entirely determined by which of these competing coalitions was successful in this seizure, initially the New Model Army and it Scottish collaborators, in this case.

Put differently, in an ancestral democratized human society, coercive power is continuously present and merely the technical tool supporting consensual choices. In contrast, in a hierarchical, elite-dominated society, decisive coercive power is no longer merely a tool, but the *sine qua non*, the only goal that can and does matter. Ethical and policy debates are the point in ancestral democratized cultures. They are merely the disposable tools of the pursuit of elite coercive power in hierarchical societies.

These simple implications of our theory will not be surprising to a few of the more sophisticated students of history. They are coincident with important empirical lessons painfully learned over the last century. However, these lessons have not become widely understood. Indeed, this perspective offends our evolved ethical sense, our self-assessment of our own public political behavior. But we *all* must grasp these insights if we are to make real progress, we argue.

Perversely, written history begins with the first elite hierarchical states. Essentially all of "history," until very recently, was written by actors embedded in hierarchical states. Virtually all of this vast record reflects the distorted elite use of the ancient ancestral human practice of public debate. We cannot use this record as data unless we interpret it for what it is—tools in the relentless scramble for elite power. We must not misinterpret this record as a discussion of the actual causes of history.

Outcomes of the uniquely human social adaptation – the past, the present and the future

All human societies everywhere and always emerge from the self-interested application of coercive threat resulting in the management of the conflict of interest problem. They can never have any other source or origin on our theory. The resulting social systems will always serve the interests of those who hold decisive coercive power. On our theory and in light of our two-million-year evolutionary history, it cannot be—and will never be—any other way.

Of course, it follows from this insight that if our goal is a humane, pan-global society, our primary concern must be the formation and survival of pluralistic, democratized, universal distributions of coercive power. We mean this statement literally. *Universal* access to decisive local coercive threat, actual control of the weapons, is what matters.

Only then can ethical and policy debates be authentically effective. Nothing else and nothing less will do. For some of us, this is a new, even startling perspective.* We will have much more to say about this crucial matter later (Chapters 15-17).

∾

---

*   How frightened we are by universal access to coercive threat is predicted to be a good indicator of how much we perceive ourselves (generally unconsciously) to be members of privileged elites or vulnerable, disenfranchises minorities. The ultimate goal, of course, is societies with neither such elites nor marginalized minorities.

Chapter 13
Elite body armor:
Hierarchical coercive power
and the archaic state

The bow-based Neolithic cultures of the American Southwest, like the Anasazi and their confreres, had established a stable existence in which each settlement or pueblo farmed its local territory. Each separate community had up to a few thousand residents and most of the adult males were co-equal warriors. [*] *Internally, the pueblos were relatively democratic, but conflict and enmity* between *them was chronic. The bow could only manage the conflict of interest problem on the scale of these local individual communities, not among multiple settlements. The relentless threat of warfare with neighboring communities was just a part of everyday Neolithic life (Chapter 12).*

However, in 1540 something new entered this world, Spanish imperialists. [†] *The contours of the famous story of this conquest are predicted by our theory and, indeed, have some illuminating lessons to teach us.*

The superior weaponry of the Spaniards rapidly subdued the local settlements of this Native American Pueblo culture (descendants of the Anasazi). The Spaniards imposed peace between these local communities. Indeed, the Spaniards claimed that the Pueblos themselves recognized the value of the peace they brought.

---

[*]   Judging from Neolithic cultures elsewhere, adult females would have been extensively enfranchised in these societies, as well.

[†]   Chapter 14.

The natives [said] that peace had reigned since we came, whereas formerly they fought continually among themselves and were never safe, and that they often said that they would support us because our presence brought them peace.[1]

However, the Spaniards held all the cards. They possessed locally decisive weaponry and, thus, coercive threat, including chain mail armor that made them relatively resistant to Pueblo arrows and thrown stones. These new weapons were very different than the last new weapon to arrive in this local world, the bow. Remember from Chapter 12 that local individuals rapidly and **universally** obtained the bow. The resulting cooperative adaptation apparently served the interests of most or all the members of a relatively democratized majority within local individual settlements.

However, the Spaniard's new weaponry produced a very different outcome. The conquerors were able to restrict access to their sophisticated weapons and used their coercive dominance in pursuit of their own interests. They were able to extract massive amounts of labor to build their large haciendas and churches. They also extorted substantial portions of the maize grown by the Pueblos—as taxes and tithes—so starvation became a constant threat to these Native Americans.

One famous and illuminating case of Spanish abuse occurred at the Acoma pueblo in early 1599. Apparently resisting depredations at the hands of soldiers, the Acoma Pueblos killed several Spaniards. Under the command of Do Juan de Onate y Salzar, the Europeans retaliated by killing four hundred to six hundred Pueblo individuals and taking another several hundred prisoner. The prisoners were tried as Spanish subjects in rebellion. The men over twenty-five had their right feet amputated and were sentenced to twenty-five years of slavery to local settlers. Women over twelve and young men were sentenced to twenty years of slavery. Children were assigned as 'servants' to the local colonialists.[2]

By 1692 chronic exploitation by the Spaniards produced a desperate uprising—the Pueblo Revolt. A well-planned massed attack by the Pueblos was able to dislodge the local Spaniards, killing many and forcing the remainder to retreat to Spanish territory further south. The Pueblos were able to hold this recaptured territory for twelve years. But they were no more able to manage their own internal conflicts than they had been before the Spanish arrived. They still lacked the necessary coercive technology suitable to such large scale policing. Their alliance of extreme necessity was not sustainable. The Spaniards were ultimately able to re-conquer Pueblo territory as a result.

The endgame of this re-conquest found Pueblo holdouts ensconced on isolated defensible mesa tops. The Spanish returned with their gunpowder artillery and chain mail armor, ultimately subduing the Pueblos, killing many and systematically executing captured Pueblo warriors by firing squad. The

Pueblo rebels left signs that are profoundly enlightening for us today. David Roberts' description as he walked through the ruins of one of these refugee Pueblos carries the message of these warriors through time to us. [*,3]

> Suddenly, before my feet, I saw something that made my throat catch in sorrow. A pile of smooth, round stones. ... I picked up one of the stones, held it in my palm, and felt, across the span of more than three centuries the weight of doom.
>
> These stones were the very missiles the defenders of the [pueblo] had used in their slingshots, or simply hurled at the Spaniards as they climbed toward the rim. I turned the round stone, about the size of a baseball, in my hand, as I tried to imagine flinging it off the cliff at a mounted armored invader. The [stones naturally on the mesa] were too light [] to serve as an effective weapon. The [Pueblos] had instead gathered hundreds of sandstone river cobbles from the streambeds below, carried them a thousand feet up the mesa [and] piled them at the top[†]....
>
> On July 24, 1694,[] arrows and stones were no match for [Spanish handguns], swords and chain mail.

The Spaniard's triumph over the Pueblo Revolt was the last straw. The formerly Neolithic settlements of Southwest would forevermore be pieces of a much bigger society. In 1912, this territory joined the United States as New Mexico, an affiliation it retains today.

This poignant story teaches us an important lesson. While the bow is not sufficient to suppress conflicts of interest between Neolithic settlements, new, later weapons are up to the task—and **the state** is the result. This is precisely the pattern our theory predicts and requires.

Contemporary descendents of the original Pueblos have mixed feelings, indeed, about their status as citizens of the state. Is it true that the Spaniards brought peace to the warring pueblos? It is. Is it true that the Spaniards used their coercive dominance to exploit the Pueblos, taxing them to near starvation and enslaving them? It is. Does each of these facts

---

*   See David Roberts' (2004) book *The Pueblo Revolt* for a well-annotated review. Roberts' highly readable book is also a richly rewarding tour of the Native American/European colonialist relationship and of the ensuing complexities for the descendents of these two groups in the American Southwest today.

†   Notice how much the piles of throwing stones on this New Mexican mesa looks like the throwing stones brought up to the plateau at Dmanisi by early *Homo erectus* nearly 1.8 million years ago (Chapter 7). These early hominids would have faced the same conflict of interest problem as modern human did, just on a smaller scale. Warfare between local groups—whose size was constrained by thrown stones rather than the bow—would have been just as likely as for the Pueblos. Perhaps, Philip Rightmire, David Lordkipanidze, Reid Ferring, and their colleagues will find that the Dmanisi site was a defensive redoubt stocked with throwing stones and that the fossil hominids they are finding are the remains of individuals who died defending themselves and their families against attack by other humans—causalities of war.

about the New Mexican past arise as a predictable consequence of the characteristically human exploitation of coercive threat in pursuit of self-interest? They do, we will argue.

Why did one weapon, the bow, become universally available while the weapons of the Spaniards remained in the hands of the few? This difference means everything when looking at the central fact about all human societies, everywhere and always, who possesses coercive dominance. These are the very issues we will explore in this Chapter.

Abusive uses of elite coercive dominance create the great tragic ironies of human history, we will claim. We will see travesties as horrible as Acoma writ large over and over in the early states that long preceded the "re-conquest" of New Mexico. This chapter will look at the birth of these first states around the globe.

In exploring this dark chapter we will not only come to understand the human condition much more completely. We will also establish the context to appreciate the courage and wisdom of our more recent ancestors who swept away these earlier, brutal hierarchical state structures, replacing them with the modern democratic state (Chapter 15). We will need to model their bravery and improve on their insight if we are to continue this vital project and shape a more humane future on behalf of our own descendants (Chapter 17).

The coming of the first states: A new adaptive revolution

In Chapter 12, we saw that the Neolithic revolutions rapidly follow introduction of the bow in Eurasia and North America, this after tens of thousands of years of stasis at levels of adaptive sophistication we called behaviorally modern (Chapter 11). Following the coming of the bow, a new period of stasis ensued at the Neolithic level of adaptive sophistication. Remember that the local level of adaptive sophistication depends on cooperative coalition size (Chapter 10).

When coalition size does not increase,[*] cultures can remain relatively unchanged for very long periods of time. For example, in Southwest Asia, this period of Neolithic stasis lasted for over six thousand years. However, beginning around 3500 BCE in Southwest Asia, something new happened. *States* arose, pulling many formerly isolated, mutually hostile Neolithic communities into single large, integrated social and economic systems. These first states were the next adaptive revolutions

---

[*]   When the maximum policeable unit (MPU) size does not increase (Chapter 12).

in the human historical story. Ultimately, vast swaths of the planet came to be dominated by them.

Written language is invented in support of this new scale of cooperation and, thus, history is born (Fourth Interlude). This history includes the local stories many of us hold dear as part of our cultural roots, our identity. To understand the rise of these first states is to take another large step toward a full understanding of how our world came to be.

Our theory makes very specific predictions about these events. First, this new state scale of human cooperation *must* result from the management of conflicts of interest on a correspondingly novel scale, produced by the self-interested application of a new form of coercive threat (Chapters 5 and 10; Third Interlude).

Second, and less obviously, we anticipate that human societies will serve the interests of those individuals who *control* decisive coercive threat—everywhere and always, no exceptions. This fact emerges inevitably from the foundational human adaptation, projection of coercive threat *in pursuit of self-interest* (Chapter 5).

The portions of the historical record we examined in earlier chapters involved behaviorally modern and Neolithic cultures where access to coercive threat was fairly broadly, democratically distributed. Such cultures will generally serve the interests of many or most of their members (up to the social scale where suppression of conflicts of interest is no longer cost-effective using the weapons at hand; Chapters 11 and 12).

Alternatively, our theory also predicts that human societies can emerge from the self-interested application of coercive threat when that threat is held predominantly or exclusively by a tiny minority, an *elite*, rather than being democratically distributed. Indeed, our theory predicts two things about such societies.

First, the technology of the decisive local weapon systems must lend itself to control by small sets of elite individuals. For example, these weapons might have an exceptionally high opportunity cost, requiring continuous training to use. They might also be expensive to manufacture. The bow is cheap and easy to use. It is not expected to produce such elite dominated societies, for example.

Second, such societies should preferentially serve the interests of the armed elites who control decisive coercive power. In the extreme case, non-elite individuals function as little more than tools or slaves of armed elites.

These central, major predictions also generate many usefully testable secondary predictions. In this chapter we will argue that the first states are precisely such elite-dominated human societies, as our theory requires. Moreover, early states arise many times independently all around the world. No single earlier theory was able to account for all these diverse events. Ours apparently does.

A word of warning and of hope

The authors are Americans of European ancestry. You may be too or you may be an Englishman of Norman descent, a Japanese of Samurai descent, a Moslem of Persian descent, an American of East or South Asian or African descent, a Mexican of Aztec or Mayan descent, and so on and so on. Our sense of where we come from is a piece of who we feel ourselves to be. Each of us has a *cultural identity*.

Moreover, the cultures each of us came from are very recently descended from powerfully intrusive and demanding societies such as the Imperial Romans, the Spaniards in the vignette above, or the Japanese Samurai, for example. These ancestral societies were almost always intensely hierarchical, ruled by earls, barons, Samurai, a'yan's, mullahs, popes, and the like, usually with overlays of religious authority. These cultures were also always male-dominated. Indeed, some early social scientists even wondered whether *hierarchical patriarchy* might be the natural ancient ancestral human biological condition, wrongly thinking that written history reflected the entire human past.

These powerfully coercive, male-dominated societies arose as distinctive new human social aggregates. Archaeologists and historians have learned to recognize these novel polities empirically as a distinctive class called *archaic states*.

The legacy of these archaic states still permeates our world, sometimes literally hanging over us, obscuring our vision, constraining our sense of options. To understand archaic states well and truly is to be liberated from some of the primary excuses powerful elites still wield in the contemporary world as they seek to manipulate us to serve their interests. These *elite licensing myths* have been fading rapidly in most of the developed world over the last several centuries (Chapters 14 and 15). Our theory's strong, simple explanation of archaic states can help us accelerate their demise, as well as giving us a new grasp of our more authentic, more ancient democratic roots.

In the process, we will also find strong new tests of our theory's capabilities to explain all of human history with simple transparency. As we proceed with this explanation, try to gradually assemble a detailed picture of how the intense intrusiveness of our archaic state heritage sometimes distorts our contemporary thinking.

As we said, the modern democratic states many of us now occupy ultimately arose from late archaic states. Thus, many of us inherit important parts of our *identifier* and *social contract* beliefs (Chapters 10 and 12) from these sources. For example, institutional Buddhism, Christianity, Confucianism, Hinduism, Islam, Judaism, and Shintoism are all products of early archaic states. Their rituals, practices, and beliefs were honed as elements of archaic statecraft (below).*

Identifier beliefs are emotionally salient to us. The make us feel like part of a potentially large, powerful group. It is, therefore, difficult for us to see them for what they are—mostly arbitrary pieces of self-referential, culturally transmitted information shaped by Darwinian selection (acting on cultural information) and historical accident (Chapters 10 and 12 and below).†

Our journey through this part of the human story can be illuminating, even electrifying,‡ but it will also be emotional and challenging (Fourth Interlude). We all carry a psychological legacy, a burden inherited from our archaic state forebears. To fully appreciate our real status and the actual logic of the origin of our world requires a monumental attitude adjustment. However, with patience and effort, this profound change in outlook is possible. The authors have spent the last decade working on exactly this personal change of perspective. This change is not merely possible, it is ethically and personally enriching.

Ironically, in adopting this increased detachment from our personal histories, we will ultimately humanize our ancestors. For example, we all know our archaic state ancestors did many violently cruel things. We will discover that they behaved this way not because they were evil, but simply because they were human. Moreover, in understanding why and how this can be, we gain the knowledge we need to place ourselves

---

*   The other major global belief system, Secular Materialism, comes into full flower later (Chapter 15).

†   As we explore the social contract beliefs and economic activities of archaic states, we will see many similarities to modern, democratized states. The *difference* lies in whose interests are being served by those behaviors and beliefs, elites or democratized coalitions of the whole (text below and Chapters 15 and 17).

‡   For example, we will find the origins of things like the Roman Coliseum, the Pyramid of the Sun, the road system of the Inca, and the Great Wall ostensibly explained.

and our descendents in the position not to repeat these terrible, all-too-human choices.

Hierarchy is a "disease" which distorts everyone it touches and brutally victimizes most of them. We will learn about how our theory predicts this pathological condition in this chapter. We will find in later chapters that democratization of access to coercive threat is the robust, effective cure to this disease, again, as predicted by our theory. Do not be discouraged by what you find in this chapter. In knowledge there is real hope for the present and the future.

How this chapter is organized

We have spent over a decade exploring the evidence from the world's archaic states. We know of *no cases* of archaic states that fail to be explained by our theory.[4] Specifically, we argue that all adequately documented archaic states are robustly interpretable as emerging from the local human application of a new and specific class of technologies for coercive threat in pursuit of self-interest. The universal, generic properties of archaic states emerge because this new threat technology is largely or entirely in the hands of a small minority of elite warriors. If this claim is right as a matter of historical fact, it is extremely powerful support for the larger theory we are exploring.

There is a massive body of empirical evidence relevant to our theory's account of archaic states. We could write volumes on this subject alone. Instead, we will cherry pick a few illuminating examples here. Skeptics are free to use cases we do not discuss to attempt to falsify our theory, of course.

This chapter is divided into three parts. Part I explores the essential and general features of our theory's account of archaic states. Part II reviews a few pithy examples of the empirical evidence supporting this picture. For those readers who want a more thorough review, several useful additional details of theory and empirical evidence can be found in Part III and in the online endnotes.

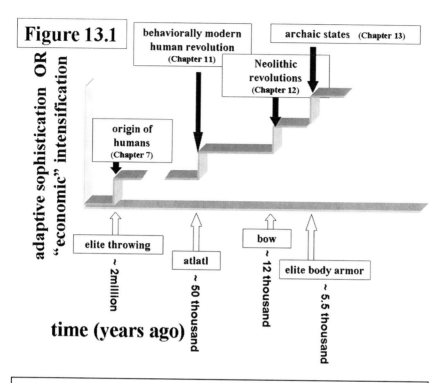

Figure 13.1: Human history consists of extended periods without improvement in adaptive capacities (adaptive stasis) punctuated by abrupt, dramatic increases in adaptive sophistication (adaptive revolutions; Third Interlude). Each of these adaptive revolutions is immediately preceded by the local development of a new coercive technology as our theory predicts.

PART I: The essential fundamentals of the archaic state

The late prehistoric planet is a massive natural laboratory for understanding the rise of the archaic state

After behaviorally modern humans spread across the planet beginning around fifty thousand years ago (Chapter 11), local human populations pursued virtually independent histories, ultimately building distinct cultures adapted to their local environments. This sequence created a massive global natural laboratory with many opportunities for testing our theory. We were able to use this natural laboratory to understand the Neolithic revolutions (Chapter 12) and it will serve us remarkably well again in understanding archaic states. We are able to use the archaeological and early historical records from approximately the last fifty-five hundred years to watch archaic states arise over and over, independently, in different places on the Earth.

For example, we will see that the Aztec state of proto-historic Central America (called Mesoamerica by archaeologists) is remarkably similar in its properties and organization to the first Mesopotamian states thirteen thousand miles away in Southwest Asia. Likewise, the Inca state in proto-historic South America is remarkably similar to the early Egyptian state of Northeast Africa. All these diverse early states are flush with informative similarities.

The archaic state is another adaptive revolution resulting from a new expansion in the scale of human social cooperation

There is substantial evidence that the archaic states represent a new *scale* of social cooperation, as our theory requires (Third Interlude). This evidence for increased scale comes in several forms.

First, some archaic states—the Aztecs and the Romans, for example—were significantly "commercial." They allowed a great deal of relatively private entrepreneurial or market activity resulting in exchange of goods from different local settlements separated by substantial distances. This kind of trade expanded dramatically in archaic states, apparently because the conflicts of interest between isolated settle-ments that would otherwise impede it were managed, suppressed.*

This new scale of commerce drove the formation of large market centers, creating the first true cities. These cities were often huge in proportion to even the largest Neolithic settlements. For example, bow-based Neolithic Cahokia and Chaco Canyon may never have had more than about ten thousand full time residents—perhaps even fewer (Chapter 12). In contrast, Tenochtitlan, the Aztec capital had at least one hundred thousand to two hundred thousand residents when the Spanish conquistadors first laid eyes on it.[5] The archaic state city of Rome, at the height of the Imperial Era, may have had as many as one million residents.[6] Moreover, population density across the landscape was higher in archaic states than preceding Neolithic cultures.[7]

Second, the settlements making up the urbanized versions of archaic states took on a strikingly new arrangement. Pre-state, Neolithic settle-ments tended to track landscape features such as river bottoms in the case of Mississippian settlements like Cahokia, for example. Moreover, these Neolithic residential arrangements consisted of single large settle-ments, sometimes with a few tiny hamlets immediately orbiting them.

In contrast, early states showed a much more extensive hierarchy of settlement sizes, with enormous cities, smaller towns, and still smaller hamlets. Moreover, these variously sized settlements tended to be orga-nized across large stretches of landscape with respect to one another in a *rational economic* way, rather than with respect to physical landscape features.[8] No longer were single settlements the fundamental unit of human cooperation. Vast stretches of landscape now became these social units. Figure 13.3 shows idealized examples of the very different settlement patterns of Neolithic and archaic state societies.

Third, the rise of an archaic state is associated with the imposition of substantial elements of common material culture over a large area. This common material culture is reflected in everything from pottery styles and religious symbolism to newly built road networks connecting cities or administrative centers with other settlements.[9]

Fourth, the written history of many early archaic states depicts them as enormous in size. Some of these claims are certainly exaggerations and elite propaganda. However, in a subset of cases these self-descrip-tions are almost certainly accurate. For example, the written record of the Roman Empire at its zenith and the archaeology of Roman architec-ture both indicate an archaic state controlling most of the vast Mediter-ranean basin.

---

*   Chapter 12

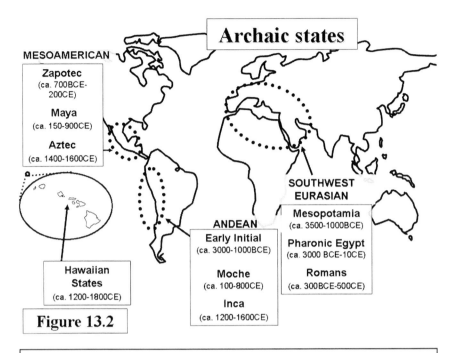

Figure 13.2

Figure 13.2: Shown is a sample of archaic states. States in each of these three regions, Mesoamerican, Andean, and Southwest Eurasian, apparently arise entirely independently of one another. Notice that there are many other archaic states in addition to these listed (text).

For now, again, what matters is that the totality of the record clearly demonstrates that archaic states are a new scale of human social cooperation.*,[10]

Body armor and supermen – elite coercion takes over the world of the bow

A scene from a famous movie has an important lesson for us.

> *Indiana Jones is searching desperately through a local Egyptian market for his lost companion. The setting is around 1941. He is being pursued by Nazi's and their local collaborators. One of these hostile locals suddenly emerges from the market crowd wielding an enormous, razor-sharp sword. The villain stands about ten meters away from Jones and brandishes the sword menacingly and with great flair*

---

*   We will see later in the chapter that the enlarged scale of social cooperation characteristic of the archaic state is not stable—it comes and goes, it *cycles*. For the moment, what concerns us is how this new scale of organization comes into existence.

and skill, making deep swooshing sounds as the weapon cuts the air. Jones pulls his side arm and, with a nonchalant shrug, hardly distracted from his goal, shoots the great warrior dead where he stands—quickly moving on with his quest.[*]

This "humorous" fictional scene makes an absolutely vital point. Brandishing a shock weapon in the presence of others possessing projectile weapons is a recipe for death, not coercive dominance. It also illustrates the converse point. If Jones' antagonist had possessed protection from Indiana's projectile weapon, our hero would have been immediately cut to pieces.

The crucial issue is what we will call *body armor*, whether it is literally worn as armor or carried as a shield. The effects of this armor are unexpectedly profound and pervasive. As we proceed, remember it is the *armor* that carries the most importance here. The shock weapons armored warriors come to use only become important because of the armor.

When the bow or the atlatl is the dominant available weapon, most people will have roughly equal access to locally decisive coercion threat (Chapters 11 and 12). However, if effective individual shielding against these projectile weapons—body armor—can be developed, the logic of self-interested, uniquely human coercion changes radically.

The crucial issues are that this body armor must be light enough to be employed by fighters in practice and provides substantial, not necessarily perfect, protection from locally available projectile weapons.

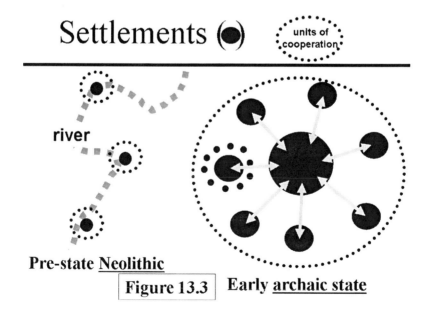

Settlements (●)   units of cooperation

river

Pre-state Neolithic

| Figure 13.3 | Early archaic state

---

[*]   Scene from *Indiana Jones and the Raiders of the Lost Ark* (1981), Paramount Pictures.

When such body armor is developed, it attenuates the role of projectile weapons in coercion and elevates the role of shock weapons—maces, swords, war clubs, and the like.* This change has a number of important effects on how the ancient, universal human trick of self-interested coercion plays out and who is in control of decisive coercive threat. Threat from a distance becomes relatively secondary and hand to hand clashes with warriors are now paramount.

Elite body armor with these properties was developed a number of times in different places around the Neolithic world (below). The records of these developments tell us a great deal about how this technology actually worked. There are three general principles that will apply to every archaic state of which we currently have sufficient evidence. It is useful and informative to designate fighters using body armor and shock weaponry as *elite warriors*.

First, for the armor to be lightweight, yet effective it must be fairly sophisticated. All effective body armor was, thus, relatively expensive to manufacture.† It was not generally possible for an isolated individual to manufacture his (or her) own elite armor at will.

Second, body armor rendered shock weapons decisive and fighting with these shock weapons was technically demanding. This property produces a very different situation from the bow whose effective use requires little training and practice (Chapter 12). Indeed, extensive training and continuous drill with these shock weapons provides such a large advantage that the local fighters using these weapons are always professional or semi-professional warriors.

One of us (Paul) took karate lessons as a middle aged father with his teenage son (foolishly, no doubt). The dojo's sensei was a world-class master in his prime with decades of experience. Paul can still remember his astonishment at the sensei's ability to flip him onto his back so quickly (and effortlessly, irresistibly) that he could not comprehend or remember precisely how it was done. Karate is ultimately derived from one feature of the training of archaic state elite warriors (Samurai, in this

---

* At close range, even an armored individual is somewhat vulnerable to the thrust of a sword or the impact of a mace at the small, but inevitable spaces between components of armor.

† Google images searches under "body armor" paired with each of the cultures we will explore here, including the Romans, Greeks, Samurai, and Aztec will yield illuminating examples.

case).* The sensei's elite abilities were the fruits of a lifetime of training, continuous practice and millennia of culturally transmitted experience.

Now imagine facing such a highly trained individual when he carries a sophisticated Samurai sword easily capable of decapitating you in a single stroke. Moreover, this individual is encased in light, maneuverable body armor[†] while you are unarmored and have only a bow or throwing stones. What do you do? To paraphrase the old joke, "Whatever he asks."

This requirement for professional status to be an effective elite warrior imposes a formidable opportunity cost. Thus, though simple shock weapons like clubs might be universally available, their elite use will generally not be universal. Moreover, elite, expensive shock weapons—Samurai swords, for example—will generally be neither widely available nor commonly used with expertise.

The combined consequence of the high material cost of good body armor and the exorbitant opportunity costs of expert employment of shock weapons is that only a small portion of the population can afford to be effective elite warriors. Moreover, these high opportunity costs and the requirement for extreme strength when using shock weapons mean that elite warriors are essentially always males. This accounts, in part, for the universal male domination of archaic states, we will argue.

Third, in addition to the cost issues above, the numbers of elite warriors is probably also *self*-limited. Once the subset of the local population that are elite warriors become large enough to exert control over the remaining *non-elites*, it is in the self-interest of each elite warrior to block access to body armor and shock weaponry by remaining individuals. Moreover, their joint coercive dominance makes it possible for these warriors to impose this block—taboos enforced by death and the like. Thus, they can control manufacture and storage of weapons as well as training in their use.

Elite warrior cadres will try to recruit new members only as needed to sustain their self-interested local dominance, but will generally exclude most members of the residue of the population from elite warrior status.[‡,11,12]

---

*   During Paul's lessons, he was told that one meaning of the word "karate" is "empty hand"—that is, the art of fighting by an elite Samurai warrior when he momentarily lacks a shock weapon.

†   A Google image search under "Samurai body armor" will produce diverse examples.

‡   As we will see later, this system has one flaw. Elite warriors often have more than one son. But what matters for now is that commoners are generally excluded from joining elite warrior ranks at will.

Though the portion of the local population consisting of elite warriors probably varies somewhat depending on the specifics of the available technology and the local terrain and economies, their fraction is generally in the range of 3 to 10 percent of the male population.[13]

Our goal is to understand how the innovation of body armor and shock weaponry shaped the adaptive revolutions we recognize as archaic states.[14]

We are elite warriors – what are our options?

In government, military matters are the essential thing.
Japanese Emperor Tenmu, 684 CE[15]

For simplicity, we can think through the idealized case of a *two class* society – elite warriors and commoners.[16] Moreover, it is useful to consider two extreme versions of the relationship between these two classes.

In the first extreme, elite warriors are *omnipotent*. They have completely unchecked coercive dominance over commoners. Under these conditions, commoners will be treated by self-interested elite warriors as just another species of domesticated animal. Notice that this harsh, ethically offensive picture is nonetheless inescapable when elite warriors have *absolute* power, in view of the ancient human adaptation to the self-interested use of coercive power.

In the second extreme, we can visualize elite warriors as having only very slight extra coercive power compared to commoners. Under these circumstances, elite warriors will generally be strongly influenced by the counter coercive threat from the numerically superior commoners. In this case, elite warriors are expected to function as well-paid "police" and "administrators," serving their own interests by, largely, serving the interests of commoners as their best available option.

In fact, we will argue that all real historical archaic states are somewhere between these two extreme alternatives. Some very early states may have approached the second, gentler extreme. However, most mature archaic states, the ones leaving large historical and archaeological foot prints, are all closer to the first, warriors-dominant extreme.

In practice, commoners in these mature archaic states had much less access to decisive coercive threat than elite warriors, but they are not quite completely without such access. When sufficiently provoked, their large numbers allow them to attack elite warriors effectively, though at huge, normally prohibitive, cost to many individual commoners.

Moreover, real historical elites were not a single class, but were generally internally stratified. Thus, sub-elites were more vulnerable to commoner coercion and, at the same time, were better able than commoners to coerce more well-armed super-ordinate elite warriors. This gives commoners a little more power—through sub-elites—than they otherwise would have.

Thus, archaic states will reflect a *balance of power* between local interest groups (Fourth Interlude), elites, sub-elites, commoners, and slaves. The outcome of this balance is predicted to reflect each group's relative access to coercive threat. With these properties in mind, we can add additional elements to our theoretical picture of these states.

Suppose we are all members of a cadre of heavily armed elite warriors in a world of farmers emerging from the Neolithic revolutions (Chapter 12). What are our best available strategic options? One is to raid the farmers' stores of *liquid assets* like grain, taking what they grow and make. However, this creates some tactical problems. We have to keep moving from destroyed village to destroyed village to make a living. Moreover, if we are not truly omnipotent—and we are not—but just relatively more powerful, coercive resistance from farmers is a significant cost of this approach.

Suppose, instead, that we impose a tax on the farmers that is substantial,* but leaves them with at least a subsistence livelihood. Moreover, while we are receiving this tax income, it is in our interests to protect the farmers from other roving bands of elite warriors, making a raiding strategy even less attractive for other elite warriors in the process. In a very real sense, the farmers would be choosing not to resist us "stationary bandits"†—aka "government"—as the best available alternative to continuing exposure to violent roving bandits.[17] We are now running a successful protection racket.

Of course, it might also be in our interests as elite warriors to promote the building of infrastructure that would increase agricultural yields and, thus, our own tax revenue. We might force commoners to build irrigation capacity, for example, or, alternatively, allow local commoner aggregates to respond to the *market forces* we create (below) by building their own irrigation infrastructure that we, in turn, will protect from theft and depredation.

---

* This tax might go by any of a variety of labels, including *tribute* or a *tithe* to the gods, for example.

† To borrow Mancur Olson's (2000) pithy descriptor of *government* in at least some of its guises.

At the same time, we will also coerce commoners to build infrastructure that supports our own elite interests, of course. This will include road and canal systems that allow us elite warriors to move easily across the landscape, projecting threat to enforce commoner cooperation. This construction work will also include the infrastructure of our elite social contract belief systems (Chapter 10; Box 13.4), such things as the Great Pyramids and the many temples, mosques, and cathedrals of the archaic world. We will publicly proclaim our role to be *keeping the peace, enforcing the law,* or perhaps, even *serving God's will.*

A related business opportunity for us elite warriors is to police public markets (Chapters 11 and 12) on a much larger scale than ever before in human history (Box 11.1). The wealth such markets can generate is spectacular and, again, we can tax it. By creating and supporting these markets we can grow wealthy as long as we remain in control. Moreover, we retain the capacity to confiscate the wealth of any unarmed "businessperson" who deigns to threaten us in any way, perhaps by attempting to hire mercenary replacements for us. Once again, "what can be exchanged can be stolen;" therefore, the merchant's wealth, alone, is not power.

As the elite keepers of commoner *domesticated animals,* it is in our interests to see that their needs are cared for to the extent that fulfilling these needs does not interfere with their generating the resources we wish to possess and consume. If we do this well, our flock can grow large. We can prosper in our elite status as *people herders.*

Notice what we are doing. As elite warriors we are suppressing and managing the conflicts of interest between formerly Neolithic communities that long prevented these communities from developing a larger scale of social cooperation—remember again the Anasazi of New Mexico (Chapter 12 and above). We can even argue to ourselves that we have brought civilization and peace to these people, as, in a very real sense, we have. Also, we can convince ourselves that we are serving the commoners by making their economic system work. Again, in a very real sense we are. But…

As this process unfolds, this new scale of social cooperation can grow very wealthy and powerful, ultimately building great cities and cultures. Think of the Athens of Pericles, Baghdad of the Abbasid Caliphate, Tenochtitlan of Moctechuzoma II, or Imperial Rome, for example.[18] We might wonder, *Is this a good thing or a bad thing?* It is the details that matter, as we will see.

Box 13.1: The "tragedy of the commons" – a useful example of the conflict of interest problem.

The Boston Commons today is like Central Park in Manhattan—a green space for universal public use. The original commons of an early Euroamerican town was a grazing area shared by all community members who had animals—sheep, for example.

The grazing commons would support a certain number of animals that would consume the ongoing productivity of the commons, but no more. A herd of this *optimal size* would trim the new grass as it grew, but leave the base of each shoot intact, allowing further growth. The grazed commons would look very much like our contemporary mowed commons.

However, as soon as this optimal herd size threshold is exceeded, hungry animals start cropping the grass closer and closer to the ground , killing it and destroying their future food supply.

Of course, as long as all the shepherds keep their individual flocks sufficiently small, the commons survives. But, each shepherd has an individual self-interest in adding more sheep to his personal flock. As soon as one of the shepherds increases his flock to the point that the threshold of optimal grazing number is exceeded, the commons is doomed. It is then in the interest of every remaining shepherd to get as many sheep onto the commons as quickly as possible before its collapse—*a race to the bottom.*

This catastrophic collapse is traditionally referred to as a *tragedy of the commons.*[19] Of course, this is just one particular instantiation of our old companion, the conflict of interest problem. As expected, this version of the conflict of interest problem can be coercively managed just like any other. If a coalition of self-interested shepherds has the means to ostracize cheaply any other individual exceeding his/her allotted number of sheep—taking ownership of those sheep and slaughtering the excess in compensation, for example, no tragedy will occur. Of course, the *IF* matters. The appropriately inexpensive means for conjoint projection of coercive threat must be available to all the shep herds just as for any other conflict of interest problem (Chapter 5). The tragedy of the commons, and the ways we can fail to manage it, will prove to be a useful visualization of some features of the conflict of interest problem in the archaic state and in later cases, as well.

Elite warriors - the tragedy of the commons and "rise and fall" of archaic states

So far we have discussed only part of the coercive logic of archaic states, elite coercion to manage commoner conflicts of interest. However, there is another conflict of interest problem. It is the conflicts of interest *between* individual elite warriors. It is vital to grasp the essential feature of when and how these elite conflicts are controlled.

First, notice that conflict based on shock weapons allows only a small number of individuals to attack a single target (somewhat similarly to non-human proximal killing animals; Chapter 5). This limitation is in contrast to the hundreds who can attack a single target with a projectile weapon like the bow (Chapter 12). Elite body armor does not provide the option to coerce using *death from a distance* (Chapter 5) *among* armored warriors. Thus, extremely inexpensive coercion is not available in policing conflicts of interest *between* or *among* elites. Only modest cost reduction is available.

However, individual elite warriors control substantial assets that can be confiscated. For example, a typical elite warrior in the Aztec, Inca, or pre-contact Hawaiian archaic states (below) might own agricultural lands farmed by commoners and/or slaves. Thus, the compensation for ostracizing an outlaw member of the warrior elite can be substantial, making it strategically sensible even when the coercive violence involved is significantly risky and costly.

In general, *individual* elite free riders (cowards in battle, for example) can be effectively policed through realistic fear of attack from their peers. Likewise *small* local aggregates of elite warriors can be overwhelmed by a numerically superior elite force.

Second, however, as free riding warrior groups grow large, coercion becomes extremely expensive. Several factors limit the scale of elite-on-elite coercion. For logistical and simple spatial reasons, it becomes difficult to achieve local numerical superiority when large numbers of elite warriors are the target. Thus, once the size of a local target elite cadre exceeds a certain limit,* elite-on-elite coercion requires extraordinarily expensive one-on-one combat. That is, roughly equal numbers of elite warriors end up effectively engaged on each side of such a conflict. This severely limits the circumstances where large-scale, elite-on-elite coercion makes strategic (adaptive) sense.

An important impact of this limitation is on the self-management of elite populations. Biological reproduction is, of course, the quintessentially self-interested act (Chapters 2 and 3). Thus, coercive threat must be used—however indirectly projected—to create a disincentive to over-reproduction.

Elite over-reproduction creates a situation in which the necessary target of elite coercion is precisely the target where the costs of elite coercion are most prohibitive, large local elite cadres. The implication of this is clear. Elite warrior reproduction presents an unmanageable conflict of interest, an inevitable tragedy of the commons.

Given this logic, we might ask how relatively integrated, massive archaic states ever arise in the first place. Should they not be destroyed in their infancy by this ungovernable basic conflict of interest? The answer to this question is apparently straightforward. To see why, we need to consider the details of the *developmental cycle* of archaic states.

When a newly dominant local elite warrior cadre initially arises and captures its first territory it will immediately begin to produce more offspring (sons, in particular) than the local social adaptation can sustain in view of high elite standards of living. This excess elite warrior population has two choices. Stay and engage in costly competition with fellow elite warriors or proceed to the *frontier* of captured territory and engage in relatively inexpensive conquest of non-elite, unarmored (or less elite, under-armored) populations. Pursuing this second choice will look like expansion or growth of the state in the archaeological record.

Indeed, existing elites have an incentive to contribute to this expansion as some of these frontier elites will be close kin. Moreover, an expanded scale of economic cooperation can be mutually beneficial to all elite parties. Further still, newly conquering elites can be coerced to return some of the booty of conquest—slaves, precious metals, and so on—to original elites to purchase their continued support.

Initially, this growth can be spectacular and incredibly lucrative. Recall, again, the massive early states we mentioned above.[20] However,

---

*   We do not currently have robust estimates of this threshold size. Moreover, it probably varies as a function of local factors like efficiency of transportation infrastructure. In spite of these serious empirical limitations, we can still make progress in understanding the overall logic of this element of archaic state coercion.

this expansion carries the seeds of its own inevitable failure. Specifically, as the empire grows, both its excess elite population and its frontiers increase. But, they increase at *very different rates*. The excess population grows much faster than the available frontier (Figure 13.4).*

Consider the effects of these different growth rates. First, elite competition for available conquest territory grows more intense as the ratio of excess elite population to available frontier grows. Second, the per capita return to resident elites in new trade and booty declines at the same time. Thus, resident elite individuals are ever less willing to underwrite new conquest, making this option still less attractive.

As a result of this inevitable developmental sequence, the incentive structure of local elites changes as the archaic state enlarges and matures. Best available individual options become competition with their local elite neighbors, seizing resources where possible and protecting what they already have at all times. By retaining what resources they have to invest in nurturing their local interests, elites are leaving the larger enterprise, the empire, bereft.

The most successful of these newly selfish local elite aggregates will be the ones most costly to attack in an act of elite-on-elite coercion. Building effective defenses against outside elite coercion, castles and the like, will become popular. The archaic state will fragment. Its local "provinces" and "communities" will increasingly go their own ways.

In some extreme cases, the local competitive pressures incentivize elites to extract maximal short-term returns from their holdings, driving over-farming, over-fishing, and so on. Thus, the elite tragedy of the commons can sometimes become a material, economic tragedy in the literal sense (Box 13.1).

Over generations, the consequences of environmental degradation and the loss of large-scale markets (destroyed by fragmentation of the original state) will further impoverish elites and commoners alike.

This new condition of relatively poor, small statelets may persist indefinitely. The Hawaiian example we will explore below is probably such a case. Alternatively, a local population may invent a new wrinkle on coercive technology in some material, tactical, or strategic manner, for example. When this innovation happens, the newly empowered elite will spread, conquering and displacing the impoverished, now-under-armored descendents of earlier elites while exterminating

---

* New excess elite population is produced as a function of the *area* of the empire—proportional, roughly, to the *square* of the diameter of the state's territory. However, the opportunities for new conquest at the frontiers of the growing state increase as a *linear* function of the diameter of the empire.

excess and resident populations and capturing others' assets as new wealth.

When this happens, the cycle can begin anew. A new empire will grow, a new elite tragedy of the commons will follow over and over again.[21]

Notice that this picture invokes nothing other than coercive self-interest in a Malthusian world. No elements of belief or culture need be considered, other than the practical fundamentals of local military and agricultural technologies.

Put differently, our theory of the management of the conflict of interest problem as the *sole* cause of the formation of human social aggregates apparently predicts a very specific historical sequence for these units when they are produced by armored elite warriors *independently of any additional considerations.*

In fact, archaic states are well known to behave in precisely the way our theory apparently requires. Archaeologists and historians refer to this property of growing large and wealthy followed by fragmentation and economic decline as *cycling*.[22] The fact that our theory apparently predicts cycling is an important element of evidence supporting the theory.*

On our theory, the *flowering* and the *decline and fall* of the Romans, the Mayans and the Islamic Caliphates are not local peculiarities. They are inevitable consequences of early successful management of the conflict of interest problem by elite warriors, inevitably leading to ensuing conflicts of interest the warriors cannot manage.

One last point is important to take away from archaic state cycling. The written history of archaic states is sometimes taken to have a very depressing message. Human societies seem doomed to flower then collapse, rather than growing and improving in perpetuity. Notice that this cyclical behavior is a very specific product of *hierarchical* states. There is no reason, theoretical or empirical, to believe that highly democratized states (Chapters 15 and 17) need be subject to this futility.

---

* Jared Diamond, in his strong, useful book *Collapse* (2005), suggests that environmental degradation was the cause of many ancient "collapses." This is true, in some cases and as far as it goes. However, environmental degradation is usually simply a tragedy of the commons. The real problem with collapses is the inability to manage the commons, to control conflicts of interest. Environmental degradation is merely a secondary effect of this primary cause on our theory, not a primary cause of collapse.

Excess elite population grows in proportion to the <u>area</u> of the state while the frontier for conquest grows as its <u>circumference</u>. Area increases in proportion to the <u>square</u> of the radius of the state ($\pi\ r^2$) while circumference increases only linearly with this radius ($2\ \pi\ r$).

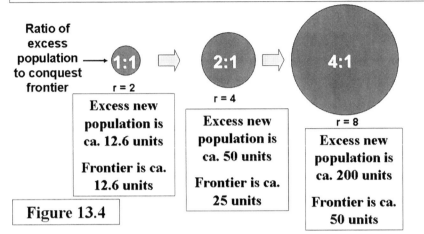

Ratio of excess population → conquest frontier

1:1 $\Rightarrow$ 2:1 $\Rightarrow$ 4:1

$r = 2$

Excess new population is ca. 12.6 units

Frontier is ca. 12.6 units

$r = 4$

Excess new population is ca. 50 units

Frontier is ca. 25 units

$r = 8$

Excess new population is ca. 200 units

Frontier is ca. 50 units

Figure 13.4

Ritual behavior and belief in the archaic state - Logic, legitimatization, and conflict

To comprehend what we see in the records of archaic states, we need to understand how human ritual behavior apparently evolves in these polities. As we saw in Chapters 10 to 12, our theory predicts that humans will have a very special relationship to culturally transmitted information, arising from our management of conflicts of interest. Unlike non-human animals, we generate, consume, and exploit vast quantities of this information. This cultural heritage is effectively stored and transmitted by our large non-kin social aggregates, our cultures, or societies. Our question now is how this feature of our behavior is exploited by members of hierarchical states.

Consider the fundamentals. Much of our heritage of cultural information serves transparently adaptive purposes such as how to farm, build, and so on. This *pragmatic* information will not concern us for the moment. Beyond the pragmatics of the external world are beliefs relevant to our adaptive relationships to other humans. We forecast that such *social beliefs* will come in two major classes (Chapters 10 and 12).

The first class we called *social contract information*. This class of information includes the publicly negotiated arrangements whereby we conjointly project coercive threat in pursuit of individual self-interest and cooperate as supported by this coercive umbrella. These are the rules of the game. Subjectively we experience this body of information as our shared ethical system. Moral, upstanding individuals, citizens, observe and coercively support the behaviors mandated by this ethical system.

The second major class of social beliefs consists of what we referred to as *identifier information* (Chapters 10 and 12). The function of this class of information is more subtle. Identifier beliefs are often arbitrary, even overtly bizarre, surreal. Their function is purely self-referential identification of the members of human social aggregates (Chapter 12).*

The substance of these beliefs is not constrained by the rules of the material universe.† This class of information allows us to identify "friend" and, by exclusion, "foe" in large anonymous polities. It also imposes a large opportunity cost (investment) for joining a social aggregate, disincentivising subsequent free riding against other members of the aggregate. Identifier information is a central part of the ancient uniquely human social adaptation.

Identifier information is everywhere. Think of gang colors or business suits. Think of forms of greeting ("g'day" versus "bro" versus "how're you?"). More deeply, recall our responses to someone who actively ridicules even arbitrary elements of something we believe in. It might be burning our national flag in public or disrespecting God, Allah, Buddha. These are all acts with no direct, tangible consequence, yet acts that might provoke us deeply.‡ Identifier information plays a conspicuous role any time competing human groups have active conflicts of interest.

With this foundation in hand, we return to the social logic of archaic states. Consider the relationship of elite warriors and commoners. Each has an urgent concern to monitor and manage the behavior of the

---

*   Of course, scrupulous observance of social contract information also serves an *identifier* function. Thus, all social contract information can also have identifier function; however, not all identifier information, in general, is also social contract information (text below).

†   In fact, identifier information often works best if it does not mirror the material universe. Mastering information that is not implicit in the material world, perhaps to the extent of being *irrational,* is challenging, making it difficult to fake being a member of the local coalition (Chapter 12).

‡   The intense impact of identifier beliefs can be a little difficult to fully experience for those of us nestled in the relative safety of the democratized, pluralistic state. But in the very different environment of the archaic state—with commoners a constant threat to revolt and elite conflicts of interest simmering relentlessly— identifier beliefs were urgent matters of life and death.

other. However, each individual also confronts a very different problem than the members of a relatively democratized Neolithic community. Remember that, in a Neolithic community, wealth will generally be broadly distributed and most community members do not object to preserving the economic enterprise (Chapter 12). They will police one another—in pursuit of self-interest—and, as a by-product, the community is sustained.

In contrast, commoners in a hierarchical archaic state have a very strong incentive to use their access to democratic coercive technology—thrown stones, bows—to build an isolated local commoner enterprise that elite warriors cannot see and tax. To suppress this strategy, elite warriors must not only monitor commoner behavior continually, they must also constantly project, or show off their ability to ostracize, massacre and pillage, commoner settlements that subvert the state in this way. Thus, it is in elite interest to bring commoners together in large groups periodically to demonstrate elite coercive threat.

Conversely, the numerical superiority of commoners means that they can overwhelm elites if sufficiently provoked. Thus, it is in commoner interests to come together in large numbers periodically to demonstrate commoner coercive threat and the limitations of elite power.

In view of this mutual interest in showing off threat, we predict that one major form of ritual behavior in archaic states will be public events that serve both these elite and these commoner objectives. These ritual occasions will include elites and relatively large numbers of commoners. Different states will develop rituals with superficial differences in detail, but the core features of these events are expected to be universal.

The first universal feature will be the demonstration of elite power. This might be overt—military games, for example. It might be slightly indirect—cruel, violent human sacrifice of outlaws or war captives by elite individuals, for example.

This first feature might be even more subtle. Non-armored elite pawns, sycophants, or priests, for example, might supervise large ritual activities requiring elaborate commoner participation. These rituals will generally make conspicuous use of local identifier information. Any local commoner population failing to participate properly will be regarded as sacrilegious, apostates, or heretics by elites. Elite warriors will wipe out or severely punish any such miscreant local commoner population, generally replacing it with a more pliable population.

The second major universal feature of archaic state social contract rituals will be commoner expression of favor/disfavor. If things are going well and the state is peaceful and prosperous, commoners may cheer and participate in public rituals enthusiastically. This scenario is most likely early in the archaic state developmental cycle (above). If things are going badly, the commoner crowd will mumble or even roar its displeasure. Elites will have been warned.

Because of the radical, persistent differences in elite and commoner interests, these public rituals will need to be very frequent, even relentless. They will be seasonal, at least, and perhaps weekly or monthly in many cases.*

It is vital to grasp that these universal features of archaic state public ritual are inescapable on our theory. They simply reflect the ongoing requirements of the central, ancient human adaptive strategy: The coercive management of the conflict of interest problem. We should find them in every archaic state we examine.

Our extensive study of the evidence indicates that this prediction is fulfilled. Indeed, one of the most striking observations of archaic state archaeology/history is that each early state—everywhere in the world, always—is associated with a vast ritual/religious apparatus.[†] Examples of this kind of public ritual from cases like the New World Aztec state and the Old World Roman state are explored in the following section.

Further, for example, Inca elite warriors took possession of *all* land in a newly conquered territory. One third of this confiscated land went to the sustenance of the Inca *religious establishment*, among whose functions was carrying out extensive, frequent rituals and, of course, monitoring the local commoner population on behalf of the *state*.[23]

A second one-third of this Inca land was farmed by the locals for the benefit of the Inca state, including production that sustained the massive Inca military capability and allowed commoner workers to be paid for building the state's infrastructure. Much of this infrastructure, in turn, was coercive in purpose. Examples include the justifiably famous Inca road system, which allowed massive elite warrior armies to move across the landscape efficiently, and the large grain storehouses that allowed these armies to be provisioned. Inca roads were not mostly used for the direct benefit of commoners, in contrast to the contemporary world where our roads are expected to serve family visits, vacations, private business activity, and the like.

Finally, the last one-third of this land went to the indigenous population of commoners and petty elites for subsistence purposes.

---

*   Notice that contemporary *religious observances* in temples, churches, mosques, and synagogues are the partially redeployed descendents of these archaic state rituals on our theory.

†   See Trigger (2003) for an excellent discussion of this remarkable observation based on a sampling of early or pristine archaic states.

Equally massive religious establishments were parts of all archaic state where we have sufficient evidence to make an assessment. Examples include the Aztecs, Maya, Hawaiians, Greeks, Ch'in Chinese, Romans, Pharonic Egyptians, and the early Mesopotamian states.[2]

In all these cases, the ritual establishment was intimately associated with the warrior elite. Kings were often "divine," god-like. Priests were commonly members of the warrior elite or their very close kin (especially younger brothers).

Religious traditions are such a pervasive element of archaic state politics that some archaeologists and historians have suggested that religious belief is a cause of the rise and/or functioning of archaic states. Our theory predicts that this is a profound confusion of cause and effect. The details of religious belief are a combination of quirky identifier beliefs (historical accidents) and technical features of local management of the rampant, intense conflicts of interest within archaic states— nothing else, nothing more.[*]

This world of warriors, priests, and commoners is a consequence of the specific distribution of access to coercive threat. How well does such a state actually work? Commoner taxes are paid—to some extent. Elite provision of sensible law enforcement and infrastructure takes place—to some extent. But commoners have a relentless incentive to underpay, depriving the state of essential revenue. Elites have an equally relentless incentive to overtax, driving commoners into starvation or semi-suicidal revolution.

These conflicting incentives tug relentlessly at archaic states. Early in their growth, when things are good, these systems work, more or less. As population densities grow, the state is under constant threat of collapse through the self-interested actions of commoners and/or elites.

Though archaic states might muddle through and even come to manage overpopulation and environmental degradation under other circumstances, they have one unavoidable and ultimately fatal flaw, as we saw above. There is no mechanism to manage intra-elite conflicts of interest consistently. These will inevitably tear the state to pieces. No *ritual practice* or *belief system* can prevent this collapse. It is simply an expression of the universal conflict of interest problem and the

---

*   Notice that identifier and social contract information will also be vital to the formation of subcultures seeking to subvert the existing order (Fourth Interlude). If such a subversive subculture ultimately overthrows the existing elite, its belief system will become a legitimate political tool of the resulting "new" archaic state. Christianity in the context of the early Roman Empire, and Islam at its birth under Persian domination, are well-known examples of this process.

limitations of its management through the existing archaic state coercive technologies, on our theory.*

An Aztec market and the Roman Forum

The gleaming white Aztec capital of Tenochtitlan was built on a large island in a lake, connected to the mainland by massive causeways. In the city's enormous market in 1500 CE, tens of thousands of people came every day to shop for almost anything imaginable, from clothing and shoes to food, medicine, and furniture. Goods were purchased with cocoa beans, sophisticated fabrics, and other *currencies*. When the first denizens of European states saw this great market in 1517, they were astonished. It equaled or surpassed the grandest market they had known at home.

On this same visit, these first Europeans watched a great public ritual at Tenochtitlan. Individuals were taken to the pinnacle of the huge pyramid in the center of the city with many thousands of commoners looking on from below. The victims were stripped to the waist and laid out on a stone table, pinned down by four priests.

A fifth priest used an impossibly sharp obsidian knife to rip a great incision in the belly of the victim and upward through the diaphragm. The priest immediately drove his hand into the incision and ripped out the victim's still beating heart. The victim's body was thrown down the stairs of the great pyramid, making room for the next victim.[25] The bodies were later apparently dismembered and the meaty parts cooked for elite consumption.[†,26]

Many were killed that day and the huge white pyramid was soaked in blood. The victims' heads were removed—the brains were apparently a delicacy—and mounted on great skull racks for all to see for many days after the ritual. Scholars estimate that somewhere between ten thousand and two hundred fifty thousand people were sacrificed by the Aztecs this way before the coming of the Europeans.[27]

The Spanish Conquistadors would besiege Tenochtitlan not long after this first visit. Several dozen of their numbers were captured during the battle and sacrificed in this same way in full view of their comrades

---

*   We will see in Chapters 14 to 17 that collapse and decay are *not* necessarily inevitable in human social aggregates where all conflicts of interest, including between elites, can be cost-effectively managed.

†   See Diaz (1963) for a translation of one of the original accounts of these events by one of the Spanish conquistadors.

surrounding the city. The Conquistadors were appalled and their ultimate sack and destruction of Tenochtitlan was, perhaps, more furious and complete than it might otherwise have been.

The Spaniards were quite confident that they were superior to these "heathen, murdering savages." They were oblivious to the profound irony of their outrage. The Conquistadors spoke Spanish, a language directly derived from Latin, the language of the Imperial Romans.

The great Forum of Rome housed a market where the citizens of the imperial capital could come to shop for many of their needs—food, clothing and shoes. Though not as large, apparently, as the Aztec market, it was impressive nonetheless. The great Mediterranean trade networks, policed by the Roman Legions, brought many richly valuable and exotic items into their capital for sale.

Nearby was the Coliseum.[28] On a day, in 100 CE, there was a great public ritual. Gladiators fought and criminals were executed, while fifty thousand commoners and some elite members watched from the stands. The unfortunate miscreants might be hanged by being suspended by the neck from a large Y-shaped timber, if they were lucky. However, they might also be burned alive—*vivus exuri* or *cremation*, to use the evocative Latin terms. They might even be fed alive and conscious to large, fierce dogs or predatory big cats.

These "criminals" are particularly interesting to us. They were almost always guilty of two crimes. One might be theft, murder, arson, or kidnapping. However, the second "crime" was being a commoner, a *humiliores*, rather than a member of the elites, an *honestiores*.[29] Indeed, honestiores would come into these fiercely cruel penalties only for serious acts of high treason. Active Roman soldiers were likewise essentially immune from such barbaric treatment unless they subverted the state.

Though the Aztecs and Romans are famous for the scale of this kind of state murder, public human sacrifice was apparently a feature of *all* early states (below).*

---

*   Historians sometimes present these acts of public cruelty as defending the interests of *elites* other than warriors—*Roman senators*, for example. However, we emphasize that there is another way to look at this. These non-warrior elite members were functionaries tolerated and enabled by elite warriors. They pronounced sentence, but the sentence was of no meaning unless the armed members of the society ultimately supported it.

To grasp this slightly subtle, but crucial point consider a judge in a contemporary pluralistic democracy. She/he passes judgment on offenders, but armed police are essential to actually restrain the sentenced individual and see that the sentence is imposed. Moreover, if a judge is attacked, *all citizens-in-good-standing* of the democratized state feel threatened and collaborate with police to apprehend the aggressors. Finally, the judge will defend the "dignity" and "prerogatives" of the court from "contempt."

Finally, it is striking that all the great institutional ritual systems, religions, surviving in the contemporary world from Old World precursors were created and refined in early archaic states, in a period religious historians have taken to calling the *Axial Age*.[30]

These ritual systems were honed to sustain massive early states. In view of this origin, it is predictable—indeed, probably inevitable— that they would be borrowed by newly emerging early modern states. These ritual systems continued to evolve to serve the *adaptive purposes* of the diverse interest groups struggling everyday to shape modern state "cooperative" systems to serve their separate interests. Having this clearly in mind will serve us well when we turn attention to modern states in Chapters 14 and 15.

Might makes right – self-justification, submission and uniquely human ethical psychology

Recall our discussion in Chapter 10 of the relationship of human ethical psychology to coercive power. An ancestral human living one million years ago, for example, will occasionally hold a piece of vast, even overwhelming coercive power. This will happen when she is a member of a large majority of individuals whose interests are confluent at that moment. Acting together these many individuals can project an awesome level of threat, originally as a result of elite aimed throwing (Chapters 5 and 7).

Thus, the feeling of possessing overwhelming power will be associated with being *right*, as defined by broad local consensus, the only definition there is. In the throes of this feeling of "proper" power, our ancestors—and we—will feel justified in committing even lethal violence against an *evil* free rider.

Conversely, our ancestors would occasionally have become the targets of overwhelming coercive threat when they did something that contravened the interests of a large majority of the members of their local social aggregate. That is, they would have become the target of overwhelming threat when they were *wrong*, again, by the only standard there is, consensus.

After two million years of Darwinian selection in the human village, this psychological association between coercive dominance of the *right*

If we look at just what the judge says and does, we might conclude that the system serves the interests of the judge exclusively, but we would obviously be mistaken. He or she is powerless without the access to the credible threat behind him/her.

and coercive vulnerability of the *wrong* is almost certainly genetically controlled, natural, inevitable.

Now consider what happens when such a creature is thrust into a fundamentally novel adaptive situation. For the first time—as a result of technical innovations in weaponry—access to overwhelming coercive dominance does *not* require a large majority consensus. Rather, it is only available to the members of small, heavily armed elites.

How will the members of this elite "feel"? Of course, they will be extremely likely to perceive themselves as being in the *right*. This is, after all, where coercive dominance has always come from throughout prior human evolutionary history. We almost certainly still carry this genetic legacy with us today.

Conversely, how are their commoner targets likely to "feel"? They are likely to see themselves as in the *wrong*, as *deserving* their inferior station. Being in the wrong is where coercive vulnerability always came from in the ancestral past.

The upshot of these predictions of our theory is clear. When coercive dominance is unequally distributed, humans will be very poorly adapted to understanding their feelings for what they are, a response to power. Instead, we will tend to see this inequality of access to power as having an ethical justification.[31] We will tend to respect and seek to join the powerful and the "right." We will hold the powerless in contempt and seek to sever our interests from theirs. Each reader who is honest with himself/herself will recognize both these impulses from his/her daily life.

We are highly adapted—under the right circumstances—to being good masters and good slaves.

Box 13.2: Adaptive novelty: Access to information and the scaling problem of the state

The central problem with the cyclic rise and collapse of archaic states is the inability to manage the escalating conflicts of interest *within* elites as population sizes and densities grow (main text above). However, there is a second effect which exacerbates this inability and which is an issue that we will confront again as we examine the emergence of the modern state (Chapters 14 and 15). Humans spent about two million years in social aggregates whose size permitted convenient, conjoint projection of threat that was inherently synchronous. All the members of a social aggregate could, in principle, project coercive threat simultaneously in one another's presence. Even if this contemporaneous action by the *coalition of the whole* was rare in practice, its potential occurrence was a constant *credible threat*.

With the coalescence of the first states, this ancient, ancestral attribute of our social coalitions was lost forever. Humans were now joined in gigantic social aggregates dispersed across vast stretches of continental land area. Individual citizens of these states might be separated by days or even weeks of traveling time from one another. Moreover, the carrying capacity of these states could not possibly feed and house the local congregation of their entire elite warrior class, never mind *all* the members of the state.

Under these conditions, locally decisive coercive threat was highly sensitive to how subsets of the individuals owning the means of coercion were distributed across the landscape and to what *information* they had access. This access to information—written documents, word of mouth from traveling functionaries, and so on—substituted for the *evidence of their own eyes* enjoyed by our more ancient ancestors. Thus, control of this information stream could sometimes be manipulated to prevent individual actors from *exercising* coercive power they might actually *possess*. Under these conditions, *potential* coercive threat was not necessarily *credible* coercive threat.

To visualize this problem more clearly call the local individuals in possession of decisive coercive threat the *local policing power.* This power will be deployed in managing the conflict of interest problem in each small area, day to day. For a state to be an effective cooperative unit, however, the local policing power must not only serve its own interests but those of the state-wide policing power.* However, they will do this only when the local policing power either has confluent interests with the state-wide policing power or when the local policing power is, itself, subject to coercion by a secondary, external policing power.

It is convenient to call this secondary policing power—a source of coercive control of the local policing power—*the elite policing power.* Notice that the elite policing power can simply be the aggregate of all the local policing powers. This condition exists, at least approximately, early in the growth of an archaic state.† However, as the scale and population densities of archaic states grow, the interests of local policing powers can diverge sharply from those of elite policing power (main text above).

This problem unfolds along the following lines. Initially for reasons of coherent control of concerted action,‡ some centralized command center(s) will be created by the elite coalition of the whole. To carry out its task, such a center must command some portion of the resources of the coalition of the whole. It is convenient to call this centrally commanded resource the "national tax." Payment of this national tax creates an immediate conflict of interest between local and elite policing powers. To the extent that the sum of all local policing powers can coercively control the actions of the elite policing power, the elite power will act in the interests of the local policing power. However, members of the elite policing power will inevitably act in their own interests, even when these are in conflict with the interests of individual members of each local power.

---

* Note that the option of the state-level policing power serving the interests of *all* the citizens of the state can exist *only* when this coalition of the whole has ultimate access to decisive coercive theat. This is an inescapable implication of our fundamental theory. Humans are adapted to project coercive threat *exclusively* in pursuit of *individual self-interest* (Chapter 5).

† Well-chronicled examples from the Mediterranean basin approximate this condition. These include the *Athenian Democracy, the Roman Republic,* and the early stages of the first *Islamic Empires,* for example.

‡ What specialists refer to as *cybernetic* or *command and control* considerations.

This conflict of interest between local and elite policing powers has diverse consequences. For example, members of the elite policing power would prefer that as much of the economic "surplus" as possible be paid as national tax. In contrast, local policing power would prefer that the national tax consist only of the minimal quantity necessary to allow the elite policing power to enforce the confluent interests of all the local policing powers. Unfortunately, the management role can be used to give the elite policing power a coercive advantage over each local policing power.

This elite advantage has at least two interconnected elements. First, the elite policing power will generally be successful initially in accruing sufficient resources to equip itself to be able to dominate each local policing power coercively. Second, the elite policing power will generally be able to restrict the access of local policing powers to the information necessary to allow *multiple* local policing powers to gang up on the elite policing power, thereby requiring the elite power to enforce local interests.

This process has the following progression in the archaic state. During early consolidation, power and access to information favors local policing powers. Elite policing powers fulfill some approximation of their nominal purpose of coordinating the activities of the entire state in the interests of all the local policing powers of that state. However, their central position allows the elite policing power to limit access to information to local policing powers. This informational asymmetry is exploited to concentrate resources, including coercive assets, in the hands of the elite policing power. This emerging power asymmetry is exploited to restrict local access to information further, creating runaway concentration of coercive power and access to information in elite hands. This initial runaway process creates a secondary arms race. Local coercive powers are severely disadvantaged by the growing elite control of coercion, information, and assets. Local coercive powers attempt to restrict elite access to resources and information, in turn. Thus, both local and elite coercive interests work to degrade access to the information necessary to coordinate both the collaborative and the coercive functions of the state. In archaic states, this phase of the process increasingly degrades the functional coherence of the state until it collapses under external invasion or some other catastrophe it can no longer manage.* This process is just another manifestation of the inevitable tragedy of the commons in the archaic state (main text above).

As we will see in Chapter 15, the distribution of control of coercive threat is different in the modern state than in the archaic state. However, these same underlying elite and local conflicts of interest exist in the modern state, including the various conspiracies by centers of power to restrict access to information. Thus, coercive centers of power in early modern states attempt to control access to information aggressively, creating tragically dysfunctional *economic systems* and brutal repression of the knowledge enterprise. This elite manipulation of information and power is broken only in those cases where a coalition of the whole has the democratized access to coercive threat allowing this coalition of the whole to enforce universal transparency (Chapters 15 and 17).

History of the archaic state: The big picture

In the sections that follow, we will review selected elements of the historical and archaeological evidence of archaic states. We will argue that this evidence represents extremely strong support for our theory of history and its application to these polities. Two fundamental insights are important to take away from this evidence.

First is how strikingly similar the Eurasian archaic states (Parts II and III), the diverse New World archaic states (Part III), and the pre-contact Hawaiian archaic states (Part II) are in their fundamental properties, even though these are each fully independent cases. They apparently had no significant contact with each other until late in their respective histories. Their language and subsistence technologies are utterly different. Even the material bases of their various coercive technologies are completely different and independent in origin. These superficially diverse archaic

---

* Local ecosystem degradation might constitute such a catastrophe in at least some cases. If centralized management of the conflict of interest problem were possible, such local ecological collapse could be avoided in the first place. Thus, ecological collapse is an effect of the failure to manage conflicts of interest, not a cause of state collapse in its own right.

An especially well-documented case of collapse of an archaic state under external threat is the conquest of Mesoamerica by the Conquistadors (Chapter 14). Cortes' ability to subjugate the Aztec elite center at Tenochtitlan was a direct result of the willingness of local coercive elites in the Aztec Empire to collaborate with the Europeans in making war on the center (Chapter 14).

states apparently have in common only and precisely the three things our theory predicts they should share: (1) the coercive dominance of armored elite warriors, (2) the intensively hierarchical political-economic-belief systems these warriors inevitably produce as they pursue self-interest in the species-typical human way and (3) an expanded scale of social/economic cooperation sustained by this elite policing apparatus. This pattern is what we predict if coercive technology is the sole and fundamental cause of the archaic state.

Second, in several major cases, including both Eurasia and the New World, we can date the invention of the elite body armor technologies permitting archaic state policing (Part III). Just as we saw with the invention of the atlatl and bow, local archaic states apparently arose very rapidly following the invention of these novel coercive technologies. Again, this is what we predict if coercive technology is the fundamental *cause* of the archaic state, with all its other important features merely being effects of this single cause.[*]

Collectively, these sets of observations represent numerous independent confirmations of the detailed predictions of our theory. We argue that this evidence, together with that from Chapters 11 and 12, constitutes very powerful support for our theory of history.

PART II: A sampling of the archaic state evidence

Case 1: The history of the archaic state: Eurasia and the "usual suspects"

As direct cultural descendents of Eurasian archaic states, the citizens of modern nations have tended to relying heavily on the written history of these earlier states, the lie made up by their ancestral winners. Thus, ironically, archaeological analyses and skeptical, detached documentary investigations are, in some ways, less well-developed for this Eurasian record than for more alien cases like the Inca, the Aztec, and the Hawaiians (below).

In spite of these limitations, a number of important, useful things about these Eurasian states are clear. The relatively large early Eurasian

---

[*]   In the hands of individual members of a species with two millions years of adaptation to the self-interested use of coercion.

states (including the Roman Empire, the Umayyad and Abbasid Islamic Caliphates, and the large Empires of early historic China) are all, without exception, hierarchical male-dominated states ruled by governments ultimately controlled by elite armored warriors.[32] Likewise, the many petty polities of the Eurasian archaic state era, such as the Greek city-states, the Japanese statelets before Tokugawa unification, the small warring states interspersed with Chinese imperial consolidation, and the toy kingdoms of Medieval Europe were, likewise without exception, controlled by armored elite male warriors.[33, 34, 35]

Indeed, all of the later Eurasian archaic states are traditionally referred to as Iron Age states and their body armor and shock weaponry was made almost exclusively of iron and its carbon alloy, steel. Armor was sometimes plate, but more commonly mail or scale. All were quite effective against the bow, the atlatl, and the sling.[*]

As predicted, all of these states, again, without exception, were based economically on the controlled labor of massive commoner/peasant/slave under-classes. These states varied in the degree to which they supported private commercial enterprise. But even when merchants were permitted to function, they were also ultimately controlled and taxed by elite warriors.[36]

The abundance of iron ore (see Part III) apparently made it possible for these elite warriors to be a larger proportion of the population than was possible with older body armor technologies like bronze. Bronze Age states may have had military elites of only a few percent of the population while Iron Age states had elites probably approaching 10 percent of males. Iron Age archaic states were, thus, sometimes self-conceived as relatively *democratic*. The Athens of Pericles, the Roman Republic, and the early Islamic Empires are good examples. Of course, these societies still contained large under-classes of peasants, merchants, and slaves. Human societies are always *democracies of the adequately armed*, so these societies looked democratic to their elite warriors.[†]

These societies accreted belief systems or institutions to serve the purposes of the competing interest groups within them. Contemporary historians from these states often focused on these institutions as the causal entities within these civilizations.[37] Subsequent historians have

---

[*]   See Roberts (1993), Freeman (1996), and Harman (2008) for three very different overviews.

[†]   This sense of democratic empowerment by elite warriors tended to deteriorate somewhat late in the development cycle of archaic states as coercive power became increasingly stratified and access to crucial information deteriorated (Box 13.2).

sometimes been fooled into believing them. However, the stories of these states and their institutions were actually written by the narrowly self-interested apologists of elite warrior interests. Any reader who is skeptical of this blunt, categorical statement should read Charles Freeman's (1996; pp. 307-357) crisp, clear synopsis of the foundation of Roman "republican institutions" and their ultimate irrelevance in the face of coercive threat by elite warriors for one example of the empirical support for our claim.

As we said, the Eurasian archaic states command our attention because our modern state cultures (Chapters 14 and 15) grew directly from them. Our religious, literary, and legal traditions are ostensibly inherited from their institutions.

As we will see in Chapter 15, these authoritarian traditions have been dramatically altered in the process of creation of the modern state. Attempts to pin our contemporary traditions to these ancestral ones are more commonly self-interested acts of attempted political/emotional manipulation than an actual statement of fact. For example, almost every day in the contemporary world, we see or hear of shared *identifier information* being used in an attempt to recruit political/coercive collaborators.[*] We may be recruited as conservatives, liberals, good Christians, devout Moslems, and so on and so on. These ancestral beliefs have subjective, emotional salience for us only for the reasons we have discussed, not because they are relevant to our contemporary adaptive challenges.

By understanding these emotional connections to the past and how they are manipulated in the present, we avoid being disingenuously manipulated. We also become better prepared to build a more humane, pan-global human future (Chapters 14-17).

The evocative stories of the late Eurasian archaic states are so well and widely known that we will not belabor them further here. Our investigations have attempted to falsify the theory using this record and have failed to do so. However, the massive record of these states stands as evidence on which others may attempt to test our theory, of course.

---

[*]   Invocation of shared religious traditions by self-interested political entrepreneurs masquerading as fundamentalist Christians in American politics or fundamentalist Moslems in South Asian politics are familiar examples. Equivalent examples would be invocation of the governmental traditions ostensibly inherited from the democratic Athenians or the republican Romans. All these invocations of the past are almost entirely fallacious and misleading, serving only narrowly self-interested contemporary purposes.

Box 13.3: Better the devil you know – the desperate stability of hierarchical politics

Humans have two million years of adaptation to pursuit of self-interest within our coercively managed cooperative coalitions. We do these things with such unconscious virtuosity that we often fail to grasp the ultimate causes of our behavior (Chapter 10). This has two powerfully toxic effects in the pathological context of hierarchical distributions of coercive power. First, we become highly risk averse when we perceive ourselves as being relatively powerless.* Second, we misconstrue our proximate motivations—*ideology* and *religion* especially—as the causes of our behavior.

The consequence of these two evolved patterns is that we are highly malleable within permanent hierarchies. If we are even one rung on the ladder above the very bottom—*white trash versus African slaves* in the antebellum American slavocracy (Chapter 15) for example—we can easily be manipulated ideologically to a violent defense of the very system that victimizes us.†

When we are relatively powerless—having access to some coercive means, but not to decisive coercive dominance—we anticipate that any change in the existing arrangement is likely to disadvantage us, not an unrealistic fear. We, thus, desperately cling to the hem of the hierarch's garment, hoping that our pittance is not interrupted and fighting with phenomenal ferocity to assure that our elite "benefactors" are protected.‡

---

* This approach to dealing with powerlessness would probably have been transitory and constructive in a democratized past. It would have allowed a person "on the outs" with his/her social coalition to adjust and recalibrate in preparation to moving back toward full enfranchisement. This approach has no such sensible consequence in a permanently hierarchical state.

† Especially toxic contemporary examples include young sub-elite males in hierarchical, male-dominated cultures from North Korea and Zimbabwe to Iran and Pakistan. These young males are the shock troops of contemporary terrorism, of course. It is unimaginably tragic that the primary effect of their violent deaths is the perpetuation of the very bondage that produced them.

‡ Of course, if we are elite warriors our *pittance* will generally be much more substantial than that of a slave or a peasant.

As we citizens of modern, ostensibly democratic states seek to democ-ratize remaining hierarchical states and our own institutions further (Chapter 17), we will need to be mindful of these toxic consequences of our evolved psychologies. Our behaviors in the face of elite coercive power create a massive inertia in hierarchical polities and institutions. We will call these phenomena back several times.

Box 13.4: Racism, sexism and ethnocentricity – elite identifier and social contract beliefs

As for all human societies—everywhere and always—the identifier and social contract beliefs of hierarchical states are designed to serve the interests of those who hold decisive coercive power. In elite-controlled polities these beliefs will justify hierarchs' domination of the larger population.* Humans have evolved to recognize projection of threat as legitimate when it is directed at evil, at free riders, cheaters, liars, and the lazy. Thus, social contract information in hierarchical states will identify non-elites as legitimate targets of repeated coercive threat—as evil. Evil doers must be recognizable. In the ancient ancestral world this recogni-tion would have involved two things—violation of social contract be-liefs and failure to evince knowledge of identifier beliefs (Chapter 10).

Our crucial question is, then, how legitimate targets of elite coercive threat are identified in hierarchical states. The two following examples are illuminating.

First, *evil* is identified as reliably associated with some recognizable physical feature. This can include markers of *class*. Thus, individuals dressed as peasants and slaves are defined to be inherently bad—shiftless, stupid, sly, devious, and so on. Of course, for this to work, patterns of public appearance—dress and behavior—must be mark-

---

* *Justification* is a crucial concept here. Getting underclass members of hierarchical states to "buy in" is useful. However, the primary goal is mutual self-justification for those elite individuals who act together to project the great coercive power they (must) hold. Elementary strategic concerns (and two million years of evolved psychology) require that elite individuals taking the risks of projection of threat will be advantaged thereby. Negotiation and clear definition of mutually agreed elite social contracts are the way, the only way, this can be arranged.

ers of class. It follows that hierarchical polities will create elaborately expensive patterns of dress and behavior to mark possessors of elite power reliably.

Of course, pricey, stylish clothing is expected and always observed. More enlightening is *manners*. Elaborate codes of elite public behavior, requiring years of training and experience to master (opportunity costs), are expected and observed.* These codes and beliefs will tend to be esoteric, even occult.

Second, congenital physical markers can also be useful in identifying "evil" non-elite individuals. Biological sex is an example. This is difficult to fake and males tend to be better elite warriors (text above) in any event. Thus, classing the female half the population as unworthy of enfranchisement is an extremely efficient elite licensing belief.

Also useful are those cases where two populations have been separated long enough to have acquired locally idiosyncratic appearance—skin tone, hair texture, and the like—before meeting again in a hierarchical state. These biological traits are quite convenient in indentifying "evil" individuals who are to be treated as domesticated animals by the holders of elite coercive power. These patterns of belief and practice not only existed in archaic states, they persist in all hierarchical states into the contemporary era. These properties of hierarchical politics will be most useful to us in confronting our future (Chapter 17). For now, sexism, racism, and ethnocentricity are always markers of elites seeking domination—no exceptions anywhere or any time. For example, it was convenient for the slavocrats of the antebellum American south (Chapter 15) to label dark skin as evil and light skin as good. It is just as convenient for the military oligarchs currently ruling Zimbabwe to reverse this relationship. When we find ourselves having racist, sexist, ethnocentric, or elitist feelings we are well advised to ask in whose hierarchical scam we are, perhaps unconsciously, playing "white trash" (Box 13.3).

---

* Ironically, early modern universities had this function (Chapter 14). They were primarily concerned with transmitting elaborate, mostly arbitrary knowledge—theology and the classics—to mark elite membership. We generally strive to make it possible for our contemporary universities do something more than this, of course.

Case 2: The history of an archaic state: A tropical "paradise" under the coercive umbrella of elite warriors is a natural experiment

Around ten million years ago, a point of hot volcanic rock peaked out of the middle of the enormous Pacific Ocean, thousands of miles from the nearest land. Over time the peak grew to a great mountain surrounded by a vast lava plain, forming the harsh beginnings of a place that would develop into the island of Kaui. After several million years, Kaui's volcano grew quiet and eventually a new island neighbor grew nearby, then another and another as the local tectonic plate slowly carried the Earth's crust across the volcanic hotspot. The entire Hawaiian archipelago ultimately unfolded. Indeed, even today, next door to the youngest of the Islands, Hawaii itself, a new sibling, Loihi, is pushing toward the surface, to be born into the world of our remote descendents.*

The relentless working of sea spray and rain slowly turned the lava of these newly born islands into a kind of soil and very, very rarely, a visitor from the incomprehensibly remote continents would struggle ashore on the islands—so saved from perishing in the vastness of the Pacific. The descendents of these impossibly lucky organisms evolved to master their initially alien environment, ultimately shaping it to their own interests while being shaped by it, in the timeless way of biological organisms.

Over eons, a bizarre collection of plants and animals came to make up this remote outpost of the life on Earth. But colonizing this outpost was not subject to patent or real estate law. Interlopers kept showing up from time to time and staking new claims. One day about sixteen hundred years ago, a most mysterious migrant peaked over the horizon, moving with purpose toward the great volcanic peaks.

This **seed** *was the designed offspring, not of a species, but of a civilization. A strange creature called "human" built massive, double hulled canoes on which they transported themselves, their domesticated plants and animals, their parasitic hangers-on, and their culture.*

This **culture propagule** *would take over the islands, utterly. Ecological dominance was one of the attributes of humans. The human population would boom, increasing a hundred times over in just a thousand years. To make this expansion possible, these humans would use their vast culturally transmitted informational repertoire to replace the indigenous forests with vast banana orchards and orderly fields of taro and yams. In the process these new creatures would drive many of the earlier immigrants to extinc-*

---

*   A Google image search of "Hawaii islands geology" will produce illuminating pictures of this process.

tion, creating a vast transported landscape. This beautiful tropical world was overrun and remade by the descendents of those that came in this tiny ocean-going canoe and their vast know-how.[38] It had taken the ancestors of these first humans thousands of years to find Hawaii. Their descendents would have the place to themselves for less than two millennia.

In 1778, British explorer Captain James Cook and his crew discovered this landscape that had so recently been transformed by its first, Polynesian human discoverers. This contact of Hawaiians with European colonialists happened over 250 years after European contact with the Aztecs.

The history of this Hawaiian society has no parallel in its value to us. The remote location of the Hawaiian Islands delayed any contact with modern colonists, so that the Hawaiian societies endured into the late 18th Century (the *Age of Enlightenment*) before being distorted and, ultimately, subsumed by the modern global political economy. Thus, arguably, the native Hawaiian civilization is our very best *direct* record of the logic and functioning of a real archaic state. Archaeologists and anthropologists have worked hard in this part of the world.[39] This rich ethnographic and material record is phenomenally useful to us.[40]

The initial factual descriptions of this civilization were relatively accurate, though the correspondents' interpretations of the significance of what they saw were often deeply confused. In the absence of good theory, it is profoundly difficult to understand what we see. The following description of traditional (mis)understanding of complex Polynesian societies by the outstanding archaeologist and Mayanist, David Webster (1998), is illuminating.

> ..Polynesian societies have, in fact, been characterized as [] "archaic" or "incipient" states by many anthropologists.... Others[] retain the term "chiefdom."[] Such confusion of terminology itself is revealing...[41]

By the time the ancestors of contemporary Hawaiians were reached by European colonialists, these societies were ruled by elite warriors. Our theory predicts that Hawaiian *chiefdoms* should, in fact, look like archaic states, though relatively young and, perhaps, in miniature.[42] As we will see, they do.

The Hawaii seen by European colonialists was orderly. It displayed vast irrigated taro fields covering, square-by-square, the great valley floors and stone fish ponds built at the surf line to farm ocean fish in huge numbers.* This spectacularly intensive landscape produced a large

---

* A Google image search of "Hawaii islands taro fields" will produce breath-taking images of this world as it still survives in some places.

surplus that was heavily taxed by Hawaiian elite warriors who, in turn, policed commoner cooperation.

Our theory is quite clear in its predictions about what this society should look like. It should be based on coercive management by elite warriors protected by some form(s) of body armor and using shock weapons to subdue and manage a large commoner population. Moreover, this elite management should have all the properties we have discussed in Part I. As we will see, this is precisely what the European colonialists, and their archaeologist/ethnographer successors, found.

Hawaiian elites were aggressive warriors. Though we have few archaeological examples of their weaponry, we have ethnographic reports indicating that they used shock weapons and wore body armor of various sorts.[*] This body armor apparently included well-designed whicker clothing as well as leather cape-shields. Shock weapons included clubs, swords, and daggers where various stones and shark teeth were used to create penetrating points and cutting edges.[43] As well, there are ethnographic indications that Hawaiian warriors cultivated techniques of elite hand-to-hand combat as all archaic state warriors apparently did. Again, recall the karate tradition handed down from the Samurai.[44]

As our theory predicts, Hawaiian society was almost entirely constructed to serve the interests of these armored elite warriors. Commoners were treated essentially as another domesticated animal along with the pigs of which elite Hawaiians were so fond. Commoner labor produced taro fields and fish ponds well beyond their own subsistence needs. They worked to pay their taxes to warrior elites.[45]

While commoner conflicts of interest were well managed, conflicts of interest between members of the elite were most emphatically not managed, as predicted. They flared continuously into open combat. These conflicts were sometimes organized around pretenders to one or another local throne—chiefs or kings. It is useful to quote Valerio Valeri's (1985) brilliant study of Hawaiian ethnography for a description of what a winner (the new king) of the latest round of elite conflict did first.

> First he sets aside certain lands for himself….Then he divides what is left among the high-ranking nobles who are his main supporters. The recipients, in turn, put aside certain lands for their use and distribute the rest [] among their own main supporters.[46]

---

[*]   The original Eurasian ancestors of the Hawaiians were almost certainly bow-using farmers (Diamond, 2000, 2001). However, sometime during a long sequence of island hoping, with small local populations, to Hawaii, the bow was apparently lost or rendered irrelevant by the development of sophisticated body armor.

Notice that the land that is distributed is owned by the nobles who were the elite warriors. They then, in turn, allowed non-owning commoners to farm the land in return for providing a large share of the produce to the warriors.

The king is apparently not paying elite warriors for services rendered in the past here. Rather, he is paying them in real time for their continued projection of coercive threat on his, and their own, behalf. Any elite warrior or noble who displeased other elite warriors risked having his assets immediately seized through armed expulsion, probably including his death (below).

It is fruitful to look at precisely how these nobles usually acquired their status. Again, Valeri is pithy.

> How is [a noble's] support obtained, then? By constituting a large faction of people who believe that their interests coincide with those of their leader.... It can be said to begin at the very birth of a noble, when his parents create a separate household for him comprising a number of clients and servants (the two are not clearly distinguished) under the supervision of a [] "guardian" who directs the affairs of the child until his is old enough. Most [of these people] seem to be adventurers who pass from one [noble] to another, often moving from district to district [] in hope of improving their situation.[47]

Finally, Valieri's description of the political world that results from this arrangement is perfectly clear. The failure to manage elite conflicts of interest is vivid.

> The rise and fall of kings, the instability of relationships of allegiance, the state of war that almost invariably follows the death of a king[], the king's attempt to maintain his faction by employing it in continuous wars of conquest, and the revolts that occur at each redistribution of land, which inevitably involves the displacement of previous landholders and the frustration of several of the king's supporters, who think they have not been sufficiently rewarded...These rivals are not simply killed; they are sacrificed....[48]

The details of this sacrifice bring home its profound violence and brutality.

> As soon as the victim is captured, he is immobilized...Sometimes his arms and legs are broken[] and one or both of his eyes are torn out.
> It also seems that the victim's penis may be partially mutilated as part of the preparation for sacrifice....
> ..Several sources say that the victims must be immolated outside the temple....[The victims] are generally placed on the sacrificial fire to singe their skin and hair; and it is thus "scorched"[] or "toasted"[] that they are placed on the alter.[49]

A second description of *elite* sacrifice is likewise illuminating about the purpose of this *ritual*.

> Pele-io-holani [chief of Ohau] cherished a feeling of enmity toward the chiefs of Molokai for the death of his daughter…and at the battle of Kapu'unonui he slaughtered the chiefs and roasted them in an oven…, Ka-hekili [chief of a different set of Hawaiian islands] sought to avenge the chiefs of Oahu [for this] slaying of the chiefs and commoners…. They had taken Ka-hui-a-Kama prisoner [] and roasted him in an oven, and they used his skull as a filth pot.[50]

These acts of brutality are directed against elites by other elites as part of competition for coercive dominance.[*] However, they also represent overt intimidation of commoners. At least elite individuals are sacrificed as a result of losing a gamble—a grab for power—that might alternatively have benefited them greatly. Their deaths are just the cost of doing business for members of archaic state elites.

Commoners have no such comfort. They may be chosen arbitrarily for sacrifice as part of the relentless need for projection of elite power. One of the early European explorers reports the following example.

> The bay had been tabooed some days on account of a large shoal of fish that appeared on the coast, at which time [an unfortunate commoner] was seen going across the entrance of it in a small canoe. [He] was immediately pursued, and when brought on shore, they broke the bones of his arms and legs and afterwards put an end to his miserable existence by stabbing his body with their [daggers].[51]

As Valeri laconically remarks about this report, "In theory, the victim of a human sacrifice is always a 'transgressor' or [a] 'mischievous" man [], but it is always possible to find a 'transgressor' when a human sacrifice is necessary." This commoner was almost certainly not infringing on the taboo, he was just in the wrong place at the wrong time, thereby unwillingly contributing to social stability by affirming elite coercive dominance.

The *belief system* associated with prehistoric Hawaiian society is startlingly rich. Indeed, some ethnographers and anthropologists have mistakenly taken the position that Hawaiian society, and, by inference, all societies, got its structure and its workings from these beliefs.

We have argued that this view is a confusion of cause and effect. Beliefs were everywhere, intensely advertised, highly detailed, indeed. They are a feature of human cooperative social aggregates (Chapters 10 and 12, above). However, recall that they are also completely susceptible to being

---

[*]   With profound irony, Captain Cook himself apparently met this fate—with his bones being distributed as prizes among various Hawaiian nobles—on a later visit to the islands (Obeyesekere, 1992).

scraped or remodeled wholesale to serve the interests of elite warriors who might seize power, today or tomorrow. Beliefs are merely the tools and ephemera of power. Only coercive power itself actually matters.

Valeri's discussion of *genealogical beliefs* about elite lineages illustrates what we mean.

> Ideally, relationships of subordination among nobles depend on genealogical distance. The closer one is genealogically to the senior line, the higher one is and the more land one receives. Thus, in theory, one cannot obtain more wealth and power than accrues by virtue of rank as determined by birth. In practice, however, it is genealogical rank that follows wealth and power, *however acquired* rather than the other way around….
>
> ..The genealogical relationships that establish and individual's rank and ultimately his right to belong to the nobility must be recognized by the king…. This, of course, allows for the genealogical claims of a strong supporter of the king to be "recognized" by him….
>
> Hence, the actual relationships of subordination and political alliance tend to be more important [] than the genealogical relationships [italics added].[52]

We argue that the record is clear. Archaic state beliefs, in practice, are merely the superficial skin and the public relations/negotiation tools of elite warriors. They are part of the contract that subordinate commoners must serve and obey scrupulously if they are to have any hope of survival. However, elite individuals may choose to ignore these beliefs almost entirely if they have the power to enforce their position.*

We argue that the Hawaiian record—the most detailed record we have of the social organization of an archaic state—is powerful in demonstrating that belief is merely symptom, effect, window dressing. Belief is never ultimate cause. For better and for worse, apparently, only coercive power determines human social organization, as our theory requires.

The one-of-a-kind Hawaiian natural laboratory, retained in "pristine" condition into the late 18th Century by the vastness of the Pacific, yields a powerful insight. Warrior elites practiced projection of coercive dominance by methodically engaging in public acts of extravagant cruelty both against cheater elites and against commoners. These rituals of human sacrifice are overtly and explicitly connected to management of the conflict of interest problem—"rebels" and "transgressors" are, at least ostensibly, the victims.[53] *Social order* is the avowed purpose and agents of violent ritual will certainly consider themselves to be the agents of the *right*.

---

* Of course, it will often serve elite interests to preserve some elements of earlier beliefs, creating an illusion of historical continuity.

Notice, again, that this is precisely what we expect if these rituals serve the function of projecting elite warrior threat as part of the ongoing self-interested negotiation of the archaic state *social contract*.

Box 13.5: Evolved human psychology and the experience of the "divine"

An evocative effect of our evolved psychological response to coercive power is apparently revealed by the archaic state and its belief systems. To grasp this point we need to understand the concept of a *supra-normal stimulus*. Such a stimulus engages an evolved psychological device designed to identify a particular cue. A version of this cue that is highly exaggerated will sometimes produce a hyper-intense response.

For example, our brains contain receptors for certain naturally produced compounds that are normally present in small amounts under specific conditions—engendering a mild, controlled "good feeling" when we execute an adaptive behavior, for example. However, drugs like cocaine can hyper-stimulate these systems producing intense, even addictive maladaptive subjective responses.

For humans, one of the most salient cues we are evolved to perceive is coercive power. For two million years our ancestors evolved to grasp this power in moderate, well-measured doses in the context of the democratized ancestral village as they managed the conflict of interest problem (Chapters 5-7 and 10-12).

However, with the coming of the archaic state, humans confronted coercive power on an entirely new, supra-normal scale. Think again of the Roman Coliseum or the Aztec Pyramid of the Sun and the massive armies of elite warriors they represent. The subjective response to such power is, predictably, like the subjective response to cocaine—intense, exaggerated, and *pathologically* addictive.

People sometimes report a *transcendental experience of the divine* in the context of rituals and religious beliefs inherited from the archaic state. We have argued that these rituals were designed, in part, for projection of elite coercive power.

We propose that the intense subjective experience of the divine is the correlate of the supra-normal stimulation of the evolved response to coercive power. The fierce experience of the divine shares elements of both possessing abnormally vast power and of being its target, on our theory.

This proposal might be directly testable through neuroimaging or experimental psychology. It would be a considerable interest to know if it is correct.

A light at the end of a short, dark tunnel – on to the modern state

This chapter is an exploration of selected details and examples of the local adaptive revolutions produced by archaic states around the world and our theory's apparently strong ability to account for them.

However, the chapter has also been an argument that hierarchical, exploitative, male-dominated societies are *not* the ancestral human condition. Rather, hierarchical patriarchy is probably a brief aberration produced by the development of elite body armor in a world where relatively democratically distributed access to threat (thrown stones, the atlatl, and bow) had probably been the rule for nearly two million years. How it feels to live in such a hierarchical state is eloquently described by the following quote used by Partick Kirch in a piece on Polynesian archaic states.

> *Side by side with legitimating ritual, there was a social control by brute force. Together with an all-pervading fear was an almost secular pragmatism….The haka'iki' [ritual human sacrifice] lay at the centre of all social life.*[54]

Archaic states were not humane places. Moreover, for all the "glory of Rome," archaic states were grimy and poor for most of their residents. You and we are the beneficiaries of the next human adaptive revolution—the rise of the *modern state*.

In Chapters 14 and 15 we turn attention to our theory's prediction of this modern revolution. We will discover how the descendants of slaves and peasants, our ancestors, seized control of the state, by capturing access to decisive coercive threat. In so doing, they created a vastly richer and potentially more humane world than all the conceited puffery of archaic state elites could even imagine.

Readers who are confident that our theory's account of archaic states is likely to be correct may wish to proceed directly to Chapter 14. For those who are still skeptical or would enjoy further exploration of the evidence, Part III below provides additional samples.

PART III: Selected additional details from the theory and empirical record of archaic states

Case 3: The Aztecs and the Inca – the ethnography of large archaic states in action

The New World archaic states are useful empirical tests of our theory. Several were in full flower when modern Europeans colonialists encountered them. For all their failings and cultural myopia, these early Europeans were sometimes competent bureaucrats, reporting somewhat accurately about what they saw to other government functionaries. These colonialists also occupied an economic system (Chapter 14) that permitted them to benefit by selling books and writing for an educated but popular audience about what they saw in the New World. This broad dispersal of information was not usually possible in archaic states.

Thus, we have a relatively extensive documentary record of how these archaic states actually worked. We have reason to think this record does not consist entirely of elite self-justification.

Further, these exotic New World civilizations have been attractive as targets of intensive archaeological investigation by the massive contemporary state scientific enterprise. These investigations are particularly valuable because many key sites of the New World archaic states were overgrown and abandoned, leaving them relatively intact and uncontaminated, rich, and ripe with potentially useful data.

Several insights from this body of evidence are particularly important.

First, both the Aztec and Inca states were run by warrior elites, as we predict in view of their vast scale and public ritual behavior we have already discussed.* Moreover, the dominant members of these elites

---

* The local ancestors of the Aztec, the Maya, are an especially poignant example of earlier scholarly confusion about the role of coercion in archaic states. The Maya were originally imagined to be governed by a ritual-mongering priestly elite, with few if any soldiers. As Mayan writing was deciphered and their archaeology better under-

wore body armor rendering them relatively resistant to the available projectile weapons such as the atlatl, bow, and sling.

This New World body armor is remarkably illuminating. It was made from quilted cotton.* The underlying principle is a little like the Kevlar of modern military/police body armor. Each of many layers of fabric formed by strong, tightly woven threads absorbs projective momentum, trampoline-like, allowing many layers in sequence to ultimately distribute projectile momentum over such a large area that no penetration occurs. Even a high-velocity penetrating projectile is experienced by the armored warrior as the punch of a fist—mildly bruising perhaps, but not seriously threatening.[55]

Though New World cultures had some experience with metallurgy, it was not used for armor or weapons as it was in the Old World. New World shock weapons often consisted of stone cutting edges, especially volcanic obsidian, and mace heads.[56]

The elite armored Aztec and Inca warriors were highly experienced. They were professional or semi-professional and could overwhelm non-armored fighters wielding local projectile weaponry under most circumstances. Even though Inca and Aztec armies often included commoner and sub-elite archers, slingers, and the like, the armored elite warriors were usually decisive, and their presence and performance determined the outcome of conflict under normal circumstances.[57] In other words, armored elite warriors were apparently coercively dominant, as our theory predicts.

Note that the Inca practiced human sacrifice as did the Aztecs (above) and as predicted by our theory.[58]

Pleasing scientifically is the observation that the Inca and Aztec states not only developed independently of Old World states, but also of one another. The fact that both these New World cases share the features we predict with one another and with the Eurasian and Hawaiian cases represents extremely strong evidence that only elite male coercive dominance—enabled by sophisticated body armor—matters.† No other features of local culture appear to be relevant.

stood, this "peaceful Maya" hypothesis was demolished. The Mayan archaic state was just as violent, hierarchical, and male-dominated as any other archaic state (reviewed in Coe, 2005). This famous case is a vivid object lesson for scholars confronting other ostensibly peaceful archaic states.

* Google images searches under "Aztec cotton body armor" will produce useful period illustrations and modern reconstructions.

† In addition to the technical references in the online endnotes, Michael Wood's beautiful, humane, readable book *Conquistadors* (2000) drives home the astonishing similarities between New World and Old World early states.

The earliest archaic states:
Elite body armor comes first

We have already seen that late, mature archaic states were apparently always and everywhere dominated by elite warriors exploiting advanced body armor to render local projectile weapons relatively unimportant and elite shock weapons decisive.

The question for us now is whether elite body armor *briefly precedes* the first archaic states as our theory absolutely requires. Again, on our theory, coercive technologies are sole *causes*, not secondary tools, or effects. We do not yet know how to predict precisely the *speed* with which archaic states will follow the first elite body armor. However, the Eurasian and North American Neolithic revolutions apparently followed very *rapidly* from the local introduction of the bow (Chapter 12), suggesting great speed. As we will see, archaic states also appear to follow local introduction of elite body armor quickly.[59]

Moreover, our theory does not predict the specific local variation of technology that will be used to generate elite body armor. However, we can predict that the technology must be suitable to some level of mass production of body armor, even though individual suites of armor remain relatively expensive. This relatively large scale production of armor is vital because an isolated individual elite warrior is still somewhat vulnerable to projectile weapon and non-elite attack.[*] Elite warriors become effective against non-elite targets in massed formations, a dozen or more individuals, depending on the size of the non-elite target population.

Thus, we anticipate that a key early development will be the discovery of some technical system/methodology that will allow routine, repeated production—over generations—of many copies of reliable elite body armor. Though the earliest development of body armor is still in need of much more extensive expert study, we have a sufficient body of evidence to make a very useful beginning here. We hope specialists in this area will give these questions renewed attention in view of their newly recognized importance.[60]

It would be ideal if we had archaeological remainders of the very first elite body armor. We probably do in some localities, but not in others. Thus, it is more useful to begin by considering whether there are reliable *indirect* indications of the arrival of effective elite body armor.

---

[*]   For example, it is difficult or impossible to produce body armor that is effective through 360° of projectile attack and a shock weapon so devastating that threat can be projected in all directly simultaneously.

As we discussed near the beginning of the chapter, brandishing a shock weapon without body armor in the presence of effective projectile weapons is suicide, not a route to elite power. Thus, indications that shock weapons are becoming locally dominant are strong, indirect indicators of the presence of effective body armor, protecting the elite warrior from projectile fire.

With this background in hand we can now examine the empirical record of the earliest archaic states.

First, the original Mesopotamian states began to show organized urbanism and the building of infrastructure on the large scale expected of the emergence of a new scale of social cooperation around 3500 BCE.[61] These first states apparently follow very rapidly after the local invention of efficient bronze metallurgy. Indeed, these polities are often referred to as *Bronze Age* states. Bronze shock weapons are common in the archaeological record at or just before the beginning of this era.[62] It is likely that bronze was used to make *scale armor* from very early in the development of routine production of the metal.[*,63]

Second, the first Egyptian state emerges from a period of rapid change and expansion of the scale of local economic/social cooperation by around 3000 BCE.[64] It appears that the first armor for professional Egyptian elite warriors might have been large shields manufactured from the hides of the domesticated cattle that had recently become much more prominent in Egyptian subsistence and ritual.[†,65] Most importantly from our point of view, period art indicates that shock weapons became very important at or near the beginning of the first Egyptian state—the late *Predynastic* and early *Dynastic* eras.[66]

Third, our theory predicts that the expense of elite weaponry should have significant consequences for the effects of this weaponry on the subsequent structure of the archaic state. Thus, the effects of the development of iron metallurgy on Eurasian archaic states are most striking. The invention of reliable iron metallurgy allowed elite body armor and shock weapons to be produced on a significantly larger scale than for bronze because of the much wider availability of iron ore than the scarce components of bronze (tin, in particular).[‡]

Iron metallurgy apparently first became well developed in present day Turkey around 1200 BCE[67], subsequently spreading outward across

---

*   A Google image search under "bronze scale armor" will produce useful examples.

†   Cattle domestication might have provided a necessary pre-condition for the invention of "industrial scale" manufacture of many copies of elite armor in this context.

‡   A Google images search under "Roman body armor" will produce useful examples of iron/steel body armor.

Eurasia and Africa.[68] Within about six hundred years, the Greek, the Persian and, ultimately, the Romans and the later Islamic Empires were founded—that is, the great archaic states we think of as ancestral to *western civilization.* Our theory predicts that the detailed political structure and scale of these states is a direct product of relatively cheap iron body armor and weaponry.

It is also extremely striking that the first great states in East Asia *follow* the introduction of iron metallurgy into these areas within a few hundred years (or less).[69]

Fourth, the New World archaic states give us a priceless *independent* opportunity to ask whether development of elite body armor briefly precedes the rise of the first archaic states as our theory requires.

The elite body armor of the Aztecs and the Inca at European contact was based on a Kevlar-like use of quilted cotton, as we have seen. Moreover, the religious art of the precursor civilizations like the Classic Maya in Mesoamerica[70] and the Moche[71] in South America shows elite individuals wielding shock weapons and wearing frocks or mantles—torso coverings—that look like representations of cotton body armor.

This indicates that cotton body armor was probably relatively ancient. Consistent with this interpretation, domesticated cotton and elite weaving technologies are available from just before the first states in both Mesoamerica and South America.[72] These observations are consistent with cotton body armor allowing the initial emergence of elite warriors in both these areas.

However, more compelling evidence comes from a different quarter—iconography or *public religious art.* Recall that the public projection of elite coercive threat is an unavoidable feature of all archaic states. Moreover, as we saw for the later Aztec, brutal public human sacrifice was an important element of this ritual behavior in the New World just as it was in the Old World and Hawaii. Likewise the precursors of late New World archaic states—earlier states like the Classic Maya and the Moche—almost certainly engaged in public execution/human sacrifice as well.[73]

Thus, iconographic representation of human sacrifice would represent indirect evidence for the onset of armored elite warriors. Remarkably, this is seen with the very earliest manifestations of presumptive archaic states in both South America and Mesoamerica.

The South American case is dramatic. The place is Cerro Sechin in the Casma Valley of Peru where one of the earliest large stone monumental buildings and its associated evidence for a state-like economic system is found. It was built between approximately 2150 to 1300 BCE.[74]

The public plaza of this early government building is surrounded by life-sized images of disemboweled individuals and severed heads.[75,*]

Reflect on this for a moment. Imagine going to the local court house in your home town and seeing life sized images of the bodies of executed criminals with their guts streaming out and their severed heads bleeding. Archaic state social systems really are different from our own. Archaeological remains from later eras strongly support the hypothesis that this Cerro Sechin iconography depicts real contemporary events.[76]

A similar kind of public display is seen with the very first small archaic state architecture in Mesoamerica. Specifically, a stair step into a public building in the Oaxaca Valley in approximately 700 BCE show a disemboweled individual strikingly like those at Cerro Sechin.[77]

We argue that such displays only makes sense if they serve the interests of an intimidating warrior elite.

In summary, the extensive, if sometimes incomplete evidence we currently possess about the very earliest archaic states is consistent with their arising as the products of self-interested coercion by elite armored warriors. These states apparently explode into the record rapidly after the invention of elite body armor. It will be of the very greatest interest to watch continuing research on these earliest states over the next several decades. These investigations will provide numerous opportunities to confirm/disconfirm the strong specific predictions of our theory.

Elite coercion from the start – Mesoamerican and South American agricultural revolutions

Our theory and the archaeological data immediately above provide an important new model for the differences between Mesoamerican/South American *agricultural revolutions* and the Eurasian/North American cases (Chapter 12).

Behaviorally modern—that is, pre-Neolithic—humans clearly domesticated plants in a number of cases.[†] This is in contrast to post-Neolithic domestication as apparently occurred in other cases, like

---

*   A Google image search of "Cerro Sechin heads" will produce vivid examples. Also, see pages 140 to 142 (and associated plates) in Michael Moseley's excellent 1992 book *The Incas and Their Ancestors*.

†   Hopewell pre-Neolithic domestication of several plants including the sunflower represents one particularly well-documented example (Smith, 1995).

Eurasia (Chapter 12). These early cases seem to include the initial stages of domestication of maize, beans, and squash in Mesoamerica and of cotton in South America and Mesoamerica.[78]

The initial stages of these domestication processes preceded local *agricultural revolutions*. As predicted by our theory, large-scale agriculture is the product of expanded social cooperation, *not* an effect of the availability of domesticated plants or animals (Chapter 12).*

Thus, we require that the South and Mesoamerican agricultural revolutions, *sensu stricto*, will follow from the local introduction of a new coercive technology, allowing a new scale of social cooperation.

As we have already seen, cotton body armor was apparently central to the archaic states in these two areas. Moreover, the early states arise very rapidly after the earliest signs of *agricultural villages*, in stark contrast to Eurasia where the first state followed Neolithic revolutions by five thousand to seven thousand years![79] (See Chapter 12 and above.) Further, the Mesoamerican evidence indicates that elite cotton spinning and weaving—necessary to make cotton body armor—arose at or just before the origins of the first agricultural villages.[80]

Thus, we propose that the South and Mesoamerican agricultural revolutions were the products of coercion enabled by cotton body armor.[†] In other words, these agricultural revolutions were simply the incipient stages of archaic state formation in this area of the world on this hypothesis. The New World archaic states arise directly from a behaviorally modern scale of social cooperation on this interpretation.

It is noteworthy that cotton cultivation apparently did not initially spread into North America with maize, beans, and squash (Chapter 12), perhaps for climatic reasons. Thus, the fact that the Mesoamer-

---

*   The small-scale use of domesticated plants is often referred to as *horticulture* to distinguish it from the larger-scale, market or state-driven practices we call *agriculture* (Chapter 12).

†   The model we propose here is, we believe, the most likely in view of the empirical evidence currently in hand. However, we emphasize that the proposals in this section could be wrong as a matter of empirical fact as to the identity of the weapon without undermining the correctness of the larger theory. For example, the agricultural revolutions of South and Mesoamerica could have been produced by the introduction of some other, currently unrecognized weapon. A hypothetical example will illustrate this alternative. These agricultural revolutions could have been driven by the bow, as in Eurasia and North America, but the local archery traditions could have used arrows that leave no archaeological traces, like fire-hardened wood, for example. On this hypothesis, we would also need to understand why this bow technology apparently did not spread into North America, of course.

ican agricultural revolutions did not extend into North America along with the Mesoamerican Triad of food domesticates is also apparently explained on our proposal—no cotton, no body armor, no agricultural revolution until the coming of the bow to North America (Chapter 12).

Roads and canals – Highways of taxation, elite coercion, and little commerce

One of the costs of elite coercion is transportation of armies of elite "police" throughout the territory of archaic states. This warrior mobility is a high elite priority. Thus, elites have historically constructed transportation systems for this purpose well beyond the needs of commoner commerce. For example, the massive Inca road systems carried elite warriors and state functionaries throughout a largely non-commercialized, centrally planned empire.[81] The Romans were likewise famous for the construction of elaborate roads to support the mobility of their legions.[82]

This insight helps us to grasp further our theory's predictions of archaic state archaeology. Both early Pharonic Egypt and early Mesopotamia are notable for the relatively modest archaeological signals of elite warrior domination.[83] Indeed, this has led some investigators to propose that *religion* (new identifier beliefs) or some other cause— rather than elite coercion—was behind these early states. On our theory such suggestions cannot be correct.

We note that efficient water transport dominated the landscape of these two states. Dramatically, in Egypt, essentially the entire state was literally within sight of the Nile.[*],[84] A relatively small elite of armored warriors could have suppressed local rebellion and tax evasion while preventing rebel movement up and down the Nile.

Likewise in Mesopotamia, the flat flood plains flanking the Tigris and Eupharates allowed local towns to be easily connected by canals to these rivers. Moreover, as new settlements were built by the state, they included canal systems very much like streets throughout.[85]

Thus, in both these cases, Egypt and Mesopotamia, elite coercive threat would have emanated palpably from the nearby water courses. Fewer *boots on the ground* per capita may have been necessary under these conditions than for other land-transportation-dominated empires.

---

* Herodotus famously called the Egyptian state the "gift of the Nile".

On to the modern state

Our ancestors, mostly the descendants of the underclasses of archaic states, went on to create a vastly richer and more aware world than the violently self-justifying, stupendously provincial elites of these hierarchical states could imagine. Chapters 14 and 15 will argue that this remarkable creative feat was the product the democratized seizure of decisive coercive threat, precisely as our theory requires.

∽

Chapter 14
Modernity emerges from the barrel of a gun, Part 1 – artillery stabilizes the hierarchical state

Kamehameha* was an intelligent, sober, industrious young member of the Hawaiian ali'i, nobility (elite warriors; Chapter 13). He was also a brave, competent military man, as all male ali'i were required to be.

Kamehameha was the nephew of the chief on a local island in the Hawaiian archipelago. He was discreetly ambitious and, in the time-honored ways of small Hawaiian archaic statelets, would ultimately take over the chiefship after his uncle's passing. But his career would be like no others before him. Kamehameha's world had recently changed in ways that mattered.[1]

In January of 1778 the famous British explorer, Captain Cook, made the first recorded land fall by a European in the Hawaiian Island group. Cook's ships were vastly larger than the Hawaiian traditional canoes and the locals were appropriately impressed. As the Europeans anchored, the first few canoes cautiously approached. They were armed with throwing stones that were discreetly jettisoned into the water as the Europeans proved peaceful and welcoming.

*In the ancient human way, market exchange began almost immediately. At first, the Europeans traded iron (rare and precious to the Hawaiians, yet mundane and functional to the Europeans) at phenomenal **exchange rates**. A single European nail could buy large amounts of local food or many sexual favors. Indeed, the wooden European ships were at risk for being stripped of every nail by the highly "motivated" sailor-traders.*

---

* *Pronounced Kah-may-ah-may-ah, with all syllables distinctly and equally articulated.*

Early in the growing relationship, a small group of Cook's sailors went ashore in a skiff. The locals immediately surrounded the boat and one became highly aggressive in trying to seize it. The British lieutenant in command of the landing party immediately shot him to death. A new weapon had suddenly entered the Islands, but the trading relationships remained and grew.

Supplies replenished, Cook's ships sailed north on other business. They returned eight months later, with the goal of a more thoroughly mapping and understanding of the island group. Word of their earlier visit had quickly spread throughout the Islands and this time royalty was prepared to greet them properly. Among them was the young Kamehameha. At one point, an armada of local vessels carrying perhaps as many as ten thousand Hawaiians welcomed Cook's returning fleet.

The Europeans were escorted around the Islands. Our first reports of the Hawaiian economy date from this visit. Hawaiian elites were carefully building relationships with the newcomers.

Cook's ships attempted to leave the archipelago in early 1779, but were forced to turn back to repair a large mast. A few locals took the occasion to steal several valuable iron tools including a large saw. Cook led a landing party to recover the items and a confrontation ensued. Cook used his side arm to kill one Hawaiian and the conflict escalated. Cook was clubbed down and stabbed to death in the surf. Several other members of his party also were killed. The remainder retreated to the safety of their ships.

Over the next few days, fighting flared repeatedly between the Europeans and the Hawaiians. The dominance of European gunpowder weaponry was clearly evident and sometimes vicious retaliation occurred. Eventually, Cook's successor, Captain Clerke, and cooler-headed members of the Hawaiian aristocracy were able to restore calm.

Cook's body had apparently been sacrificed in the Hawaiian tradition for a killed enemy chief (Chapter 13) and Clerke was able to recover only a few of his bones. Each side had a new appreciation for the other's capacity for violence. But the opportunity for mutual gain again took precedence and Clerke's fleet left the archipelago under peaceful conditions.

No European returned to Hawaii until 1786. From this date onward, Hawaii would remain a frequent port of call for European trade and war ships. They would continue to pursue the vast opportunities of Pacific basin commerce. This emerging Pacific economy involved furs and precious metals from the west coasts of North and South America and numerous manufactures, spices and other valuable goods from China and Japan. The Hawaiians had not only seen European gunpowder weapons and great ships, but they immediately began to acquire them through trade. Kamehameha was diligent, ultimately acquiring thousands of muskets, a number of artillery pieces and large ships of European manufacture. Moreover, he had recruited several European specialists—making them vested members of Hawaiian nobility—who could train his troops in the tactical and strategic use of gunpowder weapons.

By 1796, just ten years later, Kamehameha had assembled an armed interest group ready to make its move in a quest for coercive dominance. Throughout its history, Hawaii had been composed of small archaic statelets consisting of one island or a small group of nearby islands, at most, but more often just a portion of a single island (Chapter 13). Many ambitious chiefs had dreamed and attempted "unification" of the "kingdom." None had ever succeeded.

Over the next few years Kamehameha would succeed. Armed with European muskets and artillery, his coalition ultimately brought the entire archipelago under his stable control. Kamehameha administered the economy of the new Kingdom around the principles he already knew. He gave the agricultural domains with their commoner work forces to his chief allies who, in turn, parceled them out to their primary warriors. Hawaiian elite warriors owned the weapons on the ground and the Kingdom served their elite interests as before, but now on a large new scale.

How did our *modern* world come to be?

We all take for granted the absolutely astonishing world we currently inhabit. We live as none of our ancestors could even remotely have imagined. Before seeing how our theory accounts for the recent emergence of this modern human world, we need to understand exactly what it is we are required to explain.

Humans have been on Earth for nearly two million years. For most of that time, most people lived on the equivalent of about $1.50/day.[*,2] This *individual income* was probably relatively unchanged from the behaviorally modern human revolution (or earlier) through the agricultural revolutions and the rise and fall of archaic states,[†] until as little as four hundred years ago! However, beginning sometime around the 18th Century, humans in some places in the world suddenly began to grow

---

*   Note that earlier increases in adaptive sophistication (Chapters 11-13) were followed by increases in population densities, until a new Malthusian ceiling was reached. Thus, average individual wealth changed relatively little over the long haul.

†   Of course, individual members of small ancient elites were occasionally far richer than this, but we are concerned with the average human at each moment in our history for now.

richer, year-by-year, generation-after-generation—again, apparently *for the first time in our two-million-year history!*[3]

This relentless growth in access to adaptive resources continues through the present instant. Many of us reading these pages are fifty to one hundred times richer than the entire two-million-year march of our human ancestors![4] Moreover, this powerful new wealth generation gradually spread, and continues to spread, to more parts of the world.

We truly occupy a radically changed world, not just economically, but technically, too. A child born anywhere in the world just fifteen generations ago had a nearly 50 percent chance of dying in infancy or childhood. Even the lucky ones who survived to adulthood aged rapidly in a harsh, demanding environment. Life expectancy *at birth* fifteen generations ago was roughly twenty-five years. Moreover, though we do not have a lot of direct evidence from earlier eras, this figure has most likely remained constant for most of our two-million-year history. In contrast, a child born in a modern developed country today has a life expectancy three times longer than this—77.5 years in the early 21st Century US, for example![5]

Our scientific grasp of our world has likewise grown phenomenally. We understand the structure of matter down to unimaginably small, and strangely behaving, subatomic *particles*. We comprehend and routinely manipulate *chemistry* and *physics*, producing everything from block-long passenger jets to iPods, ceramic knives, internets, and antibiotics. Indeed, we have come to the truly breathtaking insight that we humans are merely a particular class of chemical device (Chapter 2), a perspective our minds were most certainly not originally designed to comprehend (Chapter 10). Moreover, we grasp the vast size and immense, yet oddly finite, age of our universe, our galaxy, our solar system.

All this insight gives us new skills. From burning wood for warmth in small fireplaces we now set off titanic thermonuclear blasts, so large they could have been seen from the Moon with the naked human eye. From scratching the ground with sticks and terracing fields to feed a few thousand, we have come to sear, plow, mow, fish, and harvest virtually the entire surface of a planet to feed billions, sometimes to the point of pathological obesity. From tiny local potentates we have truly become unchallenged, sometimes smug and thoughtless, lords of a world, transforming our planet into a massive breeding colony for a single species, our own, while altering the very stuff of its atmosphere. Over the last few centuries Earth has gone from having the odor of millions of species to literally smelling of human society.

As remarkable as these capabilities and insights are, it is their very recent and explosive origins that astonishes us even more. The vast majority of this formal scientific and engineering knowledge has been generated in only the last four hundred years! That's the same four hundred years in which we grew so wealthy.

Economics, science, and technology do not exhaust our modern world's novelty. We are a new animal socially, too. We understand that it is the height of provincial hickdom to think of those people who are different than us as *inferior*, *subhuman*, *evil*, or *heretical*. At least, most of us have acquired this insight. We have discovered that all humans alive today are the descendents of a tiny group of people who lived just a few tens of thousands of years ago (Chapter 11). Our minor, superficial behavioral differences are mostly the products of different local and *very* recent histories of the evolution of culturally transmitted information (Chapter 10). We are truly all sisters and brothers under any meaningful, biological definition of those terms. However, each of our ancestral cultures of just a few dozen generations ago considered it self-evident that they were civilized and that all other people around them were subhuman barbarians.[*]

We truly occupy an utterly novel world. Any theory of history must account for the recent, explosive emergence of this *modern* world with all of its properties, unprecedented in the 4.5 billion year history of our solar system. Moreover, such an account must predict why the changes producing this world began so disproportionately in the *West* before slowly, and not yet completely, spreading around the world. Our theory makes just these predictions—simply, directly, transparently. This chapter and the following one (Chapter 15) explores that account.

The modern world – simplicity beneath the deceptive appearance of complexity

As the astounding rise of modernity became clear over the last several centuries, many explanations were offered. It was speculated that our world was originally *caused* by the invention of modern science, the Protestant Reformation, the invention of the printing press, the development of capitalism, the invention of democratic ideologies, the establishment of modern warfare, or any of a myriad of other things. We

---

[*]   This was not a uniquely *European* delusion. The Imperial Chinese and the Japanese Samurai believed precisely the same things about themselves , as probably did every human alive at that time, and even a few alive as we speak.

will argue that all these accounts are wrong. They are confusing effect and symptom with cause.* Likewise serious scholars tried to understand our new economic power, beginning with Adam Smith, David Hume, Karl Marx, and Max Weber, among many others. As illuminating as some of these early attempts were, they were all ultimately unsuccessful as theories of the modern world. We can now do much better.

Our theory's account of the modern world will be obvious. We predict that all the rich complexity of our modern world is the simple product of a new scale of social cooperation, produced, in turn, by a weapon capable of managing conflicts of interests on this new scale. We humans deployed this new weapon as we have every other one before it throughout our two-million-year-old pursuit of coercive self-interest. The startling novelty of our world merely results from the scale and detailed properties of this weapon. This new coercive technology is gunpowder weaponry (Figure 14.1).†

It is important to be emphatically clear. The modern nation-states of which every one of us is a citizen are apparently fully predictable on our theory. They are simply new scales of the ancient, uniquely human social adaptation made possible by a new scale of suppression of the conflict of interest problem. As we have seen in earlier chapters, we humans are "one note hula hoops." We play our characteristic adaptive trick over and over again, at ever increasing scales. The modern world is just the latest installment of this story, nothing more, nothing less.

---

* More recent attempts to explain the emergence of the modern world as the product of a change in "world view" or patterns of thought have been equally unsatisfying and unhelpful—confusing effect with cause and proximate psychology with ultimate causation (see, for example, Wolin, 2004, for one perspective on this inadequacy).

† We will define the emergence of the *contemporary* world from the modern world as the currently final stage of the human story in Chapter 16. It will be convenient to date this final *contemporary* era as beginning about 1939 CE. We will turn to our theory's predictions about the social effects of modern aircraft and missile technology and of nuclear explosives in Chapter 16.

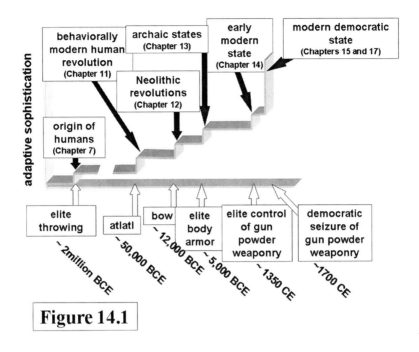

Figure 14.1

The modern state deploys the ancient human trick of massive mobilization of culturally transmitted information (Chapter 10), but on a vast new scale. As we will see (Chapters 15 and 17), with the *democratization* of the modern state (overthrowing elite coercive dominance) we are arguably returning to our ancient human heritage, the equalitarian cooperative pattern to which we are most exquisitely adapted. *Modernity* arises when we execute this ancient adaptive strategy on the scale of continents rather than villages. Individual gunpowder weapons replacing shock weapons (Chapter 13) recreates ancestral democratized control of coercive threat (Chapters 5, 11, and 12). But coercion with gunpowder artillery confers a vastly larger scale on this democratized society. These effects of gunpowder are inevitable and predictable when these weapons are employed by human minds with two million years of adaptation to the use of projectile weaponry, to the deployment of death from a distance.

Mao Zedong is reputed to have remarked that, "All power grows from the barrel of a gun."[6] We go much further. *Everything* new about our modern world emerges from that gun barrel—social humanity, wealth, science, technology, all of it. The nature and structure of the modern states humans built should be determined *entirely* by who is able to control access to these weapons, on our theory.

Many of us find this simple claim incredible or unbelievable at first glance. We may even be ethically or esthetically offended by the assertion that all our most cherished political, social, scientific, and economic beliefs and accomplishments are merely handmaidens of the self-interested projection of the threat of violence.

If we have this negative reaction, we should stop and reflect on the following logical sequence once again. First, the self-interested use of coercive threat is apparently the *only possible solution* to the problem of conflicts of interest between humans (or any organisms). There is no other solution, ever, anywhere. Second, there is no human social cooperation without this management of conflicts of interest—again, not anywhere, not ever. Thus, it is precisely this strategic use of the threat of violence that enables and protects the better angels of our uniquely human, and potentially humane, natures. Only from the threat of violence, *properly and democratically deployed*, can peace and human comfort emerge and persist.

This feature of our contemporary world is further hidden from us for other reasons. Because we have been exploiting coercion this way for about two million years, we create and navigate the resulting social world naturally and automatically, much as we breathe—with unconscious virtuosity. We are not usually conscious of our participation in and response to the coercive umbrella that overlays our every public action (Box 14.1; Chapter 10; Fourth Interlude).

Moreover, the projection of coercive threat is so deeply salient and self-interested that we are alarmed and embarrassed to discuss it explicitly in public. Just as when we are forced to discuss the most secret details of our sexual lives and thoughts in public, we deny, cringe, retreat, and change the subject.

If we wish to grasp the human condition and shape a more humane future, we must overcome these self-imposed barriers. We must understand ubiquitous, uniquely human coercion as we live it right here and right now.

The emergence of the modern world has left a spectacularly detailed historical, ethnographic, and archaeological record. This evidence will give us many opportunities to test and potentially falsify our theory in detail. In fact, we will find that the correlation—in space and time—between the development of advanced gunpowder weaponry and the emergence of the modern world is striking and very extensive, as our theory predicts.

Beginning at the end – sophisticated guns and the emergence of the modern world

It is useful to begin where our discussion here and in Chapters 15 and 17 will end—with the emergence of the full panoply of features of our world over the last few hundred years. This first, quick look will provide an overview of the compelling correlation between gunpowder and the modern world. This synopsis will ease our journey. Gunpowder weapons first reached substantial technical refinement in Western Eurasia. The first modestly effective gunpowder artillery exploded into this world in the early 14th Century. As these weapons matured, they gradually battered down and profoundly altered the last Eurasian archaic states through the 16th Century.

By the 17th Century, the personal forms of these weapons (muskets, rifles, and pistols; handguns or *guns*, for short) were refined so that they became relatively cheap, reliable and easy to use. This process of improving guns then continued through the 18th Century and beyond, into the contemporary world.

These cheap, reliable guns were decisive. Whoever controlled them controlled everything else. For example, artillery pieces are large weapons, requiring vast resources to produce and use. Who these weapons are aimed at is determined by who controls the underlying social resources. And this control, in turn, is determined by who controls the personal gunpowder weapons, the guns.

This property will prove to be good news/bad news for our world. The good news is that these individual weapons can place decisive coercive threat in the hands of vast, majority coalitions. These coalitions produce wealthy and relatively humane democratized states, protecting the interests of most of their members in the best cases. We will call cases resembling this scenario *modern democratic states* (Chapter 15).

The bad news is that production of these individual weapons has become highly sophisticated. In economic jargon, gun making is now capital- and technology-intensive. This means that small subgroups can potentially gain exclusive control over these weapons. Elite control of this coercive technology produces the *modern authoritarian state*, on our theory. Such states preferentially serve the interests of the small elites who control coercive power—inevitably and everywhere.

When gunpowder coercive weaponry first became potentially broadly (democratically) distributed, this represented a radical redistribution of threat compared to archaic states. The consequence of this redistribution was that *commoners* sometimes took control of the cooperative enterprise and forced it to serve their interests. This results from the inevitably self-interested use of coercive threat by these self-policing *electorates*, on our theory. When this happens, massive groups of individuals are now served and incentivized by their social world. The wisdom of this massively enlarged crowd (Chapter 10) and its capacity for cooperative undertakings are utterly revolutionary (Chapter 17). This democratization created our astonishing modern world, we will argue.*

For now we begin by emphasizing the spectacular, and predictable, temporal and spatial correlation between broad access to reliable guns and the rise of the modern economic world. As we will see in Chapter 17, emergence of other elements of modernity, including the Scientific Revolution, show a very similar spatial and temporal pattern, as expected.

Figure 14.2 illustrates the correlation in time and space between the wide availability of gunpowder weapons and new wealth generation. Democratized control of coercion emerged piecemeal across the global landscape beginning in, roughly, the mid 17th Century and is continuing through the present instant. There is a remarkable, yet predictable correlation between access to a novel means of coercion and the ensuing new adaptive revolution. After five thousand years of elite-dominated archaic states, decisive local access to coercive threat was returned to the hands of the majority in a few locations beginning about 350 years ago. Rapidly following this redeployment of power—and in precisely the places where it first occurred—the modern world exploded into the historical record (Figure 14.2). Empirical tests of theories of history rarely get better than this.

Our goal through the remainder of this chapter and Chapter 15 is to explore the historical record of the gunpowder-produced emergence of the modern world. In Chapter 17 we will return to other details of this story. We will look more carefully there at how democratization of access to coercive threat can produce more humane societies—with all their knowledge, wealth, and technical sophistication. These additional insights will be vital, indeed, as we turn to the human future.

---

*   Economic and scientific progress by modern authoritarian states is ultimately parasitic on democratized states, we will argue (Chapter 17).

Box 14.1: The fallacy of the bullet: Coercion and social peace

If we were to excavate a graveyard in any American or Western European city, would we find people who died of bullet wounds? Only very rarely. When we read a law passed by the parliament or congress of a contemporary democratic state (Japan, England, Australia, or the US, for example) does the law stipulate that "police armed with gunpowder weapons shall be deployed as required to apprehend fugitives and to defend the courts from intimidation"? Again, almost never. We might naively imagine, therefore, that the guns and bullets of contemporary police are secondary or unimportant. If this is our logic, we have committed the "fallacy of the bullet."

If we imagine that our economic and social systems would function for even a day without the coercive support of armed individuals such as some combination of professional police and an armed citizenry, we are fools, indeed. Without armed enforcers, merchandise and currency would be immediately looted,* indicted individuals would not show up for court dates, judges and juries would be intimidated or killed and so on and so on.

We might be thinking that this is a ridiculously grim picture of human society. We might say that most people are honest and the system would continue to work without armed coercion. If this is our logic, we would be *right* about most individual people in the short term and *completely wrong* about our social system at equilibrium. Even if only a few people "beat the system" on the first day after the armed police were gone and did better as a result, on the second day more would adopt this now-successful strategy and more still on the third. Ultimately all of us would be forced to steal and evade "the system" to feed our families and to survive.

We take the coercive umbrella under which we function so completely for granted it is like the air we breathe—essential but assumed, all-permeating but invisible.† When we exert our political power as

---

* Indeed, currency would rapidly cease to exist as un-policed counterfeiting would drive its value to zero.

† To get a better appreciation of this, do a Google image search with "courtroom." Click on the images of *occupied* courtrooms. Notice the armed woman or man standing or sitting discreetly at the margins of the court's events.

voting members of a democratic state or through many diverse non-governmental organizations, we unconsciously assume the presence of the coercive threat to enforce the policies and practices we support—almost as surely and obliviously as we presuppose tomorrow's sunrise.

Thus, if we look at the written record of our societies' great debates and billions of mundane contracts and policy decisions, we will find almost no record of the coercive means that underpin them all. If we had no theory or awareness of ultimate causation, would all these words and actions lead us to the role of coercion? Probably not. Might we be vulnerable to being misled and confused about the ultimate origins of our societies as a result? Absolutely.

Keep this in mind as we turn attention to the history of the emergence of the modern state. Many actors in these proceedings will not mention coercive policing. Instead they will talk about the same kinds of matters you and we might debate—economic policy, religious belief, and the lot. Can we be fooled into misunderstanding the causal connections between all this local *cultural activity* and the course of human history? We often are. However, our theory apparently gives us the opportunity to see through to the authentic causal relationship here and in all human societies.

Finally there is one last useful implication of the fallacy of the bullet. Without armed coercive enforcement, there is no human society—not anywhere, not at any scale, not at any time, ever. This is an unavoidable prediction of our theory and a readily falsifiable one. No one will ever find any human society anywhere that exists without coercive weaponry in the hands of some or all of its members. These weapons might be bows, atlatls, guns, throwing stones, or something we have not yet thought of, but they will always be there, no exceptions. This is a powerfully specific claim and we invite skeptical readers to scour the world, history, and ethnography to find an exception. Again, we predict that we never will—not a single case.

Wait! Something's wrong

"Never apologize, never explain." There are many attributions for this quote,[*] but our favorite is that it is advice given by the nobility of hierar-chical states to their youngsters. It is a description of how the powerful behave. Whatever its origin, the quote captures something important. People who overtly threaten physical violence are goons, thugs, banty roosters, NOT the *authentically* powerful. The truly powerful do not need to threaten. They exude a *natural authority*.

Does this not then say that threat cannot really be the source of political power in well-functioning societies? It says no such thing. Recall once again that our minds are *proximate* devices. They are designed to cause us to behave *as if* we understood the strategic logic of our behavior, *not* to actually understand that logic. For almost all of our two-million-year history true power came from a broad, well-recognized consensus (Chapters 5 and 10). Once that consensus had been negoti-ated in public, everyone was onboard. Either they agreed or they were in a small minority with no choice but to submit. Either way, consensus was completion. Power was now "natural."

Under these consistent, ancestral conditions the only time coercive violence might have been overtly threatened in discussions of policy was *during* the debates *leading to* consensus. Emerging interest groups might have projected their numbers and threat, seeking to recruit members and settle the issue in their favor. But once consensus was reached, the locus of true coercive power was unambiguous.

Thus, our evolved unconscious minds interpret *overt* projection of threat in political dialog in a very specific way. We think that someone who actively threatens is **STILL NEGOTIATING.** Such a person is *poten-tially weak*, not obviously strong. Our conscious minds do not have access to the origins of this unconscious intuition. These minds tell us threatening individuals are not *truly* powerful. We then draw the flawed conscious inference that coercive threat is NOT political power. On our theory, this is a profound mistake. Coercive threat deployed in pursuit of conjoint self-interest is the only real, *ultimate* source of political and social power.

---

[*]   Bartlett's give credit to the screenplay for the Hollywood film *Tie a Yellow Ribbon*. 1949 film *She Wore a Yellow Ribbon*.

Conjoint coercion – coalitions converge and coalesce

Humans achieve startlingly low costs of coercion by participating in coercive acts *in concert with others* sharing confluent self-interests (Chapter 5). However, no two individuals ever have *perfectly* confluent interests. Thus, part of the two-million-year old art of being human is to form and influence coercive coalitions that serve our self-interests as well as they can in spite of partially divergent interests among their members. Our minds are exquisitely adapted to this task (Chapter 10). How these processes play out in the modern state will be crucial to us.

First, we seek out sets of others whose interests are as similar to ours as possible. These *interest groups* give us potential access to political power—credible coercive threat—in proportion to their size and to their *access to decisive means of coercion*.

Second, we have evolved to fulfill the ultimate goals of replicating our design information (Chapters 3, 6, and 10). We are otherwise agnostic about how we accomplish this replication. We, therefore, unconsciously compromise in response to powerful coalitions, while simultaneously influencing those coalitions to the extent we are able.

These first two effects lead to local human cultures coalescing into large powerful interest groups whose members project coercive threat (often implicitly) in pursuit of their mutual interests.* In the context of the modern world, we thus form our stable local social units.

Modern armies, including militias, will be important to us a little later in our exploration (Chapter 15). For now, notice that they represent heavily armed interest groups embedded within a larger population. Gunpowder weapons professionally deployed make armies extremely effective at coercing cooperation among their members. Military discipline is fierce. Thus, these modern armies can be extraordinarily powerful war fighting social units, as our theory predicts.

Moreover, when a modern army possesses disproportionate access to coercive means, relative to the surrounding population, its members can extract resources from that larger population. This army

---

*   We often feel that we project political threat through our *ballots, not bullets*. We argue that this view mistakes proximate signals for ultimate causes (Box 14.1). Our ballots having meaning only as proxies for our bullets (coercive threat). This is an inescapable implication of our theory and fits well with the historical record, as we will see (Chapters 15 and 17).

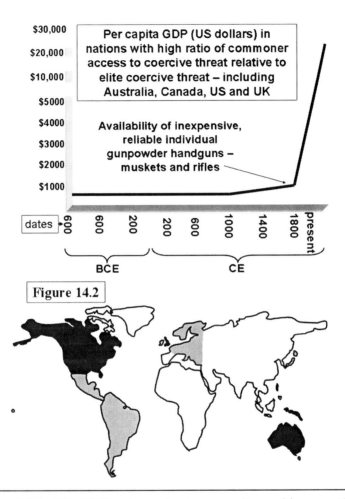

Figure 14.2

Figure 14.2 : TOP PANEL: Approximate per capita wealth generation (in 2000 CE American dollars) for the first portions of the world to show modern economic growth. Note how abrupt the onset of modern growth was and the striking correlation in time with the availability of gunpowder handgun technologies permitting commoner coercive dominance. Data from Maddison (2001) including page 42. Figures before 1000 CE are projections.

BOTTOM PANEL: Shown ***in black*** are regions of the world showing robust annual growth (roughly 2%) throughout most of the 19th Century. These regions are also those with substantial commoner access to coercive power through command of gunpowder handguns (text). ***In gray*** are regions of the world in which commoners gained partial, but significant access to coercive threat and which showed smaller, but significant economic growth throughout most (Western Europe) or the later parts (Latin America and Japan) of the 19th Century. The rest of Africa and Eurasia ***in white*** showed little commoner access to coercive power and extremely limited economic growth during this era.

can also conscript new members essentially at will—*military discipline* again. Given the highly plastic nature of the human ethical response, new conscripts quickly become dedicated to the goals of this army. As expected on our theory, armies of this sort dominate the history of huge swathes of the planet throughout much of the modern era (below and Chapter 15).

Third, the maximal size of our interest groups is limited by the coercive technologies available to manage conflicts of interest *within* them as we discussed at length in Chapters 11 to 13. In the contemporary world, we possess coercive means with ranges encompassing vast territories. Thus, our *interest groups* (our MPUs[*]; Chapter 12) can, under the right circumstances, grow large, indeed.

Fourth, because of the compromises we make as members of massive contemporary coalitions, we continuously seek, often unconsciously, to form subgroups within larger coalitions, potentially allowing us to pursue self-interest more effectively. Thus, massive modern states consist of many potentially competing subgroups.

This conclusion is not at all surprising to us intuitively. We are experienced denizens of such groups and subgroups. However, members of these subgroups always seek (again, often unconsciously) ever increasing coercive dominance over other subgroups, facilitating more effective pursuit of self-interest. Their quest for coercive dominance can even involve the seizure of coercive assets (weapons), seeking monopoly access where feasible.

Fifth, when a subgroup achieves coercive dominance, the individual members of this group have a strong interest in limiting further membership, thereby preserving their individual advantage and access to resources while projecting coercive dominance to manipulate non-dominant subgroup members. Of course, this is precisely the pattern we have already seen with archaic states (Chapter 13). However, the weaponry of the modern world produces several new effects as follows.

Modern Effect A: Contemporary gunpowder weapons make it feasible to police conflicts of interest *within* elite subgroups very effectively. This makes modern states potentially stable, in contrast to the instability produced by *tragedies of the elite commons* in archaic states. Modern states do not cycle (Chapter 13).

Modern Effect B: Modern weapons are much more effectively deployed in large numbers than the weapons of the archaic state. Thus,

---

[*]   "Maximum policeable units." (Chapter 12).

the coercive power of selectively armed elite sub-groups can become far stronger, even utterly overwhelming.*

<u>Modern Effect C</u>: With redeployment of or new access to powerful gun-dependent coercive means, political change can become blindingly rapid, often involving catastrophic short-term violence before a new coercive equilibrium is established. Such dramatic political change is what has traditionally been described as *revolution*.

Sixth, as the modern era unfolds, weaponry matures and increasingly provides *potentially* universal access to locally decisive coercive means. Thus, if and when most of the major interest subgroups within a large state-like aggregate acquire *comparable* access to coercive assets, the best available option is for all their members to reach some accommodation with other interest subgroups. No other coercive relationship is stable. This is *pluralistic democracy* on our theory.

In hierarchical polities, elite individuals seek to maintain and enlarge their control over coercive assets. Non-elite individuals seek to subvert elites and gain control of coercive assets for themselves—all subject to *rational calculations* of costs and benefits.[†] The tense, brooding peace and explosive political violence *within* hierarchical states—so characteristic of the last several centuries—are the products of this single relentless, dynamic process on our theory.[‡]

---

*   An implication of the generalized form of coercive square law, again (Chapter 5).

†   Of course, these calculations of self-interest (mostly unconscious) are executed by a proximate psychology that evolved in the ancestral past. These computations can, thus, sometimes be in error in the potentially *novel* contemporary environment. However, the evidence suggests that these calculations are often roughly accurate (subject to inevitable uncertainty and inadequate information) even in the contemporary environment.

‡   Notice again that the belief systems associated with revolutions are not their causes, on our theory. These beliefs are merely the products of negotiation of social contracts (and identifier information) within and among interest groups (Chapter 10). For example, the French Revolution was not about "liberte, equalite, fraternite." It was about the struggle of diverse interest groups for coercive dominance. Beliefs have secondary historical relevance only to the extent (sometimes very limited) that they are actually implemented by the coalition that ultimately wins the competition for coercive dominance. As we will see in Chapters 15 and 17, democratized coalitions are more likely to implement some of the social contract beliefs they negotiate during seizure of coercive dominance. This can be very good news, indeed, for the human future.

A new weapon comes to planet Earth

While we humans are much more intelligent that non-human animals (Chapter 10), we are not nearly as clever as our conceit would have it. For example, the atlatl or spear-thrower is a simple weapon, and apparently an effective agent for social coercion (Chapter 11). But it probably took humans around 1.7 million years to "discover" it.

Gunpowder weaponry is a similar story. It does not require any deep theoretical knowledge of chemistry to discover that the appropriate mixture of sulfur, charcoal, and saltpeter (nitrate salts) will explode if ignited. Moreover, once this combination is understood, the metallurgical skill necessary to build early artillery and muskets is routine iron age technology. In spite of this simplicity, gunpowder explosives apparently were not developed until about 1000 CE in what is now China. The technology was probably transferred to Europe between 1200 and 1250 CE, perhaps through the Islamic world. The initial development of more advanced artillery and guns by people in Western Eurasia occurred between approximately 1350 to 1500 CE.[7] Moreover, the then more advanced civilizations of China and the Islamic world initially lagged in developing this technology, probably as a result of short-sighted self-defense by members of stably entrenched elite interest groups.

Of course, our theory predicts that the historical accident of the early refinement of these weapons in Western Europe is the reason that all five continents today speak Western European languages, in part, and that three of the five continents speak them essentially exclusively. This cultural dominance is a product of the vast new stability and productivity of the modern (gunpowder-generated) state arising in Western Europe first.[*]

For now, our concern is the detailed properties of the weapons and how they were used to expand the scale of the ancestral human adaptive trick—individually self-interested coercion.

The world that gunpowder coercion wrought – Part 1

It is useful to group gunpowder weapons into two largely non-overlapping categories, artillery and personal weapons (guns). By artillery we will mean arms too large and too expensive to be carried as personal

---

[*]   The possibility that gunpowder's development in Western Europe was crucial to the age of European dominance has long been suspected (see Diamond, 1997, for a recent popularization). We argue that our theory is the first credible *causal* explanation for this empirical correlation (below).

weapons by an individual. Cannons, howitzers, and most mortars are examples.

Artillery has come in many forms over the approximately 650-year history of advanced gunpowder weaponry—from gargantuan bombards designed to hurl huge stones against castle walls[8] through the mobile artillery of early modern armies to the sophisticated field pieces of contemporary state militaries.[9] These weapons are expensive to build, provision and maintain. Only highly trained experts can operate them effectively. Artillery is cost-effective against infrastructure and massed infantry formations, but not against dispersed individual fighters.

Guns (personal, individual weapons) have likewise come in many forms. Early guns were, like artillery, expensive to build, provision, and maintain. These early guns also required considerable expertise. However, they gradually became cheaper to make and easier to use and maintain.

As a result of this developmental sequence, early in their history, guns were only available to coercive elites. However, over the last few centuries, highly effective guns have become readily affordable and accessible to essentially any member of any state where manufacture or sale of these weapons is not preempted by previously existing concentrations of elite coercive power.

To appreciate fully how revolutionary this widespread access is, recall from Chapter 13 the fictional warrior waving his enormous sword around. He was killed effortlessly, as a minor distraction, by Indiana Jones with his gun. Had Jones been armed with a bow instead and had this warrior been wearing archaic state body armor, Indiana, not the warrior would have perished.

In contrast, suppose that the warrior *had* been wearing archaic state body armor in the movie scene as it was actually played. The outcome would have been the same—death of the warrior, not Jones. Indiana's gun is the crucial difference.

By about the late 15th Century, gunpowder handguns were so powerful that any realistic body armor then available could often be penetrated by their bullets.[*] Thus, with the development of effective handguns, the game of social coercion became about numbers not

---

[*]   Indeed, some of the dramatic suits of armor that are favorite displays in the world's museum are so ostentatiously huge because they were attempts—ultimately unsuccessful—to render late Medieval European knights resistant to gunpowder projectiles. When you look at a museum armor display, note the dates. You will find the "big" armor mostly dates from the 15th and early 16th Centuries when guns were gaining coercive traction.

elite body armor. Moreover, as guns became cheap and widely available, most or even all people could potentially count in these coercive numbers.

The world that gunpowder coercion wrought – Part 2

We now have what we need to be very precise about what our theory predicts about the unfolding of this modern world.[10]

Late archaic states are the environments in which gunpowder weapons originally arose. Thus, the initial expensive, elite forms of these weapons should provide the means for controlling conflicts of interest *within* the elites of these states in ways that they could not be controlled before (Chapter 13). Individual hand-to-hand combat was no longer decisive—firing a gun or an artillery piece from a distance was now the meaningful event. Death from a distance, again, became possible and effective *within* elite interest groups. Early gunpowder weaponry should and did transform chronically unstable archaic states into much more stable "cooperative" social and economic enterprises. We call these enterprises examples of the *early modern state*.

Several things are important about early modern states. For example, defensive positions (castles) and the equivalent, of recalcitrant local elites became more vulnerable to reduction by gunpowder artillery. More importantly, both guns and artillery made elite policing both of other elites and of commoners outside of these isolated defensive redoubts much more cost-effective, as well. This means that day to day operations requiring elite cooperation, with commoner support, became much more reliably sustainable. Among the many different examples of such cooperative operations are the encircling and starving out of defensive redoubts. Again, it no longer paid to be a local elite holdout. Only "cooperative" state-level elite domination would work.

In addition to the material costs of early gunpowder weapons, the expertise required to use them—opportunity costs, again—meant that they were still the exclusive domain of small elites. This is quite analogous to elite body armor and shock weaponry in earlier archaic states in this one respect (Chapter 13). Thus, we expect the early modern states to still serve the interests of these armed elites. Put differently, we expect early modern states to be just as intensely hierarchical as archaic states, but with improved coercive power and increased management of intra-elite conflicts of interests. Early modern states are predicted to begin life as something like *archaic states from hell*.

Thus, though the early modern states were *adaptive revolutions*, their increased performance was still only a relatively small improvement on archaic states. They increased the scale and stability of management of conflicts of interests, but the majority of individuals (commoners) could still not protect their own interests and tended to be surly, economically dysfunctional targets of elite domination. The big adaptive pop from gunpowder was yet to come (Chapters 15 and 17).

As guns became cheaper and easier to use, a new adaptive strategy became available to commoners. Rather than submitting to elite coercion, commoners could sometimes manage to arm themselves with guns. So armed, commoners inevitably used their new coercive power. Commoner seizure of the *state* is predicted to follow.

This new commoner trick was expected to have profound consequences. Human societies were always *democracies of the decisively armed*, on our theory. That is, humans have always had *only* those rights and privileges they could coercively defend—everywhere and always. With the full development of the handgun, individual coercive power was now *potentially* returned to its ancestral universal distribution.

What can we "know" about the past?

Achieving a detached scientific grasp of our world will challenge us. Our recent history is the source of the *social contract* and *identifier beliefs* (Chapter 10) every one of us, authors included, have held dear. We will be looking *directly* at ourselves; our immediate families; the founders of our cultures, religions, and political systems; our fellow countrymen; and our mortal enemies. To see all these people with objective detachment will involve repeated, sometimes jarring shifts in the perspectives we take for granted.

Indeed, much of what we think we "know" about history is actually not knowledge. Rather it is a set of social *beliefs* masquerading as knowledge. Our evolved proximate social psychologies tend to confuse and defeat our analytical capacities.

None of us is completely free of this confusion. However, by collaborating and sharing our different perspectives on a global scale, we can help one another correct our local provincialisms.

Indeed, if we do not find our historical beliefs rattled and even, sometimes, destroyed utterly by a new theory of history, we can be very confident that such a theory is inadequate.

Gunpowder weaponry comes to the global natural laboratory

Our next step must be to ask whether we can falsify these precise predictions of our theory on the historical, ethnographic, and archaeological records. This record is massive and multi-faceted. We cannot possibly review it all here. As with the archaic state in Chapter 13, we will review a few specific examples—chosen to test the theory and illustrate its simplicity and power. We emphasize that we investigated the record of gunpowder's effects on human social units extensively and we know of *no cases* that contradict the predictions of our theory. Of course, the cases we choose not to discuss remain available for skeptics to attempt to falsify the theory.

We will proceed with a few examples of the dramatic changes in human social cooperation, adaptive revolutions, produced by gunpowder-based coercion as follows.

The social effects of gunpowder weaponry *and* the continued evolution of the weaponry itself occur concurrently in early modern Western Eurasia. This complexity can sometimes obscure the causal processes over time, somewhat complicating our task. Thus, it is simpler to get our empirical sea legs elsewhere and then return to Eurasia.

As the early modern states of Western Europe consolidated as a consequence of gunpowder coercion, their new adaptive power allowed them to extend their coercive, and thus economic, reach.[*]

Therefore, these early modern states sometimes introduced well-developed gunpowder weaponry into "pristine", still archaic state settings as they expanded. Members of these recipient archaic state cultures suddenly—literally overnight—found themselves confronted with fully developed, sophisticated new coercive means. These cases are elegant natural experiments. They give us beautiful opportunities to test our theory against the empirical evidence.

In the following three sections we discuss three specific cases of pristine introduction of gunpowder weaponry. These specific cases were chosen for two reasons. First, the evidence is strong and detailed for each. They are empirically well attested. Second, each of the three cases is as completely independent of the other two as it can possibly be. These three archaic states into which gunpowder weapons were introduced *de novo* share no common cultures, languages, religious beliefs, subsis-

---

[*]   As always, *states* do not do these things, strictly speaking. Rather, self-interested collections of individuals within states are the actual agents—often exploiting the coercive and economic resources sustained by the coercive umbrella of the state environment to pursue their private goals.

tence technologies, or any other identifiable cultural antecedent. One is a Polynesian archaic state, the second is a New World archaic state, and the third is an island East Asian archaic state.

As we will see, in spite of the differences in cultural details, each of these archaic states is transformed in the same way by the coming of gunpowder weaponry—precisely as our theory predicts.

"The end of the world as they knew it" Gunpowder coercion transforms the last archaic states Case 1: The Kingdom of Hawaii

In the opening vignette to this chapter we saw the role of gunpowder weaponry in the consolidation of the Hawaiian kingdom. This temporally compressed example of the formation of an early modern state let us see some of the essential features of the process clearly.

The elite warriors who controlled the endlessly cycling archaic statelets of the pre-gunpowder Hawaiian archipelago (Chapter 13) had the same psychology and self-interested goals after they possessed gunpowder weaponry as before. However, because of gunpowder's superior capacity for magnifying the coercive power of a numerical advantage, the strongest of these chiefs was immediately able to consolidate a larger state, the *Kingdom of Hawaii*.

Notice that the gunpowder-enabled advantage is just another way of saying that the individuals in this numerically superior interest group were now capable of enforcing their own interests and keeping one another in line. Elite conflicts of interest became manageable in a way they never were before.

The initial stage of consolidation involved several events that can be called *wars of conquest* as these new powerful elites expanded their control. However, one particular feature of this story is crucial. One of the archipelago's major islands, Kauai, was added to Kamehameha's kingdom simply as a result of the *prospect* of war, by coercive *threat*.

Kamehameha's state was still intensely hierarchical, with underchiefs who were the local administrators and second-order economic beneficiaries of the resources of the state. These lesser chiefs were a constant threat to rebel, fragmenting the kingdom again. However, this fragmentation never occurred. A few violent local police actions against rebels were necessary as demonstrations of overwhelming coercive power early in the history of the kingdom, but needed to be only limited and transient. This is a very different scenario from the constant and almost daily public intimidation of the earlier archaic state phase of

Hawaiian culture (Chapter 13). The new Hawaiian state remained stable thereafter through its formal incorporation as the 50th US state in the mid-20th Century.

The vital lesson from these details is that the early modern state is not predominantly a product of war in the sense that some military historians have traditionally believed.[11] Rather what is called war in the early consolidation of a state is better thought of as the overt demonstration of the *credibility of coercive threat*. The ongoing presence of this umbrella of threat, year after year, is what creates and sustains the state. Two million years of evolutionary adaptation to such social settings have shaped us to contribute and respond to such coercive umbrellas with virtuosity (Box 14.1; Chapter 10).

It is not possible to grasp our theory's account of the modern world without fully internalizing this point. If we are distracted by brief, catastrophic occurrences of war, we are focusing on a tiny, idiosyncratic element of a much larger and more important process. To say that overt war creates the state is tantamount to saying the law enforcement consists primarily of those rare moments when a police officer actually shoots a suspect. It fundamentally misconstrues the issues (Box 14.1). Our attention is distracted by minute moments of noisy gore rather than focusing on the larger causal picture.

Finally, returning to the Hawaiians, they did not have the capacity to manufacture gunpowder weapons during the era in question. They purchased them at high prices from European merchants. Thus, only elite individuals could initially acquire them. Even though these weapons were cheap and potentially widely available in the larger world of the colonial powers, they were expensive elite weapons in 19th Century Hawaii. As expected, they sustained the continuing domination of Hawaii's warrior elites.

Unfortunately, from a scientific point of view, this elegant natural experiment in early modern state formation was destroyed after just a few generations. The kingdom of Hawaii was quickly pulled into the emerging global economy driven by more powerful European colonial powers. The kingdom was too small and poor to resist the overwhelming coercive dominance of the larger British, French, American, and Spanish states. Their massive warships began frequenting Hawaii's harbors in the late 19th Century, making their insistence salient. When the Imperial Japanese attacked the American base at Pearl Harbor Hawaii 145 years after the founding of the kingdom, the American aircraft carriers were larger than a typical ancestral Hawaiian village and the destroyers had artillery pieces capable of reducing any point on the Islands to dust.

"The end of the world as they knew it" Gunpowder coercion transforms the last archaic states
Case 2: The Aztec-Spanish state of New Spain

In his wonderful book *Conquistadors,* Michael Wood quotes the German artist Albrecht Durer's reaction to Aztec materials returned to the court of Charles V in 1520 by Spanish conquistador Hernan Cortes.

> I saw things which have been brought to the King from the new land of gold, a sun of all gold a whole fathom broad [6 feet], and a moon all of silver of the same size, also two rooms of armour of the people there, with all manner of wondrous weapons, harness, spear, wonderful shield, extraordinary clothing, beds and all manner of wonderful objects of human use, much better worth seeing than prodigies…..All the days of my life I have seen nothing that touches my heart so much as these things, for I saw amongst them wonderful works of art and I marveled at the subtle ingenia of men in foreign lands.[*]

The Aztec civilization was an archaic state in full bloom (Chapter 13). It had artistic and technical traditions that rivaled those of the contemporary early modern states of colonialist Europe. To Durer, this meant wonder, enlightenment, enrichment. To the more worldly members of European elites they meant wealth—for the taking.

Cortes landed on the east coast of what is now Mexico in 1519. He was fully prepared to take extravagant risks in pursuit of potentially huge rewards—ultimately adaptive rewards in the form of extra mates and vast quantities of resources.[12] He was a member of the growing Spanish presence in the Caribbean basin. He had reason to believe, from earlier, furtive contacts, that a great and wealthy civilization might lay inland.

Cortes was *licensed to kill* in pursuit of conquest for his king and in spreading of the faith, the Spanish elite's social contract and identifier belief systems. He was also possessed of a quick, subtle political mind. Through interpreters, he discovered that local provincial cities greatly resented domination by elite Aztec warriors—quite predictable in an archaic state. He understood that these locals could be recruited as allies to conquer the wealthy inland center of the state.

What was required was to show these potential collaborators that they could win if they threw in their lot with the Spaniards. To parade his superior coercive power, Cortes exploited his gunpowder weaponry

---

[*]   Quoted in Wood (2000), pages 15-16.

and other assets to defeat, and in at least one case massacre, some local Aztec subjects. As good humans, the balance of the locals understood the implications of overwhelming coercive threat perfectly well.

Moreover, it was not the actual guns and tools of war possessed by Cortes that were the major message. To understand what we mean, suppose an extraterrestrial named Greep showed up tomorrow with a ray gun side arm and proceeded to melt the Eifel Tower with it. Would our primary concern be that little weapon, itself? Hardly. The issue would be the alien civilization implied by that weapon—and their interests and intent. The Mesoamerican locals meeting Cortes were just as smart as we are and they reacted to his weapons as we would react to Greep's. The point of Cortes' demonstrations of power was taken and the Spaniards had the indigenous foot soldiers they needed to march on the Aztec capital of Tenochtitlan.

The Spaniards were initially welcomed by Moctezuma II, the reigning Aztec emperor. However, Cortes and his men soon took Moctezuma captive, demanding and receiving extravagant ransom. While initially the Spaniards were driven out, they returned some months later with more native troops and successfully demolished the Aztec capital, seizing everything of value they could carry away.

The battle for Tenochtitlan is illuminating. The Spaniard's artillery and guns were certainly useful. However, the bulk of the fighting was done by the Spaniard's local collaborators with their own archaic weaponry. These local collaborators fought with their own weapons in knowledge that they were on the winning side—a product of the Spanish state's vast resources *as well as the coercive environment created by the Spanish weapons*. Again, the capacity of gunpowder coercion to sustain the "cooperative" state apparatus is just as important, or even more important, than the weapons used in any specific local act of war.[13]

However, the Spaniard's Aztec collaborators, as expected, would ultimately be sold out. They did not actually hold coercive power; they merely fought under its umbrella. The European gunpowder weapons were manufactured elsewhere and remained in the nearly exclusive control of the Conquistadors. They used their coercive advantage to seize the Aztec state, including its commoner work force, and simply replaced the Aztec elite with themselves.

As with the Hawaiians, these initial acts of overt warfare by the Spaniards were merely the opening coercive gambit, not the definitive event. Cortes the conqueror had to put down indigenous conspiracies, both real and imagined, over the next few years. Equally importantly, in the wake of Cortes' conquest, armies of Spanish immigrants followed—all supported by Spanish gunpowder-armed troops. The Spanish state, itself, was stably integrated through gunpowder policing. Its citizens cooperated to mount vast projects, taking over the Aztec agrarian economy and building large new enterprises to grow and export sugar, coffee, chocolate, and other products.

This sustained gunpowder coercive umbrella was owned by the Spaniards and they used it to shape the new Spanish-Aztec early modern state in their interests.[14] A few elite Aztecs—including females of royal blood married to Spanish males—benefited from this new state. But most surviving Aztecs lived as peasants in this hierarchical new world, not unlike their role in the pre-Spanish Aztec state. The intimidation of public execution by Aztec priests (Chapter 13) was replaced by the public intimidation of firing squads—"meet the new boss, same as the old boss" once again.

As similar as this new hierarchical early modern state was to its archaic precursors, it differed in one salient way. Gunpowder weaponry allowed the persistent management of elite conflicts of interest. This new Spanish-Aztec early modern state was quite unlike its unstable Mesoamerican ancestral states (Chapter 13). It endured. The racist, hierarchical organization of this state continues to reverberate through its present in contemporary Mexico and Guatemala.

"The end of the world as they knew it" Gunpowder coercion transforms the last archaic states
Case 3: The Tokugawa Shogunate of Japan

The early modern states of Hawaii and Mexico are very useful in understanding the immediacy of the impact of gunpowder and the social consolidation it supports. However, these two cases are limited as empirical evidence by the extent to which the resulting states quickly become enmeshed in the larger global economy. Thus, we can see the initial effects of gunpowder in isolation, but not its ongoing effects.

In contrast, our last case lets us see these longer-term consequences of gunpowder. This third archaic state, Samurai Japan, took in European gunpowder weapons, rapidly making them their own. They then relentlessly excluded almost all other outside influences for the next approximately 250 years. This gives us a spectacularly robust natural experiment.

Japan was occasionally, transiently unified between the coming of iron around 400 BCE and the coming of gunpowder weaponry in the 16th Century CE. Indeed, Japan developed an ostensibly common culture, language, and religious tradition—even a common ruler in the form of the emperor.

However, in reality, Japan was almost always composed of sets of small local archaic statelets, occasionally unified briefly by conquest, only to fragment again—classic archaic state cycling (Chapter 13).[15] Indeed, the period just before the introduction of gunpowder is referred to in traditional Japanese history as the *warring states* era.

However, in 1543, a European colonial trading ship ran aground on the small island of Tanegashima just off the southern coast of the large island of Honshu.[*] The Portuguese sailors demonstrated their firearms. These included artillery pieces and arquebuses, the earliest well-developed gun or musket. Tanegashima was part of the warlike Satsuma statelet whose warriors immediately appreciated the unprecedented opportunity these new weapons presented.

Similar to the Hawaiians, the Japanese initially traded for European weapons. However unlike the Hawaiians, the Japanese were skilled metallurgists and chemists by the standards of the era and quickly began manufacturing their own gunpowder weapons. By 1575 the battle of Nagashimo was won by the side having two thousand arquebusiers. Gunpowder was firmly entrenched in Japanese combat technologies.[16] Arquebusiers came to be the most numerous component of late 16th Century Japanese fighting forces. These weapons swept away the one-thousand-year-old dominance of body armor and shock weapons in elite Japanese coercion.[†] Our theory predicts that a specific, dramatic social change should ensue.

Within a few years, these weapons were relatively widely distributed among Japanese elites. Ambitious, self-interested political entrepreneurs began consolidating coercive coalitions, incorporating both traditional Samurai cavalry and large numbers of gun toting soldiers.

A first major step toward unification was engineered by the armies of Oda Nobunaga, who unified central Japan around 1560, just seventeen years after the first coming of Portuguese weapons. Nobunaga was killed in 1582, forced to commit seppuku after being captured by a rival. His successor, Hashiba Hideyoshi inherited Nobunaga's gunpowder-armed forces.

Hideyoshi's forces continued to expand their domain of control at the expense of weaker rivals. In 1588 Hideyoshi took a most interesting

---

[*]   Tanegashima island is the site of the contemporary Japanese spaceport.

[†]   As Japanese commoners would remain almost entirely disarmed during and after this period, the Samurai sword would remain a feared weapon to commoners, but no longer to elites with access to gunpowder weaponry.

step.* He issued an ordinance requiring all weapons possessed by commoners or by samurai not associated with his fighting forces to be confiscated—Hideyoshi's famous *sword hunt*. This began a policy of systematic concentration of coercive power in the hands of Hideyoshi's forces. This policy would continue for centuries to come, setting up a strong feedback process concentrating decisive coercive power in the hands of what came to be called the *Tokugawa* elite.

Hideyoshi was succeeded by Tokugawa Ieyasu who ultimately killed Hideyoshi's young son, and intended successor. By the great battle at Sekegahara in 1600 and the shelling of the castle as Osaka in 1614 the issue was settled. Japan was unified under Ieyasu's forces. An early modern state, referred to as the *Tokugawa Shogunate* in traditional Japanese history, was established, and would persist as a stable entity for the next 250 years!

There was limited, and generally local, civil strife during the Tokugawa era, but no real threat to the stability of the state, until later (Chapter 15). Of course, this is precisely the outcome our theory predicts. Gunpowder weapons should allow stable management of elite conflicts of interest and end cycling. The fantastic speed of the unification of the Tokugawa state with the introduction of gunpowder weapons, after more than a millennium of cycling, is rather dramatically consistent with the predictions of our theory.

Moreover, the Tokugawa state has some other important lessons to teach us. First, given that gunpowder weapons were monopolized by the Tokugawa elite, members of this elite are predicted by our theory to possess coercive dominance and to use the resulting state to serve elite interests as they, in fact, did.[17]

This dominance included rather substantial rates of commoner taxation and very limited levels of economic growth.[18] This economic profile will be absolutely vital for us in Chapter 15. It is one of the many pieces of evidence that *peace and stability* are *not* sufficient for *modernity* or even merely for healthy economies. We will need something more if we are to explain the *modern economic miracle*.

Second, in spite of the fact that the Tokugawa Shogunate was ever further surpassed in per capita wealth by contemporary European states, this early modern Japanese state remained stable. Indeed, the Tokugawa social contract was only changed rather violently much later

---

* We adopt the traditional simplification of speaking as if Nobunaga or Hideyoshi himself was the agent and decision maker for this large armed force. Of course, in actuality he represented the interests of those who held coercive power.

by another importation of new coercive technology in the mid-19th Century (Chapter 15).

This insight is profoundly important for us. It confirms our theory's prediction that the *only* thing that matters in considering a *state*, or any other human social aggregate, is *who holds decisive coercive power*. The state may serve the interests of almost all its members poorly. It is of no *a priori* consequence. If a tiny elite holds decisive coercive power and is adequately served by the state, the state and its hierarchical social contract beliefs will endure—perhaps forever, in principle.

On to real modernity

We have several strong empirical confirmations of the essential predictions of our theory about the effects of gunpowder coercion. We possess what we need to understand how the modern world we think of as our *cultural heritage* explodes into existence, on our theory. Chapter 15 will explore these astounding events.

෨

Chapter 15
Modernity emerges from the barrel of a gun, Part 2 – handguns democratize the modern state

An introduction to the "original case" – the modern European state

> *Man will never be free until the last king is strangled with the entrails of the last priest.*
>
> Diderot[*]

The shocking brutality of Diderot's famous remark reflects the ferocious hatred provoked by the authoritarianism—the concentrated elite coercive power—of the early modern European state (Chapter 14). Both the direct agents of coercive power (the king is their representative) and the managers of elite social contract and identifier beliefs (the priests) were equally visible actors in these fiercely hierarchical social enterprises—warrior popes and divine right kings alike.

But, by the late 18th Century, Diderot could not only say in public, but actively publish his subversive statements without being beheaded or shot. Something new was afoot.

In the following section we will look at the long, slow, painful emergence of the early modern state in Europe—the rise of Diderot's kings and priests—at the barrels of the first generations of gunpowder weapons. In the sections after that, we will look at their downfall with the rise of the modern democratic state—at the barrels of still newer gunpowder weapons.

---
[*]   18th Century Enlightenment philosopher.

The European record of these events is rich. The rapid follow up of the *early modern state* with the *democratic state* in Europe is predicted by our theory to produce a large uptick in adaptive sophistication (below and Chapter 17). This new level of adaptive sophistication also produced legions of publishers, writers, and thinkers to preserve and interpret the record of this once-in-a-planet event.

However, this record is also fraught with the interpretive complications we know so well—many winners telling many lies. Moreover, every one of us alive on Earth today has been touched, with furious intensity, by these original European states, whether we live in their direct descendents in Europe, North America, South America, or Australia/New Zealand, or in their indirect descendents, the states of the rest of the world.

Not one of us alive anywhere in the world today is a detached observer of the early European states. We may admire their social, scientific, economic, and technical accomplishments or deplore their extravagant brutality. Indeed, we may harbor *both* these attitudes. But we are, none of us, objective.

At this juncture, it is more important than ever to remember our extraterrestrial anthropologist. We each must struggle every moment for as much detachment as we can muster. As fellow citizens of an emerging pan-global human culture (Chapters 16 and 17), we can be of great assistance to one another in overcoming our local provincialisms and preconceptions about these crucial historical events.

The early modern world emerges in Western Europe

Gunpowder technology was introduced into the Islamic and European worlds around 1250, as we saw in Chapter 14. Western Europe was distinguished from most of the rest of Eurasia at this time as a hick backwater of tiny statelets—small kingdoms, earldoms, and so on, each looking to parasitize its neighbors where possible. No large-scale central authority had existed since Charlemagne over four hundred years earlier. In contrast, the richer parts of the Islamic world were dominated by a few large archaic states, including those that would ultimately produce the Ottoman Empire. These extra-European states were relatively stable at that time and managed by conservative, self-protective elites.

The fragmented, competitive European environment provided a natural laboratory in which a great deal of the experimental refinement of gunpowder weaponry would occur.* Throughout the first

---

* These innovations rapidly diffused across Eurasia, so that Western Europe often led, but did not have a monopoly on these technologies.

approximately two hundred years of its evolution—through the early 15th Century—this technology remained relatively primitive. It played a supporting role in what still looked a lot like archaic state warfare.[1]

However, by the late 15th Century, recognizable early modern states began to consolidate in Western Europe. Traditional *warfare and the state* scholarship sees this consolidation as the product of wars of conquest and of the selection for entities that could mount such wars.[2] Though not entirely wrong, we believe this view is profoundly distorted and misleading. These states emerged when coalitions that could mount decisive local coercive threat coalesced in pursuit of their mutual self-interests.[*,3]

Once these locally successful coercive coalitions formed, they were well positioned to police cooperative behavior, enforcing tax payments and labor from commoner populations, as well as waging war to protect their positions and expand their domains. In other words, early modern states did not consolidate *because* of the demands of war. Rather, their consolidation under gunpowder policing allowed them to wage war on a new scale. Waging "modern" war is merely a particular case, among many, of kinship-independent social cooperation on the scale of modern states.

Many of these early modern states were ostensibly ruled by kings. Kings had existed since the rise of the first archaic states five thousand years earlier. However, these earlier kings had always been, first and foremost, mouthpieces for dispersed elites. In contrast, with the coming of gunpowder weaponry, it sometimes became practical for kings to hire large *standing armies* composed of self-interested merchants of coercive threat, available to the highest bidder.

The former elites of the archaic states were useful members of the earliest of these standing armies. They were able to use their martial and managerial skills to assist in the projection of violence and coercive threat and, thereby, to derive the economic (adaptive) benefits from managing the cooperative enterprise. Their original wealth represented part of the seed money that made the new armies possible.

At first glance, we might imagine that the elites of early state standing armies were no different than the armed elites of archaic states. However, there are important differences.

First, improved policing made possible by gunpowder weaponry allowed the construction of improved production and supply

---

*   The so-called War(s) of the Roses leading to the foundation of the Tudor dynasty and the early modern British state is an especially well documented and illuminating, yet typical example of this process.

capabilities—what professional merchants of coercion call *logistical support*. This made the local assembly of ever-larger forces possible.*

Second, even heavily armed *archaic state elites* could be over-whelmed by peasant mobs if these mobs grew large and angry enough. In contrast, early modern armies—armed with cannon and massed musket fire and better supported logistically—were much less vulner-able to commoner discontent. Their power was virtually unchecked in the presence of unarmed commoner populations.

Finally, these well-provisioned, powerful military elites were supported by bureaucratic state tax systems that were quite detached from the peasants and merchants who were simply their *tax base*. In contrast, most members of earlier archaic state elites were in more direct contact with the peasants and merchants, who generated the wealth they consumed.

As a result of this set of properties, early modern states were even more coercive and repressive, in general, than their brutal-enough archaic state ancestors. The early modern states were hierarchical, elitist enterprises like earlier archaic states. They were simply better at it.

The coercive power of early modern state elites was absolute enough to make them extremely domineering. Indeed, the more authoritarian early modern states, ironically referred to as the *ancien regime* in the propaganda of later democratic subversives, could be well thought of as outstandingly efficient *protection rackets*.[4] They charged extravagant fees called taxes to purchase protection from the very violence of which they were the agents. The primary difference between most of these early modern states and Al Capone and his goons on Chicago's South Side in the early 20th Century was scale.

As we have mentioned, hierarchs ostensibly tracing their status to the elite licensing beliefs of earlier archaic states, continued to play a central role in many of these early modern states. However, maturing gunpowder weapons were increasingly subversive of their status and ultimately destroyed these early criminal enterprises *cum* states (at least in some areas).

We can get a hint of this emerging subversive power by listening to what two of the nobles of the early modern state elites thought of the improved guns and the commoners who wielded them. First is the great mid-16th Century warrior Blaise de Monluc, whose military

---

*   Note again the direction of causation. These early states had improved logistics *be-cause* of their better-policed cooperation. They then went on to win conflicts as a result. These logistics did not emerge for some unknown reason, allowing the benefactors of such accidental "magic" to prosper selectively, as many traditional military theories of the rise of the state would have it.

career was ended when a portion of his face was blown away by a musket ball, requiring him to wear a mask in public for the remainder of his life.

> Would to heaven that this accursed engine had never been invented. I had not then received those wounds which I now languish under, neither had so many valiant men been slain for the most part by the most pitiful fellows, and the greatest cowards; poltroons that had not dared to look these men in the face at hand, which at distance they lay dead with their confounded bullets.[5]

Don Miguel de Cervantes was a European noble who also fought to help establish early European states, even suffering a wound at one point. He later put his feelings about guns held by commoners into the mouth of his famous fictional knight, Don Quixote.

> Blessed were the times which lacked the dreadful fury of those diabolical engines... whose inventor I firmly believe is now receiving the reward for his devilish invention in hell; an invention which allows a base and cowardly hand to take the life of a brave knight, in such a way that, without knowing how or why, when his valiant heart is full of courage, there comes some random shot—discharged perhaps by a man who fled in terror from the flash the accursed machine made in firing—and puts an end in a moment to the consciousness of one who deserved to enjoy life for many an age.[6]

These quotes hint at the end of traditional elitehood produced by advanced guns, as we will see in the following sections. However, notice the other lessons embedded in these impassioned self-revelations.

First, these individuals had no difficulty whatever seeing themselves as "brave" and "true-hearted," and the gun-toting commoner adversaries as "base" and "cowardly." Yet, these brave knights would not have had the slightest compunction in beheading an unarmed insubordinate commoner on the merest pretext. At work here is the infinite human capacity for ethical self-justification— the fundamental adaptive goal of most of our public political discourse (Chapter 10).

Second, for purposes of rhetorical impact we have violated our own dictum about detachment in this section. At one level our indictment of early modern states as brutal criminal enterprises is perfectly accurate. However, at another level, our *own* ethical overlay is also profoundly misleading. The cast of characters in this historical process, Richelieu, Louis XIV, and tens of thousands of others are just humans doing what humans have always done throughout their two million years on the planet. They simply used whatever means were at their disposal to project coercive threat in pursuit of their own self-interest. They used public ethical justifications as recruiting tools to garner the collaborators necessary to make the enterprise work just as humans have always done.

As much as we might wish to, we cannot indict early modern state oligarchs for being evil, only for being human. If wish to have the privilege to consider their actions evil—as we, the authors, often do—our challenge is first to put ourselves and our descendents in the position to do better with the fundamental tools of humanness. Through the rest of this and the remaining chapters we will gain the last few insights we need to be in a better position to be not only human, but also more humane.

Creeping commoner subversion of the early modern state – The evolution of the gun and the dawn of modernity

As individual gunpowder weapons, guns, became ever cheaper and more plentiful, a larger fraction of the population could, in principle, acquire and own them. As a result, coercive coalitions increasingly contained sub-elite armed individuals, including illegitimate offspring of royalty, middle class individuals, and even some peasant commoners.[7] As always, these individuals used their access to coercive threat to pursue self-interest. Over time and with abrupt fits and starts and local accidents, this unfolding process sometimes produced states that began to serve the interests of somewhat larger fractions of their citizens better, the *modern democratic state*.

Two issues are crucial. First is the continued improvement in guns in the 17th and 18th Centuries. Earlier matchlocks and wheelocks were a first step, but the big breakthrough was the flintlock.[*,8] This simple firing mechanism was relatively cheap to make. A local gunsmith could reliably manufacture the device for general sale. The flintlock mechanism was relatively insensitive to moisture and dirt and would fire on demand most of the time. During roughly this same time interval, gunpowder manufacture also became much more efficient and productive, reducing powder prices dramatically.

The original flintlocks were mostly *smoothbore weapons*. They fired a spherical ball out of a smooth metal tube—the musket barrel. These weapons were loaded from the muzzle of the gun so that the ball had to be slightly smaller than the diameter—the bore of the musket. As a result, the musket ball rattled slightly as it flew out of the gun on firing.

---

*   See results of a Google image search under "flintlock." Also, see online endnote 8 for additional details of gunpowder personal weapons.

This banging against in inner walls of the barrel impart a spin orthogonal to the direction of flight or, sometimes, no *net* spin at all. These effects produced a lot of movement away from the intended trajectory of the ball similar to a curve ball or a knuckle ball thrown in American baseball.

The net effect of these design limitations was that smoothbore flint-locks had very poor accuracy beyond 50 to 100 meters. It had long been known that if the bore of the musket was *rifled* (had helical groves) and if the ball or bullet fit very tightly, the accuracy of the musket could be improved dramatically. This improvement worked because the tight fit meant that the bullet always exited the barrel at the same trajectory and the rifling meant that it had a spin precisely perpendicular to the path of the bullet. This spin produced *gyroscopic stabilization*, like the tight spin of a well thrown pass in American football, making the flight of the bullet extremely true and reproducible.[*,9]

The problem with this approach was the tight fit of the ball or bullet. It required the slug to be jammed down the barrel against resistance. This took time and was not practical under fire during policing and military applications. This last problem was solved in the first half of the 19th Century. A projectile, called a *Minie ball*, after its inventor, was developed with the conical bullet shape and hollow concave back.[†] Moreover, it was slightly smaller than the bore of the rifled barrel so that it dropped easily down the gun at loading. However, when the gunpowder exploded behind the Minie ball, its hollow back of soft lead expanded so that the bullet grabbed the rifling on its way out.

The new practical rifled muskets or *rifles* allowed by this technology, were spectacularly accurate. Infantry could reliably hit targets several hundred yards away and trained snipers became lethal out to nearly a mile. These weapons made massed infantry even more dominant. *Boots (rifles) on the ground* were now absolutely decisive.

This innovation became nearly universal by the middle of the 19th Century.[‡] However, these early rifles were still single shot weapons. Moreover, the powder, bullet, and firing cap had to be loaded separately, a complex, demanding sequence of manual maneuvers. Under fire even trained infantry could deliver no more than about twenty to thirty shots per hour in practice.

---

[*]   See results of a Google image search under "gun rifling."

[†]   See results of a Google image search under "Minie ball."

[‡]   The spectacular lethality of infantry battles in the American Civil War—approximately twenty-eight thousand killed at Gettysburg in three days in early July of 1863, for example—resulted in large measure from the terrible accuracy of these weapons.

During the latter 19th Century, supported by the *industrial revolution* (Chapter 17), breach-loading weapons were developed. These new guns were loaded from the back like modern weapons and used self-contained ammunition consisting of complete cartridges in the sense of modern bullets.* Breach loading allowed the bullet to fit tightly as it exited a rifled barrel as before. Moreover, proper design of the mechanism and the cartridge casing made the velocity or momentum of these bullets potentially very high.

Finally, spring, gear and wheel-based mechanisms were developed to allow rapid, repeating firing before reloading. The *six shooter Colt revolver* and the *repeating Winchester rifle* were among the earliest mass-produced versions of these weapons. They were cheaply available in time to help "win the West" in the conquest of Native American, Spanish and French lands to complete the modern United States between the 1870s and the 1890s.

All this technical progress had many effects beyond its direct impact on the costs of coercion. For example, many of the features of assembly line production, mass casting, and machine tooling that we think of as constituting the *industrial revolution* in the 19th Century were actually initially developed to permit the large scale, inexpensive production of gunpowder handguns.[10]

Through the 20th Century these weapons continued to be refined until today we have access to highly reliable, extremely deadly individual weapons of diverse sorts—pistols or handguns, including Glocks, Colts, and the like and assault rifles including the famous Russian Kalashnikov or AK-47 and the American M-16 are among the many.†

These assault weapons retain the high accuracy and killing power of earlier rifles while allowing an individual to fire up to six hundred rounds per minute. The large, spring-fed clips for these weapons also allow very large numbers of rounds to be fired before reloading.

In overview, the tremendous fire-power of these weapons makes those who deploy them with numerical dominance the locally decisive coercive threat, from the 18th Century (at least) through the present instant. Even the larger modern coercive weapons—artillery, aircraft, and the like—are ultimately controlled by those who control these individual weapons.

Our theory is absolutely clear in its predictions. The modern world, through the present instant, should be a product of the human use of

---

*   See results of Google image searches under "breach loading rifle," "bullet," and "cartridge."

†   Especially useful are Google image searches under "AK-47," "Glock handgun," and "Colt handgun."

these weapons in self-interested coercion. Our social systems should be shaped decisively by them and these systems will serve the interests of those who control these weapons. All other factors shaping the contemporary world should be secondary or derivative of this primary cause, *the redistribution of access to decisive coercive threat.*

We will spend the rest of this chapter arguing that the history of the contemporary world looks remarkably precisely as it should if this prediction of our theory is correct.

Modernity explodes from the barrel of the handgun –Introduction to the empirical evidence

As we have seen, the availability of cheap gunpowder handguns rendered all earlier individual coercive weapons largely irrelevant. This transition changed utterly the logic of the self-interested coercion central to the ancient, uniquely human social adaptation. The introduction of these weapons ignited a planetary fire that still burns around us today. The hierarchical ways of life inherited from the archaic state are being consumed in totality by this global conflagration. Everything about our world—our social, religious, artistic, economic and scientific lives has been utterly remade as it emerges from the ashes of this blaze.

The melodramatic prose of the preceding paragraph is a way of emphasizing two vital, interrelated points.

First we predict that the modern world will be profoundly different than what came immediately before it. But this profound difference is, nevertheless, perfectly understandable on our theory.

Second, this modern world of ours is so different than the past *not* because humans have invented something truly new. We have not. Instead, over the last several centuries we have merely exploited the ancient human trick of coercive management of conflicts of interest. However, new weapons have made it possible for us to prosecute this ancient adaptive strategy on a vastly larger scale than ever before in our two-million-year history.

In a very real sense, we all still occupy the ancestral human village, doing what our ancestors have done for thousands of millennia. It is just that our village now has hundreds of millions of members rather than merely hundreds. This difference in scale has some rather striking effects (Chapter 17).

Pause to remember what we have already forecast. Whatever we may feel subjectively about power growing from the barrel of a gun ala Mao—he understood modern politics perfectly. Our world grew from the barrel of a gun and from nowhere else. The contemporary world emerges from the human exploitation of these weapons and from nothing else, on our theory.

We will spend the rest of the chapter exploring key pieces of evidence that this strong, simple claim is correct. The evidence is clear and compelling, we will argue. As we proceed, remember that these insights, however much they may trouble us at first glance, will ultimately uplift us. They give us the real, substantive possibility of becoming agents of our own more humane future (Chapter 17).

Box 15.1 Sam Colt and Mikhail Kalashnikov – coercion and equality

Sam Colt was one of the most successful American manufacturers of revolvers and repeating rifles in the late 19th Century. His outstanding salesmanship included a slogan of particular interest to us here, "God made men free, Sam Colt made them equal."

Seventy years later the Russians either stole or independently invented the identical slogan. "God made men free, Kalashnikov made them equal," they said.

Aside from the rich irony of this slogan used by an authoritarian (and ostensibly atheist) modern state, the underlying point is crucial. With a two-million-year history of using projectile weapons for self-interested coercion, humans everywhere have an immediate intuitive grasp of the social implications of these weapons.

Box 15.2 Money is not power – power is power and money is its food

Money…not only affords you no protection, but makes you the sooner fall a prey…it is not gold, as is acclaimed by common opinion, that constitutes the sinews of war [coercive power], but good soldiers [armed fighters]; for gold does not find good soldiers, but good soldiers are quite capable of finding gold.

Niccolo Machiavelli*

---

*   *The Discourses*, Book 2, pages 300-302. Bernard Crick, editor (1970) London, UK: Penguin Books Ltd. (Penguin Classics.)

The empirical evidence – *Democratization*

The written history of the emergence of the central core of the modern world over the last approximately four hundred years is monumentally vast. We could write tens of thousands of pages of non-redundant discussions of our theory's accounts of specific local empirical evidence. Instead, as before, we will briefly review a few selected elements of this record. Again, our extensive studies have revealed no evidence that contradicts the simple empirical predictions of our theory and, as always, the appropriate and valuable response of skeptics would be to produce cogent counter-examples.

In pursuit of the goal of empirical testing of our theory we must return again to the nature of the evidence. *States* are the terms we use for the temporary winners of local skirmishes in the world-wide struggle for coercive dominance initiated by the coming of cheap guns. Dominant coalitions within these states create identifier and social contract beliefs, as human cooperative coalitions have always done (Chapters 10-14). We have all been trained by state-sponsored educational systems to think of these beliefs as "history." If we are to grasp the empirical record of these events, we need to have a jaundiced, skeptical, critical eye—to say the least—in reading this "historical" record.

For example, terms like Glorious Revolution, American Revolution, French Revolution, American Civil War, Meiji Restoration, Bolshevik Revolution, the rise of the Nazi Party, Maoist Revolution, the Khmer Rouge takeover of Cambodia, and thousands of other designators of local events fundamentally mis-focus our attention at the wrong scale. These are all merely local flairs in the global fire of exploitation of gunpowder-based coercion, on the one hand, and local consequences of the actions of heavily armed local interest groups, on the other.

We will date the beginning of this era of modernizing the state with the opening of the Dutch Revolt against the Spanish crown in 1572.[11] Any single starting date is inevitably somewhat arbitrary. But, arguably, the first large scale, successful deployment of commoner, gun-armed coercive forces begins here. As we said, this process continues around us today. We optimistically assume that the worst of this political violence is behind us and will continue to wind down and end by late in this 21st Century (Chapter 17). Perhaps our remote descendents will call this era in human history something like the *500 Years War of Global Democratization*.

Before taking another step it is vital that we stop and confront terms like *democracy* and *democratization* more carefully. These terms, of course, have quasi-religious connotations for many of the people in the contemporary world. These words can connote aspirations most of us, authors included, would share. These ostensible goals are parts of the social contract and identifier beliefs of virtually every modern state regime.

Of course, the verisimilitude of these terms in state propaganda is extremely doubtful. Regimes calling themselves democratic include those we have come to think of as *relatively* gentle like the 19th Century European outlier states in North America (Native American genocide and African slavocracy notwithstanding) and Australia (aboriginal genocide notwithstanding). These also include regimes that we tend to think of as murdering scum like the Stalinist and Nazi states in Western Eurasia and the Maoist and Khmer Rouge states in Eastern Eurasia, for example.

One of the most demanding changes in perspective we must make if we are to understand our recent history is to grasp that the differences between Stalinist Russia and the late 19th Century United States, say, are *quantitative, not qualitative.* Very few of us, indeed, can do this easily or readily. Every one of us holds cherished identifier and social contract beliefs that cry out against such thoughts. We have work to do.

To be more specific, all modern states have the same funda-mental underlying structure. They are dominated by a heavily armed, internally quasi-democratic dominant set of coalitions or interest groups. These interest groups use this coercive dominance to shape the state in pursuit of their member's self-interests, subject, of course, to much enforced compromise and log-rolling. This pursuit of interest often includes disadvantaging—even enslaving and murdering—mem-bers of interest groups who lack access to adequate levels of coercive threat to defend their interests. This violent pursuit of self-interest is, of course, precisely what our theory predicts.

The difference between an *authoritarian state* like Stalinist Russia and a *democratized state* like the late 19th Century United States is the *relative size* of the armed, powerful interest group set in comparison to the disenfranchised set. When the dominant coalitions are a large frac-tion of the population, we think of these states as *democratized* and, when the dominant coalitions are a small minority, we think of these as *authoritarian.*

The level of ongoing coercive violence and overt threat necessary to sustain these states, their social contracts, and their levels of economic productivity varies strikingly and predictably as a function of these proportionalities (Chapter 17). But, for now, we are more concerned with the underlying similarities between all modern states, whether authori-tarian or democratic.

Three brief vignettes will help us grasp these similarities among all modern states.

First, in the early 19th Century, North Americans of European descent used superior armed force to disenfranchise Native Americans, seizing what is now Illinois, Indiana, Ohio, and the American South by killing or displacing Westward non-Europeans who occupied valuable agricultural lands.* Some of these seized lands, in what became Mississippi and Alabama, were excellent for raising cotton on a massive industrial scale for sale into a booming global market. This gave fresh life to enslavement of African American labor, a practice requiring the continual projection of massive armed coercive threat.[12]

Even the relatively poor European elements of this Southern cotton economy—*white trash*—benefited to some degree from African slave labor. Moreover, even though Africans were a majority in some intensive farming areas in the South, Europeans retained highly disproportionate control of gunpowder weaponry.[13] Thus, this society was benevolent and democratic by the lights of even its less wealthy European members but oppressive and aggressively authoritarian to its African members.

Second, in what is now Russia, in the early 20th century, armed groups eventually coordinated by Trotsky, Lenin and, later, Stalin committed murder on a massive scale during the civil war that ultimately produced *the Soviet Union*.† The major coercive force they managed came to be called the *Red Army* and its elite *secret police* created the *Communist Party* as its front. This dominant Red Army/Communist coalition was always a small minority of the Soviet state. The commoner population managed by the Red Army was vast, but impoverished, consisting mostly of poor, unarmed peasants. This majority of the residents of the Red Army state were actively enslaved and manipulated.

Thus, though the Soviet elite called itself democratic, most members of Soviet society did not agree and would have called the state authoritarian.

Third, a group of armed thugs organized around Adolf Hitler were able to use a combination of discreet intimidation, assassination of political rivals and appeals to the economic misery of Weimar citizens to gain initial control of the German state. This control was consolidated and maintained by a continued program of political murder by this group calling its political face the Nazi Party and its elite coercive

---

*   Of course, this continued throughout the 19th Century until the "West was won" and "manifest destiny" was fulfilled (McDougal, 2004).

†   See pages 141-159 in Ferguson (2006) for a powerfully succinct recent synopsis of these events.

core the Gestapo and SS. This elite coercive core was able to co-opt and control the army (Reichswehr, later Wehrmacht).*

This coercively dominant group was superficially similar to the Red Army/Communist interest group in the Soviet Union. However, German citizens had access to considerably more coercive threat—that is, political power—than Russian peasants. Thus, the SS/Nazi coalition had to be more mindful of the interests of its majority of citizens if it was to avoid ouster. The Nazi elite coalition pacified these citizens by an aggressive program of deficit spending to fund a welfare state in which public works generated increasing employment and which yielded many other benefits—cars, the Autobahn, vacations, and so on.

Unfortunately, this deficit spending could not be supported on the basis of realistic future German economic productivity, as contemporary United States deficit spending (ostensibly) is, for example. Thus, the Nazi system was propped up by the systematic plunder of wealth from relatively disarmed domestic minorities—especially German Jews—and of conquered territories throughout Europe and North Africa.†

Among other things, the pretexts used to seize much of this wealth (redeemable securities issued in exchange for material wealth) made it expedient for these creditors of the SS/Nazi state to disappear before redemption. The SS/Nazi death camps were ultimately set up, in part, to solve this financial problem.

The SS/Nazi welfare state was very popular with the majority German coalition/electorate, and the few dissidents were easily dispatched by the SS/Gestapo. Moreover, the SS/Nazi state was sufficiently discreet in its genocide that a happy, well-fed electorate could arrange to pretend to be plausibly ignorant.

Thus, the majority of the denizens of the SS/Nazi state considered their country to be benevolent and democratic well into the early 1940s. Indeed, the SS/Nazi state was remarkably successful in capturing the idealism and commitment of many of its most talented citizens. Jewish, Slavic, and Romish members of this state took a very different view.

Before we move on, stop and consider one last point. Suppose that the slavocracy had survived the American Civil War or that the SS/Nazi regime had survived World War II. Does anyone seriously doubt that the dominant coercive coalitions in these states would have successfully written a benevolent, democratic "history" of themselves?

---

*   See pages 221-269 in Ferguson (2006) for an excellent recent review of these events.

†   See Aly (2007) for the scholarly work behind this discussion.

Late modern states – The nature of experimental evidence in practice

Even in very sophisticated experimental sciences—molecular genetics, for example—there is never a *perfect experiment*. Any experiment, no matter how well controlled and elegant, might be interpretable in some way the experimenters have not thought of. Moreover, sometimes it just is not practical to do completely controlled studies. We must settle for studies that give us the potential to falsify our hypothesis, but which—when successful—still leave several different potential interpretations in play.

Thus, in the real world of practicing scientists, reliable insight comes from having multiple experiments each interpretable, at least in principle, in multiple ways, but each with different design and approach. If all of these independent approaches converge on the same answer, we begin to have high confidence in this answer.

The historical records of the emergence of late modern states represent a set of natural experiments, each with its own idiosyncratic ambiguity. In any individual case, there are multiple possible primary causes of the historical changes we see.

For example, we might imagine that the rise of the North American state (the US) was the product of the *Pilgrim's pride*, the *Protestant ethic*, the efficiency of slave labor, the rise of frontier capitalism, or one of many other things. However, our theory predicts that the emergence and structure of this state should be determined by which interest groups held (and continue to hold) decisive local coercive power. We can ask whether the evidence is consistent with this prediction.

Likewise, the emergence of the modern Russian state could, in principle, have been caused by the emergence of *proletariat awareness*, the diffusion of *Marxism* into tsarist Russia, the inevitable contradictions of quasi-feudal tsarist rule, or one of many other factors. However, again, our theory predicts that the emergence and structure of this state should be determined by which interest groups held (and continue to hold) decisive local coercive power.

Finally, the emergence of Meiji Japan in the mid-19th Century could have been caused by the penetration of Western beliefs, the desire to catch up with the West, the redesign of *bushido* doctrine or one of many other things. However, again, we can ask whether the evidence is consistent with the Meiji state emerging from the deployment of decisive coercive power newly captured by a different interest group than before the Meiji state.

Notice that in each of these three cases, most of the possible explanations for the historical processes in question are idiosyncratic to the individual local case. In contrast, the causal role of capture of coercive dominance predicted by our theory is common to all three.

If the evidence is consistent with this coercive cause in each case, we will have much more confidence in our theory of history and in our larger theory more generally. We will argue that this is precisely what we see. This insight, in turn, sheds powerful new light on the emergence of the modern world.

Late modern states, Case 1: A hot house plant in an accidental incubator – The British Empire produces the North American state

Self-interested individual members of any coalition fight to retain and expand their access to coercive power. This is universal, ancient human behavior (Chapters 5 and 10). This property has an important effect in state-level social units. If coercive means are broadly distributed, members of the empowered majority will fiercely resist elite capture of these means. Conversely, if coercive power is monopolized by a small elite, like an authoritarian army, members of that elite will violently resist disbursement of coercive means more broadly.

Access to coercive means remained relatively broadly distributed during the rise of the early modern state in the British Isles.* In contrast, coercive means were more concentrated (including large standing armies) in some continental European early states, like the France of Louis XIV.†,14

This *relatively* less hierarchical British state pursued a hands-off colonial policy in North America, largely for budgetary reasons. Britain used its maritime power and extremely limited North American infantry deployments to protect the colonies from outside state intruders, France and Spain. However, the colonists were largely left to their own devices for local law enforcement and the common defense against hostile Native Americans.‡ As a result, a large fraction of colonists owned

---

*   This may result from the long tradition of a heavily armed electorate dating from the era of the English long bow through early "gaming" handguns (Malcolm, 2002). Further study of late pre-gunpowder English politics will be of the very greatest interest and importance.

†   See Skocpol (1979) for a discussion, including the French case.

‡   Even before the North American revolt, elitist British monarchs—James II, for example—occasionally tried to police the North American colonies more aggressively.

guns and were competent in their use."[*,15] Moreover, the geopolitically decisive handgun of this era, the flintlock, could be manufactured locally by individual gunsmiths.[†] Finally, the British coercive umbrella also helped forestall formation of strong local elites *within* the North American colonies. No colonial warlords allowed.

Thus, the best available option for the North American colonists was to mutually police their own economic systems and focus on agriculture and/or commerce with trade into the global economy. This focus included commercial production and export of raw materials like timber and cotton into the robust European market.

By the late 18th Century, this novel constellation of circumstances produced a situation unprecedented in the history of any state up to that time. The entire eastern seaboard of North America was inhabited by a massive, democratically armed coalition of several million people and no entrenched, professionally armed elite.

Moreover, local magnates like plantation owners George Washington and Thomas Jefferson or a wealthy merchant like John Hancock could help fund militias in defense of their own very local interests (and those of the local militia members). However, no local concentration of wealth comparable to a European monarch and capable of funding a large standing army existed.

This unique historical accident produced an extreme natural experiment in democratization of access to coercive threat.[‡] In pursuit of mutual self-interest, the North American colonists declared their

---

However, they were defeated in this effort by the events of domestic British politics—including the *Glorious Revolution* in James' case.

*   See, for example, John Galvin's 1989 book *The Minute Men* for a review of early colonial American policing and warfare. Also, see Joyce Malcolm's 1994 book *To Keep and Bear Arms* for a discussion of internal American political response to a broadly armed electorate. The contrary argument that colonial Americans were poorly armed (Bellesiles, 2000) has now been discredited (endnote above).

†   This unique *inherently* universal availability of decisive handguns did not exist before the invention of the flintlock. Neither did such access survive the dominance of the flintlock into the contemporary era. Both before and after the flintlock, gun manufacture was so expensive and sophisticated that entrenched elites *could* monopolize and control it. Before and after the flintlock, democratized access to handguns required the *previously existing* democratized control of the means of their production. We will be well advised to recall this when we turn to the contemporary world and the future (Chapter 17).

‡   Also important were the related cases like the areas that would become Canada, Australia, and New Zealand.

independence and revolted against their British patrons and protectors, ultimately producing the United States.

It is noteworthy that many of the British taxes that sparked the North American revolt were extremely light.[16] These taxes were offensive not because they were quantitatively onerous, but because they were designed to fund increased local British policing, including the beginnings of enlarging professional *standing armies* opposing the interests of the colonists.

The initial conflicts with British forces were executed by democratic militias. As the Revolutionary War dragged on, the Continental Army increasingly consisted of professional soldiers recruited on the basis of financial incentives. However, this army remained under a tight financial leash held by the Continental Congress and the potential for policing by the large non-professional militias of the whole*, if it became necessary. Indeed, the American Constitution was written in the crucible of local post-Revolution insurrection. The key issue was to achieve coercive stability while retaining control of the state's coercive means by a democratically armed electorate.[†]

This historical accident had one last element. The North American colonies may not have had the resources or expertise to defeat even the limited professional military assets the British were willing to contribute to subduing a remote popular revolt.[17] The American Revolution might have failed on its own. However, in pursuit of its own economic interests, contra the British, French elites were willing to help fund the American revolution. The French government also provided crucial training personnel and finally naval support.[‡]

This unique set of accidents produced a particularly novel outcome. A massive, radically majoritarian commoner interest group gained essentially uncontested control of a vast, ultimately continental, state. The reverberations of this extreme and unprecedented experiment continue to rumble around us through the present moment.

The government of the North American state (the US, henceforth) has remained constantly under assault by diverse internal elite interests,

---

*   See Martin and Lender (2006) for an especially pithy discussion of the American Revolution's military establishment.

†   See Szatmary (1980) for a strong description of the early post-Revolution coercive instability. See Williams (2003) for one view of the shaping of the American Constitution (especially Article I and the Second Amendment) in response (Chapter 17).

‡   Ironically, the French monarch Louis XVI would perish under the guillotine in a quasi-democratic revolution against elite domination in France less than two decades later.

from its inception through the present instant. This process has often produced large concentrations of highly manipulative, self-interested wealth.[18]

However, the electorate has remained aggressively surly and intensely jealous of its prerogatives, including private ownership of a huge number of personal gunpowder weapons. Americans have been tolerant of local corruption when it serves their interests. At the same time, they have consistently insisted on governmental transparency—rewarding and protecting an aggressively disrespectful press, for example. Further, through the present instant, suspicions of elite domination and self-interested malefaction permeate the public American political dialog. Any and all politicians are routinely, publicly, daily disparaged in the most vitriolic possible terms.[19]

To date, this approach appears to have been generally successful in maintaining a significant level of popular sovereignty, initially of white males and ultimately of others, still unfolding. It appears that Lincoln was right about the wisdom of crowds—"you can fool some of the people all of the time, and all of the people some of the time, but you cannot fool all of the people all of the time."[20] The heavily armed North American crowd has, so far, remained pushy and, in the long run, mostly uncowed.

Several additional details of the development of the radically democratized North American state are illuminating for us here.

First, the small-time American elites—southern slavocrats like Patrick Henry and ambitious northern merchants/financiers like Alexander Hamilton—were able to influence the original American Constitution to nominally serve their interests, in part. Their vision was a republic governed by a "natural aristocracy."*

However, the members of the heavily armed citizenry of the young US had their own goals. Over the next approximately eighty years, the citizens of the US redefined property law, the rights of citizens in a radically democratized state and the role of government as a servant of a democratic electorate.[21] This process included trampling some of the intentions of the elite authors of the original American constitution, reinterpreting the meaning of some of its Articles and Amendments.†

---

* See McDougall (2004) for a beautifully rich, acerbic, eminently readable review of the massive body of scholarship on the early emergence of the US. Also, see Chapter 17 herein.

† We Americans should remember this history when we hear someone invoke "strict constructionist" interpretations of the American Constitution. We need to ask if this advocate might be looking to defend elite privilege, consciously or otherwise.

Second, one pivotal event in the development of this radical democratization was the American Civil War. The gunpowder weaponry of the later modern state was deployed in a massive conflagration, including approximately two hundred thousand killed by direct combat action and another four hundred thousand from the consequences of the war.[22] This war asserted the unchallenged power of a radically equalitarian, "free labor" Northern coalition over a more elite dominated Southern slavocratic coalition. Indeed, it is arguably the American Civil War and its aftermath, rather than the American Revolution, that established the pattern for all radically democratized states through the present.[*,23]

It is striking that the American Civil War played out on a global stage following the *Crisis of 1848* in Europe, where democratic risings were put down with substantial preservation of entrenched elite coercive power.[24] The self-serving conventional wisdom of surviving European elites was that radical democracy was a goofy failed experiment—destined to collapse wherever it existed. The North American state would be the next domino to fall, on this view. It did not turn out that way.

European observers watched the American Civil War with keen interest including Karl Marx and his daughter, who sent letters of advice to Lincoln. The American electorate understood how the world looked at their experiment. There is excellent reason to believe that the electorate as a whole grasped the global issues at stake perfectly well.[†]

When Lincoln wrote the Gettysburg Address for delivery at the massive cemetery for the soldiers (all volunteer) fallen at this battle—twenty-eight thousand killed in three days—he was speaking to all the (white male) American people and to the world. He clearly considered that the issue was whether radical democracy—"government of the people, by the people and for the people"—would "perish from the Earth." This crisp powerful address is arguably one of the most illuminating documents in all of human history and is reproduced in its entirety in Box 15.3.

For reasons we will explore in detail in Chapter 17, democratization is expected to produce dramatic increases in the adaptive capabilities

---

*   Ironically, but predictably, both the Northern mercantile/free labor interests and the Southern slavocratic interests considered that they were fighting for "freedom" (McPherson, 1988).

†   James McPherson's book *For Cause and Comrades* (1997) uses the vast correspondence from Civil War troops—the first mass armies to be highly literate and the last to be largely free of military censors—to show that even the foot soldiers of the American Civil War knew very well what the global stakes were.

of human coalitions. The North American case is an excellent example of this expectation. The radically democratized North American state rapidly became the most powerful economy on Earth—a status it retains through the present.

Before leaving the North American natural experiment, one last issue is vital. The US has apparently remained largely commoner controlled. However, its national government also acts external to its boundaries. Its actions affect millions of individuals who do not hold coercive power over it, in particular, *foreign nationals*.

Do we expect the international actions of the US government to be constrained by the interests of these powerless foreign nationals? Of course, we do not. Do we expect American citizens to be highly sensitive to the costs of US international actions to powerless foreign nationals? Again, of course, we do not. Indeed, we expect most US citizens to be no more mindful of the interests of powerless foreign nationals than citizens of the Nazi state were to their government's actions against disenfranchised Jewish and Romish residents, for example.

Thus, our theory predicts that the US state should act mostly humanely toward its own majority citizens and less humanely—sometimes even brutally—toward non-citizens. Arguably, US government actions over the last two centuries have largely conformed to these expectations.

In spite of this somewhat grim assessment, there is authentic hope for a far better international future (Chapters 16 and 17).

The American experiment has many other lessons to teach us. Reinterpreting US history in view of our theory will be of considerable interest.[25] We will examine a few of these important additional details in Chapter 17.

Box 15.3 – The Gettysburg Address

Four score and seven years ago our fathers brought forth on this continent, a new nation, conceived in Liberty, and dedicated to the proposition that all men are created equal.

Now we are engaged in a great civil war, testing whether that nation, or any nation so conceived and so dedicated, can long endure. We are met on a great battle-field of that war. We have come to dedicate a portion of that field, as a final resting place for those who here gave their lives that that nation might live. It is altogether fitting and proper that we should do this.

But, in a larger sense, we cannot dedicate—we cannot consecrate—we cannot hallow—this ground. The brave men, living and dead, who struggled here, have consecrated it, far above our poor power to add or detract. The world will little note, nor long remember what we say here, but it can never forget what they did here. It is for us the living, rather, to be dedicated here to the unfinished work, which they who fought here have thus far so nobly advanced. It is rather for us to be here dedicated to the great task remaining before us—that from these honored dead we take increased devotion to that cause for which they gave the last full measure of devotion—that we here highly resolve that these dead shall not have died in vain -- that this nation, under God, shall have a new birth of freedom— and that government of the people, by the people, for the people, shall not perish from the earth.

Late modern states, Case 2: Faux democracy – The soviets, Trotsky, Lenin, Stalin, and the exploitation of guns by authoritarian elites

Tzarist Russia in 1917 was an archconservative bastion of the authoritarian early modern state. It was a classic *ancien regime* with a small wealthy elite supported by a strong coercive policing and war-fighting machine, paid for by a heavily taxed underclass. Russia was fighting elite German forces in its western territories as part of World War I. Germany was winning against the Russians, if not against France, Britain, and the US on its other front. Not only was the war going badly for Russia, but it cut the primitive Russian economy off from the European markets vital to its functioning. All interest groups within Russia, both elite and commoner, were suffering and disaffected.[26]

Most importantly, the economic assets to compensate the infantry were depleted while the risks of death skyrocketed. In pursuit of self-interest, many members of the tsarist army withdrew their coercive support. Moreover, local worker groups armed themselves, forming militias. They were able succeed in this arming with the help and tolerance of the disenchanted military who controlled armories. Together, these forces represented decisive coercive threat in crucial urban localities, including St. Petersburg and Moscow. Predictably, they collaborated to take over the Russian state.[*]

These new concentrations of coercive means were sufficiently powerful that they initially swept everything in their path. Individuals who later acquired some leadership roles, like Trotsky, Lenin, and Stalin, were initially just along for the ride. This armed coalition of disaffected soldiers and workers was ultimately able to establish control over crucial parts of Russian infrastructure, including the rail system. This acquisition allowed them, in turn, to take control gradually of most of the country and to feed themselves.

On behalf of this concentration of coercive force, Trotsky and Lenin negotiated peace with the Germans, surrendering massive amounts of territory temporarily, but stabilizing the situation in what remained of Russia. This new armed coalition (now called the Red Army) then prosecuted a three year civil war against other armed interest groups, ultimately winning that war. Lenin's original political group was referred to as the *Bolsheviks* and this civil war came to be called—for public relations purposes—the *Bolshevik Revolution*.

The Bolshevik Revolution was to the modern authoritarian state what the American Civil War was to the modern democratic state, a defining event. The crucial point for us here, though, is that these seemingly different cases actually reflect *precisely the same logic*. In both the Russian and North American cases the very same process played out. A well-armed, self-policing coalition seized coercive dominance. It then used that dominance in pursuit of self-interest, producing local economic systems that served the interests of these armed interest groups.[†] The only difference between the two cases was the fraction of the population with ownership of a stake in decisive coercive power.[‡] This observation is, of course, precisely what our theory predicts.

---

[*]   See Wade (1984) for an excellent description of the redistribution of coercive assets at the outset of the Russian Civil War.

[†]   In fact, we argue that ALL modern revolutions have precisely this logic and structure. If read for its empirical (not theoretical) content, Theda Spockol's *States and Social Revolutions* (1979) is a wonderfully insightful account of other famous cases, including the French Revolution, the Great Rebellion and Glorious Revolution, and the Chinese Civil War.

[‡]   This "modest little difference" between the North American and Russian cases turns out to have substantial economic, ethical, and social consequences, of course (below and Chapter 17).

The following details of the Russian Civil War are valuable and illuminating.

As we discussed in Chapter 13, elite control requires the systematic, continuous intimidation of non-elites. Thus, we predict that elite coercive coalitions controlling large commoner populations would be forced to employ the same strategies. As predicted, the Red Army used coercive threat on an unprecedented scale during the Russian Civil War. Gunpowder-based intimidation as a political weapon had been applied on a small, less industrial scale before—the Terror of the French Revolution is a well-known example.[27] However, terror on this new scale required the development of gunpowder weapons that were cheap and easy to use in very large numbers. These had just been invented a few decades before (above). Trotsky, Lenin, and the Red Army used them in the service of acquiring elite domination on a truly massive scale.[28]

The Red Army used terror in two illuminating ways. First, they established an authoritarian military. This involved conscription of foot soldiers, policed, in turn, by elite military units (below). Further, elite military units were deployed with machine guns behind the lines during combat so that those conscripts who deserted or refused to fight were killed.

Of course, this type of military discipline enforcing cooperative war making is just the fundamental human trick put to narrow elite ends. This approach was extraordinarily successful. The Red Army rapidly became the most formidable of the combatant interest groups. As the Red Army became more dominant, conscription became ever easier and the level of overt coercion necessary to sustain military discipline declined.[*] The Red Army was now formidably capable of enforcing the taxation necessary to sustain itself.

Having become the locally decisive elite coercive power, the Red Army was now in a position to control and domesticate the commoners who originally helped bring them to power, as predicted by our theory. Peasants were given land to own and cultivate, though this was temporary as the Red Army and its administrative elite, the *nomenklatura*, would ultimately re-seize the land through *nationalization* and *collectivization*. Likewise, the representative political bodies initiated at this time were almost immediately suppressed—never being a real, functional factor in Russian politics.

Unarmed Russian peasants and workers were among the targets of another major application of terror as a political weapon. They were

---

[*]   This is just one more example of the universal human pattern. See Box 14.1.

effectively re-enslaved by the Red Army—"meet the new boss, same as the old boss," yet again. The Red Army made liberal use of firing squads in every locality, killing people publicly on the flimsiest pretext to establish complete coercive dominance over commoners.* You will note this is exactly the same logic we have seen before in smaller scale elites, for example in the cases of the Hawaiians and Aztecs (Chapter 13).

The Trotsky/Lenin model for the exploitation of modern weapons on a large scale by authoritarian elites would be well noted by many to follow into the contemporary era.[29]

Before leaving Russia, we note that its dysfunctional Bolshevik era authoritarian economy (Chapter 17) ultimately led to massive defection by factions of the Red Army. This happened when the collapse of oil prices deprived them of their primary source of income in the late 1980s. What followed was an ostensibly *democratic velvet revolution*, as various elite factions again recruited commoner support in their grabs for power.

Initially, coercive power was captured by wealthy oligarchs having private police capability. These were rather nakedly organized criminal enterprises. However, with the assistance of residual coercive power in the military and state police apparatus, a faction led by Vladimir Putin was able to build another elite coalition to seize stable control of the Russian state.

As Russian chess grand master Gary Kasparov recently stated, contemporary Russia is, once again, *a police state.*† The present condition of the Russian state is remarkably similar to its pre-1989 condition. Its dysfunctional economy is fundamentally parasitic on the functioning economies of the democratized West, running off the profits from the sale of vast natural gas and oil reserves to functional economies. As in 1989, when and if gas and oil prices sharply decline, the Russian economy may once again fold. This may be beginning as we write in early 2009. It is a worthy goal of the global community to find a way to democratize this chronically failed state (Chapter 17).

---

*   The Red Army faction had no monopoly on this practice. The so-called *White counter-revolutionary forces* behaved in precisely the same way, as predicted. James Palmer's (2009) account of parts of the White forces in the Russian Civil War is most illuminating. It is also quite general. This is how elite concentrations of coercive power behave, everywhere and always.

†   CBS News interview, *60 Minutes*, September 23, 2007.

Late modern states, Case 3: The Meiji Restoration and modern Japan – A natural experiment in the control of coercive dominance[30]

Japan's relative physical isolation from the Asian mainland has made it a valuable natural experiment for us throughout this book. It will now serve this valuable role again. We saw previously that the 16th Century importation of first-generation gunpowder weapons rapidly produced the consolidation of the expected hierarchical early modern state, locally referred to as the Tokugawa Shogunate (Chapter 14).

The Tokugawa elite continued to pursue a policy of radical isolation from the rest of the world, in part, as a strategy to maintain relatively unquestioned elite domination. This isolation policy was amazingly successful for nearly 250 years! However, the modern economic miracle (Chapter 17) was well underway throughout other large parts of the world by the mid-19th Century.

The newer weaponry produced by this global economic revolution ultimately penetrated Tokugawa isolation. It is traditional to date this intrusion to the visit of the American Navy in its huge, heavily armed *Black Ships* under the command of Commodore Matthew Perry.[31] Perry's ostensible goal was to coerce the xenophobic Japanese to desist from abusing and killing shipwrecked American sailors. His more important goal was to begin the process of forcing open Japanese markets to American trade.

Perry's flotilla sailed into what is now Tokyo Bay in 1853. Not only were American ships armed with artillery vastly beyond anything the conservative Tokugawa state had developed, American soldiers also had modern rifled, flintlock muskets capable of relatively high rates of massed fire with stunning accuracy and range.

Before 1853, coercive power was distributed across the Tokugawa countryside in a nominally samurai elite. Local elite oligarchs—including *daimyos* (roughly governors)—ruled *hans* (roughly provinces), on behalf of the central ruler, the *shogun*. Among these han were Choshu and Satsuma. Both of these han were in the southern part of the county, well removed from the central government. They were also coastal provinces with extensive illicit commercial contacts with Western merchants.

After Perry's visit, Choshu and Satsuma rapidly acquired large numbers of modern guns and artillery from Western merchants. Their forces radically reinvented Japanese war fighting, basing it increasingly on massed peasant/commoner militia (*shotai*) armed with Western muskets. Other han emulated Choshu, producing wide distribution of this new coercive capability.

By 1863 a civil war/revolution was underway. The earliest social contract beliefs on which the revolt was ostensibly based was ejection of foreigners, a superficially conservative, even reactionary message. However, the commoner militia responded better to a different appeal. Commoners who felt abused by the hierarchical Shogunate were told that the Emperor, marginalized to symbolic irrelevance under the shoguns, had always had their interests at heart and that restoring the Emperor to his rightful place would address their concerns. Hence the nominal goal of the revolt became restoring the emperor. The ultimate benefactor of this public relations gambit was the new young emperor, Mutsuhito. His formal name later became the Meiji (enlightened) emperor and this successful revolt came to be called the *Meiji Restoration*.

As will see below, the Meiji Restoration completely remade the Japanese state—its leaders, its beneficiaries, its formal forms. Thus, within *fifteen years* of Perry's first visit to Japan bringing elite modern handguns with him, a massive social revolution had taken place! This is one of the most remarkable natural experiments in all of human history. The introduction of a new level of coercive weaponry arguably produced a complete remodeling, even recreation, of a modern state, *overnight*. It is enlightening to pause and consider some of the details of this particular transformation.

First, the coercive threat of the heavily armed militias of the Choshu and Satsuma han and their allies gave them decisive control over the state. Many of members of the Choshu/Satsuma leadership were middle level samurai or nobility and even, in some cases, commoners. They used their control over local finances (agricultural taxes mostly at the beginning, taxes on commerce later) to sustain their military assets and control until the state was remade according to their wishes.

Second, early in their takeover, the Restoration forces appealed to commoner interests, forcing the Emperor to sign the Five Articles Oath (Box 15.4) which ostensibly committed him and the government to respect commoner rights. Its subtext was also to reassure the new Meiji elite that their rights would be protected.

Third, the extensive national Restoration coalition exploited its coercive dominance to remake the logic of the state completely. This included the de facto exclusion of most of the members of the former ruling samurai class from their traditional state subsidies. Indeed, this expulsion from the bounty of the state was so severe that samurai revolts had to be put down, ironically by commoner militia, on several occasions.

Fourth, the early Meiji militias supported the construction of an ostensibly democratic state with representative, judicial and executive institutions. Moreover, since the Meiji militias were extensive and diverse, this emulation of the institutions of radical democracy was more realistic than in the Red Army's Russia, for example.

Fifth, in spite of these ostensibly democratic institutions, the military descendents of the Meiji militia never surrendered coercive power. In fact, with a largely disarmed electorate, there was no credible competing source of coercive power. Over the late 19th and early 20th Centuries, the Japanese military became increasingly ambitious in its control of the Japanese state and its use of that control to pursue its interests. By the onset of the Pacific War in the mid-20th Century, Japan was a police state little different from the Russia of the Red Army.* This brings us to the next coercive transformation of the Japanese state.

Beginning in Manchuria and Korea early in the 20th Century, aggressively ambitious Japanese military cliques successfully conquered large areas of Asia, creating a fictitiously independent state dependency Manchukuo.[32] These successes were extended, including conquests in Southeast Asia and throughout the western Pacific Basin.

The major competing military power in the western Pacific was the US. Japanese military planners hoped to force US planners into a negotiated settlement granting Japan hegemony in large parts of the western Pacific basin. One of the tactical moves toward this goal was the famous attack on American naval assets at Pearl Harbor Hawaii in early December 1941.

This attack was a desperate gamble that ultimately failed. After establishing its dominance in the European theater against the Nazi/SS state—ironically with the indispensable collaboration of the Red Army's Russian state—the US was able to turn the massive technical and economic superiority of the radically democratized state to the Pacific War (Chapter 16). Within several years, the US was able to cut off all overseas Japanese assets and take extensive control of the air over Japan.

This air superiority allowed the US to burn to the ground many Japanese cities, initially with massive conventional incendiary raids and later with nuclear weapons, inflicting hundreds of thousands of civilian

---

*   Notice that social contract and identifier beliefs, including *radical nationalist, national socialist, racist, benevolent imperialist,* received extensive public airing in early 20th Century Japan. However, as always, these are merely the recruiting tools of coercive power, not causes of historical processes, on our theory (Chapter 10).

deaths. At the same time the island-hopping, air-supported infantry campaigns by US forces took and inflicted appalling casualties.*

Shortly after the nuclear attacks on Hiroshima and Nagasaki in August 1945, the Japanese surrendered. Supreme Commander, Allied Powers (SCAP) Douglas MacArthur led an occupation including approximately eighty thousand American infantry. SCAP staff oversaw the complete demilitarization of Japanese society. Returning Japanese soldiers were sometimes reduced to begging on the streets. The creation of a modern constitutional democracy with traditional representative institutions was the end result. This was, yet again, a radical reconstruction of the Japanese state; this time at the barrel of American rather than Meiji guns.

It is noteworthy that the Japanese constitution emerging from the American occupation includes a *Peace Article* in which Japan disavows future militarization. Under the umbrella of American protection/coercion this article continues to be observed sixty-five years later. The Japanese electorate remains relatively unarmed. However, in the absence of internal concentrated professional coercive power, and with relative American indifference, Japanese democratic institutions have, thus far, survived intact.

Notice how stunningly rapid and radical each of these two transformations of Japan were. Notice also that the Meiji Restoration was based on elite control of coercive power and ultimately—inevitably, on our theory—produced an authoritarian state. Conversely, the radically democratized North American state used its coercive domination to de-militarize the Japanese police state, producing an apparently stable democratized Japan.

Japanese culture, language, religion, traditions remained essentially the same—only the distribution of coercive threat changed. Yet this single change apparently trumped all other forces and interests. This is, of course, precisely the result our theory predicts.

---

* For example, the infantry assault on Iwo Jima and Okinawa—islands near the Japanese mainland— killed approximately fifty thousand and one hundred and seven thousand Japanese with about fifty thousand and forty-nine thousand American deaths, respectively (see page 651 in Jansen, 2000).

Box 15.4 – The Five Articles Oath (Meiji Japan)

Article I : Deliberative assemblies shall be widely established and all matters decided by public discussion.

Article II: All classes, high and low, shall unite in vigorously carrying out the affairs of the state.

Article III: The common people, no less than the civil and military officials, shall each be allowed to pursue his own calling so that there may be no discontent.

Article IV: Evil customs of the past shall be broken off and everything based upon the just laws of nature.

Article V: Knowledge shall be sought through the world so as to strengthen the foundations of imperial rule.

Authoritarian state militaries are ponzi/pyramid schemes while democratized state military personnel are civil servants

It is vital to grasp how the ancient universal human trick of self-interested coercion actually plays out in modern states. We begin with a look at the worst that human coercion has to offer followed by the alternative, a radically democratized and more humane solution.

The Russian Red Army and the Japanese Meiji and Imperial Armies illuminate something fundamental about elite gunpowder coercion—everywhere and always. The armies of authoritarian states exert their coercive force in the interests, preferentially, of military elites whose individual members generally bear little of the day to day risk of dying in warfare. Yet, the real coercive power of these armies is their millions of foot soldiers, individuals who benefit only modestly from the authoritarian state. Why do these soldiers agree to take these risks for, mostly, someone else's benefit?

Historically, many answers have been offered. For example, new Imperial Japanese conscripts were heavily indoctrinated being forced to recite endlessly, "Duty is weightier than a mountain, while death as light as a feather."[33] Recruits were required to believe, or at least publicly pretend to believe, such foolishness. On our theory, indoctrination is secondary. It really is just a cue of actual threat and the required response to that threat. In effect, the Imperial Japanese recruits were being told to behave *as if* "death were as light as a feather" if they did not wish to be shot. Military discipline, again.

Imagine you are a new, young conscripted foot soldier in Stalin's or Hirohito's army. Typically, you would have come from a commoner, even an impoverished background. Moreover, you are individually powerless to resist initial conscription. And, of course, you wish to eat. Under coercive threat—remember Trotsky's machine guns aimed at the backs of his own troops—you fight. It is your best available option at the moment.

You survive and you are promoted. Each promotion increases your pay and puts you in a better position to reduce your risks in combat, provided new recruits are conscripted to take your former place.

What are your choices?

First, you can perpetuate the system. You contribute your coercive threat to recruiting and training the next round of conscripts, gaining better pay, less risk. As a recipient of tax revenue, your condition is improving. Moreover, once local dominance by your forces is established, your combat risks often become small indeed.

Second, alternatively, you can retire, return to your poorly paid civilian existence, and surrender your hold on a share, however small, of the only coercive power in your world. In other words, you could voluntarily change status from predator to prey.

Third, you could try to organize your fellow soldiers to break the cycle, challenging the current social contract and refusing to carry out orders designed to enforce elite interests and to coerce new conscripts. If you are successful, a coercive power vacuum will have been created. You could then work toward a more democratized system, but with little confidence of success and even less assurance of reward if you do succeed. Moreover, unless you are successful in recruiting a large enough number of your fellow soldiers to your cause—you are powerless alone or in small numbers—you will be shot/hung/guillotined for treason.*

Which option do we choose? Be honest!

---

* Precisely such revolts from within authoritarian militaries initiated both the Russian and the French Revolutions (above). They occurred when the capacity of the existing paymasters (ostensible elites) could no longer make the cost-benefit ratio of service acceptable. Of course, these armies-in-revolt still hold decisive coercive power. They never democratize access to threat. They merely transfer their coercive assets from one elite to another—from Louis XVI to Napoleon, from the Tsar to Lenin, Trotsky, and Stalin.

Thus, we see once again that authoritarian military organizations are really organized criminal enterprises. They share some features of pyramid schemes, though even the lowliest conscript is usually a little better off than the workers and peasants he "protects." Moreover, these military organizations always run aggressive protection rackets. They extract a massive resource flow from the larger population often under the guise of a *centrally planned economy* or *urgent national defense* needs.

In Chapter 17 we will return to the history and future prospects of disassembling such criminal enterprises.[*]

Contrast this with the military of a democratized state. First, coercive conscription generally occurs only when an electoral majority supports it in pursuit of their coercively enforceable interests. This, in turn, generally requires a real, consensually recognized external threat—actual war. During extreme danger—the US in World War II, for example—this military approximates a coalition of the whole. More commonly the "need based" military of a democratized state involves relatively small numbers of combat infantry.

Second, these properties of democratized militaries result from the fact that the electorate has sufficient access to coercive threat that the professional infantry is vulnerable to majority coercive threat, on our theory. Under these conditions commoners are not managed by the military elite, but vice versa.[†] The military of a modern democratic state has the same status as the local constabulary. As long as it serves its function, it is supported and respected. When it fails (or threatens) it is disassembled and rebuilt.

---

[*]　Notice that we have deliberately allowed ourselves to take an ethical position here. Of course, in the broader view, the members of authoritarian militaries are just doing what two million years of humans have always done, projecting coercive threat in pursuit of self-interest, and justifying it.

[†]　At first glance, we might think that this is a poor description of many modern democratic states—Japan and Western Europe, say. Their electorates are very lightly armed. We will return to this issue in Chapter 17. For now, note two things. First, Japan, Western Europe, Canada, Australia, and others also have relatively small, weak militaries. Thus, the low level of civilian coercive threat in many of these states is arguably sufficient to manage their militaries. Second, more importantly, all these states developed their contemporary political and military structure under the umbrella of global American military power. Thus, the coercive balance that matters most is between the *American* electorate and its infantry. The contemporary American electorate is still extraordinarily heavily armed (Chapter 17). Moreover, the number of infantry on American soil at any moment in time is highly limited.

If these implications of our theory for the maintenance of democracy disturb you, be patient. There is authentic humanity and great hope for the future here (Chapter 17).

Box 15.5 – Elite cadres and control of information in the modern authoritarian state

The preceding section of the text describes how authoritarian militaries share properties with pyramid schemes. The question arises as to why the foot soldiers of the authoritarian state put up with their relatively inferior status. Collectively these soldiers hold decisive coercive power within the state. They could overthrow the elite dominated system and install themselves as an internally democratized, large ruling group. Why do they not?

The answer to this question will be extremely important to us as we examine the human future in Chapter 17. We argue that the answer has two parts, both crucial.

First, the main body of authoritarian state military is under constant, mostly secret monitoring by elite armed units. Call them hyper-elites. These include the Gestapo in Nazi Germany and the KGB and related agencies in Stalinist Russia. These hyper-elite police help run the state and are very well paid. They have the power to summarily imprison or execute a member of the military for "treason" without due process. They are able to intimidate local components of the state military, unless the state military can bring its larger force to bear against them. These hyper-elites also police conflicts of interest within their own ranks.

Second, the enormous size of modern states means that access to the necessary information to enable the millions of soldiers of the state to coordinate coercive action against a hyper-elite is not assured (Box 13.2). Hyper-elites jealously control the flow of such information. Indeed, in general, they interpret any attempt by a non-hyper-elite member to address the members of the larger military as prima facie evidence of treasonous intent.

Thus, authoritarian militaries *must* have secret police. They *must* also be highly secretive about their functioning and decision making. Otherwise, they are immediately overwhelmed by the self-interested members of their necessary legions of cannon-fodder. The militaries of democratized states have neither of these properties. These insights will be extremely useful to us as we turn to the human future (Chapter 17).

Where are we going?

The empirical record is powerful. After five hundred years of archaic states, gunpowder arrives and the cooperative human coalitions we call *modern states* consolidate rapidly in its wake. This is precisely what our theory of history predicts. Moreover, when access to individual gunpowder coercive threat is broadly and relatively uniformly distributed, these states are spectacularly wealthy (Chapter 17). These democratized states tend to use their wealth to support the well-being of their enfranchised citizens. Under the best of circumstances, democratized modern states can be relatively humane, rewarding environments. This, too, is what our theory predicts.

However, we have two large problems for the pan- global community to solve:.

First, disenfranchised individuals in the modern state—in the minority in democratized states and in the majority in authoritarian states—are not fulfilled. They are impoverished and desperate. Now that we understand the human condition and its origins, we must respond to this unacceptable state of affairs. In Chapter 17, we will return to this crucial ethical obligation.

Second, the properties of gunpowder weaponry make it inadequate to the task of cost-effective, self-interested management of conflicts of interest on a global scale. In other words, individual states have conflicts of interest with one another that they cannot manage with gunpowder weaponry. The ineffectual brutality of World War I illustrates this limitation, for example.*,[34]

This creates perfectly predictable failures of cooperation, producing chronic warfare and tragedies of the commons on a planetary scale (Chapters 5, 13, and 17). Even highly democratized states that behave relatively humanely toward their own citizens are capable of profound brutality outside their own boundaries.

To forestall chronic conflict between states, and their amoral international exploitative behavior, new coercive capabilities are required, on our theory. We must have the technical means for coalitions of multiple states to relatively cheaply coerce individual states. We have recently come into possession of precisely such technologies and this is the subject of Chapter 16.

൭

---

*   Modern states are just the MPUs (maximum policeable units, Chapter 12) for gunpowder weaponry. They are expected (and observed) to war with one another relentlessly, just as Neolithic settlements did, for example (Chapter 12).

Fifth Interlude:
Never again - pathological deployment of human coercive threat and the brutalization of the powerless

In Chapter 15 we touched on some of the horrendous human tragedies inflicted by modern states. In Chapter 16 we will briefly encounter
some of these yet again. In order to be prepared to confront our common
task in building a humane, pan-global human world we must examine
and understand these things. But it is equally important that we do not
allow the overwhelming horror of some of these events to either numb us
into inactivity or repel us into turning away from the crucial challenge.

To better inoculate us against this kind of self-defeating revulsion it
is helpful to recall what our theory says about how such terrible things
come to pass. Throughout most of our two-million-year evolutionary
history, access to coercive power resulted from being a member of a
broad, majoritarian consensus. Anyone outside of that consensus, and
therefore powerless, was, by definition, a free rider on the common
enterprise. There was no other route to powerless.

As a result, our adapted minds are extremely, irresistibly oriented to
react to powerlessness *within our social units* with contempt, even with
revulsion (Chapter 10).

In the ancestral environment, this response was one of the very
strong incentives for individuals to avoid free riding or other kinds of
anti-social behavior toward the consensual coalition. However, in the
context of modern authoritarian states this response has a profoundly
dark consequence. Coercive power is extremely hierarchically distributed in these states. The vast majority of members of their populations

are powerless relative to the elite armed policing apparatus universal to these states (Chapter 15).

Thus, almost everyone is held in contempt by the elite policing machinery of the modern authoritarian state. Moreover, this elite policing machinery can facilitate its management of the large commoner population by publicly subdividing that population to create extremely powerless small subgroups, the hyper-powerless. These hyper-powerless subgroups are then held in contempt even by the relatively powerless larger commoner population.* These hyper-powerless individuals can be used for public display of the coercive power of the state to the larger commoner population in a form that commoners will tolerate, even support. Thus, the desperate stability of the hierarchical state (Box 13.3) is established.

The supreme contempt toward the hyper-powerless natural to the evolved minds of authoritarian hierarchs not only opens the floodgates to their own basest impulses. It also creates an environment in which pathological, sadistic personalities can succeed and prosper (Chapter 16). This, we argue, is the simple recipe for the unthinkable monstrosities of the modern authoritarian state.

This is the (unspeakably) bad news. But there is good news hidden here. By systematically empowering all members of the global community and democratizing the local political cultures in which they live, these monumental inhumanities need never occur again. These tragedies are *not* the products of belief systems or economic/political systems, per se. They are not the products of specific cultures and not others. Rather, they are always and everywhere the products of only one thing, the hierarchical distribution of coercive power.

Universal pan-global democratization of access to coercive power is an achievable goal, perhaps by the end of this 21st Century (Chapter 17). Our great-great-grandchildren, living on such a planet, will look back on the brutalities of the 20th Century as something alien and unthinkable.

When we go on to say "empower" all the citizens of the pan-global human society, we will mean this *literally*-give them access to locally decisive coercive threat. We will *not* be using this word in the utterly inadequate, metaphorical, illusory sense in which it has so often been used in the past. We will have much more to say about this in Chapter 17.

∽

---

*   This reaction to powerlessness also explains how even democratized states can be inhumanly brutal toward the truly powerless—the genocide against Native North Americans by the early Euroamerican colonialists, for example.

Chapter 16
The pan-human population takes full ownership of planet Earth – aircraft, nuclear explosives, and the emergence of the pan-global human village?

*Now, I am become Death, destroyer of worlds**

On the morning of August 20, 1680 the Pueblo Revolt had it first great success. The Spanish governor of Santa Fe evacuated the town, leaving what is now New Mexico in the hands of Native American forces for the next twelve years (Chapter 13). Constantly watching its flanks, this retreating column of Spanish refugees followed the Rio Grande River south to safety at El Paso, on what is now the US/Mexican border

165 years later the era of the great modern states, each acting as a relatively autonomous agent (Chapters 14 and 15), was drawing to a close. At 5:30AM local time, on July 6, 1945, just a few miles east of the Spanish governor's escape route, the scientific power of these states unleashed a force never before seen on Earth's surface. The first human nuclear detonation—the blast code-named Trinity—*unleashed the explosive power equivalent to twenty million pounds of TNT.*†

---

* From the Hindu scriptures in the Bagavad Gita. Robert Openheimer reported this phrase to have been his first thought in the New Mexico desert on the morning of the first atomic explosion in Earth's history—the Trinity blast (text below).

† TNT is one of the modern descendents of the original *black powder* gunpowder (Chapters 14 and 15). It is a chemical explosive. One pound of TNT can demolish a typical small North American suburban house, for example.

This new nuclear high explosive was much more than just a potent scientific curiosity. One month later—August 6 and 9—bombs similar to Trinity were dropped from American aircraft on the Imperial Japanese cities of Hiroshima and Nagasaki. Approximately one hundred thousand to one hundred fifty thousand people were killed immediately by these explosions with many more perishing over the next few days from acute aftereffects.[1] Others continue to die prematurely through the present from radiation-induced cancers. This catastrophic loss of life was produced without the loss of a single American pilot.

A few months earlier, British and American bombers dropped tons of chemical explosive and incendiary devices on the German city of Dresden, engulfing much of the city in an unimaginably intense firestorm. Two young children ran in panic from their home with the first flames. Wandering through the city, they survived the inferno, by luck alone. Afterward, they saw thousands of dead people and animals littering the streets. Returning home, they inspected the bomb shelter in the basement of their home. There they found a delicate replica of a human being seated along the wall. It was so fragile that it collapsed into a small pile of fine, talc-like ash when touched. Within the pile was a piece of partially melted jewelry, allowing the children to identify the person who had been flash-incinerated by the bombing. She was their mother.[*]

This incendiary raid took the lives of tens of thousands of German civilians.[2] A few tens of Allied pilots were lost in the raids, mostly to anti-aircraft fire from the ground and mechanical malfunction. Similar area bombing by the Allied forces was used to destroy many other German and Japanese cities in the end stages of World War II.[†] Total deaths from all such raids, conventional and nuclear, are not precisely known ,but figures exceeding six hundred thousand are reasonable.

The predictions of our theory about the emergence of the contemporary world from the earlier modern world are clear. The limitation on the scale of cost-effective coercion—and thus, cooperation—allowed by gunpowder weaponry determines the scale of the modern state. Once this inherent limitation on the size of the state, imposed by gunpowder's capacity to manage conflicts of interest, is reached, states will have unmanageable

---

[*]   This fictionalized story is realistically derived from various survivor accounts. See Murray and Millet (2000) and Ferguson (2006) for examples.

[†]   These raids were carried out essentially exclusively by the British and the Americans as the Soviets, the other major Allied power, had limited strategic bombing capability (see Murray and Merritt, 2000, and Bergerud, 2000, for reviews).

conflicts of interest among themselves. Gunpowder states will be in chronic, recurring, potentially endless conflict with one another (Box 16.1). The logic is precisely the same as for the chronic conflict between Neolithic settlements policed by the bow, for example (Chapter 12; Box 12.1). On our theory, this grinding perpetual conflict between gunpowder states is the late modern world for the last several centuries through the end of World War I.

In contrast, today, we live in a world where militarily equal nation-states relatively rarely go to war directly with one another. Moreover, different states now routinely engage in extensive, though not yet fully "free," trade.[3] We occupy an increasingly global economy.* Further still, the UN and international agencies like the IMF and the World Bank play relatively central roles in the planet's economy. Previously, such transnational agencies either did not exist at all or were even more ineffectual and marginal than their contemporary equivalents. Think of the League of Nations in the early 20th Century, for example.

*This contemporary **global cooperation**—incomplete as it is—raises the next historical opportunity to falsify our theory. We predict that this cooperation will be like all human cooperation, everywhere and always. It is ultimately limited exclusively by the conflict of interest problem. The existence of this cooperation implies the existence of a new coercive means allowing the <u>**relatively**</u> inexpensive management of conflicts of interest <u>**between**</u> states. We argue that modern aircraft—sensu lato, including planes and missiles of all sorts—delivering high explosives is this predicted new coercive means (Figure 16.1).*

We have two goals in this chapter. First, we will explore how well the empirical evidence conforms to the predictions of our theory. Second, we will continue to use the massively detailed record of the modern and contemporary eras to illuminate further the implications of the ancient human adaptation, coercive management of conflicts of interest. This second goal will give us further insights crucial to thinking about the human future in Chapter 17.

---

*   The global financial liquidity crisis beginning in late 2008 and the international response to it bring this planetary relationship into sharp focus (Chapter 17).

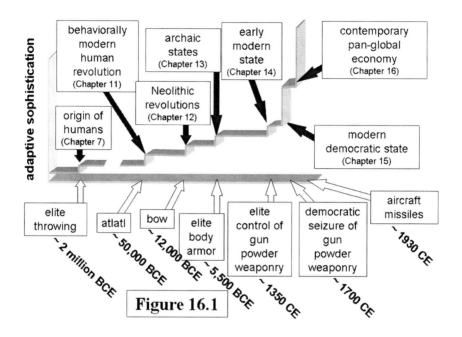

Figure 16.1

We global citizens – thinking big

This chapter deals with the most technically complex coercive technologies in human history to date. These new technologies allow humans to "think" with new scope, projecting inexpensive coercive threat and reaping its by-products on the vast scale of our entire planet. In spite of the enormity of a "village" with six billion citizens, the fundamental story remains as it has always been—and as simple, too—on our theory.

No matter how colossal a weapon might be, all that matters is whose interests it is used to defend. Moreover, enormous modern weapon systems are like the gunpowder artillery (Chapters 14 and 15). They require a vast network to supply the capital, expertise, and labor to make and use them. Thus, they are always controlled, in turn, by those who control this social supply network.

Of course, our theory tells us precisely who will control these local supply networks. These networks will be owned and deployed by those who hold decisive coercive power **on the ground**. This coercive power can only emerge from the control of locally decisive individual weaponry. In our world, this decisive local weaponry is still the same gunpowder handgun of the last several centuries.[*,4]

---

*   Of course, modern gunpowder is several technical generations removed from the original black powder that allowed the initial formation of the modern democratic and authoritarian states (Chapters 14 and 15). However, the principle remains the same.

Thus, we can have relatively high confidence that the new weapons of planetary range will be deployed in the interests of coercive elites by modern authoritarian states and in the interests of majoritarian coalitions in modern democratized states (Chapter 15; Box 16.1).* We can realistically think about the strategic logic of these weapons of mass destruction as long as we keep this crucial feature in mind.

We will use two historical terms with specific meaning. First, we called the era that gunpowder wrought *modern* (Chapters 14 and 15). Second, we will refer to the era engendered by aircraft and missile technology as *contemporary*. The contemporary era begins with the run up to World War II and it is convenient to use this war as the start point of this distinct period.

My country right or wrong – the social logic and human psychology of contemporary war making

The carnage of modern gunpowder wars like the American Civil War or World War I was appalling. Yet contemporary aircraft-based warfare takes death and destruction to a new level. For example, six hundred thousand died in the American Civil War. They were mostly soldiers; whereas roughly that number of innocent civilians died in World War II area bombing alone and probably another *twenty to fifty million* through other actions and effects of the war! Moreover, the majority of the *deliberate* military killing of civilians by specifically planned area bombing was carried out by the US and Great Britain, ostensibly democratized states.

These details of the Second World War can outrage our ethical sensibilities. However, as always, we must seek the wherewithal to look at these events with the dispassion of our extraterrestrial anthropologist. This perspective is easier to adopt if we pause to reflect again on the role and evolutionary origins of ethical outrage in our public political behavior.

Our theory predicts that these moral feelings are not what we consciously believe them to be. Rather, public expression of ethical outrage is merely a recruiting action designed to bring others to share and potentially enforce the social contract (and identifier) beliefs that serve our interests (Chapter 10). If we aspire to have our ethical reactions taken as more than this we need to understand them better and deploy them in the pan-human interest.

---

* In Chapter 17, we will return to the crucial questions of the ongoing relationships between authoritarian and democratized states.

World War II area bombing lets us see how situational our ethical sensibilities have traditionally been. The Allied planners of these actions were mostly not psychopaths or megalomaniacs. Rather, they were military planners *ethically outraged* at the actions of their targets—the Nazi and Imperial Japanese states. The vast inhumanities perpetrated by Nazi and Japanese troops helped stimulate this moral outrage. For example, the summary execution of resistance fighters in Europe and the savagery against civilians in the Philippines and China were among the atrocities well known to Allied planners.[*]

Indeed, soldiers fielded by the pyramid schemes of authoritarian modern state militaries tend to commit individual war crimes more often than soldiers fielded by more democratized states (Chapter 15). Thus, the Allied planners' moral outrage was not entirely self-deceptive. Yet, its consequence was millions of horrendous individual experiences like our two small children in the aftermath of the fire-bombing of Dresden.

We can go even further. Most military planners for the Nazi and Imperial Japanese state were not psychopaths either, though some certainly were.[†] Many were ethically outraged at the historical treatment of the German state at the end of World War I, for example. Some of this outrage was arguably justified. The Versailles Treaty ending World War I was a testament to short-sighted vengefulness by the victors. Other sources of outrage were racist and delusionally paranoid, with people imagining themselves to be the victims of international conspiracies by "inferior" ideologies or "races." Again, however, many of the unspeakable outrages of the Nazi state were committed by individuals nursing a feeling of entitled ethical outrage.[‡]

Our own moral outrage, no matter where we stand on an issue, is not the point. The strategic *role* of this moral outrage *is* the point. Ethical outrage is merely a proximate psychological phenomenon evolved as a social/political tool. It has no *a priori* moral standing. For our ethical psychology to be *authentically humane* it must first be better educated.

---

[*]   Ostensibly, the stupendous scale of the racist Nazi Holocaust was not yet known to Allied planners.

[†]   Authoritarian modern state hierarchies tend to select for sadistic personalities (Fifth Interlude). Thus, some of the local brutality of World War II resulted from full-blown sociopathic behavior. The moving PBS Point of View Documentary *Inheritance* poignantly illustrates one of thousands of examples of this phenomenon (2008; PBS Home Video).

[‡]   Imperial Japanese military planners were equally motivated by ethical outrage. They saw their country—a late comer to modernity (Chapter 15)—as having been unfairly excluded from the fruits of international imperialism by the older Euroamerican colonial countries.

To move toward this education we must first set our subjective ethical concerns aside and view the second half of the 20th Century with detachment. Chapter 17 will explore what might empower us to have a pan-human moral perspective. We will assume this detached attitude and stance about the early stages of the contemporary world—as well as we can—throughout the rest of this chapter.

Aircraft and the contemporary world – the first fundamentals

The famous cynical statement of the Golden Rule says that "He who has the gold makes the rules." As we have seen, "gold" is not what determines human history. Rather, "He who possesses coercive dominance makes the rules." (The gold comes later and as an effect.)

It is essential for us to begin exploration of the contemporary world by looking at how aircraft changed the scale of coercive dominance. Our theory's fundamental prediction about the new features of the *contemporary* world is that they emerge from a single, simple cause. All the diverse novel properties of our world are merely effects of that single cause.

Specifically, the contemporary world emerges from the fact that sophisticated aircraft allow the formation of massive concentrations of coercive power capable of self-interested policing of cooperation between nation-states. The result is enforcement on a planetary scale. Earlier gunpowder projectile weaponry, in contrast, was fundamentally incapable of cost-effectively supporting such global coercive threat.

We are aware of no evidence in the massive historical record of the second half of the 20th Century that contradicts this simple claim. A survey of all the evidence is far beyond our scale here. Rather, in what follows, we will use several selected details and events to illuminate this empirical test of the theory. As always, we welcome attempts by skeptics to use any or all of the record to attempt to falsify our theory's predictions.

Aircraft versus gunpowder artillery – a new scale of coercive threat

It is vital to grasp how aircraft expand the scale of cost-effective coercion (Figure 16.2). In massive state-level warfare, high density use of gunpowder artillery creates "fronts" that are relatively stationary, as

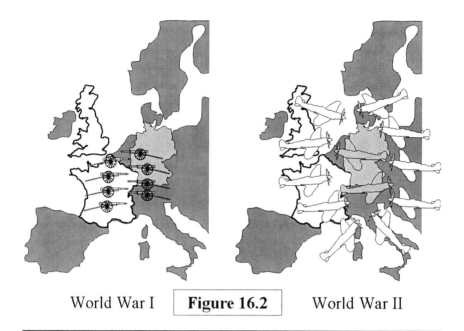

World War I   **Figure 16.2**   World War II

Figure 16.2: Shown is the cost logic for the coercive technology of the First and Second World Wars in the European theater (text).

Consider World War I (LEFT PANEL). Gunpowder artillery creates *fronts* that are relatively stationary. Thus, each side can only bring a limited, and similar, number of weapons to bear at any moment. Throughout most of the conflict each side experiences the same (very high) costs and losses. This situation is similar to proximal killers in non-human animal conflicts as discussed in Chapter 5.

Contrast the World War II condition (RIGHT PANEL). Aircraft allow the conflict to be spread across the entire landscape limited only by range, pilots, and planes. Numerically superior forces can bring their advantage to bear at once and throughout the conflict. This creates an approximation of the coercive square law that governs remote killing in general (Chapter 5). This allows the numerically superior side to seize control rapidly of the air at relatively low costs. The ultimate outcome of the war is then no longer in doubt (text).

opposing sides fortify defensive positions that can survive high intensity artillery fire in a way that moving formations cannot.[5] Under these conditions, for much of the confrontation, the stronger force, with more personnel and weaponry, nonetheless takes initial losses at about the same rate as the weaker force. This happens because only so many men and artillery pieces can be arrayed along a stationary one-dimensional front.

As men and weaponry are destroyed, new assets replace them in a drawn-out, bloody war of attrition. Only very briefly at the end of the conflict, when the weaker side can finally no longer replace its losses, does the stronger side gain numerical superiority and, thereby, the capacity to exploit the coercive square law (Chapter 5) to rapidly end the conflict at relatively low cost to themselves. Overall, the costs to both the winning and losing sides are virtually identical and extremely high. In other words, gunpowder artillery is a poor, costly weapon for policing cooperation *between* nation-states.

In contrast, the availability of movement in the vertical dimension with aircraft allows the fight to be spread across the landscape at will limited only by range, pilots, and planes (Figure 16.2). Thus, the combatant coalition with numerical superiority can immediately assert that dominance to seize control of the air at relatively low cost. Again, notice that all a state's planes can be thrown into battle at will while only that fraction of a state's artillery that fits along the front can be brought to bear at any one moment.

Under the new conditions that aircraft create, again, a reasonable approximation of the coercive square law applies (Chapter 5). Numerically superior aircraft formations can inflict horrendous losses on the numerically inferior side with relatively limited costs to themselves. The war of attrition for control of the air works rapidly and relatively cheaply in favor of the superior side—in contrast, again, to the artillery case.*

Finally, once air superiority is so achieved, the military position of opposing artillery, armor, and infantry formations on the ground become untenable. Ground forces can be strafed and bombed at will. Moreover, ground forces can be starved for supplies and material by cutting transportation lines from the air. The ultimate military outcome is then inevitable, provided the side with air superiority has the capacity and incentive to press its advantage with its own infantry and artillery on the ground.†

---

* Note that these costs are extraordinarily low when averaged over the entire population of the nation-state with numerical superiority in aircraft number. These costs are high to individual pilots, of course. However, these pilots are enmeshed in a coercive environment (military discipline; Chapter 15) requiring their investment in coercion, almost no matter its individual cost to them (text below).

† Note that long-term control of the civilian population is not an inevitable consequence of military victory. This control requires collaboration of the civilian population, their coerced cooperation. How well this can be achieved depends on how heavily armed the civilian population is and the willingness of the winning military force to invest in the costs of coercive management. Democratized states generally find the costs of managing an armed con-

Overlord – aircraft come to the world of gunpowder projectile weaponry

Operation Overlord was the code name for the Allied invasion of the Nazi-controlled Normandy Coast in June of 1944. Comparing Overlord to the World War I trench warfare in this same area twenty-seven years earlier is a good way to illuminate the impact of aircraft on coercive violence between state-level social units (Figure 16.2).

By the inception of Overlord, the Nazi air force (Luftwaffe) had been battered by two and half years of relentless attrition. As Allied aircraft production came up to capacity the more limited capacity of the Nazi state was overwhelmed.[6]

As a result, the Allies had effective air superiority in Normandy. Range considerations allowed extensive fighter protection of heavy Allied bombers against the residue of Luftwaffe fighters. (See Figure 16.3 and online Endnote 7 for examples of these weapons.) In other words, the Allies could bomb and strafe infrastructure essentially at will in the area of the invasion because they possessed coercive dominance in the air.

Exploiting this air superiority, the Allies systematically degraded road and railroad networks on the continent. This bombardment was spread throughout Western France, both to avoid revealing the intended location of the invasion and to compromise the ability of Nazi planners to move heavy equipment in tactical response to the invasion.

When Allied troops hit the beaches of Normandy, a massive mechanized infantry assault, complete with air support, was poured on locally thin Nazi defenders. The German Army (Wehrmacht) could not bring its Panzer mechanized divisions or large infantry formations to reinforce the points of the initial Overlord incursions effectively, as a result of Allied air power. Moreover, throughout the Allied march across Europe to Berlin, columns of Nazi men and materiel were relentlessly harassed by Allied aircraft with tremendous losses.[*]

As an ultimate result of Allied air superiority, Allied infantry and mechanized divisions progressed more or less relentlessly across Western Europe and into the heart of Germany, ultimately helping to end the war in the European theater.[8]

---

temporary civilian population prohibitive in the long term, while authoritarian elites have a stronger incentive to pursue such coercive dominance (Box 16.1; Chapters 15 and 17).

[*]   World War II fighter pilot Quentin Aanenson's interview, sprinkled throughout Ken Burns and Lynn Novick's film *The War* (2007; PBS Home Video), provides a profoundly human and evocative perspective on the terrible work of these Allied pilots.

Contrast this scenario with the trench warfare of World War I. The capacities of infantry and artillery were roughly comparable to those of World War II troops. However, World War I aircraft were much more primitive and of very limited strategic or tactical impact.[9] Thus, artillery fire was only useful to soften up a point on the enemy line before an infantry push—a local tactical act. However, artillery bombardment could not destroy the remote infrastructure behind the enemy lines that would allow enemy planners to mass its own artillery and infantry behind the "bulge" to resist the coming incursion.

The costs of doing business in the global village

The crucial issue is to understand the self-interested logic of all the players making up the contemporary state in view of these properties of aircraft in combat. How does an animal with two million years of adaptation to projection of coercive threat reason—not necessarily consciously—about these new weapons?

First are the pilots who put their lives at risk to allow a state to project such massive coercive power. At first glance, their behavior appears to be insane from the perspective of a self-interested individual. Death rates among pilots on all sides were extremely high throughout the early stages of World War II,[*] until Allied aircraft production ramped up and ultimately allowed the seizure of air superiority. In return for

---

[*]   Joseph Heller's famous novel *Catch-22* (1961) captures this perfectly in the following dialog:

Yossarian says, "You're talking about winning the war, and I am talking about winning the war and keeping alive."
    "Exactly," Clevinger snapped smugly. "And which do you think is more important?"
    "To whom?" Yossarian shot back. "It doesn't make a damn bit of difference who wins the war to someone who's dead."
    "I can't think of another attitude that could be depended upon to give greater comfort to the enemy."
    "The enemy," retorted Yossarian with weighted precision, "is anybody who's going to get you killed, no matter which side he's on."

The position pilots are put in creates Heller's famous "catch" as follows:

There was only one catch and that was Catch-22, which specified that a concern for one's safety in the face of dangers that were real and immediate was the process of a rational mind. Orr was crazy and could be grounded. All he had to do was ask; and as soon as he did, he would no longer be crazy and would have to fly more missions. Orr would be crazy to fly more missions and sane if he didn't, but if he was sane he had to fly them. If he flew them he was crazy and didn't have to; but if he didn't want to he was sane and had to. Yossarian

Figure 16.3: Joanne's father (Cesare Monastra, second from left, front row) is shown with his fellow crew members in front of the strategic B-24 bomber they flew in the Pacific theater during World War II. *Photograph courtesy of Cesare Monastra.*

taking such outrageous individual risks, these pilots were clothed, fed, and paid a small salary for the brief duration of their service/survival. Those who were not killed and did exceptionally well might be honored as heroes or aces but this reward translated into relatively little tangible value.*

---

was moved very deeply by the absolute simplicity of this clause of Catch-22 and let out a respectful whistle.

    "That's some catch, that Catch-22," [Yossarian] observed.

    "It's the best there is," Doc Daneeka agreed.

The real Catch-22 for World War II pilots was that the only way to go home was to fly the required number of missions (twenty in some cases), but the probability of surviving more than five to seven bomber missions in the early days of the Allied bombing campaign over Nazi Europe was low. The choices for the individual air crewmembers were starkly awful. Only the massive coercive power of modern state military discipline can make such behavior common (Chapter 15).

*   Of course, acting to provide ultimate protection to close kin at home—especially parents, siblings and potential spouses of young pilots—was also a part of the rational incentive set.

So why were pilots willing to do what they did? In World War II many pilots were either conscripts or they volunteered early in the war before the huge risks of being a combat pilot were clear. Once in the military, by whatever route, the pilots were subject to military justice. Refusing to obey an order would produce results depending on the context and the state, ranging from imprisonment and having ones economic future ruined to being shot or hanged.[*] Thus, the individual pilot weighed this massive personal coercive threat against the benefits of evading the risks of combat. The (unconsciously) rational, self-interested pilot often elected to fly.

Second are the "funders" who supported the pilots' work, paying for the aircraft and infrastructure as well as the Selective Service and the coercive military justice system. In the case of large states like the US, these costs are distributed among many taxpayers, so the individual cost for most was low. Thus, all those who are not close relatives of combat pilots would regard the costs of achieving the air superiority for Operation Overlord to be quite low and acceptable. Moreover, the benefits of military victory are reasonably expected to accrue to many of the members of a democratized electorate. In general, these benefits exceed the small average individual costs of achieving air superiority when the choice of even cheaper constructive negotiation was not available.

In authoritarian states the cost-benefit logic of those funding the pilots is essentially identical—except that well-armed elites replace democratized electorates as those benefiting from that air superiority (Box 16.1).[†]

We can now ask how we expect geopolitics to unfold in the wake of the invention of aircraft. When the weapons are first refined to the point of battlefield effectiveness, we expect authoritarian states to build large numbers of them and use them for attempted conquest. This prediction results from the much higher individual potential rewards of conquest for members of authoritarian elites than for members of majoritarian coalitions (Box 16.1). Arguably the history of the Nazis and Imperial Japanese in the run up to World War II during the 1930s is consistent with this prediction.[‡,10]

---

[*]   These threats were enforced by MPs (military policy) wielding gunpowder side arms.

[†]   Self-interested pilots cannot use their aircraft to capture the state. Coercive control of the essential social supply network by others precludes such a takeover.

[‡]   The other large authoritarian state of this era, Russia, was not yet up to the industrial challenge of producing these aircraft in large numbers. Likewise, authoritarian China was both industrially primitive and racked by civil war between warlords, as well as Imperial Japanese incursion.

As the threat of belligerent authoritarian states becomes clear, democratized states are expected to attempt to catch up in aircraft production. If time and circumstances permit, the higher productive capacities of democratized states (Chapters 15 and 17) should allow these states ultimately to overwhelm the initial belligerents. Again, the history of World War II looks as we expect. The relatively isolated position of North America protected the US from the brunt of the initial Nazi and Imperial Japanese expansion, leaving the massive productive capability of the US economy in the position to be the *arsenal of democracy*, in FDR's famous phrase.

The US ultimately produced stupendous numbers of aircraft, allowing the Allies to sweep Nazi Messerschmitts and Stukas and Japanese Zeros from the skies.[11] As we have seen, once a force establishes air superiority, its infantry is irresistible, provided it is sufficiently large, well-funded and incentivized.

Once the strategic logic and credible threat of aircraft as coercive weapons is clearly established, we expect all states to seek to have access to these weapons at sufficient levels to render hostile seizure of air superiority costly. We also expect a persistent arms race to increase the efficacy of these weapons, rendering highly productive and knowledge-tolerant democratized economies coercively dominant. Arguably, the contemporary world looks very much as we would predict.

Thus, a quasi-equilibrium of coercive threat is expected to emerge and our theory predicts increased pan-global "cooperation" as a result. We will return to this prediction in more detail in a moment. However, we need to resolve a few matters first.

Box 16.1: Democracy and hierarchy – When modern states choose war

Modern states are human social units whose scope is limited by the capacity of gunpowder technology to manage conflicts of interest, on our theory. Such states tend to behave as individual units in international politics and economics. These units behave as if they have interests that can be in conflict (Chapters 14 and 15; text above). However, these states are actually aggregates of many millions of self-interested individuals. The behavior of states will be determined by the behaviors of the individuals making them up. Moreover, these individuals will behave in ways determined by the proximate psychological mechanisms designed to serve their interests in the ancient, universal human world of coercion and cooperation.[*]

These features of the modern state are important in many ways. At the moment our concern is how these states will make war (Box 12.1). The answer to this question is fundamentally different for the modern authoritarian state and the modern democratic state (Chapters 14, 15 and 17).

For both democratic and authoritarian states the costs of war are primarily born by commoner populations, both soldiers and civilians. However, the rewards of potential victory are distributed very differently. In authoritarian states, small elites can avoid most of the costs of war while reaping a very disproportionately large share of the rewards of victory. In contrast, in democratized states the rewards of victory are more broadly distributed or diluted. Moreover, elites are coercively empowered to make decisions for the state in authoritarian polities, while the decisions of leadership in democratized polities usually require commoner consent (Chapters 15 and 17). Commoners are the ultimate war-making decision takers in democratized polities.

Thus, the cost-benefit ratio of war is far more attractive for authoritarian state decision makers than for the decision makers of the democratized state. Authoritarian states will be more likely to initiate international warfare than democratized states.[†]

Initiators of the all the great wars of the 20th Century were authoritarian states, as expected. Other contemporary examples are illustrative. Authoritarian North Korea constantly threatens democratized South Korea, mostly not vice versa. Democratized neighbors Canada and the US have many conflicts of interest, but are most unlikely to go to war over them. In contrast, authoritarian late 20th Century neighbors Iran and Iraq lost five hundred thousand to one million dead in a war that netted nothing but continued survival of small authoritarian elites.

---

[*]   Actually, the interests of their design information, of course (Chapters 3 and 10).

[†]   This conclusion applies to the contemporary world where all states are at least somewhat well-armed. In contrast, democratized electorates will pursue self-interest through military adventure when the target of war is helpless to resist. The Euroamerican conquest of the Native American societies during the building of the US is an example.

These very different tendencies to warmongering by authoritarian and democratized states will be vital for us to keep in mind as we contemplate the human future (Chapter 17).*

Do nuclear hyper-explosives change pan-global coercion?

There is a massive, if internally contradictory, scholarly literature on the role of nuclear weapons in geopolitics.[12] Our theory apparently contributes some clarification to these issues. Specifically, cost-effective, self-interested projection of threat is the only criterion for weapons-related action/policy. Note again that cooperative, even humane, outcomes occur *only* when the pursuit of self-interest by the coercively dominant allows them, as always in the human world (Chapter 17).

This calculus has and will determine the use and non-use of nuclear weapons, on our theory. There is much well-meaning but poorly thought-out posturing about the inconceivability of nuclear weapon use. Nevertheless, we believe that a human future in which these weapons are never used is, indeed, achievable, even likely. However, this desirable outcome is best served, in our view, by acquiring the clearest possible pragmatic understanding of these weapons in the hands of an animal like us.

First, on our theory, nuclear weapons use has a cost-benefit calculus just as with any other weapon. We do not need new ways of thinking about these weapons; we merely need to think about them clearly.

A science fiction example will illustrate this point. In an imaginary world with all of today's conventional weapons—**including aircraft and cruise missiles, but no nuclear weapons**—let us say the Americans launch a massive attack on China. Suppose further that this attack was motivated purely by American self-interest rather than some previously negotiated conjoint interest with other countries. The target country, China, would counter with all the military assets at its disposal. Under these conditions, both the US and China would suffer massive economic losses and casualties. It is unlikely that either country could thoroughly subdue the other in isolation.

---

* Weart (1998) reviews the early scholarship on democracy and war. This initial work is valuable, but seriously limited, especially by its naïve and uneven definitions of democracy. We are now in a position to do much better and a new generation of scholarship on war and the state would be extraordinarily valuable.

More importantly, this unilateral US action might incentivize other states to enter the conflict on China's side. Members of the winning coalition of states could carve up the formidable US asset base, rather than the more limited Chinese asset base, among themselves. This base would include military and economic resources as well as the possibility of extracting "reparations" from the massive, relatively efficient US economy.

What is the rational course for the US, or any other hypothetical country, under these conditions? Of course, the answer is not to launch the attack in isolation in the first place. More generally, even a great power would experience difficulty acting parasitically on the global market in any persistent and substantial way. Such economic behavior, too, would incentivize a concerted counterattack or, equivalently, global economic boycott not breakable by great power military action.

Now introduce nuclear weapons back into our science fiction world. These weapons shorten the roll out to conflict (hours rather than months) and increase the total amount of destruction (higher explosive yields and radioactive after effects are both significant); however, the underlying logic is apparently unchanged. The costs to the fictional US of launching a unilateral nuclear attack on another country would be prohibitive.

We suggest that this pattern is general. As we think through the logic of international coercion, nuclear hyper-explosives do not qualitatively alter the logic of the contemporary world created by effective aircraft.

A global coercive equilibrium enforcing international "cooperation" is expected on this picture. The properties of our world indicate that such an equilibrium is consolidating and beginning to function (Chapter 17).

The contemporary world – law enforcement in the pan-global community

All states (nations) are not coercively equal, of course. There are powerful major coercive players. These include the US, China, and Russia, states already heavily armed.[*] Major sources of *potential* coercive power also include the European Union, Japan, Canada, and

---

[*]   Recall from the discussion above that the decisive essential element of this armament is air power, in all its diverse manifestations from fighters and drones to stealth fighters and strategic bombers to cruise missiles and intercontinental ballistic missiles.

Australia—countries not now heavily armed but with the economic and technical capabilities to rapidly arm as needed. These states can be called, collectively, the "global powers." Many other countries, including India, Pakistan, Iran, and South Korea can project significant military power in their neighborhoods. We will call these states *regional powers*. Finally, almost every country on the planet has some military air power and armored infantry, allowing significant projection of threat at and within its borders. These are *local powers*.

Global powers have conflicts of interest with one another. The coercive power of these states is so large that they will come into direct, active military conflict only when the stakes are relatively large. Stakes rarely, if ever, become this high in the day-to-day policing of the contemporary global economy. Moreover, no individual global power can afford to become the target of coercive violence by a coalition of the other global powers. Even the US, the single most powerful of the global powers, would be highly vulnerable to a concerted attack by China, Russia, the EU, and Japan, for example.[*]

Thus, a global economic system that serves the mutual interests of the global powers is policed by them. There is an effective strategic equilibrium among them. As the global legal system and police, the global powers effectively control the UN, World Bank and International Monetary Fund, for example.

The regional and local powers have significant capacity to resist coercion by the global powers. For example, the US was easily able to subdue the regional power of Baathist Iraq, but only at the costs of thousands of US causalities and of more than one trillion dollars over about the first five years from a ten trillion dollar annual economy. These were formidable costs for the US, not to be undertaken lightly or often.

Thus, the aggregate capacity of the regional and local powers to resist policies imposed by the global powers is very substantial.

Moreover, mutually policed requirements of *fair play* among the global powers establish a global market that all states can exploit.[†]

---

[*]   Equally importantly, US military power is not sufficient to pre-empt concerted economic action by the other global powers.

[†]   Note that any individual global power seeking to exploit a regional or local power in that individual global power's exclusive interest and in conflict with the interests of the other global powers would find itself targeted. Thus, the weaker states often benefit from the mutual policing of the global powers.

Moreover, as we will see in Chapter 17, the incentives for democratized electorates to engage in severe economic exploitation of the populations of other states can be less than their incentives to help those other states grow into robust trading partners. As several of the global powers are significantly democratized, this factor also contributes to policing of a functional global economy.

An example will illustrate one of the important ways this system works. Several of the global powers would prefer to depress global oil prices artificially, as the US, Japan, the EU, and China all import substantial fractions of their hydrocarbon requirements. However, another global power, Russia, benefits from elevated oil prices.[*] Moreover, many regional and local powers are heavily dependent on oil exports for their economic survival.[†] These regional and local powers are willing to use their coercive means to resist the global powers. Again, the US experience in Iraq illustrates that this resistance is strategically important to the global powers.

The net effect of these factors is the establishment of a global oil price structure that is at least *minimally* acceptable, survivable, for both the global powers and the oil producers.[‡]

Likewise, under these moderately well-policed conditions, it can become in the interests of many states to integrate with the global economy, allowing their currencies to float and managing property rights so as to attract foreign capital investment.

Slowly, sometimes painfully, the coercive units of this new scale of social cooperation, consisting of all nations to greater and lesser degrees, will work out the other elements of a *global legal system*. World courts are being explored. International trade and intellectual property agreements are being developed—piece by piece. A global economy is emerging—slowly, incrementally, without a *central plan*. This scenario is how human cooperative systems always emerge from individually self-interested acts within a new scale of cooperation—a new scale emerging from the exploitation of a new means of coercion.

We are just beginning to feel the effects of this new global economy. Our grandchildren will take it for granted much as we take vast national markets for granted. For both good (tens of millions of people have been lifted out of abject poverty in the early stages of this process and billions more may eventually benefit) and ill (the cooperative global community

---

[*] Indeed, at this writing, the relatively dysfunctional Russian economy is directly and indirectly dependent on gas and oil sales into the global market to maintain the military coercive power essential to its global power status.

[†] That is, without these revenues, their militaries would be impoverished and political elites would risk being ejected and replaced (and, often, killed) by competing oligarchs.

[‡] Technological assaults on equilibrium oil prices through development of alternative energy sources are probably beyond the capacity of oil producers to control.

is now capable of environmental mayhem on an unprecedented scale) the die is cast.[*]

The planet has begun the next stage of becoming a massive incubator for the production, care and feeding of human animals. It remains for us to manage this inevitable process with as much insight as the global wise crowd can muster (Chapter 17).

The contemporary world and the fallacy of the bullet (again)

As we saw in Chapter 14 (Box 14.1), the absence of large numbers of bullet-riddled bodies in the cemeteries of modern democratized states most emphatically does NOT indicate that gunpowder projectile weapons are irrelevant to coercive enforcement of social cooperation.

Likewise, some of us do not immediately see the relevance of the preceding section's argument. After all, the US, the Russians, and the Chinese are most unlikely to attack one another directly with thermonuclear weapons. Indeed, the late 20th Century is a litany of careful avoidance of confrontations with this potential. Surely the global powers, and others, are not usually projecting active coercive threat toward one another.

With a little thought, however, we immediately see that all the global powers do project threat toward the other members of the global community continuously. For example, the very public "secret" testing of weapons achieves this projection of threat quite nicely as did the Space Race, the many "military exercises" and even local armed "interventions" (like Korea, Vietnam, Iraq, and Afghanistan).

Moreover, all humans, including the electorates and decision makers of contemporary states, have a virtuoso's elite intuitive grasp of the logic of coercion. It is, after all, apparently the central element of our two-million-year-old uniquely human birthright. We navigate a coercive environment—even a global one—with the same elite, often unconscious, command with which a raptor in flight navigates the air.

Finally, we each spend almost all our conscious effort in exploiting the myriad opportunities the presence of our coercive umbrellas provides. We create the umbrella with little conscious thought, then consciously obsess over how to use it. Thus, when we think of our lives

---

[*]   The global financial crisis ongoing at this writing (Spring 2009) does not negate the net long-term trend toward increasing global GDP and increasing peace. We will discuss these issues in Chapter 17.

we almost never think about coercion and almost always think about creative endeavor.

This state of affairs is how it should be. But if this astonishingly powerful human creative trick is to persevere and encompass more people more completely and humanely, we must not forget the unconsciously natural part—the very important coercive umbrella.

When might nuclear weapons be rationally used in the future by desperate elites and for recruiting violence?

Of course, nuclear hyper-explosives might be rationally used in the same way that conventional explosives are sometimes used. The specific case of the September 2001 attacks on the World Trade Center and the Pentagon in the US illustrates what we mean.[13]

These attacks were arguably highly *rational*. They had very high risks coupled with very high potential returns for a small group of self-interested players. The logic is as follows. These attacks were a public performance for purposes of fund-raising and recruiting. The funds were to be raised from well-heeled fellow would-be oligarchs, mostly wealthy Saudis. The foot soldiers were to be recruited from the massive groups of frustrated and powerless young males throughout intensely hierarchical states like Egypt, Saudi Arabia, and Syria.*

With these foot soldiers and funds under the influence of the organizing would-be oligarchs, primarily al Qaeda leadership, including Osama bin Laden and Ayman al-Zawahiri, this military enterprise would have been in a position to seize control of the Afghan state from the Taliban and Northern Alliance. From this base of operations, seizing control of Pakistan, including its nuclear arsenal, might have been achievable over a few years. From this larger base, seizing control of Saudi Arabia and, perhaps, other Sunni strongholds might have become feasible, expanding manpower and asset bases at each step. From such an expanded base, overwhelming and seizing traditionally Shiite areas might then have become feasible. Thus, the mythical *Caliphate* might have been achieved. Rulership of such an entity, or even some small part of it, would have brought stupendous wealth and power to the individual winning oligarchs.

---

* The foot soldiers are not our focus here, but see the logic of the infantry of hierarchical authoritarian modern states in Chapter 15.

This was the potential payoff for this original small group of aspiring oligarchs—huge, indeed. The equally huge risk was that the US would retaliate for the 2001 attacks, pre-empting al Qaeda seizure of Afghanistan and short-circuiting the enterprise, perhaps at the cost of the lives of those oligarchs who planned the attack. This latter scenario is—in part and so far—what has transpired. The best-case scenario of the original attacks was not fulfilled and the worst-case scenario may still play out, though the ultimate outcome remains in doubt at this writing.

Now imagine this same scenario in which the 2001 attacks made use of pirated nuclear weapons. The number of US causalities would have been substantially higher—perhaps millions rather than thousands. But would the ultimate outcome have been any different? Probably, not. Would the *net* gamble have been any greater? Again, probably not. The likelihood and magnitude of US retaliation most certainly would have been increased, but so also would have the fund-raising and recruiting impact.

More importantly, this case illustrates where we can expect nuclear weapons to be used in the future. They will be used by desperate, high-staking-gambling would-be oligarchs with potential access to funds and to a large set of potential foot soldiers.

The logical requirements for use of nuclear weapons in this way are extremely specific. Their use for this purpose can only work if either of two conditions is fulfilled. One is if the attacked power has a significant tendency not to retaliate rapidly. The second condition is if the perpetrator lacks a location to which state-level retaliation can be easily addressed.

In view of this logic, it is uncertain that nuclear weapons would ever be deployed in this way. Here is the problem. If the fund-raising recruiters are initially successful, they immediately acquire an address. For example, consider again the hypothetical scenario for al Qaeda success above. These oligarchs find their new force in possession of Afghanistan and Pakistan, for example. Now, suddenly, they own a great deal of valuable real estate, real estate that can be targeted, including by the global powers. Are these new property owners likely to escape preemptive retaliation in response to their initial nuclear attack? Probably not. The likelihood of success in this endeavor is probably higher with the measured use of conventional weapons of mass destruction than with a novel use of nuclear explosives.

Finally, we must consider the horrendous possibility that would-be oligarchs actually do use nuclear explosives in this way and somehow escape initial retaliation by the global powers. What is likely to happen then? These formerly *stateless bandits* would probably become more conservative *stationary bandits*, stakeholders in the global economic system. Further international adventures would now carry the large new downside risk of losing gains already in hand. We live in a world that already has a number of such bandits, Robert Mugabe and Kim Jong Il with their military oligarchies are familiar examples.

Such new states would thus likely begin life as highly hierarchical, like Maoist China, Stalinist Russia, or contemporary North Korea, say. However, we can realistically anticipate their oppressive structure gradually melting under international and domestic sources of democratizing coercion (Chapter 17).

Here we are – Now what?

With this chapter, we have completed our acquisition of all the insights we need to see what our theory has to say about our lives at this moment and into the future. Following up on these insights will be the goal of Chapter 17.

Can we realistically try to predict the human future? We will argue that we can. Notice how our theory predicts the contemporary world with the properties we recognize. Our current world emerges simply and transparently from our ancient adaptation to the self-interested exploitation of all available coercive means.

Notice, further, that we had no need to invoke isolated individual decisions. For example, the specific choices made by Churchill, Roosevelt, Hirohito, Stalin, or Hitler played no role in our discussion, and are unnecessary to understanding the essential features of the outcome of World War II.

Specific choices did, indeed, influence *details* of the outcome—the precise boundary of the "Iron Curtain" at the end of the war, for example. However, the larger global outcome is affected only marginally and relatively negligibly by specific individual choices, on our theory. Only the aggregate of millions of choices matters to the larger picture.

We might also worry that we cannot be confident of this approach because the 20th Century does not provide redundant tests of our theory against the evidence. We cannot, after all, rerun the 20th Century under conditions where Roosevelt's early polio might have killed him or Hitler might have died in the trenches of World War I.

However, each Chapter from 11 through 16 approaches different elements of history in this same way and our theory is equally successful in accounting for each era. Moreover, the structures of completely independent agricultural civilizations (Chapter 12) or archaic states (Chapter 13) are strikingly similar and predictable.

Thus, we argue, the historical record *in toto* gives us the power to test our picture of historical causation. History is the product of all of us—not just a few great men or key actors. It is, of course, precisely for this reason that we can claim to have a *theory of history* and to pursue a *science of history* (Third Interlude).

This perspective matters to every one of us very much. As we look into the human future we obviously cannot predict the individual choices that will shape the superficial patina of our unfolding common destiny. A UN Secretary General may perish in a plane crash…or not. The next potential Bernie Madoff may suffer an early heart attack…or not. A North Korean military leader may mount a coup attempt…or not. *However*, we need not be overly concerned about our inability to anticipate such minutia. As long as we react to them as a global community, the broad course of the river of our future can arguably be anticipated on our theory. Understanding this predictable, common future will be our final goal in Chapter 17.

ᏉᏉ

Chapter 17
Final considerations – a humane future from our evolutionary past

We begin this chapter with great hope and equally great trepidation.

We turn first to the hope. We have built a robust new theory of the evolution of our properties as a fundamentally new kind of creature, a **human** *creature (Chapter 5-10). Moreover, this theory of our origins also gave us a powerful new theory of our history from two million years ago through the present instant (Chapters 11-16). As a result, we know things our ancestors could not even have imagined.*

We live in a world where the welfare of all six billion of us is daily at risk from political violence and economic deprivation and uncertainty. But, knowledge is opportunity. We can now realistically contemplate seizing control of our common future with profoundly new confidence. The benefits to the billions of our descendants can be enormous, perhaps even beyond our contemporary capacity to comprehend. Moreover, in using our new knowledge in this way, we would be redeeming the heroic contributions and sacrifices of millions of our ancestors.

More specifically, we argue that our theory illuminates the fundamental processes underlying contemporary human politics and economics. We are now in the position to see formerly daunting, mysterious plagues, like war or economic depression, as relatively simple public health problems. We mean this analogy literally. Our medieval forebears saw bubonic plague as a chaotic evil, but we now understand it as a manageable bacterial infection. Likewise, we have historically seen political violence and economic dysfunction as chaotic "evils," but we are now positioned to understand their origins with pragmatic clarity.

Indeed, we will argue that **all** our collective social traumas result from two simple, sometimes interconnected practical problems— elite domination of authoritarian polities or narrowly self-interested manipulation of more democratized ones. Democratization is the antidote for the first problem. Aggressive knowledge seeking by the global wise crowd under well-defined conditions is the cure for the second.

Moreover, we now understand that achieving democratization requires broad distribution of access to coercive threat. Nothing less or else will do. Likewise, we know that in democratized societies, we must have transparent access to information in order to drive our cooperative knowledge seeking. The knowledge enterprise, in turn, will allow us to eliminate ever more opportunities for free riding on our common economic and political systems.[*]

We are not only empowered to understand our political and economic public heath, but we also have the knowledge to implement pragmatic, effective therapeutic actions. These actions **must** result from pan-global democratic consensus and, thus, they will be slow, halting, and generational.[†] But we have realistic hope that the results of these actions can be a vastly wiser, richer, and more peaceful world for our descendants.[‡]

These are the reasons for hope. Now we explain our trepidation. In a very real sense, the authors have returned from a journey of many years across a very unfamiliar landscape. This book has been our travelogue. As a result of our journey, the world looks and feels very different to us than to many of our readers before you picked up this book. Communicating a fundamentally new perspective on our shared condition is challenging. We worry that our communication skills were not be up to the task.

We believe that the new theory we have developed and described here gives us fundamentally new insight into the human future.[§] We will proceed on that basis. But, some of the most important things we need to re-examine

---

[*]   Equally importantly, transparency allows us to identify and remove *perverse incentives* to economically destructive behavior. The contemporary global economic crisis (2008-2009) is apparently, in large part, a consequence of such perverse incentives (see Box 17.3, below).

[†]   To paraphrase Winston Churchill's view of the 'inefficient' democratized path to progress, "Democracy is the worst form of government...except for all the others...."

[‡]   Indeed, we concur with those who argue on empirical grounds that human history has a direction toward ever larger scales (ultimately pan-human) ethical inclusion and cooperation (see, for example, Singer, 1981, Wright, 2000, and Dennett, 2003). Our theory accounts for and predicts this directionality (Chapters 11-16).

[§]   Of course, as always, it remains for the global wise crowd to decide if we are correct or to make needed refinements in our picture.

are so familiar, so seemingly settled in our shared consensus, we know it will be difficult for us to avoid being misunderstood.

For example, the political labels so traditional and seemingly full of contextual content for Americans and Europeans, like left, right, or center, have lost all meaning for us. As you read through this chapter you may be variously tempted to classify us as leftist or rightist, or, perhaps, even utopian or anarchist, at one moment or another. If you make these categorizations, you will not only be wrong, you will miss our most important argument.*

We are, in fact, "centrists", but not in the traditional sense. Rather, we believe that the only possible source of authentic insight comes from something like the mean or consensus of large democratized wise human crowds (Chapter 10).† It is this mean that is the "center" we believe in as individual political actors.

Indeed, we believe that recognizing the unique, decisive role of the wise crowd in the human world is the single most important practical insight from our long journey. Learning to listen attentively to one another's contributions to our collective wisdom, even when we do not understand them at first, is the most fundamentally human thing we can do.

In other words, all wisdom comes from patient listening under the democratically enforced conditions where we are each free to speak as we wish and all required to hear and criticize one another, without trivial social categorization or preconception. There is simply no other place where authentic knowledge and understanding **can** *come from, not anywhere, not ever. It is in this process, we believe, that a truly great and real hope for a more humane future lies.*

Against this backdrop of both hope and trepidation we are going to speak to you with the greatest candor we can muster. The theory we are exploring really does reorder all of our experiences and world views. You may not always accept or understand what we say on your first pass through this chapter. We ask only that you listen with the possibility in mind that our views might be a viable alternative to the unconsciously narrowly self-interested political intuitions we each originally absorbed from our local, provincial culture. We acquire these older intuitions much as we learn our native language. Their origins are preconscious, yet their hold on us is pervasive. Mastering a new language takes time and effort.

---

*   Indeed, we believe these labels mostly reflect arbitrary identifier information for interest groups (Chapter 10; Fourth Interlude), beliefs invented by our evolved psychology for purposes of local political competition, not coherent sets of intellectual or policy beliefs.

†   Of course, in view of the fundamental logic of human social cooperation (Chapter 5) this democratization requires universal access to coercive (political) power (Chapter 15 and below).

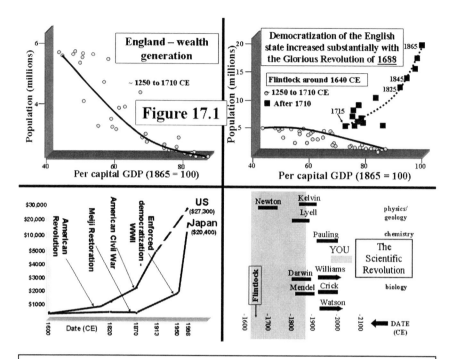

Figure 17.1: TOP LEFT PANEL: Population (vertical axis) is plotted against per capita GDP for England from the 13th to the early 18th Century. Per capita GDP grows only when population declines, for example, after a plague epidemic. This is the classic Malthusian pattern (text) and it persists throughout English history until the early 18th Century.

TOP RIGHT PANEL: The same variables as in the top left panel are plotted, but values for the later 18th and 19th Centuries are included. Notice that the scales are changed to accommodate a larger range of values; thus the points from the top left panel are now clustered in the bottom left corner of this panel. Beginning in the 18th Century—immediately after the early civil wars of democratization in the UK—the values for *both* population and per capita GDP grow dramatically and enter entirely new ranges (compare the solid black line for pre 1710 CE to the dashed line after that date. This change reflects a profound adaptive revolution (text). Data redrawn from Chapter 1 of Bernstein (2004).

BOTTOM LEFT PANEL: Per capita GDP over the last four centuries in the US and Japan is shown. Notice the correlation between democratization of the US and Japan and their economic productivity. American economic growth began to be significant after the Revolution; however, the massive democratization produced by the Civil War apparently resulted in further acceleration of economic growth (Chapters 15 and 17). The partially democratized Meiji Restoration in Japan apparently produced a small uptick in economic productivity. However, the large, modern growth in productivity did not occur until the enforced, more complete democratization of Japan

at the end of World War II (Chapter 15). Data redrawn from Maddison (2001) pp. 43, 90, 126, 215, and 264.
BOTTOM RIGHT PANEL: Shown are a few selected life times from among the more well-known scholars having built the contemporary scientific knowledge enterprise. The lifetime of an average reader of this book is indicated by the small horizontal bar to indicate time scale. Notice how recent all these scientific careers are and how they all follow, rapidly, the early democratization of the state from the English Civil War through the American Civil War (large vertical grey bar) (Chapter 15; Malcolm, 2002).

What is our big picture now?

We stand at an unprecedented moment in the four billion year history of our solar system and, possibly, in the approximately thirteen billion year history of our universe. We are one of Earth's millions of animal species; but we are also a completely new kind of animal. We approach biological adaptation in an unprecedented way. We each pursue self-interest through projection of coercive threat, which, in turn and under the proper conditions, enforces collaborative, mutually beneficial adaptive activities of spectacular scale and power. Put differently, we are the first of Earth's animals to be able to manage the conflict of interest problem on a substantial scale. This seemingly mundane new attribute is utterly revolutionary (Chapters 5-16).

Our unprecedented biological adaptation continues to drive a highly directional process. We are turning our planet and solar system into a single integrated unit directed at nurturing us and serving our confluent interests, for better and for worse. The die is cast. There is no going back. The shape of the future of our solar system, and possibly of the universe, will almost certainly be dominated by this single process.

Our ancient evolved proximate human psychology was designed to allow us to navigate this human social world, not to understand the ultimate causal origin of that world. Only our uniquely human, adaptive relationship to knowledge can give us this ultimate grasp of our condition. Indeed, at times, our proximate psychology can work at cross-purposes to our pursuit of insight. Our evolved political psychology often actively interferes with our ability to understand our ultimate origins (Chapter 10; Fourth Interlude). We sometimes anticipate a past that no longer exists and our best intentions are defeated.

Thus, our subjective sense of the world can sometimes become a powerful impediment as we examine together the present and the future. Such discussions engage the compelling, immediate *life and death* individual interests of each of us. We must try hard one last time to think with the dispassionate detachment of an extraterrestrial anthropologist. The dividends of success will be substantial.

In the four sections immediately following, we will tour some vital details of how our contemporary world came into existence. We each have learned local theories and conventional wisdom about these events, their empirical features and their causes and relationships. Our theory reformulates many of these causes and relationships. We will find that some elements of conventional wisdom—like robust empirical patterns in event sequences—will come into crisp new focus. Other elements of conventional wisdom, particularly traditional interpretations of causal processes and subjective motivations, will be mostly demolished.

The only meaningful causal processes in the public life of humans are conflicts of interest and their coercive management. Everything else, including all our history, politics, economics and knowledge-seeking, is consequence and by-product of this single central process. This simple law of the public human world has been with us for the last two million years, it is with us now, and it will be with us two million years in the future.

Thus, we begin by looking directly at how the contemporary world emerged from the unprecedented scale of our coercive management of conflicts of interest. This exploration will also represent one last set of powerful new empirical tests of our theory. We then turn to how we can retain what our ancestors gained, often at a terrible individual cost to them. Finally, we look at how we might regain some things they transiently lost, thereby establishing a trajectory toward a more humane, pan-global human future.

Humanity and hope, enabled by coercion – The "democracy effect" on wealth generation

One of the most startling features of the emergence of the modern world is its sudden discovery of the ability to generate adaptive assets, what we call *wealth*, on a previously unimaginable scale (Chapter 15). For example, using data from several sources we can estimate the per capita wealth generation for a typical Englishman for the last two thousand years.[1] This value remains essentially unchanged for the first seventeen hundred years of this time period and probably long before that, as well.

Then, beginning around the 18th Century, per capita GDP begins to grow. This process accelerated in the 19th Century and has continued for approximately the last three hundred years. This growth has resulted in a contemporary world where an average Englishmen, or other *Westerner,* is enormously richer than any of her/his ancestors throughout our two-million-year history.

This *modern economic miracle* is just one symptom—stunning as it is—of a broad, pervasive transformation of human existence. Looking at the details of wealth generation shows us something deeper. Per capita GDP *before* the modern miracle displays small fluctuations, mostly reflecting population densities. Thus, for example, when populations crashed as the result of a plague epidemic, surviving people grew a little richer. However, as population rebounded, people once again grew poorer (Figure 17.1, top left).

This population/wealth relationship reflects the fact that increasingly marginal agricultural lands are forced into cultivation as populations grow. This reduces the average per capita productivity of agricultural workers. At the same time as food prices go up pay rates for now-abundant labor go down. People begin to go hungry. Over the long haul, population bounces up and down, constrained by this Malthusian ceiling.*

In contrast, beginning during the 18th Century and accelerating dramatically in the 19th, this population/wealth relationship is shattered, apparently for the first time in human history (Figure 17.1, top right). Population now began to grow far beyond anything that had ever been seen in two million years of human history. However, instead of people starving, they were better fed and prospering (on average) in the parts of the world undergoing this dramatic change. The Malthusian ceiling had been raised, dramatically.

This radical change tells us something simple, but utterly profound. The increase in per capita wealth occurred at exactly the same time as people also began to learn how to produce more food *per unit land area.* In other words, increased wealth generation is not just an increased ability to manufacture gewgaws, to provide heretofore unknown luxuries or to increase the money supply. It reflects a fundamental increase in know-how, in the ability to manipulate the material world.[2] The human cooperative (economic) enterprise was suddenly able to feed millions more people on the same limited amount of agricultural land that formerly constituted the iron lid on our Malthusian box.

---

* Observing this historical relationship between wealth and population was the basis of Malthus' original work that was so important to Darwin and to our picture of the broader biological world (Chapters 2 and 3). See Fischer (1996) for a useful contemporary perspective on these historical cycles.

We will see this large, abrupt improvement in proficiency, capability, competence reflected in other ways below. For now, this increase in *productive ability* is extremely well known and famously documented.[3] This new economic growth is the consequence, most fundamentally, of the accumulation of human know-how, of a vast reservoir of personal experience shared and stored among the members of the human population.[4] Beginning around three hundred years ago, people in some parts of the world appeared to begin thinking, creating and behaving fundamentally differently than they had throughout the preceding two million years.

Our theory predicts this profound revolution in the human condition. Specifically, we predict that democratization of the state, caused by redistribution of coercive power, should produce this effect. Specifically, we have argued that this profound reconfiguration of the state returns humans to the ancestral, democratic condition. This democratic condition is the social environment to which we are particularly well adapted (Chapters 5, 7, 11, 12, and 15).

This fundamental social change, in turn, apparently occurred as a result of modern gunpowder handguns. These are quintessentially democratic weapons (Chapter 15). Moreover, the ensuing effects recreate the ancestral democratized social environment *on a vastly larger new scale* allowing the ancient human tricks of self-interested coercive enforcement of cooperation to produce collaborative results of a spectacular new magnitude (Chapters 10, 15 and 16; Third Interlude).*

What precisely do we expect the outcomes of democratization of access to coercion in large, formerly hierarchical authoritarian states to be? This is a question whose answers will benefit enormously from further analysis by well-informed economists. But we already know enough to suggest the outline of some of the most important answers.

It is useful to begin simply. *Unchecked elite coercive power*—whether in the archaic state, the hierarchical early modern state, or the authoritarian late modern state—*yields, inevitably, a fundamentally criminal enterprise from the perspective of the larger commoner population.* Such elite-dominated states are massive extortion and protection rackets.†

---

* The massive *scale* of the modern enterprise is sustained by larger-scale weaponry (artillery and aircraft; Chapters 14-16). However, these larger weapons *can* remain under democratic control through deployment of personal gunpowder weapons (Chapter 15 and below).

† Again, recall that our theory predicts this condition to be inevitable. All people project threat in pursuit of self interest, everywhere and always. When access to coercive threat is concentrated in the hands of the few, these elites coercively shape the local

Suppose that you are a commoner in such a state. What will happen if you invest your effort in creating some innovation that yields new value to the larger economy? In general, of course, you can expect coercive elites to seize that added value rather than seeing it go to your own, commoner, enrichment. Even worse, any resulting enrichment of coercive elites will be partially invested in improving their weaponry and infrastructure increasing their capacity to dominate you, a commoner.*

Under these conditions, what is your best strategy as a commoner? To invest effort and risk in only those few enterprises that exclusively benefit your immediate family and your highly local social unit; your clan, extended family, village. You and your local colleagues can use non-elite coercive means to police this internal cooperation and its products can be rendered partially invisible to elite thieves.

Contrast this with what happens in a radically democratized state. Any local interest group seeking to engage in extortion or protection is vulnerable to being threatened by the coercively powerful self-interested cooperative individuals surrounding them. This encircling democratized interest group possesses the means to kill or ostracize such "criminal" individuals, seizing their assets in compensation.† These enforcement actions also enhance *secondary returns* (see following section) on future cooperative enterprises.

What is the net effect of this democratized coercive enforcement? The members of the coercive majority coalition are in a position to create economic rules (social contract beliefs; Chapter 10) that serve their confluent individual interests. This includes the formulation of *property law* which protects the assets generated by individual members of the coercive majority. This also has the indirect effect of making individuals more effective contributors to public economic output as they are incentivized to produce more for market exchange.‡ For example, dramatic

---

cooperative enterprise to serve their own (elite) interests. Paternalist elite rationales and propaganda are irrelevant, self-justifying rubbish, everywhere and always.

* Elites will justify this behavior as being required by their greater intelligence or richer perspective. The human capacity for delusional self-justification is infinite when unchecked, of course.

† Recall that this coercive ostracism will generally be immediately self-interested as the assets of the target criminals will be seized, directly or indirectly, by ostracizing individuals. As the democratized state matures, this asset seizure takes the form of formal fines and forfeited gains contributing to the funding of policing agencies, though these agencies, in turn, remain under the vigilant supervision of the larger armed population.

‡ Put differently, Adam Smith's *invisible hand* only works with real effectiveness in democratized economies. See Smith (1776) *An Inquiry into the Nature and Causes of the*

increases in potent individual specialization resulting in modern professions and careers become adaptively rational, which is to say economically sustainable.

There is substantial empirical evidence that this democratized creation of *property law*, and the disinterested *rule of law* more generally, is absolutely indispensable for the miraculous economic growth in the modern world.[5] Peruvian economist Hernando de Soto's exploration and discussion of this issue in his thoughtful book *The Mystery of Capital* (2000) is highly illuminating on the unfolding and consequences of democratized property law.* His discussion of the emergence of property law in the early, radically democratized North American state in his Chapter 5 is rewarding. No state without such property law has ever been able to produce sustained, endogenous economic growth.†

Humanity and hope, enabled by coercion – Modernity and the very essence of being human

The preceding section addresses the specific question of how the classic economics of the modern miracle emerge from the democratization of coercive threat. However, careful analysis illustrates that these effects are part of a broader, more pervasive change in our societies, a change predicted by our theory.

To understand this larger insight, it is useful to imagine ourselves in the highly democratic ancient ancestral environment. Most of the time we (our distant ancestors) would have engaged in cooperation with others we expected to frequently cooperate with again in the future. These groups would have been small, based on the then-current coercive means, and recurring interactions between specific individuals would have been common.

In this ancestral condition, our coercive means would have allowed us to forestall the toxic effects of conflicts of interest and participate in mutually enriching projects (Chapter 5). These projects would yield rewards—food, shelter, water, mates, and other resources. Let us call these rewards the **primary return** on cooperation.

Wealth of Nations.

*   Though he does not grasp the central causal role of access to coercive threat.

†   Notice that authoritarian states showing some economic growth largely funded by selling oil or other naturally occurring resources to developed democratized economies—Russia and Iran are examples—do not falsify this claim (below).

However, our fellow cooperators would also prosper as a result of our combined efforts and, thus, they would have survived to cooperate with us again next week or next year. When potential colleagues survive to cooperate with us in the future, in part, because of our efforts today, the returns from these *future* collaborations constitute a **secondary return** on our *current* cooperation.

Indeed, it would be rational for our ancient ancestors to have invested extra effort and risk at any one moment, beyond the minimal amount necessary to produce the best immediate primary return, so they could enhance the secondary return on their efforts in the future.

Thus, we are expected to have inherited the ancient adaptive psychology that would cause us to consider both primary and secondary returns in our public economic behavior. In a highly democratized coercive, and thus cooperative, environment, we unconsciously anticipate not only that our efforts and property will be respected. We also expect future secondary returns from our current cooperative efforts. This evolved psychology produces some very characteristic modern behaviors. For example, we not only work *for a paycheck* (primary return) we also work to *build institutions* (secondary return). Indeed, we are not subjectively happy and fulfilled unless our efforts are producing both returns.[*]

Remarkably, detailed analysis strongly suggests that the modern economic miracle results precisely from this ancient human psychology. We are rich because so many of us pursue secondary returns, not just a paycheck. Placed in a democratized environment, we not only deliver the immediate efforts required to execute some task or job, we are also motivated to innovate in ways that build the larger economies and institutions of which we are a part.[†]

To be specific and emphatic, we propose that it is precisely this pragmatic and psychological effect of democratization of access to coercive means that drives the immense scale of the modern miracle. No future hopes of building this miracle to new heights can be successful without a firm grasp of this insight.

---

[*]   It is vital to notice that anticipating secondary returns does *not* depend on (dubious) adaptation to what we called *social selection 1* in Chapter 10 (feeding effort into a cooperative enterprise because of the inevitable, automatic adaptive echo of that effort). Rather, it primarily reflects the anticipation of ongoing *social selection 2*, our ability to enforce coercively the cooperative contributions of others in the future.

[†]   Of course, these expectations, and their benevolent effects, are dramatically reduced when we are trapped *inside* compartmentalized, hierarchical, arbitrary institutions nested within democratized societies (below).

A powerful natural experiment indicates how strong the evidence for this claim is. In mid 19th Century England the first wave of modern industrialization was in full throat. This economic success generated enormous amounts of capital that could be invested in the colonies of the British Empire. This investment included India, where textile mills were built as literal clones of mills on the home front. Thus, levels of technology and capitalization were equivalent. We can now ask whether economic productivity or wealth generation per hour worked was similar or different.

The answer is *profoundly different*. Indian workers were nearly *fivefold* less productive than British workers.[6] Though many racist Victorian explanations for this were offered at the time, we now know these are rubbish. Indians are just as bright and hard working as Englishmen.

What was the difference? English workers labored under a significantly *democratized* coercive umbrella that they themselves created and maintained. They saw themselves as *Englishmen* and their larger society as a potential source of future protection and personal benefit. Their (partial) control over their state gave them confidence that the state would use some of its assets on their behalf in the future.* They were working for both primary and secondary returns.

In contrast, 19th Century Indian society was intensely hierarchical. Indian workers worked only as hard and as efficiently as they needed to in order to produce the stipulated, short-term primary return. The effort that an English worker put in to produce secondary returns, the Indian worker withheld for investment, instead, in his immediate family and friends.

Think now about your own psychology. Is your desire to work merely *for a paycheck*? The vast majority of us living in relatively democratic states will answer, "No," authors included. We do, indeed, need primary returns. However, we also desire to invest in *something larger* represented in our companies, our universities, our professions and, indeed, in the larger societal enterprise.

We are members of democratized societies that are inherently more *humane*. We are living similarly to the way two million years of human evolution have designed us to live. When sustained by *democratized* threat, our modern crowds are wiser, richer and more technically powerful than those of our ancient ancestors *only* because they are so much larger.

---

* Of course, the capacity of 19th Century British society to deliver *social services* was pitifully poor by our standards; however, neither was it irrelevant. Being an Englishman had some benefits.

Box 17.1: On the fundamental poverty of radical "conservatism," American style

Conservatism is in quotes because this word has different meanings in different places and at different times. The contemporary American version of this policy package purports to disempower government and empower individuals, a superficially seductive message. However, in practice, this approach disempowers democratized monitoring and empowers compartmentalized, hierarchical organizations and corporations nested within democratized societies, at the expense of individuals.

This approach has two predictable, highly toxic features. First, individual members of the leadership of these hierarchical institutions are incentivized to sacrifice the interests of the larger culture for their own—extravagant compensation, golden parachutes, and so on. Long term investments in infrastructure, education, research, and regulation look like "government inefficiency" from this perspective. Leaders of large corporate entities play the economic game much like authoritarian oligarchs always have (Chapters 13 and 14).

Second, as they acquire unchecked power, hierarchical entities subvert the most important individual behaviors producing economic growth. Insulated hierarchical institutions disincentivize their individual workers' support of long-term societal investments generating secondary returns. The voices of individual workers are never heard. Equally importantly, leadership in hierarchical organizations is incentivized to steal all credit (financial and otherwise) for any successes the institutions might have.* Workers within such institutions have the same incentive structure as medieval peasants. They contribute to the primary return but refrain from nurturing the institution that exploits them in ways that share).

This is a toxic double whammy—poorly managed conflicts of interest once again. These private institutions work to pauperize the larger

---

* Big annual bonuses need to be explained to shareholders in private institutions, for example. Likewise, continued control of decision-making power by bureaucrats needs to be justified (generally to other self-interested bureaucrats). When bureaucrats become insulated from public scrutiny, even public institutions become effectively private.

culture. At the same time they themselves are impoverished by clashing interests *within*.*

Solutions to this systematic pathology are urgently important to a more humane and wealthier human future. Increasing *transparency* to shareholders and more sensible regulation of extravagant pursuit of narrow interest will be part of the solution. However, ultimately, the coalition of the whole must be attentive. For example, it would be relatively easy to compile data systematically on the level of toxic hierarchy in various institutions. These compilations could become an important contribution for academic economists to make. We all might then make our job-seeking, purchasing, personal investment and tax-supported public investment decisions accordingly.

Humanity and hope, enabled by coercion – The "democracy effect" on the ancient human knowledge enterprise

The generation and exchange of culturally transmitted information on a new scale has apparently been a central element of the uniquely human adaptation throughout our two-million-year history (Chapter 10). This cooperative use of information among non-kin presents the conflict of interest problem, and the scale of this information sharing is limited by the scope of our ability to manage conflicts of interest (Chapters 9 and 10; Box 10.2; Third and Fourth Interludes). Thus, humans always achieved new scales of adaptive sophistication, new adaptive revolutions, only through new scales of social cooperation. The behaviorally modern human revolution (Chapter 11), the agricultural revolutions (Chapter 12) and the rise of archaic (Chapter 13) and modern (Chapter 15) states are all examples.

These earlier adaptive revolutions gave us wonderful opportunities to observe the predicted correlation between scale of coercive management of conflicts of interest and adaptive sophistication. But, the rise of the modern democratic state (Chapter 15) gives us the opportunity to

---

* Some institutions have apparently had some success at avoiding this morbidly blind hierarchy (see Collins and Porras, 1994, for example). It will be of the very greatest interest to have a more well-developed understanding of effective solutions to this endemic, highly destructive pattern.

see these effects in unprecedented detail and to grasp more fully their inner workings.

It is useful to begin with an evaluation of the quality of the empirical historical evidence. The correlation between democratization of the modern state and the explosive modern growth of human knowledge, the *Scientific Revolution*, is spectacular (Figure 17.1, bottom right). Of Eurasian states, England was the most democratized the earliest (Chapter 15; above). As a result of this cultural change, Englishmen played a very disproportionate role in the early Scientific Revolution.[*]

For example, Newton, Boyle, Lyell, Darwin, and Dalton were all English citizens. Indeed, Newton was born in the same year as the beginning of the English Civil War and lived through the Glorious Revolution, key events in the long process of radical English democratization (Chapter 15).[7]

Moreover, the *European Enlightenment* was strongly influenced by partially democratized British culture. Indeed, Porter has argued persuasively that the Enlightenment was fundamentally English, with continental contributions being largely secondary and derivative.[†]

The central role of democratized states in the ongoing knowledge enterprise has continued through the present. Radically democratized states contribute Nobel Laureates in the sciences disproportionately per capita. Watson was an American and Crick an Englishman, for example. Likewise, Bill Hamilton and Richard Dawkins were Englishmen and George Williams and Edward O. Wilson were Americans.

Again, the temporal and spatial correlation between the explosive modern growth of the knowledge enterprise and democratization of the state is dramatic. Our challenge now is to capitalize on this correlation to penetrate more deeply into the detailed mechanisms whereby democratic management of conflicts of interest drives dramatic growth of knowledge. Much remains to be learned, but some things already seem clear.

First, and most importantly, the fundamental tool of the knowledge enterprise is public doubting based on transparent access to reliable information (Preface, Introduction, and Chapter 10). This universal

---

[*] As we have already seen, racist "genetic" explanations for the disproportionate early contributions of Englishmen to science are highly dubious.

[†] See Porter (2000) *The Creation of the Modern World.* Porter's lively, detailed description of the Enlightenment illuminates its origin in the critical, doubting public dialog that promotes the wisdom of crowds (Chapter 10). It follows that those who emphasize the continental contributions are seduced by elite propaganda (and, perhaps, wannabe emulation).

doubting imposes a severe discipline on developers of knowledge, ultimately improving the quality of the product. However, public doubting is also highly toxic to all narrow self-interested pretense and self-serving rationalization. Elite free riding is almost immediately destroyed by democratized, transparent doubting. Dominant hierarchs in authoritarian social systems—archaic states and authoritarian modern states—are relentlessly subverted by this public questioning (Box 15.5). They cannot tolerate it—anywhere, ever—and the historical record shows that they do not. Oligarchs universally strive to restrict public doubt and access to reliable information.

In contrast, no member of a democratized majority has the power to forestall public doubting of his/her actions and motivations.* Indeed, almost all members of a democratized social/economic system have strong incentives to pursue as much reliable information as possible.

The implication of these considerations is that democratized majorities will actively support the knowledge enterprise, whereas authoritarian elites will generally subvert and suppress it by all available means.

Second, the highly formal parts of knowledge generation that we think of as professional science are extremely and continuously dependent on uncounted thousands of pieces of know-how possessed by non-scientists—everything from glass jar manufacture to machine tool design and electricity generation. This know-how is largely generated in democratized economic systems through the actions of individuals sharing information, including in anticipation of secondary returns. Thus, essential skills diffuse rapidly throughout democratized economies further supporting formal knowledge generation.

In contrast, individual commoners in authoritarian states get little in secondary returns from their contributions of know-how. Indeed, they often merely enhance their own subjugation by elites. Thus, the same non-growth of know-how that prevents robust economic performance in authoritarian economies also prevents efficient formal knowledge generation.

Third is the self-interested psychology of knowledge seekers in democratized cultures. Discovery and publication of scientific insight opens the door to personal reward. For example, this includes salaried positions at—and research funds from—the knowledge-oriented

---

* For example, in a democratized culture like the US, wealthy and powerful individuals are always subjected to viciously skeptical, penetrating treatment by the press. Ongoing, relentless elite attempts to defeat this process in democratized cultures have been only transiently successful *to date.*

universities democratic cultures endow.* Further, private investors in a democratized market often provide support for researchers of demonstrated competence. Moreover, the educated citizenry of democratic cultures buy books written by scientists. Many of the prominent scientists listed previously, including Newton, Darwin, Lyell, Dawkins, Williams, and Wilson, have received income from this source.[†,8] The crucial point is that the market in democratized, knowledge-hungry cultures is an important agent of sustained knowledge growth.

Finally, democratization of our cultures liberates the better angels of our evolved psychologies. We saw these effects on economic productivity in the preceding section. Knowledge generation is comparably effected. Scholars (amateur or professional) are motivated by immediate self-interest as allowed by democratized economies/societies (primary returns). They are also motivated by the awareness (often unconscious) that what they do can change the world around them in ways that benefit them and their children in the future, shaping the scholarly environment all will inherit (secondary returns). These secondary returns on knowledge generation benefit the rest of us, as well.

This somewhat technical discussion of the modern knowledge enterprise tends to obscure something profoundly important to the human future. Humans are spectacularly gifted users of the wisdom of crowds to uncover the workings of our world. Ultimately, this uniquely human adaptive trick is the only thing that distinguishes our awareness of the world from that of any other of Earth's animals. We understand the world so much more completely than any other animal for this reason and *only* for this reason.

Though contemporary democratized cultures have been vastly more efficient at knowledge generation than hierarchical ones, there is very good reason to believe that even democratized cultures can still improve their efforts dramatically. We should remember that *better* is not necessarily *good*. We have a long way to go in fulfilling the true human potential for knowledge generation. The following brief discussion introduces a few of the many remaining problems we face here.

First, as we said, authentic knowledge is caustically subversive of *all* privilege, entitlement, and hierarchy. Moreover, even in relatively

---

* It is striking that the number of universities and colleges per capita in radically democratized cultures like the US is very high. For example, the massive *land grant colleges* of the American Midwest were endowed by Congress in the immediate aftermath of the Civil War (Burton, 2007).

† The authors of this book have also been supported in various ways by these sources of income.

democratized states various individuals will sometimes find their interests undermined by knowledge built on transparency and doubt. For example, the white majority in the North American state long sought to protect itself from the corrosive effects of a rigorous analysis of racism. Likewise, members of some economic elites consistently resist efforts to make banking, investment, and business activities more transparent to public scrutiny and to scientific analysis.* Knowledge about how our economic system is working is just as important as knowledge about how climate changes works.

Of course, these are just two of many, many global examples of resistance to knowledge generation. These examples were also chosen to illustrate how democratized cultures chip away at such pretense and presumption. Both racism and lack of financial transparency continue to recede in the face of the onslaught of knowledge-hungry citizens capable of conjointly, coercively protecting their interests.

A second problem we face derives from the internal logic of large, bureaucratized institutions,† including those specializing in education and research. These large organizations are nurtured by the enormous flow of research funds from governments, foundations, and private sources. Their elite, dominant members have substantial incentives to redefine the knowledge enterprise to consist of those things that most efficiently bring in resources they can subsequently control. These are not necessarily the activities that have the best chance of producing authentically new insight.

An example from the basic science research enterprise is instructive. The human genome project consumed billions of dollars; but it was a massive engineering and clerical project, not science.‡ Such projects enhance resource access and stature of elite managers, independently of their real scientific impact.

The early 20th Century physicist Arthur Eddington is reputed to have remarked with contempt, "There are scientists, and there are stamp

---

*   Obviously, authoritarian elites are much more aggressive in blocking knowledge generation. For example, the military/political leadership of China and North Korea has long sought to deflect realistic analyses of *state socialism*.

†   See below for a more detailed discussion of such institutions.

‡   Some clerical and map-making projects, like the genome project, can actually be quite valuable. Our argument is merely that they should not be conflated with science. We share the expectation of many biologists that the massive body of clerical information generated by the genome project will ultimately be used by scientists to generate authentic insight and novel clinical and pragmatic capability. There are already many encouraging signs that this will happen.

collectors." *Institutional* science constantly tends toward degradation into an enormous stamp collecting endeavor. It is important that technical description, rather than innovative science, is attractive to the middle managers running the major funding agencies (NIH, DOD, NSF, etc.). Description is predictable and, thus, an attractive product for self-justifying bureaucrats. Authentic science, in contrast, is highly unpredictable, progressing in explosive bursts from sources no bureaucrat can hope to anticipate. (After all, Darwin was a gentleman farmer/investor and Einstein was a patent clerk.) A community of investigators who can reliably generate a strong stream of description can form a reciprocally self-rewarding echo chamber with funding agency bureaucrats. This pattern of *social capture* of resources actively (if generally unconsciously) parasitizes the knowledge-seeking coalition of the whole.[*]

Thus, the knowledge enterprise is under constant threat and attack, even in democratized states. The only defense for its integrity is the broadest possible public involvement in relentless public doubting, transparency again. For example, a global and domestic consensus has progressively forced white Americans to begin to recover from their racist history. Skeptical democratic electorates can likewise force bureaucratic institutions to demonstrate authentic, new insight, rather than endless description, as a product of spending the state's research budget.

Humanity and hope, enabled by coercion – The *humanities* and the *democracy effect* on art, joie de vie

In Chapter 11, we first examined the social/economic origins of *art* as predicted by our theory. In short, the availability of a sufficiently large market for esthetic objects of all sorts is probably sufficient to drive the evolution of these "odd," uniquely human activities. Moreover, the size of this market is expected to determine the sophistication and scale of artistic production. The empirical record supports this prediction. For example, the birth of modern literature and music correlates well in space and time with the first emergence of the large markets of the modern democratized states.[†]

---

[*]   See Gina Kolata's story on the front page of the June 28, 2009 issue of the New York Times (Playing it Safe in Cancer Research) for one commentary on this toxic pattern. This bureaucratic problem is not unique to institutional science, of course. It is, arguably, a systematic problem with all large institutions, including government and private business.

[†]   It is convenient to distinguish between *intensive* and *extensive* art. Intensive art includes painting and sculpture. These activities generate unique, non-counterfeitable

These strong effects of the scale of cooperation include autocatalysis of creative production. One of millions of examples illustrates this effect. The massive growth of global markets in democratized states in the mid-20th Century allowed recording technology to become economically viable. So routine was recording, in fact, that cheap records made by poor African American artists in the southern US could find their way across the Atlantic into the modest houses of British working- and middle-class youngsters with names like John Lennon, Mick Jagger, Eric Clapton and Pete Townsend. The new musical tradition these artists helped launch, in turn, reverberated back across the Atlantic to influence Americans, including a young African American artist of astonishing virtuosity and originality named Jimi Hendrix.

These various market-size effects probably result in large part from primary returns on artistic achievement. However, as with production of goods and knowledge generation, the liberation of our evolved cooperative psychology by the potential for secondary return is also crucial here. Again, artists—like economic actors and scholars—are not only motivated by ambition/avarice but also by the knowledge that what they do can change the world around them in ways that benefit them secondarily by shaping the esthetic environment in which they and their children will live.

These secondary returns on individual artistic achievement produce music, and literature and art, of true beauty and variety. Massive democratized coalitions do not merely make us richer and wiser, they also flood us with wonder and joy.

The view from high altitude – The limits of human potential

As we think ethically about the human future, *the modern economic miracle*, the *Scientific Revolution* and the development of modern art forms remind us of a profound, vital lesson (Chapter 11; Box 11.3). The human potential to know and to do is *not* limited by our genetic patri-

---

objects that serve as pieces of very large denomination currency (Chapter 11). Large-denomination currency is extremely valuable to elite individuals. Thus, supremely sophisticated intensive art is also developed and funded by elite-dominated cultures. The artistic heritage of Classical Greece, Imperial Rome, the Islamic Empires, or the early Renaissance are examples of this from the Mediterranean basin. In contrast, massive democratized cultures preferentially support art that can be easily copied and marketed on a massive scale. We call this extensive art and it includes literature and music.

mony. Rather, it is limited by the scale and internal logic of our social cooperation.

We briefly return to a few issues that merit our special attention and careful reflection.

First, humans living ten thousand years ago had the *potential* to put footprints on the Moon or discover Darwinian evolution. They merely lacked the right social scale and organization for *actualizing* that potential.

Second, there is no sound reason to believe that the current status quo, even in democratized states, approaches an optimal environment for fulfilling human potential. Thus, the very exciting possibility exists that our modern adaptive revolutions have hardly begun to yield what humans are ultimately capable of. Our goal now must be to see how we can "just do it."

The ancient ancestral adaptation and the human future – The contemporary snapshot

*People have **only** those rights they can defend.*[*]

In the preceding sections we slowly swam out into very deep waters. Now we must learn to confront the remaining implications of our theory for the present and the future. For many of us this part of the discussion will not be comfortable. It will even be actively frightening or appalling to a few of us. Be patient. The potential for a peaceful, wise, wealthy, transparent, just pan-global human future lies at the end of this leg of our journey.

On our theory, all things human, including our present condition, are the consequences of a single ultimate cause: The individually self-interested projection of coercive threat, performed conjointly and cost effectively. It simply is not possible to think productively without this insight being perpetually at the center of our focus, politically, economically, socially. This process is what made us human in the first place and it will remain at the center of our existence forever.

The distribution of access to this fundamental human tool, coercive threat, is the single most important thing for us to be aware of. Yet, it tends to be hidden from us (Chapter 10; Fourth Interlude). Our human

---

[*] Various searches indicate that this is a widely held popular view, showing up in screen plays, novels, and blogs of diverse origins. We have not been able to find a reliable original attribution. Our theory predicts that this traditional wisdom is literally, inevitably correct.

minds are designed to take democratized access to coercive threat for granted. After all, it was thus for nearly two million years and we tend to presuppose (unconsciously) our own access to coercive threat just as we anticipate the air we breathe. Therefore, our evolved minds profoundly misconceive the single most important issue in the contemporary world—**who will own coercive dominance and, thus, the human future.**

Our evolved minds are designed to see human ethical and pragmatic debates as causal. We fail to recognize the *implicit assumption* of democratized access to the threat necessary to police our consensual decisions. We insist on believing that if we make the right ethical and practical arguments, a benevolent outcome will ensue. In fact, it matters what we discuss and decide *only* to the extent that our decisions also mobilize *decisive* coercive assets—always and everywhere.

Our inability to grasp the coercive logic of our politics is as great a threat to the human future as a planet-busting asteroid. Failing to understand truly the logic of humane cooperation, we invite the risk of handing over control of the global human enterprise to powerful, hierarchical elites. Once entrenched, such elites would never surrender their control and our remote descendents would be domesticated animals in a fully toxic and very much more profound sense than we are (Box 10.3).

Every other goal and hope we have must begin with avoiding this fundamental intellectual error and its pervasive, threatening consequences. We must capture democratized control of decisive coercive means for all the world's children and for all time. Nothing else matters if we fail in this endeavor. Our theory is quite clear about this requirement, and the empirical support for this requirement is legion (Chapters 13-15; below).

Crucial issues hang in the balance. Military oligarchs control Russia, China, and many smaller states (Iran, Zimbabwe, and North Korea, for example). Moreover, continued democratized control of military power in the US is uncertain. At present the heavily armed American electorate has apparently retained effective coercive control over its ever-more-sophisticated military. If this control were to be lost, the consequences would be unimaginably bleak. The resulting American military elite would pursue domination of their state just like Russian, Chinese, and other oligarchs. In aggregate, these now-unchecked oligarchies would create a brave new world making Orwell's *1984* look utopian by comparison. Notice that this is apparently an inescapable prediction of the universally self-interested use of coercive power (Chapters 5 and 11-16). It is not a vague, intuitive claim.

We did not mention democratized states other than the US in the preceding paragraph. We omit them because there is a case to be made that these other states are not currently carrying their global weight. The other major democratized polities have lived under the umbrella of American coercive power since the end of World War II (Chapter 16). They are derivative dependents of American coercion, not self-sustaining autonomous states.* The other democratized states need to grow into counterweights to US power and help-mates to the global democratization project we will discuss shortly.

Our theory seems clear about future pragmatic actions we must take if humane, pan-global democracy is our goal.

First and most immediately, the US electorate *must* retain civilian coercive control of its terribly powerful military. This is both a pressing political issue and a crucial area for immediate scholarly attention. Has the US electorate already lost coercive control of its various professional police and military groups? We suspect not, but the issue cries out for careful study. What exactly is required to retain coercive control by the US electorate in the future? We need a good answer (below).

Second, the citizens of the other democracies must look to take a stronger coercive position toward their own militaries. Where these militaries are strong, they arguably remain subservient to domestic civilian interests primarily because of the indirect effects of US threat. This places far too many eggs in the single American basket.

The essential third and fourth goals are inextricably intertwined. We must patiently, but relentlessly help to democratize the remaining authoritarian states (below). The ultimate democratization of these states will reduce the level of military threat each state must retain for defense and enforcement of pan-global cooperation.†

This reduction in standing professional military force, in turn, will reduce the amount of firepower individual members of electorates must retain to assure coercive control of their countries. The world resulting from this gradual, partial demilitarization would be much less threatening than ours. These goals can probably be substantially achieved by the end of this century (below).

---

* Imagine the results if we remove the US entirely from the global picture. What would the status of Japan, South Korea and Taiwan be, vis a vis China? What would be the status of Europe vis a vis Russia?

† Recall that democratized states are unlikely to engage in expansionist warfare against polities with any significant ability to resist militarily. In contrast, elite-dominated states are habitually expansionist. This universal rule derives from the different distributions of cost and benefit from *imperialism* to the coercively decisive groups in each case (Chapter 16; Box 16.1).

If we, the pan-global people, are successful in these simple, but challenging steps, we can realistically expect our descendents to inherit a planet that is vastly richer, wiser, more aware, and more humane than at present. This hope is not naïve. These benefits arise as an inevitable by-product of the way humans manage our cooperative enterprises when coercive threat is democratically distributed (above). To enable these outcomes, we aspiring citizens of a pan-global democracy are well advised to understand the ultimate logic of coercive power as well as aspiring oligarchs always have.

The American Constitution debate and the human condition

The preceding section will strike many readers as too strong and too simple. Geopolitics is surely much more complicated than we have made it out to be. If you reacted this way, we urge you to keep an open mind, to consider the possibility that you might be wrong. Several points might inform your thinking.

First, we believe our view of the asymmetric American role in the geopolitics of the last half-century arises from a realistic assessment of American coercive power and political structure. We do not believe it is a distortion resulting from the authors' perspective as American citizens. Of course, skeptics might look for robust, reliable ways to falsify this claim.

Second, empirical evidence argues that our starkly simple view is fundamentally correct. One of the richest sources of such evidence is the natural historical experiment represented by the rise of the US itself (also see Chapter 15). A few details of the American story will illustrate what we mean.

Before we begin, recall that coercive threat is like sex, nationalism, and other emotional topics. Our interests and unconscious evolved minds are engaged in powerfully evocative ways. Moreover, the ultimate logic underlying these evolved responses is generally hidden from our proximate social psychologies, the parts of our minds whose job is to manage our public behaviors (Chapter 10).

Thus, the contemporary public dialog on democratized access to coercive means tends to be deeply emotional. Much of what passes for scholarship is actually political polemic, often unbeknownst even to its authors. However, strong new studies have recently emerged. Currently this debate flies under the flag of the Second Amendment to the American Constitution*,9—its meaning, and its purpose, the intent of its authors and the evolving history of its application.†

This new work gives us fresh insight into what the political actors who wrote the American Constitution were thinking. David Williams' 2003 book *The Mythic Meanings of the Second Amendment* is one rich source. We quote below from Williams' book where the original period documents are referenced, in turn. The point of these quotes is that the Framers of the Constitution knew very well that access to coercive means was the central issue.‡ We argue that they had the real world experience against which our theory's predictions might be compared. Most of us today lack this direct experience. We urge readers who are intuitively skeptical about democratized access to threat to reflect carefully on these period remarks.

The two sides of the debates producing the American Constitution and its Second Amendment came to be called the *Federalists* and the *Anti-Federalists* (Williams, 2003). Members of the first group favored a relatively

---

* Readers searching for writings on the "Second Amendment debate" in the US will be flooded with books and short pieces (formal and popular). Much of this material is emotional, narrowly (if often unconsciously) self-interested, and hyperbolic, in our view—on all sides of the debate. Analyzing the details of all this intellectual chaos is beyond our scope here. However, we propose several working interpretive hypotheses for future investigation. US citizens who favor broad citizen access to coercive means are predicted to include individuals who perceive themselves as members of a potentially powerful majoritarian interest group. In contrast, those who favor restricting citizen access to coercive means include those who perceive themselves either as elite benefactors of the contemporary status quo (government officials, university professors, mainstream clergy, law enforcement administrators, corporate executives, etc) or as vulnerable minority targets of a potentially hostile majoritarian interest group (some African Americans, for examples).

On a related issue, we emphasize that we are agnostic about legal restrictions on the public transportation of guns by non-professionals (Lott, 2000; reviewed in Kahan and Braman, 2003). Private weapons deployed in everyday crime control are *not* crucial to our discussion. Only private access to weapons for the ultimate ability to "police the police" (and the military) is important.

† In a sea of questionable scholarship, several books stand out as sober, thoughtful, and historically grounded, in our view. We particularly recommend Malcolm (1994) and Chapter 1 to 3 Williams (2003). (We disagree with some of Williams' theoretical suggestions, but his review of the history is outstanding.) Galvin (1989) and Martin and Lender (2006) are illuminating about the pragmatics of civilian coercive means in shaping the incipient US.

‡ The Framers also struggled mightily with the nature of the human cooperative enterprise, itself, often invoking the intuitive construct of the "Body of the People" as the real unit. These period thoughts and actions are rich data against which to test our continuing theoretical refinement in the future.

strong central government, but were *moderately* alarmed about the threat of elite concentrations of coercive power. Members of the second group were dubious of a strong central government and *highly* alarmed about the threat of elite coercive power. We will quote from the first group (with two exceptions) to illustrate the *universal* attention to democratized access to coercive means among the Framers (page numbers from Williams, 2003).

Noah Webster:
> *The supreme power in America cannot enforce unjust laws by the sword; because the whole body of the people are armed , and constitute a force superior to any band of regular troops that can be, on any pretense, raised in the United States.* (p. 51)

Thomas Jefferson:
> *And what country can preserve its liberties, if its rulers are not warned from time to time, that this people preserve the spirit of resistance? Let them take arms…The tree of liberty must be refreshed from time to time, with the blood of patriots and tyrants.* (p. 50)

Patrick Henry:
> *Guard with jealous attention the public liberty. Suspect everyone who approaches that jewel. Unfortunately, nothing will preserve it but downright force. Whenever you give up that force, you are ruined.* (p. 51)

George Mason:
> *To disarm the people;…[I]t was the best and most effectual way to enslave them…* (p. 51)

James Madison:
> *[Against any standing army] would be opposed a militia amounting to near half a million of citizens with arms in their hands, officered by men chosen from among themselves, fighting for their common liberties…* (p. 51)

Elbridge Gerry (Anti-Federalist):
> *What, sir, is the use of the militia? It is to prevent the establishment of a standing army, the bane of liberty.* (p. 55)

James Madison:
> *Besides the advantage of being armed, which the Americans possess over the peoples of almost every other nation, the existence of subordinate governments, to which the people are attached and by*

which the militia officers appointed, forms a barrier against the enterprises of ambition... (p. 65)

Alexander Hamilton:
In a single state, if the persons entrusted with supreme power become usurpers...The citizens must rush tumultuously to arms, without concert, without system, without resource; except in their courage and despair. (p. 66)

Joseph Story (early 19th Century Supreme Court Justice and interpreter of the Constitution):
The right of the citizen [note singular] *to keep and bear arms has justly been considered as the palladium of the liberties of a republic, since it offers a strong moral check against the usurpation and arbitrary powers of rulers....* (p. 86)

Joyce Malcolm's outstanding 1994 book *To Keep and Bear Arms* also harvests useful period writings. The following two examples are illuminating.

Zachariah Johnson (Virginia Constitutional Convention):
...the people are not to be disarmed of their weapons. They are left in full possession of them. The government is administered by the representatives of the people, voluntarily and freely chosen. Under these circumstances should anyone attempt to establish their own system, in prejudice of the rest, they would be universally detested and opposed, and easily frustrated. (p. 157)

Newspaper article explaining the 2nd Amendment, published in Philadelphia, New York and Boston in 1789:
As civil rulers, not having their duty to the people fully before them, may attempt to tyrannize, and as the military forces which must be occasionally raised to defend our country, might pervert their power to the injury of their fellow-citizens, the people are confirmed.....in their right to keep and bear their **private** *arms.* (emphasis added; p. 164)

It is worthwhile to read and reread these period quotes.* They appear to reflect practical, empirical awareness of precisely the issues

---

* We are unqualified to comment definitively on the controversial subject of the *original intent* of the Framers. See Malcolm (1994) and Williams (2003) for discussions illustrating how complex this scholarly debate has grown. However, our theory clearly indicates that collective human action (formation of the Body of the People in the Framer's our theory predicts to be important[*]. Notice the clear pragmatic recognition of the role of coercion. The Revolutionary leadership, including Washington, recognized the danger of a standing professional army. For example, the Newburgh Conspiracy in 1782 was an incipient military coup of "disgruntled" continental army soldiers and officers. If the civilian militia had not retained the capacity to resist the standing Continental army, the outcome of the Revolution might well have been very different.[†]

Box 17.2: A world awash in Kalashnikovs – do we have democracy yet?

How we react to the world's tens of millions of copies of the brilliantly simply engineered Russian assault rifle—the Kalashnikov or AK-47[‡]—is a rich litmus test of how we view the human condition. At first glance, the impact of the AK-47 seems unambiguously evil. This is the weapon of the Taliban and of every two-bit military oligarch in the Congo basin. It has been used by child soldiers and the Janjaweed to murder

---

terminology) can only emerge from the self-interested actions of individuals (Chapter 5). Thus, the *individual right* school of thought on the 2nd Amendment (supposing that the right to keep and bear arms resides with each individual) is the only theoretically coherent approach to the formation of a viable democratized polity, in our view. Some of the Framers may have grasped this fact; others may not. Note especially that some of the Framers were elite slavocrats, certain to have mixed feelings about broad access to coercive threat.

We also note that the early North American colonists would have had extensive experience with the local Native American communities, originally policed by the highly democratized access to coercive threat represented by the bow (Chapter 12). We agree with those scholars who argue that this Native American model might have facilitated the early Euroamerican comprehension of the feasibility of radically democratized societies (reviewed in Johansen, 1998).

[*]   Of course, elite oligarchs like Stalin, Hitler, Mao, Pol Pot, and others were just as well aware of the crucial role of access to decisive coercive means in determining political outcomes (Chapter 15).

[†]   See Martin and Lenders (2006) for a discussion of these issues, including the Newburgh Conspiracy.

[‡]   See Kahaner (2007) for a description of this weapon. The more elaborately engineered assault rifles of the American military are serviceable with strong logistical support, but not so useful to a rural insurgent without similar support.

innocents by the hundreds of thousands (at least).* Surely stemming the flow of these weapons must be one of our highest humane priorities.

We suggest that we should also consider precisely the opposite position *depending on context*. Our problem might sometimes be *too few* Kalashnikovs, not too many. AK-47s are now primarily in use by local oligarchs and criminal enterprises not by the local democratized crowd as a whole. They are being employed in the longstanding tradition of the Nazi SA and the Red Armies of the 20th Century (Chapter 15). AK-47s are incredibly effective weapons of elite intimidation of unarmed commoners.

When acting geopolitically, the democracies have traditionally managed these local armed oligarchies badly. We have allowed ourselves to be trapped into supporting one elite oligarch who will pretend to serve our interests rather than another who pretends to serve the interests of an enemy state. For example, FDR reputedly remarked about the brutal, but pro-American dictator Anastasio Somoza that "He may be a sonofabitch, but he's our sonofabitch."

This approach may have been temporarily expedient, even necessary in the past. For the future it is self-defeating. We have different options now. First, the global power of the democratized states is sufficient that we can begin to block the local competition between powerful states that ultimately helps empower local oligarchs. Second, where military oligarchs are not yet entrenched—Afghanistan and rural Pakistan come to mind at this writing—we must consider what might happen if the AK-47 were universally available. If we really believe in universal democracy, perhaps we should put our money where our mouths are. Do we really believe that the universal human condition should be democratized access to the coercive threat that inevitably shapes our societies? We, the authors, do.

Of course, this approach would present a host of practical problems, including ways to get our service people out of harm's way while blocking local attempts at elite seizure. However, at this writing an attempt to use this approach in rural Pakistan is ostensibly under way. It will be of the greatest interest to see if it is competently executed and what its long-term effects will be.

---

* According to Wikipedia, translations of Janjaweed include "a man with a gun on horseback."

Equally importantly, this approach will almost certainly not work in places with existing, entrenched elite concentrations of coercive power—Iran, the Palestinian territories, and North Korea come to mind. Large state-level professional military forces can often pre-empt large-scale democratization of access to threat. Different approaches will be necessary in these cases as we will discuss in the main text below.

The conflict of interest problem *within* democratized states

Conflicts of interest are a fact about the universe, like gravity (Chapter 3). They were with us at the birth of *Homo* around two million years ago (Chapters 5 and 7) and they will be with us two million years hence. The question will *never* be, "How do we *eliminate* conflicts of interest?" It can only and always be, "How do we *manage* them."

Thus, no culture can ever avoid this problem. Each society will constantly confront new approaches to free riding by small subsets of their citizenry. But these social cheaters will generally be revealed by the agents of transparency and doubting like the press and rest of us. Democratic polities will meet each new case with legal and regulatory responses, closing off some approaches to future free riding. Still newer free riding tactics will be discovered and the process will be repeated, ad infinitum. We will discuss specific historical examples of this process in the following section and in Box 17.3. But some general points are especially important.

First, this never-ending, cyclical process of responding to free riding can be executed well enough that democratized economic systems function relatively well most of the time (above). Moreover, there is good reason to hope that we will get better at this monitoring and insistence on transparency, so that the systemic effects of free riding—like the global financial crisis beginning in 2008, for example—will become milder and rarer.

Second, even in relatively democratized contemporary states, the larger coercive umbrella (the legal system) supports creation of smaller institutions (businesses, universities, hospitals, governmental bureaucracies) that are often internally hierarchical and compartmentalized. Humans pursue individual self-interest (often unconsciously), everywhere and always. In compartmentalized, hierarchical institutions, self-interested individuals are commonly in the position to (discreetly) subordinate the ostensible purpose or function of the institution to their individual self-interests. To the extent this happens, our institutions work badly and we are all impoverished. Indeed, many of us are also individually oppressed and intellectually abused as members of such hierarchical institutions. Such institutions are not *wise crowds* (Chapter 10).

This pattern of institutional dysfunction is epidemic and pervasive, as most readers will be very well aware from their own personal experiences. Again, this toxic institutional culture is merely another reflection of the ancient, universal conflict of interest problem. Our institutions can almost certainly work far better than most of them do now *if* we find better ways to manage these conflicts of interest. Creating a better theory of institutions presents an enormous humane opportunity.

For the moment we can make one strong statement. Such self-serving hierarchies in democratized cultures represent a malicious, self-interested exploitation of the legal system. These hierarchies rely on a lack of transparency, on restricting access by the democratized wise crowd to their ongoing functioning. Shining increasing daylight on the functioning of all our institutions will be part of the antidote to the systemic pathology of bureaucracies.

The North American state, again – the day-to-day functioning of democratized states is a massive, slow-motion brawl

As we have seen, our theory gives us new insights into the contemporary human condition and its historical origin.[10] Our theory's picture of democracies as immortal arms races and simmering cauldrons of managed pursuit of self-interest may sound ethically discouraging, but it need not be.[11] We now have the insight to recognize this clearly as just another fact of life. Moreover, it is a feature of human biology with which we have two million years of experience. We are highly adapted to dealing with this, the natural human public world.

As a result of our fundamental human biology, part of our being citizens of democratized states is to participate in the constant monitoring of our public (non-kin) social and economic system, empirically patching our social contract beliefs, including writing new laws, as need arises. When Thomas Jefferson said that "the price of liberty is eternal vigilance" he had something else in mind, but his maxim applies perfectly well to the day-to-day functioning of all democratized human societies for all time.

The nascent US is useful to us once again in this context as empirical evidence. The partial democratization of the US occurred early, as we have seen (Chapter 15; above). The gradual unfolding of this state is well documented and studied. We argue that US history illustrates how democratized human entities work. Looking at how this state grew and changed over time is, thus, both a test of our theory's perspective and richly illuminating.

A brief survey is as follows. The early American "white" male oligarchy used its disproportionate access to coercive threat to disenfranchise other recognizable subgroups, including women and individuals of African, Asian and Native American descent (Chapter 15). However, this attempted coercive disenfranchisement was much more general than racism and sexism. Self-interested subgroups form relentlessly within all societies, including democratized ones. These subgroups represent a common form of the recurring pattern of self-interested exploitation which it to say, free riding.

Our concern in the next few paragraphs will be how the white majority population subdivided itself into warring interest groups during the 19th and early 20th Centuries. What did the North American society producing revolutionary economic growth (Figure 17.1) look like to its white citizens?*

A broad democratized subgroup (white males) struggled through the early 19th Century to establish property law and a larger legal system that served its members' interests.[12] The early stage of this process is traditionally referred to by American historians as the *Age of Jackson* followed by what one recent historian referred to as the *Age of Lincoln*.[13] Economic growth was robust (Figures 17.1 and 17.2). However, these social contact beliefs left many openings for self-interested individuals to exploit the system for sometimes extravagant personal gain.

Continuing, partially democratized monitoring imposed some constraints on this exploitation, however. Thus, for example, confiscatory taxation of a permanent underclass—characteristic of elite-dominated states (Chapter 13)—was infeasible. Self-interested would-be elite actors were therefore constrained to serve the larger market, at least to some degree, as the best available self-interested option. This

---

*   There is some wonderfully sharp, opinionated recent scholarship in this area. We particularly recommend Wilentz (2005), Beatty (2007), Burton (2007), and McDougall (2008).

constraint required them to contribute to continued rapid economic growth. These strategies involved massive transportation enterprises including railroad and transoceanic shipping as well as mass-produced commodities like hydrocarbon products and iron/steel. These enterprises were spearheaded by individuals with names like Carnegie, Rockefeller, Morgan, Gould, Vanderbilt, Frick, and others, of course.

As these initial service enterprises grew larger, their capacity to exploit holes in social contract practices grew. In particular, they used large amounts of money to manipulate all three branches of government substantially.[*]

One detail of this struggle for elite control of government is salient. It is the creation of the *National Guard*. This entity was designed to produce federal control of what had been state-controlled militia. This was intended, by some of its sponsors, to make the Guard a more easily controlled coercive agency to manage labor. Indeed, the Seventh Regiment Armory still stands just east of Central Park in New York on Park Avenue between 66th and 67th Streets.[†,14] This armory is notable for having been built with private money. It was explicitly intended to support armed suppression of labor unrest.[15] Indeed, railroad

---

[*]   Beatty's account in *Age of Betrayal* (2007) is especially noteworthy for its review of the extreme corruption of the court system—at least as egregious as the more well-known corruption of the legislative and executive branches. The courts were directly, actively responsible for some of the most hideous abuse of working people in all of American history. It will be educational to contemporary left-leaning citizens to see how toxic *judge-made law* from the right can be. These events throw a whole new light on contemporary discussions of *judicial restraint*.

[†]   See Google image search results at "Seventh Regiment Armory" with "New York."

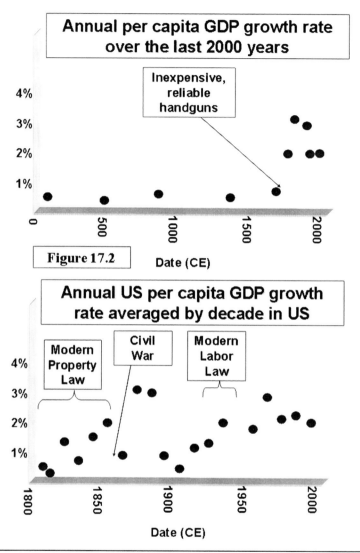

Figure 17.2 : This figure extends the picture of the modern economic miracle we developed in Chapter 15 and above (Figure 17.1). Notice in the TOP PANEL that per capita growth rates remained essentially stagnate until the democratization of the enormous cooperative units represented by the modern state. The BOTTOM PANEL shows a blow up of these events in the specific case of the nascent US (text). US productivity growth begins to become obvious in the early 19th Century. This growth continues to the end of the century. By 1900 growth rates subside. However, with the continuing democratization of the North American state, productivity growth rates rebound and have remained relatively robust through the present. Most American readers are dramatically wealthier than their early 20th Century forebears.

monopolist Jay Gould is reputed to have claimed that "I can hire one half of the working class to kill the other half."[16]

Fortunately for the majority of American citizens, Gould was wrong. Over the late 19th and early 20th Centuries large numbers of Americans reshaped social contract beliefs, including the form of modern labor law, to rein-in some of the more extravagant elite exploitation of the democratized market economy.[17] As our theory predicts, these goals of the democratic majority sometimes required violent action. Of course, this violence constitutes coercive threat in conjoint defense of individual interest. Moreover, contrary to contemporary elite propaganda, much of this action was disciplined and responsible, as expected given the long human adaptation to controlled, strategic projection of threat.*

The consequence of this labor movement was extensive revision of the American social contract. This revision resulted in improved control of monopolistic manipulation of markets and government expenditures, hence creating a substantially more effectively democratized system.

It is also noteworthy that Carnegie, Frick, Rockefeller, and other wealthy individuals from this era ultimately donated huge amounts of wealth back to the public good, perhaps to placate a coercively dangerous, outraged electorate (below).

Finally, it is potentially important to note that economic growth slowed and grew rockier as elite control of the economic system increased in the late 19th Century (Figure 17.2). Moreover, economic growth again became more robust and consistent in the wake of improving management of elite manipulation of the American economic system in the early 20th Century (Figure 17.2). Our theory predicts this relationship— improved economic performance from refined social contract beliefs enabled by conjoint control of the social enterprise by a democratized consensus.

It would be of great interest for professional economic historians to test these proposed causal relationships in more detail. Understanding well and truly how democratized economic systems work over time is vital to the long term prosperity and health of the pan-global human enterprise.[18]

The larger point from the American experience, again, is that relentless monitoring and renegotiation of social contracts, policed by democratized coercive power, is the way of the human world, now and forever. Because we are so highly adapted to this world, this insight is actually very hopeful for the human future.

---

* See Zinn's *A People's History of the United States* (1995) for a sympathetic account of labor's use of coercive threat.

Box 17.3: Financial crises, wealth, the conflict of interest problem, and the human future

Practical men, who believe themselves exempt from intellectual influences, are usually slaves of some defunct economist. Madmen in authority, who hear voices in the air, are distilling their frenzy from some academic scribbler of a few years back.

The social object of skilled investment should be to defeat the dark forces of time and ignorance which envelop our future. The actual, private object of the most skilled investment today is ..to outwit the crowd, and to pass the bad, or depreciating, half-crown to the other fellow.

[T]he position is serious when enterprise becomes the bubble on a whirlpool of speculation. When the capital development of a country becomes a by-product of the activities of a casino, the job is likely to be ill-done.

John Maynard Keynes (1936)*

Economics faces very special challenges. Scholars are attempting to do science on the very individuals and interest groups who fund them. This is a little like doing genetics and biology under Stalin (producing the notoriously deranged Lysenkoism). Unremarkably, economics as a science is deeply confused. Professional economists, working in insular, severely distorted academic environments, will never fix this problem alone. Only support from relentless doubting by the global wise crowd has a real chance of improving our vision, we argue, and the need is urgent.

Our theory clarifies three issues useful to improving our comprehension of economics. First, the conflict of interest problem and its often imperfect management is the central underpinning of every action at every moment in human public life (Chapter 5). Economics to date has focused on what happens when this management is presupposed, mostly unconsciously. Coupled with the distorting influences described in the preceding paragraph, this myopia is sufficient to account for the lack of progress.

---

*   From Keynes' *The General Theory of Employment, Interest and Money* (1936). The three quotes are from pages 395, 167 and 171, respectively, of the 2008 reissue by Management Laboratory Press, Hamburg, Germany. It is telling that Keynes' analysis remains precisely as relevant now as when it was written over seventy years ago.

Second, the human management of conflicts of interest is two million years old (Chapter 7). Our minds are exquisitely and deeply adapted to our uniquely human solution to this problem (Chapter 10). We are so well adapted to being human, in fact, that we are generally not aware of where our economic behaviors come from.

Third, our ancient human public cooperative enterprise has now expanded to a vast, never-before-seen scope (Chapter 15 and above). This creates novel opportunities for hostile manipulation, for self-interested behavior on an unprecedented scale (Chapter 10; Box 13.2).

We need to recognize the actual strategic, adaptive challenges ancestral human minds were designed to deal with; the historically idiosyncratic psychological devices that evolved to meet these challenges; and how these particular devices act in the new and massive social enterprises of modern state and global economies.

The science of economics will be improved in response to public doubting in ways no one can anticipate. However, the relatively new approach of behavioral economics is likely to be important[*]. Likewise, skeptical thought about how massive contemporary economic systems work in aggregate will certainly be vital, as well. Three examples of this kind of iconoclastic thought are Hyman Minskey's *Stabilizing an Unstable Economy* (1986), Robert Shiller's *Irrational Exuberance* (2005) and George Soros' *The New Paradigm for Financial Market* (2008). We need many more examples of such fresh explorations from different perspectives.

At this writing the global economy is mired in the largest contraction of equity prices and credit availability in living memory[†]. Where democratized access to coercive power exists[‡], new insight can

---

[*]   See Taleb (2007) and Ariely (2008) for recent discussions of this approach to our public economic behavior. A very useful videotaped discussion of some key issues between outstanding psychologists Daniel Kahnemann and Nassim Taleb can be found at http://video.dld-conference.com/watch/aj4OXAg.

[†]   We are writing in June, 2009. See Partnoy (2003), Cooper (2008) and Cohan (2009) for readable discussions of the current financial crisis from perspectives inside the financial system.

[‡]   Manipulation of the global economy by elite-dominated actors (Russia, China and the OPEC countries, for example) will be an adaptive challenge in the future. However, the current financial crisis appears to reflect dysfunction *within* democratized economic systems.

generate practical and enforceable responses while we await a stronger science of economics.

Some fundamental points are obvious. As we said, all elements of our economy, including the financial system, are direct outcomes of our ability to manage the conflict of interest problem. These institutions fail and malfunction almost *exclusively* as a consequence of inadequacies in managing this problem. Important insights from the current experience include the following:

While overt free riding (ponzi schemes, say) is a perennial problem, the current crisis emerged from more subtle forms of free riding. Our continued inability to support *enterprise* rather than *speculation* through our financial markets is the central expression of this problem. Enterprise, properly conceived, is the economist's term for the ancient, uniquely human trick of collaboration to produce outcomes that individuals acting alone cannot (Chapters 5 and 10). In contrast, speculation is a rarified form of gambling on what other gamblers will think about the short term future of enterprise. Our financial markets often reward (in the short term) the creation of artificial bubbles among the gamblers rather than the capitalization of real productive enterprises. One time-honored approach to restraining speculation is restricting *leverage* – borrowing to speculate. Of late, professional regulators have dropped the ball badly on this count.

The preference for speculation over enterprise can be further restrained by regulating the creation of complex financial derivatives. Most recently, these novel instruments supported high-risk, high-return lending surreptitiously masquerading as low-risk lending. Would-be investors in enterprise are conned into gambling on the persistence of a bubble. Notice that when financial instruments – or any institutional practices – are claimed to be too complicated for any but experts to understand, we can be quite confident that we are being maliciously manipulated. The universality of the conflict of interest problem allows no other conclusion. Hiding behind the supposed inscrutable complexity of an institution or a process is the last refuge of a scoundrel, always and everywhere. The antidotes are two. Professional regulation of speculative practices too arcane for the public wise crowd to monitor day-to-day is the first. Transparency and democratized doubting of both this regulation and the ongoing functioning of our financial markets is the second.

Finally, free riding on the financial system through generation of misleading practices vividly illuminates the fundamentals – and fundamental problems – of human non-kin cooperation. When one small group of individuals (a few bankers, say) begins to have success with a new free riding strategy (misleading financial instruments) others are increasingly forced to either join in the scheme or find another line of work. The short-term benefit from these schemes draws others (naïve or self-serving capital) and leaves no viable alternatives. Honest bankers cannot generate enough profit to retain capital investment in competition with the mavens of the new instruments. This sequence is perfectly general. *Anytime* a free riding strategy becomes successful, ultimately everyone is forced to free ride and the larger enterprise comes crashing to earth. Only coercive preemption of the free riding strategy to begin with can prevent this perennial cycle – but, again, such preemption is the essence of the ancient human adaptation.

Most importantly, these considerations illuminate a magnificent opportunity. The modern economic miracle (text above) clearly shows that democratized human cooperation can be astoundingly productive of human welfare. However, as each of us knows very well from personal experience, most of our institutions work *much* more poorly and inefficiently than they might.* Moreover, our theory emphasizes that all this institutional dysfunction has a single source, conflicts of interest and our failure to manage them. The good news is that we can get better at this management, knowledge is opportunity. If we do improve our institutions, the modern economic "miracle" may ultimately prove to be just a faint first hint of the colossal, as-yet-unrealized power of massive democratized human coalitions to improve our lives.

---

* See Fox (2009) and Matt Taibbi's July 2009, Rolling Stone piece "Inside the Great American Bubble Machine" [http://www.rollingstone.com/politics/story/28816321/the_great_american_bubble_machine] for recent reviews of the dysfunction of our financial system. Keep in mind that *all* our contemporary institutions (governmental, corporate, educational, philanthropic, professional) work just as poorly as our financial institutions.

"All power corrupts, absolute power corrupts absolutely"

The title's paraphrase of Acton's well-known Dictum is intended to emphasize a simple, but crucial implication of our theory. Poisonous concentrations of power can occur in many places and in many guises.[*] Two examples (again, among many) are illuminating.

First, with the appropriate support of governmental coercion, labor unions can become powerful, narrow interest groups, much like private corporations. This threat seems remote in the present climate of indifference to organized labor; however, it has been important in the past and may be again.

Second, our institutions of higher education are constructed to give elite power to faculty over students. Instructors are the sources of grades, letters of recommendation, and other assets students need to move up and be professionally successful. The ostensible goal of this power asymmetry is to incentivize hard work by students and to give the faculty the opportunity to train and evaluate. Indeed, these beneficial consequences often ensue.

However, this power asymmetry can also have toxic side-effects. For example, there is an incentive for faculty to exaggerate their *elite expertise*. This can be done (often unconsciously) by making the simple seem arcane and esoteric. Another way this elite self-enhancement can be achieved is by presenting a topic area as a massive string of details, laboriously memorized—a "reading of the phone book"—rather than as a coherent, readily comprehensible, relatively easily mastered conceptual whole. In other words, our system creates individual incentives to teach badly.

These practices are doubly toxic. Faculty come to see teaching as disconnected from the larger intellectual enterprise, eliminating one of the enterprise's key elements, students as the knife edge of democratized doubting (Chapter 10). Moreover, second rate scholars can prosper in this artificial environment. Likewise, students can be marginalized and educated into learned helplessness.[†] The most confident students see through these distortions and take from the educational experience what they need and want. But a substantial fraction of students

---

[*]   He actually said, "Power tends to corrupt, and absolute power corrupts absolutely" in a private letter written in 1887. We would argue that a more accurate statement would be *"Elite* power corrupts…." Only elite power can be absolute in Acton's sense. A Google search with this quote produces a wealth of interesting context.

[†]   Being coercively forbidden to doubt is to be enslaved at the most fundamental level.

are simply not yet mature or experienced enough to react to this power asymmetry productively. Members of this second group are trained to believe they are less capable than they can actually be.

As in many other cases, the solutions to these problems lie in the vigilance of the democratized coalition of whole. For example, it would be useful to collect data systematically on the real-world performance of colleges and universities. A suggestion for a useful measure would be to poll the job satisfaction and economic success of graduates ten, twenty, and thirty years after graduation in relation to the circumstances of their parents. Again, professional social scientists might have exceptionally valuable careers carrying out such measures.

The *humanities* and how we might think about the human future

The knowledge enterprise is just as vulnerable to corruption by elite power as any other human endeavor. What is now thought of as the *humanities* was originally the sole content of higher education in the hierarchical medieval and early modern European states.[*] The humanities were actually the necessary tools (and identifier information; Chapter 10) for aspiring young oligarchs, while science was then seen as the vulgar practical tools of commoners, not suitable for a university. The contemporary humanities continue to wrestle with this elite legacy.[†]

We illustrate this ongoing struggle with a story from our own experience. As you read the story, recall that the issue is *not* the *humanities* versus the *sciences*. Rather, the issue is *democratized doubting* versus *elite manipulation* of the knowledge enterprise. *Both* the humanities and the sciences are vulnerable to elite manipulation.

We were sitting in on a conference hosted by a major American research university and attended by scholars from around the US. One scholar gave a presentation on the liberalism and tolerance of early 20th Century British novelist/essayist E. M. Forster. Forster's essays through the traumatic years of World War II—directed at the war's genocidal

---

[*]   See C.P. Snow's famous 1959 essay *The Two Cultures* for one perspective on this history.

[†]   See Wolin's 2004 book *The Seduction of Unreason* for a discussion of this issue, so vital to the future of the knowledge enterprise.

racism in particular—are, in the speaker's view (and ours), thoughtful and humane.[*]

One of the speaker's important observations was that Forster receives extremely little attention from contemporary humanists, in spite of the humanity of his views. Remarkably, the speaker came under attack from academics in the audience – ostensibly on the grounds that he, Forster, was a privileged member of the Imperial British class structure. It was argued that the universal, modest, self-doubting tolerance he advocated was only possible to someone holding the kind of elite power Forster possessed as an Imperial Brit.

This attack is profoundly important and illuminating for several reasons.

First, the attackers were three academics who held positions in major American universities but had cultural origins elsewhere. Two were obviously upper class descendents of the highly class-conscious Indian society. Contemporary Indian society, with its substantial impoverished underclassses, is considerably less democratic and considerably more power-conscious than Forster's England. Adding further irony, both of these individuals spoke English with ostentatiously cultivated British accents.

Adding yet another layer of irony, the third attacking scholar was a white male who grew up in South Africa before the end of apartheid, a society rivaling the antebellum American South and contemporary Zimbabwe for the most racist societies in all of human history.

So, what were these attacking scholars, unconsciously, saying? "We grew up in status-conscious, racist societies but are capable of humane insight; **however**, Forster is not." We need not belabor the obvious suspicions this position invites.[†]

The more important points are what this little event and billions like it occurring all across the planet every day, tell us. Every one of us responds to what we imagine power to be. It seduces us and obscures our self-awareness. It is frightfully easy for us to imagine that we are

---

[*]   These essays are bound in Forster's 1951 classic—still in print—*Two Cheers for Democracy*.

[†]   One hypothesis springs to mind and would be interesting for political psychologists to try to test as follows. These scholars seem likely to be elite power-seekers jealous of what they imagine to be the power of Forster's class and unconsciously wishing to capture some of that power for themselves. Of course, as a tiny, self-referential academic community, their capacity to seize any real power in a democratized society like the US is probably negligible. But they can bureaucratically hijack some local assets that could be much better deployed.

benevolent and that everyone who seriously challenges us is either ignorant or evil.

The lack of self-awareness of his critics is symptomatic of precisely the mind set Forster argues against. The inability of self-imagined elite individuals to accept the Forster view of civil, open society is a predictable product of the ancient human evolved social psychology (Chapters 10 and 13).*

When seduced by coercive power we gravitate to whatever beliefs advance our chances of sharing in that power. When that power is aggressively democratized our only choice is the one advocated by Forster—public questioning, self-doubt in the presence of wise crowds. Humans are rather good at this and democratized societies work well as a result (above). Our challenge is to assure that this democratization penetrates and ultimately governs the workings of all our institutions.

Elite power – The human drug of choice

We suspect that the conclusions of the preceding two sections will be relatively straightforward to those readers who have taken the entire journey of this book.† We have chosen to use mostly the professional knowledge enterprise for purpose of illustration because it is an area we both know very well from personal experience. However, essentially identical critiques could be written about the everyday behavior of investment bankers, corporate managers, organized labor leaders, or government bureaucrats.

The deeper, universally applicable punch lines are vital. If acquisition of the limited, shared power we can exercise as members of a democratized local coalition results in internal release of the natural "endorphins"‡ that help us experience the small satisfactions of everyday life, then elite power is the "heroin"§ that hyper-stimulates that evolved psychological mechanism. Elite power is extraordinarily difficult for us to resist. Once we have held any type of elite power, it is nearly impossible for

---

* Also, see Popper (1945) for a different statement of a very similar view.

† Though not necessarily to readers who have flipped to the back of the book to begin at the end.

‡ Naturally produced hormones that engender pleasant subjective feelings. We use them here as a metaphor for internal natural reward systems.

§ A drug that apparently hyper-stimulates the same good feelings produced by some kinds of natural compounds.

us to give it up. Kicking heroin or elite power addictions are equally challenging.

We suggest that this pathological addiction to elite coercive power feeds the astonishing lack of self-awareness of our humanists from the preceding section just as surely as it did the racism of many 19th and 20th Century American whites or of Nazi ideologues. It is so easy for us to imagine that our expert or privileged position gives us insight that no one else has—utterly unaware that our faux insights are actually designed to feed our own elite power needs. We are—every one of us, authors included—extraordinarily vulnerable to this delusion.

This kind of (unconscious) self-deception is a huge problem for the knowledge enterprise. It is the equivalent of *crack epidemics* in local neighborhoods. We need to be able to distinguish *proposals* of fact by honest brokers from *assertions* that are actually the drug-induced delusions of those seduced by the dream of elite power.

Of course, the solution, we suggest, will already be apparent. Apply the judgment of the democratized doubting wise crowd (Chapter 10). Thus, we propose a rule of thumb. Someone who claims the power to control public decision-making based on *obscure expertise* or *esoteric revelation* is, beyond all reasonable doubt, one of our elite power-druggies. In contrast, if the most we aspire to is modest, democratically shared power, our assertions about public policy—or anything else, including human uniqueness, for example—must be maximally transparent and as vulnerable as possible to falsification by doubting wise crowds.[19]

The drug of elite power distorts human judgment—always and everywhere. This powerful narcotic drives us to do otherwise insane things. These things can be manipulated to feed the ambitions of others. It is this drug high that drives both the leadership and the shock troops of elite power.

In local environments, such shock troops—Nazi brown shirts, Mafia goons, armed Strangelovean bureaucrats, KKK/Maoist/Islamist/fascist militia, or Abu Ghraib prison guards—can be overpowering. If the material/financial means are available to arm and sustain them and if there is no sufficiently powerful democratized counter-force, the violent druggies of elite power win, everywhere and always. Modest Forsterian doubters apparently do not often become storm troopers and suicide bombers. The democratized portions of the global wise crowd have an obligation (and a self-interested incentive) to provide their coercive means to manage elite power addicts wherever they are.

The end of the world as elite power addicts knew it – The human future

Is the human future doomed to be dominated by addicts of elite coercive power? Possibly, but we have been suggesting that this outcome is NOT inevitable— probably not even likely.

As we discussed in detail in Chapter 15, radically democratized modern states apparently arose as a result of local historical accidents allowing a coercively powerful democratized majority to grow up before sources of elite domination could overwhelm them. This happened to a limited degree in the UK, but most radically in the northern half of what became the US and in a few other places.

Moreover, the northern half of the early US ultimately took a unique additional step. It created a powerful professional military *under the control of its democratized electorate*. This powerful coercive force was used to crush the military of the southern slavocracy blunting its local growth and forestalling its spread into the western half of the emerging US.[*]

This experimental North American state grew to enormous wealth over the next century, predictably, we argued above. Its democratically controlled, powerful military decided the outcomes of the great wars of the 20th Century imposing relatively radical democratization on Europe, Japan, Taiwan and South Korea while preserving it in the UK, Canada, Australia and New Zealand.

There is a cogent argument to be made that contemporary global democratization is, in large part, the product of this accidental formation of North American democratized coercive power.

Self-serving American political entrepreneurs are fond of pointing this out and speaking of *liberty*, *freedom*, and *democracy* as American values. On the contrary, democratized social organization appears to be the ancient, universal human heritage originating, most likely, 1.8 million years ago in Africa. The historical accidents that produced the North American state merely allowed the reemergence of this most successful ancient human trick and its expansion to the vast new scale of the modern state. Large parts of the global human population are still recovering from the relatively brief (roughly 5500 year long) interval of elite domination under archaic and early modern states (Chapters 13 and 14).

The ember of this ancient human adaptive trick—the wisdom of radically democratized crowds—still burns powerfully. It remains for all

---

[*]   See Chapter 15 for additional discussion.

of us over this century to nurture this fire. If we are successful we will give birth to a radically democratized *global* society. If we fail we will allow the druggies of elite power to snuff out a hopeful human future, perhaps irretrievably. How our remote descendents will live will almost certainly be irrevocably decided before the year 2100.

Knowledge is opportunity in facing this world-shaping challenge, perhaps more than in any other. We suggest that we can now see ourselves clearly enough to no longer be diverted and deluded by the corner pushers of the additive drugs of elite power. But the larger task will not be easy. We will have hard choices to make. Perhaps most difficult of all is the realization that those of us who hold elite power now will be required to surrender it. The global community will insist upon it and our difficult choice will be enforced.

Each and every one of us needs to aspire to being neither slave nor slave-owner but, instead, to being members of the global human coalition, each with her/his own small share of democratized power, no more, no less. Moreover, we must be willing to accept the real-but-modest pleasures of Forsterian wisdom, not the orgasmic, but fake high of elite domination.

Much patient work remains to be done. Many steps may need to be taken, some ancient and time-tested, others radically novel, perhaps, but all and only as revealed by our reasoning together as the global wise, self-aware crowd empowered to protect our interests. We propose a few important candidates for such steps in the following two sections.

The democracy effect and the quandary of elite – dominated contemporary states

We know from the record of Nazi Germany and Imperial Japan that elite-dominated states can be turned into democracies quite effectively. The problem is the stupendous loss of life and the cost of the approach used in these cases. South Korea and Taiwan suggest an alternative model for transforming militarized hierarchies into democracies: Patient coercion over several generations.* This, we suggest, is the model for transforming the planet's remaining elite-dominated states.

It is useful to consider, first, what is not enough. Paper institutions are of no value or significance, like the futility of the anti-slavery 13th, 14th

---

*   More peaceful coercion was possible here because South Korea and Taiwan were dependent on the American protective umbrella to defend them against the malevolent, expansionist elite power of Maoist China. Political change supported by American foreign policy was preferable to Maoist domination.

and 15th Amendments to the American Constitution in the white-dominated Jim Crow South following the American Civil War, for example. On our theory, such social contract beliefs/institutions are relevant *only* as they serve the interests of those who hold decisive coercive power. When this power is democratized, democratic institutions work. When power is controlled by small elites, democratic institutions are farcically irrelevant.

Thus, economists who imagined that paper economic institutions alone could create democracy—for example, in the former Soviet Union—were ineptly naïve. Coercive power remained in the hands of small elites—organized criminal enterprises and the residues of the Red Army and KGB. It was predictable, indeed inevitable, that these coercive powers would determine the short-term course of the Russian state, as they have.

Against this background, Russia and China represent similar problems for the future of the planet. It is useful to begin by considering how elite-dominated *economies* can work in the contemporary pan-global world.

The logic of elite coercion *within* the state is little affected by the availability of aircraft/missile technology (Chapter 16). However, the capacity of oligarchs to export their domination is effectively constrained when other states also possess these weapons. The global response to self-interested aggression would be far too costly (Chapter 16). Under these conditions, the power of the democracy effect on economic productivity (Chapter 15 and Figure 17.1) must be dealt with by oligarchs in some way if they are not to become helplessly poor and vulnerable.

One approach is to sell products to functioning democratized economies that do not require democratized economies to produce. Oil, diamonds, and narcotics are well-known examples. Selling oil and natural gas to the functioning democratized economies is currently the central pillar of the feeble, elite-dominated Russian economy, for example.

A second approach is to try to create a managed democratized economic system within a "containment vessel" maintained by the elite-dominated state.* This is the contemporary Chinese model. Chinese elites pirate know-how, innovation, and the locally hierarchical corporate institutional structures for producing manufactured goods. These

---

* Another useful approach is to extort economic assistance from the functioning democratized economies. This is the North Korean strategy. However, this strategy seems unlikely to produce more than tiny, temporary revenue streams for the elite state and we will not consider it further here. Of course, democratized states may be forced to use military force against such states at some point in the future.

goods are produced by slave labor and sold cheaply to wealthy democ-ratized states.*

Both these models seek to retain elite control of the state, predict-ably. The broader toxic global effects of these approaches on the welfare of workers and citizens of democratized states are obvious. The global labor market is grossly distorted and our ability to deal with environ-mental problems is severely hampered, among other dangerous conse-quences.

The question for us at the moment is what our theory has to say about whether Russian, Chinese, and other oligarchs might succeed. Of course, purely internal processes within these oligarch-dominated states will never undo entrenched elite power. However, in a pan-global world, there are alternatives.

On the Russian model, the economy will contract radically when oil prices fall again, as they have in the past and are doing again at this writing. When this next radical contraction comes, there may be another opportunity to exploit international coercion to improve the ratio of citizen to state coercive power within Russia, and, thus, Russia's democ-ratization.

On the Chinese model, the Red Army (the locally decisive coercive interest group) will continue to tax the economy and sustain its domi-nance. However, this capacity of the Red Army to fund its existence is ultimately dependent on the willingness of the citizens of high-func-tioning democratized economies to buy what China sells. We do this now, but abuse of the Red Army's power could bring this to a screeching halt, impoverishing both the citizenry and the elite of the Chinese state. Again, this is an opportunity to exploit our global coercive power indi-rectly to increase the access to power by the non-elite citizens of the Chinese state.

Under both these models, the citizens of democratized states are lending their coercive power to the otherwise powerless citizens of elite-dominated states. This loan is self-interested on our parts as we are creating new economic collaborators and de-fanging elite mili-tary threats. This loan of our power is probably a sustainable strategic approach. However, it will require considerably more finesse than we have shown to date and our theory helps us see why.

---

*   If you doubt that Chinese workers are essentially slaves, begin by reading Chapter 9 in Ellen Ruppel Shell's 2009 book *Cheap*. Also see page 20 in the July/August, 2009 *Harvard Business Review*.

Political psychology in the global village

Elite oligarchs can benefit from manipulating commoners as we have seen (Chapters 13-15). An effective way to enhance elite control is to create an external threat whose resistance requires commoner submission to elite supervision. This is, of course, a centuries-old gambit by kings and military rulers (Orwell, 1956; Kim, 2008).

In the past, this commoner manipulation required very little false propaganda. For example, elite-dominated Nazi Germany and elite-dominated Stalinist Russia were, in fact, very real, imminent, mortal threats to one another (Box 16.1). However, in the present, the massive coercive power of functional democratized states including North America, Western Europe, Japan, Canada, Australia, and New Zealand, means that expansionist military action is not a viable option for *any* state (Chapter 16).

Of course, authoritarian elites will use democratized global coercive dominance as a *bogey* to scare their commoner populations. However, we need not play along. Though the first American administration of the present century was quite incompetent at this approach, the electorates of the US and other democracies are considerably more sophisticated. We need not return spit-ball for spit-ball from elite manipulators (Putin, Ahmadinejad, Mugabe, Kim, Chavez, and others). Moreover, contemporary communications technology enables the global community of citizens to speak directly to one another over the heads of their leaders.

The combination of these long-term processes can deprive oligarchs of some of their ability to manipulate their citizenry. The effect of this, in turn, can leave elite authoritarian militaries quite isolated, no match for the economic and technical mastery of the global democratized states. Such elite militaries are then more fragile threats, open to the continuous, patient pushing of the global democratic consensus toward relinquishing ever more of their local coercive control. Moreover, the ability of hyper-elites to control their elite foot soldiers (Chapter 15) can likewise be severely eroded by improved access to information.

When surrendering to this democratic push, oligarchs would be foregoing elite control in return for the best self-interested option we leave open for them. We might slowly, patiently, over decades make their continued puny local dominance irrelevant, not worth having.

We can give the elite military establishments of contemporary authoritarian states two choices. One option is to continue attempting elite control of their domestic populations. If they continue this practice, we can ultimately crush them under the relentless piecemeal manipulation—readily enforceable—of the overpowering global democratized consensus. They will be remembered by history, if at all, together with the likes of Hitler, Stalin, Mao, and Pol Pot.

But we can also give them a second option, and a very attractive one at that. Authoritarian elites can morph into the necessary protectors of their domestic citizenry against asymmetric domination by the global democratic consensus. A source of such local coercive power will be necessary and these erstwhile elites are in a perfect position to provide it.*

For former oligarchs to take on this new, essential role of *national protector* will require the gradual empowering of their citizenry and surrender of their elite control of coercive power. Of course, they would not normally ever contemplate such a thing. However, the overwhelming power of the democratized global consensus transforms this choice into their best remaining option. Moreover, if they pursue this option successfully, they can be remembered as creative elements of the emergence of their nation-state into wealthy, peaceful modernity—not as the final, tragic dregs of the five-thousand-year-old human digression into authoritarianism.

This really is an offer they cannot refuse if we are able to make it with sufficient wisdom and finesse. It remains to those hundreds of millions of us who are members of the global democratized consensus to have the good judgment to make this choice as easy for them as we can.[20]

Box 17.4: Elite-dominated states – a list of symptoms

Our theory makes very clear what the properties of elite-dominated, hierarchical states *must* be, everywhere and always. Moreover, many of these features *must* be public and obvious. Thus, these properties become *symptoms* of elite-domination that the citizens of such states (and the rest of us) can use to diagnose them unambiguously, irrespective of their self-serving public propaganda. These states are the most serious disease of the contemporary pan-global world. We need to be able to recognize them.

---

* Remember that democratized majorities are not "naturally humane." They merely grant the rights and privileges local populations can defend. Recall the fate of Native Americans in the face of the radically democratized North American state, for example. Another chillingly clear case is the violent exploitation of Africans in the Congo by the democratized Belgian state in the late 19th and early 20th Century (Hochschild, 1998). Another telling example is the current behavior of American corporations in abetting oppression of Chinese labor merely to keep the prices we pay for consumer goods lower (Chapter 9, Shell, 2009).

This list of symptoms includes the following

(1) **Individual coercive means (personal gunpowder weapons) will be held preferentially or exclusively by professional military/police organizations.** Of course, this is the first and most fundamental law of the modern authoritarian state (Chapter 15). An important and equally diagnostic tributary principle is that professional military/police organizations in elite-dominated states will have strongly hierarchical relationships, while these organizations will be pluralistic and horizontally related in democratized states (Chapter 15).

(2) **Active public intimidation by elite military/police power, often including overt public cruelty, will be continuous.** This relentless display of coercive threat is required to suppress the pursuit of self-interest by the many members of massive disenfranchised commoner groups. This intimidation can take the form of public executions or the anonymous knock on the door in the middle of the night followed by people being ostentatiously *disappeared* without recourse.

(3) **Governmental self-justification will emphasize arbitrary, obscure identifier beliefs over pragmatic, empirically testable social contract beliefs.** Examples of such emphasis on identifier beliefs are legion. They include things like *communism, national socialism, laissez-faire capitalism, sexism, racism,* and, of course, the various *theisms*.

(4) **Elites will sharply limit the human knowledge enterprise.** This action will have several crucial motives. The knowledge enterprise is subversive of arbitrary identifier beliefs and of elite privilege more generally. Equally importantly, the massive sub-elite professional armies of the modern authoritarian state are a lethal, perennial threat to the smaller hyper-elite military/police cadres that manage them. This hyper-elite management depends on sub-elite military/police forces being too poorly informed to act on their very real, practical potential to overwhelm hyper-elite cadres coercively (Chapter 15). Restricting the knowledge enterprise—sanitizing it of "sedition"—is crucial to keeping all this subversive firepower uninformed.

Our theory predicts that no autonomous, democratized polity will have *any* of these symptoms. Conversely, we predict that all elite dominated polities will always have *all* of these symptoms.

Pax Pangaea – what might the human future look like?

Pangaea means, roughly, *one earth* or *whole earth*. It is a term geophysicists use to designate a period around two hundred and twenty-five million years ago when all the major continental landmasses were joined to form a single super-continent. We use it metaphorically as a designator for the pan-global, universal human coalition, emerging all around us now (above and Chapter 16). What does our theory say about how Pangaea might look one thousand years in the future?

Each human adaptive revolution throughout our two million year history has also represented a new level of complexity in the universe, an unprecedented scale of social cooperation (Figure 16.1). Pangaea represents the most revolutionary transition of all; the resources of an entire planet will come to serve the interests of a single, massive, coherent cooperative enterprise. As with each human adaptive revolution, this one presents stupendous danger and spectacular potential. This book has been one long argument that we now have the knowledge to tame the danger and tap the potential of Pangaea.

We claim that we now know what we need to know in order to begin, patiently and wisely, to systematically transform our world into a home for our descendants that is vastly richer, wiser, more peaceful, and more humane than we or our ancestors could ever have imagined.

Many of the more thoughtful among us may have been discouraged or depressed by what we have said in the sections immediately above. But we want to argue that this attitude is profoundly mistaken. The fundamental human adaptation uses the inexpensive *threat* of violence to suppress violence itself. If we put ten non-kin adult top predators—lions, for example—in a small space, they will immediately explode into lethal violence (Chapter 3). In contrast, humans—also top predators—sit down in non-kin groups of tens, hundreds, or thousands in rooms all around the world millions of times every day, with hardly ever a scuffle and nearly never a homicide. We use the threat of violence to produce peace – and much more.

Never forget the *fallacy of the bullet* (Box 14.1). Once we understand a coercive technology, we rarely need to use it overtly. Merely its credible threat is sufficient to sustain our cooperative enterprises. With these thoughts in mind how might democratized Pangaea unfold?

The emergence of the US over the last two centuries is a useful small-scale analogy. The US began as a mutually suspicious, wildly diverse groups of states. A struggle of ninety years, including an incredibly violent civil war, established the coercive logic and social contracts of the new unit, the *United States* as a singular noun. A long reconstruction after this period—still underway—gradually produced a functioning cooperative polity capable of protecting and sustaining many of its members in relative peace and prosperity.

Pangaea has arguably already been through its *civil war*—World War II. We have paid our (terrible) dues to establish the coercive logic of the pan-global coalition (Chapter 16). Our *reconstruction* is well underway, with the *Cold War* and the *War on Terror* being notable episodes in this unfolding global struggle and negotiation.

The American Congress began as a bungling caricature of a government, but it can be argued that it is sometimes less ridiculous now. The UN is likewise an institution with dubious abilities and track record. However, it can grow into something more and something better.

A pragmatically achievable vision for the Pangaea of a mere century hence would include most or all of the Earth's component human social units—nation states—being highly democratized. Each nation-state would probably possess global coercive assets, mega-weapons, contributing to maintenance of global peace and fair economic behavior, while retaining, cooperatively, local influence over the global central government.

If universal democratization is achieved, the levels of armament, both individual weapons and mega-weapons, could gradually become relatively low, though, of course, never zero (above and Chapters 5, 15 and 16).

This dry, technocratic discussion of a possible Pangaea has a richly human implication. A peaceful, pan-global coalition of billions of people can generate knowledge for all of us on a scale we can barely imagine at this moment (Chapter 10). Our descendents might see clearly things we now glimpse dimly or not at all.

Moreover, there is not the slightest reason to believe that the modern economic miracle of the last three centuries is the last step on the staircase of human adaptive revolutions (Figure 16.1; Figures 17.1 and 17.2). It is quite conservative to suppose that the citizens of a highly democratized Pangaea might enjoy a per capita planetary GDP orders of magnitude higher than the richest contemporary nation.

This level of wealth and knowledge may enable our descendents to realistically contemplate the next step in the human story, escaping our planet and, ultimately, our solar system. A blind, purposeless, amoral universe awaits our voice.

We apparently now have in hand what we need to build a wiser, richer, more humane world, patiently, democratically, and pragmatically over the next several generations. The crucial tool is our shared coercive means deployed in service of the capacity of democratized wise crowds to acquire insight. If we do this well, we may receive the greatest possible reward—having all the world's children one thousand years hence look back on us with admiration for our courage, but with pity for our ignorance and poverty.

Where to now?

The suggestions for action above must be informed by ever growing insight and relentless pragmatism. We have argued that we can now provide many long-sought answers. But if our theory is right—and we believe it is—the answers we have explored will not be what are most important. The real value of new science always lies *not* in what it *finishes* but in what it *starts*. It remains for you, the members of the global human wise crowd, to decide what this work will start.

We invite you to join one set of dialogs about the future and the human knowledge enterprise at www.deathfromadistance.com.

⚬

Postscript
Meaning and materialism – fulfilling our hearts and minds

The existential implications of what we think we know – Overview

Ah, love, let us be true
To one another! For the world, which seems
To lie before us like a land of dreams,
So various, so beautiful, so new,
Hath really neither joy, nor love, nor light,
Nor certitude, nor peace, nor help for pain;
And we are here as on a darkling plain
Swept with confused alarms of struggle and flight,
Where ignorant armies clash by night.
> final nine lines of the poem "Dover Beach," Matthew Arnold (1867)

Arnold's famous poem was written eight years after the publication of Darwin's *Origin of Species*. It expresses a terrible emptiness many people felt in the face of reductionist, materialist explanations of life. We are sometimes tempted to think that such explanations sweep away all sources of comfort, meaning, and purpose. They do not. The theory of human uniqueness explored in this book refines and extends Darwin's materialist picture. If we choose to look at the world as Arnold did, our theory can only deepen our existential despair. But the opposite choice is much more attractive.

We want to argue that Arnold did not really understand the ultimate implications of a materialist perspective. Rather, he was misled by the effects of unchecked elite power in the world he knew. First he saw the destruction of an earlier set of beliefs without yet seeing where new and better beliefs would come from. New insight into our condition does sweep away many illusory sources of meaning and purpose, but not real ones. Second, Arnold still saw coercive means (his "ignorant armies") as the tools of predatory hierarchs, as they had always been in the history he knew—mostly the history of elite-dominated archaic and early modern states (Chapters 13 and 14). Arnold did not see clearly that coercive means can be made to serve us, all of us, wisely and humanely.*

From a strong, clear materialist view of our universe can come true hope, peace and help for pain. Let us be clear about our argument. Many feel at first glance that a materialist view of how humans came to be unique can only provide a pale, wan imitation of real meaning. We argue that this perspective is a choice, not an inevitability and it is utterly mistaken.

To feel our lives have meaning, we need to experience a sense of larger or higher purpose.† We need something immortal and beyond the immediate interests of ourselves and our families. We suggest that the existential failure of materialism to date results, in part, when it artificially fails to recognize such a higher purpose. This failure is completely unnecessary. Humans, conjointly, can and do have a higher purpose, a purpose requiring the combined input of insight, experience, and expertise from all of us. As we have seen, humans probably own the galaxy and, perhaps, even the universe (Chapter 2). Human insight, comprehension, and adaptive capacity are something fundamentally new in that universe, resulting from our unique management of the universal

---

*   Arnold was also a member of a privileged elite. To the extent that he grasped the implications of the global tsunami of democratization, he would have had mixed feelings about it.

†   This sense of purpose reflects an evolved psychology that leads us to seek to be a part of the larger kinship-independent, uniquely human social adaptation surrounding us (Chapter 10). However, this proximate psychology can also be viewed as *pre-adaptation* to building a pan-human universe. An analogy helps grasp this perspective. We have recently come to understand that we eat food because of the requirements of the Second Law of Thermodynamics (Chapter 2). Our esthetic and sensual response to food reflects psychological mechanisms evolved to encourage us to take in enough of the right kinds of food. However, knowing these things in no way reduce our joy and pleasure in great cooking. Likewise, understanding how our desire for a higher purpose arose in our ancestral past in no way undermines our ability to choose knowingly to build our pan-global public lives around this desire.

conflict of interest problem (Chapters 10 and 17). We humans represent an opportunity—perhaps the only opportunity—for a self-aware cosmos. A cosmos *intelligently designed* by the *wise human crowd*. From this unprecedented position might come a true higher purpose.

The existential implications of what we think we know – Proximate social psychology, conflict, and hierarchy

We are highly adapted to negotiate, enforce, and live by social contract beliefs (Chapters 5, 7, and 10-17). Our survival and prosperity within a local, uniquely human social adaptation are utterly dependent on all the important details of these beliefs and have been for about two million years.

The cosmic, supernatural, institutionalized theistic elements of these social contract beliefs often also form the fictive source of ultimate coercive threat. A god will punish us if we commit evil. This is a useful fiction when actual coercive power is held by small elites with dubious claim to a larger social mandate (Chapters 13 and 14). Whatever its immediate strategic rationale, attempts to replace this fictional source of social coercion is a psychologically traumatic threat to all members with an investment in an existing social order, elite or commoner.

These fictional elements of social belief notwithstanding, our social contract beliefs also include the behaviors and strategies that are allowed and disallowed in everyday life, including many necessary and humane behaviors. These include the terms of the social contract (Chapter 10). These are the real adaptive pieces of our belief systems.

Thus, the total package of our social contract beliefs are absolutely vital to us both pragmatically (ultimately) and subjectively (proximately). We cannot feel fulfilled, or even merely safe, without confidence in the viability and shared acceptance of these beliefs.

Thus, when materialist theories of ultimate power and moral stricture attempt to displace supernatural theories, local believers in those earlier theories are inevitably threatened. The threat is not abstract or delusional. It is real and immediate. The fabric and rules of the local human social adaptation are potentially at risk, even the humane ones. A hostile, defensive reaction is reasonable and expected.* However, it is

---

* At this writing the following URLs have useful examples of such reactions: http://amnap.blogspot.com/2007/08/as-skeptical-as-pat-robertson-or-billy.html; http://mediamatters.org/items/200603200013; http://www.achristianblog.com/2005/11/blog-

possible to rebuild our social contract beliefs around pluralism of belief without the need for fear. We can not only retain our sense of humanity, but we can expand on it.

This process has mostly failed to happen in some parts of the world. This failure has traditionally been ascribed to the fundamental existential poverty of materialist belief systems. We believe this explanation is simply wrong. Rather, we suggest that materialism has been rejected by many for the same reason that committed Christians usually reject Islam and vice versa. Taking a moment to understand what we mean by this statement will pay crucial dividends.

Social contract beliefs include rules for ownership and use of coercive power. Moreover, in hierarchical social/economic systems. these rules include rationales for elite power. Institutional religions and contemporary theistic social contract belief systems were reshaped profoundly by their use (arguably perversion) by hierarchical early state elites (Chapters 13 and 14). In contrast, the modern Scientific Revolution was apparently a product of increasing democratization of early modern states (Chapters 15 and 17). Indeed, scientific belief systems were also potent political psychological tools in the early democratization of the state.* However, the problem arises because this democratization remains incomplete.

More specifically, individually self-interested agents are equally capable of exploiting materialist *or* theistic belief systems in support of elite power. Professional scientists and scholars have almost always succumbed to the temptation to allow themselves to be portrayed as priests of arcane, esoteric "knowledge."† In this role, they become useful,

---

connection-materialism/. Also, Google searches with terms like "materialism" and "Christianity" or "Islam" are fruitful.

* See Chapter 11 in Desmond's powerful 1997 biography of Thomas Huxley, Darwin's bulldog, for an especially clear, poignant account of an example of this political use of scientific belief systems, in Victorian England. Huxley provided massive public education in science while mobilizing working class and middle class (Marx' bourgeoisie) individuals in overt subversion of the traditional Anglican English hierarchy.

† To get a feeling for how pervasive this is, think back on the high school and college courses you have taken. How often does the instructor present herself/himself to you as an elite expert who is conveying to you specialized information? Your role was to write down and remember what was said, like a catechism. Regrettably, the answer is, "very often." In contrast, how frequently did you have instructors who communicated information to you in the context of active skepticism? Moreover, how often were the tools of effective doubt as much a part of the course as the nominal factual content? How frequently was your role to doubt everything you thought you knew including what you were hearing from this particular instructor? You then take away with you only what

convenient collaborators for elite interests. They are, in fact, modestly well compensated for their services in this regard.

Most professional scholars are not fully conscious of the various elite social schemes into which we are sometimes co-opted. We speak from personal experience. One of the authors (Paul) is a traditional *white coat* natural scientist and the other (Joanne) has worked at times as a research psychologist. We never consciously conceived of ourselves as serving *privileged elites* in that work. However, directly and indirectly, we sometimes have served such elites, especially when we contribute to the cultural practice and belief that professional scientists are "experts" whose views on public policy should be given decisive weight (Chapter 10). As a result of this history, our claims as scholars are seen by some to represent elite-justifying social contract beliefs, not potentially true claims about the material world.

Thus, when a scientific materialist argues for her (or his) view of the world, she naively, consciously imagines that she is merely defending an intellectual position supported by evidence, surviving falsification. However, she is also seen by some members of her audience as a self-interested apologist for a particular elite social system.

In other words, her claim, or our claim, that a materialist view of the world might lead to a superior set of social contract beliefs is no more credible, on its face, than the corresponding theistic claim by an Iranian mullah, say. A *materialist* takeover of the state is seen by many as essentially indistinguishable from the foundation of a hostile *theocracy*.

Fortunately, we are not doomed to continue a futile dialog between *science* and *religion*—between theism and materialism.* Science, properly practiced is ultimately a democratic enterprise. It works by exploiting the wisdom of crowds (Chapter 10). Professional scientists are not priests who know things others do not or cannot know. Rather, we are to science somewhat like professional players are to the game of basketball. We demonstrate in public how the game of science can be

---

survives you own personal, individual doubting (Chapter 10). Tragically, few, indeed, of the courses we take are of this second form.

*   The fact that so much of the theistic criticism of materialist views of the world—and of Darwinism in particular—is logically and scientifically inept leads many scientists to dismiss it as either malevolent or merely incompetent. (See Johnson, 1993, and Yahya, 2000, for two characteristic examples of such criticism from the Christian and Islamic traditions, respectively.) Indeed, such theistic attacks often are (unconsciously) malevolent when they are designed (again, often unconsciously) to preserve narrow elite privilege. However, what materialists generally fail to grasp is that these critiques get their persuasiveness from their ethical and political content. Under these conditions, their intellectual incoherence and factual accuracy are irrelevant.

played, but we do not control the game, its rules, or its ultimate product. The global human community has this control.* Notice that this is in sharp contrast to elite theistic systems where priests, mullahs, Popes and the like attempt to dictate the social contract.

By continuing to open up, to democratize the scientific enterprise[†] and by morphing the roles of professional scientists from priests to player/coaches, we can ultimately transform the pan-global dialog profoundly. The authors expect the core features of our reductionist, materialist view of humanity to survive this democratized pan-global doubting. If they do, materialism can emerge as a common consensus, no longer a narrow doctrine serving, or appearing to serve, parochial, privileged elites.

Of course, science is always about doubting. Thus, even when and if humane materialism emerges as a majority pan-global consensus, other belief systems will inevitably be tolerated and protected. Only the use of belief systems—*any* belief systems—as a pretext for elite manipulation of human societies need be disallowed and ostracized.

The existential implications of what we think we know – Life and Death

Our uniquely human grasp of the logic of all life gives us something more that no non-human animal will ever have. We can see our fear of death for what it is, merely a proximate psychological process designed to support our reproduction-related individual survival. As informational creatures, the most valuable and important parts of us are immortal or, rather, co-mortal with the long term survival of the human lineage and of our own cultural traditions.

First, obviously, genetic design information that builds animals who fear and resist death does better than design information that builds

---

*   The analogy of science to basketball is surprisingly rich to contemplate at length. The professional basketball game, played in public, influences the play of all the world's amateurs. These many amateurs, in turn, experiment and improvise around the professional model with which they grew up. Occasionally, on the world's many playgrounds and gym floors, talented amateurs make radical new contributions to the game. These innovations ultimately find their way into the professional game brought as gifted youngsters grow into the professional ranks or when pros copy an amateur they see and even with whom they compete. And the cycle continues. Science is just like this, at least when it is working the way it should.

†   Concurrently with continued democratization of economic and political systems (Chapter 17).

animals who have a reckless indifference to death. Hence we fear death and seek to replicate our genetic design information through reproduction when we can, directly or indirectly through close kin. We care very much about our children and families.

But each of the diverse pieces of genetic design information that builds us is also present in hundreds of millions of others. We need not fear the demise of this information if individual human lineages are not targeted for destruction by others. If genocide is abolished, only our narrow, artificial kin-selected psychologies remain to make us afraid (Chapters 2 and 3). And this is a small fear we can overcome.

Second, the same logic applies to culturally transmitted information. Cultural information that builds us to regard its preservation and transmission as vital, does better than cultural information that builds human minds to be indifferent to that information's survival (Chapter 10). Thus, our minds are also built by self-regarding culturally transmitted information. Hence we fear the death of our cultural traditions, and seek to replicate this information through cultural transmission in all modalities.

These insights teach us something profound, transcendental. Once we have arranged to respect our individual lives and cultures—our pan-global genetic and cultural heritage—we each gain the informational immorality that is reflected in our evolved proximate goals. We can then caste aside the illusion of individual despair—and the fear of informational extinction that is its underlying reality.*

As long as *any* of us seek the annihilation of others of us, death will continue to haunt the footsteps of *all* of us. But, if we are successful as a pan-global coalition in forbidding and suppressing ethnocentric murderousness, each of us can live with confidence and peace. We can serenely enjoy the time of individual consciousness each of us happens to possess.

It could hardly be more transparent or simple, we argue. It remains for us to act (Chapter 17).

෧෧

---

* More precisely, we have argued that the only humans who should experience this fear are the small elites in remaining hierarchical states (Chapter 17).

References

(Endnotes numbered in the text are available online at www.deathfromadistance.com)

Agur, A.M.R., Lee, M.J., and J.C.B. Grant. 1999. *Grant's atlas of anatomy*. Philadelphia: Lippincott Williams & Wilkins.

Aiello, L., and C. Dean. 1990. *An introduction to human evolutionary anatomy*. London; San Diego: Academic Press.

Alberts, B., A. Johnson, J. Lewis, M. Raff, K. Roberts, and P. Walter. 2008. *Molecular biology of the cell*, 5th ed. New York: Garland Science.

Alexander, R.D. 1987. *The biology of moral systems*. Hawthorne, NY: A. de Gruyter.

Altman, J.C., and Australian Institute of Aboriginal Studies. 1987. *Hunter-gatherers today: An Aboriginal economy in north Australia*. Canberra: Australian Institute of Aboriginal Studies.

Aly, G. 2007. *Hitler's beneficiaries: Plunder, racial war, and the Nazi welfare state*. New York: Metropolitan.

Ames, K.M., and H.D.G. Maschner. 1999. *Peoples of the Northwest Coast: Their archaeology and prehistory*. New York: Thames and Hudson.

Angier, N. 1999. *Woman: An intimate geography*. Boston: Houghton Mifflin Co.

Ariely, D., 2008. *Predictably irrational: the hidden forces that shape our decisions*. New York: Harper.

Arnold, M. 1994. *Dover Beach and other poems*. New York: Dover Publications.

Arnold, J.E. 2004. *Foundations of Chumash complexity*. Los Angeles: Cotsen Institute of Archaeology, University of California.

Avital, E., and E. Jablonka. 2000. *Animal traditions: Behavioural inheritance in evolution*. Cambridge, UK: Cambridge University Press.

Baker, R.R. and Bellis, M.A. 1994. Human *Sperm competition: Copulation, masturbation and infidelity*. New York: Chapman & Hall.

Baker, R. 1996. *Sperm wars: The science of sex*. New York: BasicBooks.

Bean, L.J., and T.C. Blackburn. 1976. *Native Californians: A theoretical retrospective*. Ramona, CA: Ballena Press.

Beatty, J. 2007. *Age of betrayal: The triumph of money in America, 1865-1900*. New York: Alfred A. Knopf.

Bellesiles, M.A. 2000. *Arming America: The origins of a national gun culture*. New York: Alfred A. Knopf.

Bergerud, E.M. 2000. *Fire in the sky: The air war in the South Pacific*. Boulder, CO: Westview Press.

Bernstein, W.J. 2004. *The birth of plenty: How the prosperity of the modern world was created*. New York: McGraw-Hill.

Biederman, I. 1987. Recognition-by-Components—A theory of human image understanding. *Psychological Review* 94: 115-47.

Bingham, P.M. 1997. Cosuppression comes to the animals. *Cell* 90: 385-87.

Bingham, P.M. 1999. Human uniqueness: A general theory. *Quarterly Review of Biology* 74: 133-69.

Bingham, P.M. 2000. Human evolution and human history: A complete theory. *Evolutionary Anthropology* 9: 248-57.

Bingham, P.M. 2009. "On the evolution of language: Implications of a new and general theory of human origins, properties and history" In *The Evolution of Human Language: Biolinguistic Perspectives*, ed. R.K. Larson, Viviane Deprez, and Hiroko Yamakido. Cambridge, UK: Cambridge University Press.

Bogin, B. 1999. *Patterns of human growth*. Cambridge, U.K: Cambridge University Press.

Bramble, D.M., and D.E. Lieberman. 2004. Endurance running and the evolution of Homo. *Nature* 432: 345-52.

Burton, O.V. 2007. *The age of Lincoln*. New York: Hill and Wang.

Buss, D.M. 1994. *The evolution of desire: Strategies of human mating*. New York: BasicBooks.

Buss, D.M. 2003. *The evolution of desire: Strategies of human mating, 2nd ed*. New York: Basic Books.

Calvin, W.H. 1983. *The throwing madonna: Essays on the brain*. New York: McGraw-Hill.

Carr, C., and D.T. Case. 2005. *Gathering Hopewell: Society, ritual, and ritual interaction*. New York: Kluwer Academic/Plenum Publishers.

Cech, T.R. 1986. RNA as an enzyme. *Scientific American* 255: 64-69.

Choi, J.K., and S. Bowles. 2007. The coevolution of parochial altruism and war. *Science* 318: 636-40.

Coe, M.D. 2005. *The Maya*. New York: Thames and Hudson.

Cohan, W.D., 2009. *House of cards : a tale of hubris and wretched excess on Wall Street*. Waterville, ME: Thorndike Press.

Collins, H., 2009. We cannot live by skepticism alone. *Nature* 458, 30-31.

Collins, J.C., and J.I. Porras. 1994. *Built to last: Successful habits of visionary companies*. New York: Harper Business.

Cooper, G. 2008. *The origin of financial crises: Central banks, credit bubbles and the efficient market fallacy*. New York: Vintage Books.

Corballis, M.C. 1991. *The lopsided ape: Evolution of the generative mind*. New York: Oxford University Press.

Corballis, M.C. 2002. *From hand to mouth: The origins of language*. Princeton, NJ: Princeton University Press.

Cordell, L.S. 1997. *Archaeology of the Southwest, 2nd ed*. San Diego, CA: Academic Press.

Crocker, W.H. 1990. *The Canela (Eastern Timbira), I: An ethnographic introduction*. Washington D.C.: Smithsonian Institution Press.

Crocker, W.H., and J. Crocker. 1994. *The Canela: Bonding through kinship, ritual, and sex*. Fort Worth, TX: Harcourt Brace College Publishers.

Daly, M., and M. Wilson. 1988. *Homicide*. New York: A. de Gruyter.

Dancey, W.S., and P.J. Pacheco. 1997. *Ohio Hopewell community organization*. Kent, OH: Kent State University Press.

Darwin, C. 1859. *On the origin of species by means of natural selection*. London: J. Murray.

Darwin, C. 1871. *The descent of man, and selection in relation to sex.* D. New York: Appleton and Company.

Dawkins, R. 1976. *The selfish gene.* Oxford: Oxford University Press.

Deacon, T.W. 1997. *The symbolic species: The co-evolution of language and the brain.* New York: W.W. Norton.

Dennett, D.C., 1995. *Darwin's dangerous idea : evolution and the meanings of life.* New York: Simon & Schuster,.

Dennett, D.C., 2003. *Freedom evolves.* New York: Viking.

Dennett, D.C., 2005. *Sweet dreams : philosophical obstacles to a science of consciousness.* Cambridge, MA: MIT Press,.

Desmond, A.J. 1997. *Huxley: From devil's disciple to evolution's high priest.* Reading, MA: Addison-Wesley.

Diamond, J.M. 1997. *Guns, germs, and steel: The fates of human societies.* New York: W.W. Norton.

Diamond, J.M. 2000. Taiwan's gift to the world. *Nature* 403: 709-10.

Diamond, J.M. 2001. Polynesian origins - Slow boat to Melanesia? Reply. *Nature* 410: 167.

Diamond, J.M. 2005. *Collapse: How societies choose to fail or succeed.* New York: Viking.

Díaz del Castillo, B. 1963. *The conquest of New Spain.* Baltimore, MD: Penguin Books.

Dickerson, R.E. 1978. Chemical evolution and origin of life. *Scientific American* 239: 70-74

Ferguson, N. 2006. *The war of the world: Twentieth-century conflict and the descent of the West.* New York: Penguin Press.

Fischer, D.H. 1996. *The great wave: Price revolutions and the rhythm of history.* New York: Oxford University Press.

Fisher, H.E. 1983. *The sex contract : the evolution of human behavior.* New York: Quill.

Forster, E.M. 1951. *Two cheers for democracy.* London: E. Arnold.

Fox, J., 2009. *The myth of the rational market : a history of risk, reward, and delusion on Wall Street.* New York: Harper Business.

Freeman, C. 1996. *Egypt, Greece, and Rome: Civilizations of the ancient Mediterranean.* Oxford: Oxford University Press.

Galvin, J.R. 1989. *The minute men: The first fight: Myths & realities of the American revolution.* Washington D.C.: Pergamon-Brassey's International Defense Publisher.

Gazzaniga, M.S. 2000. Cerebral specialization and interhemispheric communication - Does the corpus callosum enable the human condition? *Brain* 123: 1293-1326.

Gazzaniga, M.S. 2005. *The ethical brain.* New York: Dana Press.

Gazzaniga, M.S., 2008. *Human: The Science Behind What Makes Us Unique.* New York: HarperCollins.

Gould, J.L., and W.F. Towne. 1987. Evolution of the dance language. *American Naturalist* 130: 317-38.

Gould, T. 2000. *The lifestyle: A look at the erotic rites of swingers.* Buffalo, NY: Firefly Books.

Hamilton, W.D. 1996. *Narrow roads of gene land: The collected papers of W.D. Hamilton.* New York: W.H. Freeman/Spektrum.

Harman, C. 2008. *A people's history of the world.* Chicago: Bookmarks.

Hauser, M.D. 2006. *Moral minds: How nature designed our universal sense of right and wrong.* New York: Ecco.

Heinrich, B. 2002. *Why we run: A natural history.* New York: Ecco.

Heller, J. 1961. *Catch-22, a novel.* New York: Simon and Schuster.

Hewes, G.W. 1992. Primate Communication and the gestural origin of language. *Current Anthropology* 33: 65-84.

Heyes, C.M., and B.G. Galef. 1996. *Social learning in animals: The roots of culture.* San Diego, CA: Academic Press.

Hobbes, T. 1651/2003. *Thomas Hobbes Leviathan*. Ed. G. A. J. Rogers and Karl Schuhmann. New York: Thoemmes Continuum.

Hochschild, A. 1998. King *Leopold's ghost: A story of greed, terror, and heroism in Colonial Africa*. Boston: Houghton Mifflin.

Hölldobler, B., and E.O. Wilson. 1990. *The ants*. Cambridge, MA: Belknap Press of Harvard University Press.

Hrdy, S.B. 1999. *Mother nature: A history of mothers, infants, and natural selection*. New York: Pantheon Books.

Huxley, A. 1935. *Brave new world*. Garden City, NY: The Sundial Press.

Issac, B., 1987. Throwing and human evolution. African Archaeological Review 5, 3-17.

Jackendoff, R. 2002. *Foundations of language: Brain, meaning, grammar, evolution*. New York: Oxford University Press.

Jansen, M.B. 2000. *The making of modern Japan*. Cambridge, MA: Belknap Press of Harvard University Press.

Johansen, B.E. 1998. *Debating democracy: Native American legacy of freedom*. Santa Fe, NM: Clear Light Publishers.

Johanson, D.C., and B. Edgar. 1996. *From Lucy to language*. New York: Simon & Schuster Editions.

Johnson, P.E. 1993. *Darwin on trial*. Downers Grove, IL: Inter Varsity Press.

Justice, N.D. 1987. *Stone age spear and arrow points of the midcontinental and eastern United States*: A modern survey and reference. Bloomington, IN: Indiana University Press.

Justice, N.D. 2002. *Stone Age spear and arrow points of the Southwestern United States*. Bloomington, IN: Indiana University Press.

Kahan, D.M., and D. Braman. 2008. The self-defensive cognition of self-defense. *American Criminal Law Review* 45: 1-65.

Kahaner, L. 2007. *AK-47: The weapon that changed the face of war*. Hoboken, NJ: Wiley.

Kennett, D.J. 2005. *The island Chumash: Behavioral ecology of a maritime society*. Berkeley, CA: University of California Press.

Keynes, J.M., 1936. *The general theory of employment, interest and money*. London: Macmillan and Co., ltd., Reissued, 2009, Hamburg: Management Laboratory Press.

Kim, M., 2008. *Escaping North Korea : defiance and hope in the world's most repressive country*. Lanham, MD: Rowman & Littlefield Pub. : Distributed by National Book Network.

Klein, R.G. 1999. *The human career: Human biological and cultural origins*. Chicago: University of Chicago Press.

Klein, R.G., and B. Edgar. 2002. *The dawn of human culture*. New York: Wiley.

Kolata, G., 2009. Playing it Safe in Cancer Research, *New York Times*, June 28, pg. 1.

Laumann, E.O., J.H. Gagnon, R.T. Michael, and S. Michaels, eds. 1994. *The social organization of sexuality: Sexual practices in the United States*. Chicago: University of Chicago Press.

Leakey, R.E., and R. Lewin, R. 1992. *Origins reconsidered: In search of what makes us human*. New York: Doubleday.

Lloyd, E.A. 2005. *The case of the female orgasm: Bias in the science of evolution*. Cambridge, MA: Harvard University Press.

Lott, J.R. 2000. *More guns, less crime: Understanding crime and gun-control laws*. Chicago: University of Chicago Press.

Machiavelli, N., and B. Crick, B. 1970. *The discourses*. Harmondsworth, UK: Penguin Books.

Maddison, A. 2001. *The world economy: A millennial perspective*. Paris: Development Centre of the Organization for Economic Co-operation and Development.

Maher, J. 1996. *Seeing language in sign: The work of William C. Stokoe*. Washington, D.C.: Gallaudet University Press.

Malcolm, J.L. 1994. *To keep and bear arms: The origins of an Anglo-American right.* Cambridge, MA: Harvard University Press.

Malcolm, J.L. 2002. *Guns and violence: The English experience.* Cambridge, MA: Harvard University Press.

Malthus, T.R., and J. Bonar. 1965. *First essay on population, 1798.* New York: A. M. Kelley, Bookseller.

Martin, J.K., and M.E. Lender. 2006. *A respectable army: The military origins of the republic, 1763-1989,* 2nd ed. Wheeling, IL: Harland Davidson.

Maynard Smith, J. 1982. *Evolution and the theory of games.* New York: Cambridge University Press.

McCullough, M.E. 2008. *Beyond revenge: The evolution of the forgiveness instinct.* San Francisco, CA: Jossey-Bass.

McDougall, W.A. 2004. *Freedom just around the corner: A new American history, 1585-1828.* New York: Harper Collins Publishers.

McDougall, W.A. 2008. *Throes of democracy: The American Civil War era, 1829-1877.* New York: Harper.

McPherson, J.M. 1988. *Battle cry of freedom: The Civil War era.* New York: Oxford University Press.

McPherson, J.M. 1997. *For cause and comrades: Why men fought in the Civil War.* New York: Oxford University Press.

Metzinger, T., 2009. *The ego tunnel: the science of the mind and the myth of the self.* Basic Books, New York.

Milner, G.R. 1998/2006. *The Cahokia chiefdom: The archaeology of a Mississippian society.* Washington, D.C.: Smithsonian Institution Press.

Minsky, H.P., 1986. *Stabilizing an unstable economy.* Yale University Press, New Haven.

Moseley, M.E. 1992. *The Incas and their ancestors: The archaeology of Peru.* New York: Thames & Hudson.

Muller, J. 1997. *Mississippian political economy.* New York: Plenum Press.

Murray, W., and A.R. Millett. 2000. *A war to be won: Fighting the Second World War.* Cambridge, MA: Belknap Press of Harvard University Press.

Napier, J.R., and R. Tuttle. 1993. *Hands.* Princeton, NJ: Princeton University Press.

Obeyesekere, G. 1992. *The apotheosis of Captain Cook: European mythmaking in the Pacific.* Princeton, NJ: Princeton University Press.

Okada, D., and P.M. Bingham. 2008. Human uniqueness-self-interest and social cooperation. *Journal of Theoretical Biology* 253: 261-70.

Olson, M. 1965. *The logic of collective action: public goods and the theory of groups.* Cambridge, MA: Harvard University Press.

Olson, M. 2000. *Power and prosperity: Outgrowing communist and capitalist dictatorships.* New York: Basic Books.

Orwell, G. 1949. Nineteen Eighty-Four. A novel. New York: Harcourt, Brace & Co.

Palmer, J. 2009. *The bloody white baron: The extraordinary story of the Russian nobleman who became the last Khan of Mongolia.* New York: Basic Books, a Member of the Perseus Books Group.

Paradis, J.G., and G.C. Williams. 1989. *Evolution & ethics: T.H. Huxley's "Evolution and ethics" with new essays on its Victorian and sociobiological context.* Princeton, NJ: Princeton University Press.

Partnoy, F., 2003. *Infectious greed : how deceit and risk corrupted the financial markets.* New York: Times Books.

Pepperberg, I.M. 1999. *The Alex studies: Cognitive and communicative abilities of grey parrots.* Cambridge, MA: Harvard University Press.

Pinker, S. 1994. *The language instinct.* New York: W. Morrow and Co.

Pinker, S., 1997. *How the mind works.* New York: Norton.

Popper, K.R. 1945. *The open society and its enemies.* London: G. Routledge & Sons.

Porter, R. 2000. *The creation of the modern world: The untold story of the British enlightenment.* New York: W.W. Norton.

Potts, R. 1988. *Early hominid activities at Olduvai.* New York: A. de Gruyter.

Pringle, H.A. 1996. *In search of ancient North America: An archaeological journey to forgotten cultures.* New York: John Wiley & Sons.

Raab, L.M., and T.L. Jones. 2004. *Prehistoric California: Archaeology and the myth of paradise.* Salt Lake City, UT: University of Utah Press.

Ratnieks, F.L.W., Foster, K.R., and T. Wenseleers. 2006. Conflict resolution in insect societies. *Annual Review of Entomology* 51, 581-608.

Ridley, M. 1996. *The origins of virtue.* London: Viking.

Roberts, J.M. 1993. *History of the world.* New York: Oxford University Press.

Roberts, D. 2004. *The Pueblo Revolt: The secret rebellion that drove the Spaniards out of the Southwest.* New York: Simon & Schuster.

Shea, J.J. 2006. The origins of lithic projectile point technology: evidence from Africa, the Levant, and Europe. *Journal of Archaeological Science* 33, 823-46.

Shell, E.R., 2009. *Cheap : the high cost of discount culture.* New York: Penguin Press.

Shiller, R.J., 2005. *Irrational exuberance.* New York: Currency/Doubleday.

Shreeve, J. 1995. *The Neandertal enigma: Solving the mystery of modern human origins.* New York: Morrow.

Singer, P., 1981. *The expanding circle : ethics and sociobiology.* Farrar, Straus & Giroux, New York.

Skocpol, T. 1979. *States and social revolutions: A comparative analysis of France, Russia, and China.* New York: Cambridge University Press.

Smith, A. 1776. *An inquiry into the nature and causes of the wealth of nations.* New York: A. M. Kelley.

Smith, B.D. 1995. *The emergence of agriculture.* New York: Scientific American Library: Distributed by W.H. Freeman.

Snow, C.P. 1959. *The two cultures and the scientific revolution.* New York: Cambridge University Press.

Soros, G., 2008. *The new paradigm for financial markets : the credit crisis of 2008 and what it means.* New York: Public Affairs.

Soto, H. de 2000. *The mystery of capital: Why capitalism triumphs in the West and fails everywhere else.* New York: Basic Books.

Stringer, C., and C. Gamble. 1993. *In search of the Neanderthals: Solving the puzzle of human origins.* New York: Thames and Hudson.

Stuart, D.E. 2000. *Anasazi America.* Albuquerque, NM: University of New Mexico Press.

Surowiecki, J. 2004. *The wisdom of crowds: Why the many are smarter than the few and how collective wisdom shapes business, economies, societies, and nations.* New York: Doubleday.

Symons, D. 1979. *The evolution of human sexuality.* New York: Oxford University Press.

Szatmary, D.P. 1980. *Shays' Rebellion: The making of an agrarian insurrection.* Amherst, MA: University of Massachusetts Press.

Taibbi, M., 2009. Inside The Great American Bubble Machine. *Rolling Stone,* July 9, online at *http://www.rollingstone.com/politics/story/28816321/the_great_american_bubble_machine.*

Taleb, N., 2007. *The black swan : the impact of the highly improbable.* New York: Random House.

Tattersall, I. 1995. *The fossil trail: How we know what we think we know about human evolution.* New York: Oxford University Press.

Tilly, C. 2006. *Why?* Princeton, NJ: Princeton University Press.

Tilly, C. 2008. *Credit and blame*. Princeton, NJ: Princeton University Press.

Tooby, J., and L. Cosmides. 2001. Does beauty build adapted minds? Toward an evolutionary theory of aesthetics, fiction and the arts. *Sub-Stance*, 6-27.

Trigger, B.G. 2003. *Understanding early civilizations*. Cambridge, UK: Cambridge University Press.

Valeri, V. 1985. *Kingship and sacrifice: Ritual and society in ancient Hawaii*. Chicago: University of Chicago Press.

Von Neumann, J., and Morgenstern, O., 1944. *Theory of games and economic behavior*. Princeton, New Jersey: Princeton University Press.

Vrba, E.S., D.H. Denton, T.C. Partridge, and L.H. Burckle. 1995. *Paleoclimate and evolution, with emphasis on human origins*. New Haven, CT: Yale University Press.

Waddington, C.H. 1960. *The ethical animal*. London: G. Allen & Unwin.

Wade, R.A. 1984. *Red guards and workers' militias in the Russian Revolution*. Stanford, CA: Stanford University Press.

Walker, A., and P. Shipman. 1996. *The wisdom of the bones: In search of human origins*. New York: Knopf: Distributed by Random House.

Weart, S.R. 1998. *Never at war: Why democracies will not fight one another*. New Haven, CT: Yale University Press.

Webster, D. 1998. "War and status rivalry: Lowland Maya and Polynesian comparison." In *Archaic states*, ed. G.M. Feinman, and J. Marcus. Santa Fe, NM: School of American Research Press, 311-51.

Wenke, R.J. 1996. *Patterns in prehistory: Humankind's first three million years*. New York, Oxford University Press.

Whiten, A., J. Goodall, W.C. McGrew, T. Nishida, V. Reynolds, Y. Sugiyama, C.E.G. Tutin, R.W. Wrangham, C. Boesch. 1999. Cultures in chimpanzees. *Nature* 399: 682-85.

Wilentz, S. 2005. *The rise of American democracy: Jefferson to Lincoln*. New York: Norton.

Williams, G.C. 1957. Pleiotropy, natural-Selection, and the evolution of senescence. *Evolution* 11: 398-411.

Williams, G.C. 1966. *Adaptation and natural selection: a critique of some current evolutionary thought*. Princeton, NJ: Princeton University Press.

Williams, D.C. 2003. *The mythic meanings of the Second Amendment: Taming political violence in a constitutional republic*. New Haven, CT: Yale University Press.

Wilson, E.O. 1971. *The insect societies*. Cambridge, MA: Belknap Press of Harvard University Press.

Wilson, E.O. 1975. Sociobiology: *The new synthesis*. Cambridge, MA: Belknap Press of Harvard University Press.

Wolin, R. 2004. *The seduction of unreason: The intellectual romance with fascism: From Nietzsche to postmodernism*. Princeton, NJ: Princeton University Press.

Wood, M. 2000. *Conquistadors*. Berkeley, CA: University of California Press.

Wright, R. 1994. *The moral animal: The new science of evolutionary psychology*. New York: Pantheon Books.

Wright, R., 2000. *NonZero : the logic of human destiny*. Pantheon Books, New York.

Yahya, H., and T. Mossman. 2006. *Atlas of creation*. Istanbul: Global Pub.

Zinn, H. 1995. *A people's history of the United States: 1492-present*. New York: Harper Perennial.

Endnotes

Numbered endnotes in text can be found online at
www.deathfromadistance.com

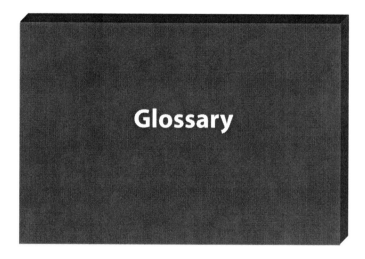

Glossary

Acheulean tools: This set of stone tools was initially produced by **Homo erectus** beginning around 1.6 million years ago. They include the hand axe, an elaborately crafted bifacial cutting, boring, and chopping tool, resembling a giant stone arrowhead. These tools require sophisticated manufacturing techniques and are far beyond the capabilities of non-human animals. However, they are much less sophisticated than those made later by **behaviorally modern humans** (Chapter 11). Moreover, these tools continued to be made in essentially unchanged form for well over 1 million years by early humans! Such a period of **adaptive stasis** is profoundly foreign to us today as modern humans.

AD (dating convention): See **CE** and **BCE**.

adaptive revolution: Refers to the fact that the two million year interval of human history has consisted of very long periods of **adaptive stasis** during which human capabilities improved very little followed by periods of dramatic increase in adaptive sophistication. These periods of dramatic increase in adaptive sophistication are referred to as adaptive revolutions. Apparently, these revolutions result from the following sequence of events (Third Interlude): A novel weapon technology allows cost-effective projection of **coercive threat** producing an expanded scale of **kinship-independent social cooperation**. Human **adaptive sophistication** is determined by this increased scale of social cooperation (Chapter 10). All the major adaptive revolutions of the human story, including the behaviorally modern human revolution (Chapter 11), the various Neolithic/agricultural revolutions (Chapter 12), the rise of the archaic state (Chapter 13) and the rise of the modern state (Chapters 15-17), are apparently examples of adaptive revolutions that follow this causal sequence.

adaptive sophistication: Refers to the capabilities of any **organism** to interact with the world in ways that serve its interests as a **vehicle**. We will use this term almost exclusively to refer human capacities. Human adaptive sophistication can be measured in many different ways, though each of these measurements gives us a very similar picture when we examine our history (Third Interlude; Chapters 11-17). One easy measurement is the sophistication of tools. An **Acheulean handaxe** is less sophisticated than an iPod, for example. As well, we can measure changes in sustainable population densities or economic productivity (see **GDP**).

adaptive staircase: The graphic depiction of all of human history as consisting of many instances of abrupt, stair step-like increases in **adaptive sophistication** (referred to as **adaptive revolutions**) separated by flat intervals of **adaptive stasis** with no increase in adaptive sophistication (see Figures in Third Interlude).

adaptive stasis: Refers to the tendency of human **adaptive sophistication** to remain largely unchanged over an extended period of time (even millions of years; Chapters 10 and 11). We explore the evidence that adaptive stasis results from a prolonged period during which human cooperative coalition size remains unchanged. Periods of adaptive stasis are ended by **adaptive revolutions** whose origin and details are predicted by our theory (Third Interlude).

adolescence: This **life history** segment is unique to humans (Chapter 6). As we use the term, it refers to the time when humans delay reaching full adult size. Following an adolescent bodily growth spurt, humans continue to mature, socially and sexually, over a period of several additional years. This adaptation probably reflects several things, including providing additional time for the acquisition of the uniquely large amount of culturally transmitted information that human cooperative coalitions possess and share (Chapter 10).

africanus: see **Australopithecus afarensis**.

agriculture/agricultural revolutions: This phrase is applied to dramatic changes in human **adaptive sophistication** occurring at different times in different parts of the world, ranging from roughly eleven thousand years ago to about thirteen hundred years ago. These changes are associated with increased reliance on domesticated plants and animals, in contrast to living by hunting and gathering of wild foods as apparently humans did for the preceding 1.8 million years. Traditionally, agricultural revolutions have been thought of as *resulting from* the domestication of plants and animals. However, the last several decades have produced increasing evidence that this is not the actual causal sequence. As our theory predicts, improvements in weaponry precede expanded social cooperation which then apparently engenders local **adaptive revolutions**, including agriculture and other **Neolithic** adaptations (Chapter 12).

aircraft: We will use the term to include all contemporary weapon systems dependent on powered flight. These systems include military fighter and bomber aircraft, ballistic and cruise missiles, and all forms currently classified as precision munitions. These weapons make possible the (relatively) inexpensive projection of coercive threat against an entire nation by any sufficiently overwhelming coalitions of other nations or powerful individual nations (Chapter 16).

AK-47: See **handgun**.

Alexander, Richard D.: One of the greatest 20th Century thinkers regarding evolution. Alexander's monumental book *The Biology of Moral Systems* (1987) is one of the important foundations on which this work is based.

allele: Refers to different versions of the same piece of **genetic design information** produced by mutation (Chapter 2). For example, two different versions of a 3000 **DNA** base long eye color gene might be identical in all but one position, say position 1357.

At this single position one version might have a C base and the other a G base in the DNA molecule. One version might produce green eyes and the other blue eyes. Each of these two forms would be called a different allele of this gene.

altricial: The opposite of **precocial**. Altricial youngsters or juveniles remain highly dependent on social and/or parental support for a relatively long period after birth. This prolonged period of dependence has the benefit of allowing the youngsters to acquire more information from others, ultimately becoming more successful as adults. Therefore, altricial animals tend to have more sophisticated behaviors. Dominant animals like carnivores are usually relatively altricial. Humans are, by far, the most altricial animal that ever lived. (The noun form of this adjective is altriciality.)

American Civil War: A crucial element of the ongoing battle to democratize the **modern state** (Chapters 15 and 17). Elite Southern slavocrats and more democratized Northern electorates fought to the death for control of the emerging North American state. Arguably, the contemporary United States was more shaped by the Civil War and its aftermath than by the **American Revolution**.

American Revolution: For a variety of reasons - especially lack of concentrated mineral wealth and remoteness from Europe - the North American colonial population remained relatively free of elite domination from Europe. As a result, North America constituted a massive natural experiment in state formation. Moreover, North American colonists lived in an environment with potentially hostile native populations and abundant game for hunting. Several centuries of this set of environmental factors produced a population of highly independent, heavily armed individuals free of elite military domination. These individuals initiated the American Revolution in 1776, ultimately breaking free of European control and founding one of the early, relatively democratized modern states (Chapters 15 and 17).

amino acid: See **protein** and **DNA**.

Anasazi: One of several terms, in addition to "ancestral Pueblo" and Hitsatsinom, used to describe one of the major Neolithic cultures of the American Southwest (Chapter 12). This culture spread throughout the "Four Corners" region, where Utah, Arizona, New Mexico and Colorado meet. It was sustained by maize agriculture and arose around 600-900AD. Among other things, the Anasazi are famous for building the Great Houses of Chaco Canyon and the cliff dwellings of Mesa Verde.

anatomically modern humans: These were the first members of our contemporary anatomical species, *Homo sapiens*, who evolved (apparently from local *Homo heidelbergensis* populations in eastern and southern Africa) around 150,000-250,000 years ago. They are called anatomically modern because they look virtually exactly like us in skeletal morphology (with minor quantitative differences) yet they did not display the full panoply of capabilities that would have made them **behaviorally modern humans.** They had a brain size indistinguishable from ours (averaging around 1350 cc and ranging from 1100-1500 cc with occasional individuals outside this range). However, these animals continued to make tools similar to older human populations and to the contemporary **Neandertal** population in Eurasia. For this reason their behavior, in contrast to their anatomy, remains distinctly non-modern (Chapter 11).

Annie Oakley: She rose to fame in the last decade of the 19th Century as the greatest marksman of her era. She traveled North America and Europe giving shooting demonstrations and was an international celebrity. In her biography she claims to have trained over 10,000 American and European women to use firearms. The influence of the women of her generation on extension of the political franchise to women was very substantial (Chapter 15).

ants: See **hymenopteran**.

apes: The group of **primates** including chimps, bonobos and gorillas (African great apes) and orangutans (Southeast Asian great apes). Humans are also African great apes as determined by evolutionary relationship (Chapters 6 and 7). Indeed, chimps and bonobos are more closely related to humans genetically than they are to the other African great ape, the gorilla. The lesser apes include gibbons and siamangs. Apes differ from more distantly related monkeys (like macaques) in generally being larger and more ground dwelling rather than arboreal (tree dwelling).

apical tufts: The enlargements at the tips of the outermost finger bone. Their enlargement in humans is probably part of our adaptation to elite throwing, including the powerful yet supple and controlled griping of thrown projectiles, like stones. See Figures in Chapter 7 for details.

arboreal retreat: The capacity to escape ground living threats by very rapidly climbing high into trees. Modern savanna baboons use arboreal retreat to escape the big cats (lions, for example). Some of the skeletal adaptations of the **australopithecines** like Lucy appear to be designed to permit this kind of elite, high-speed tree climbing. With the rise of the first humans, these elite tree-climbing adaptations were lost in deference to other capabilities. This suggests that early Homo had some strategy other than arboreal retreat for dealing with the big cats. The human capacity to throw with elite skill, thereby injuring and intimidating the big cats, is an excellent candidate for such an alternative strategy.

archaic state: The archaic state, and its transiently enlarged form, the **empire,** reflect a new scale of **kinship-independent social cooperation** characterized by extensively integrated networks of towns and cities organized across a relatively large landscape. Archaic states follow from the local introduction of elite **body armor** and the effects of this technology on the distribution of access to decisive coercive threat (Chapter 13). **Armored warriors** dominate the politics of all archaic states. As a result, archaic states are hierarchical and male-dominated in their social/political structure (Chapter 13).

archaeology/archaeological: The study of the material and technological remains of ancient human cultures and civilizations. Archaeology most commonly focuses on the human story after the **behaviorally modern human revolution** about forty thousand to ninety thousand years ago (Chapter 11) but before the era when recorded history becomes available beginning around five thousand years ago (Chapters 13-17).

armor: See **body armor**.

armored warriors: Individuals who posses several characteristics as follows. First, they have body armor which significantly reduces their vulnerability to the projectile weaponry of their era. Second, they exploit shock weaponry to overwhelm non-armored individuals and in conflict among themselves. Third, this armor and weaponry is not only expensive but a great deal of time in training is required to use it effectively. Thus, armored warriors must pay a substantial opportunity cost for their status. However, when these costs can be overcome, armored warriors are the decisive source of local coercive power. This dominance of elite warriors has profound effects on local human social organization (Chapter 13). Examples of armored warriors include the hoplites of Athenian Greece, the Samurai of pre-state Japan, the military elites of the Aztec and the Inca and knights of Medieval Europe.

artillery: We use this term to refer to all large (not hand-held) gunpowder projectile weaponry, including what are now called howitzers, cannons and mortars. These weapons were originally developed in advanced form in Western Europe beginning in the late 13th Century. Their introduction was rapidly followed by the consolidation of the **early modern state** (Chapter 14).

atlatl: The Aztec word for the spear-thrower or throwing stick (see images in Chapter 11). This device allows bolts (small spears) to be thrown at a greater distance than they can be thrown by hand. The device acts as an extension of the hand, greatly increasing the leverage of the hand during the throwing motion. This device is first seen in **behaviorally modern humans** (Chapter 11).All modern humans possessed the atlatl (or a more advanced weapon) and all non-modern humans lacked it.

australopithecines: These are now-extinct **hominids** that include our immediate ancestors after the divergence of our lineage from our **last common ancestor** (LCA) with modern chimps and bonobos. These australopiths were bipedal (walking up on two legs), much like we are, rather than being relatively quadrupedal "knuckle walkers" like modern chimps and bonobos. The genus name for these animals is Australopithecus. A well-known early species is Australopithecus afarensis (includes the famous Lucy fossil) and a more recent one is Australopithecus africanus. **Australopithecus garhi** will also be of particular interest to us. See figures in Chapters 6 and 7 for additional details.

Australopithecus afarensis/africanus: See **australopithecines**

Australopithecus garhi: The name given to a provisionally identified new species by Tim White, Owen Lovejoy, and their colleagues who discovered the fossil evidence. This animal lived in East Africa about 2.5 million years ago. This species appears to have begun a much more extensive commitment to scavenging (or possibly hunting) compared to other earlier **australopithecines** (Chapter 7). Its time and place strongly suggest that this australopith might have been an immediate ancestor of the first humans - the first members of the genus **Homo**. These animals probably made very primitive stone tools that included some traditionally referred to as "early **Oldowan**".

australopiths: Short for **australopithecines**.

Aztecs: An **archaic state** in what is now central Mexico. The Aztec state/empire was still in existence at European colonial contact so that we have extensive records of how it was organized (Chapter 13).

babyhood: We use this term specifically to refer to the uniquely human period of extreme dependency for the first nine to eleven months after birth (Chapter 6). Among other things, babyhood is associated with the extension of extremely fast fetal-like rates of brain growth through most of the first year of life. This pattern is a uniquely human **life history** feature. All other animals cease rapid fetal-like brain growth at or immediately after birth. The extensive social support necessary to evolve babyhood became available only with the evolution of uniquely human social cooperation (Chapters 6 and 7) allowing protection and support from the human village.

baseball grip: Refers to the three-fingered grip humans typically use to throw small round objects like stones or baseballs. This grip is an apparently unique capability of the human hand. The other living great apes cannot use their hands in this way. It is part of our adaptation for throwing with characteristically human virtuosity (Chapter 7).

BC (dating convention): See **BCE** and **CE**.

BCE (dating convention): Also see **CE**. An older dating convention created a transition point (ostensibly around the birth of Christ, hence BC for "before Christ"). Earlier dates counted backward into the past from that point. Hence, the Roman Julius Caesar was born around 100 BC and humans evolved around 2,000,000 BC. More recently the trend has been to internationalize this dating system by referring to BC as BCE, translated as "before the common era". Thus, Caesar's birth year becomes 100 BCE.

behaviorally modern humans (BMH): These humans arose from an **anatomically modern human** (AMH) population, probably in Africa, and they are apparently ancestral to all of us alive today. These moderns are anatomically essentially indistinguishable from the older AMH population. However, their behavior is dramatically different on the basis of a number of criteria (Chapter 11). These differences include more sophisticated and/or complex, composite tools as well as production of complex art and the use of a profoundly new weapon called the **atlatl** or spear-thrower. These very dramatic behavioral changes emerge explosively, probably over a few thousand years. [Recent controversy about the speed of this revolution probably results from confusion of early weaponry development with the subsequent adaptive revolution as discussed in Chapter 11.] As a result, these changes between around forty thousand to fifty thousand years ago are often referred to as the behaviorally modern human revolution or just the *modern revolution*, for short. Among their many other unique achievements, BMH were the first humans to reach a number of remote places, including Australia and the Western Hemisphere. These particular early BMH migrants were ancestral to contemporary Australian aboriginals and Native Americans.

biological father: This refers to the male whose sperm actually fertilized the egg producing the individual child. This status is potentially distinct from **social father** or from co-father (Chapter 8).

bipedal locomotion: Refers to animals that walk on their two hind limbs or legs rather than on all fours like quadrupeds. Humans are bipeds as are most birds, for example.

BMH revolution: See **behaviorally modern human** revolution.

body armor: Armor that will resist available projectile weaponry, including atlatl bolts and arrows projected by bows during the eras that will concern us (Chapter 13). It must also be light and maneuver-friendly enough to be worn in battle. These constraints mean that body armor requires sophisticated manufacture. It must either be light-weight, robust metal or some other composite material with the appropriate properties. In practice, in Eurasia metal-based armor was the most common. In contrast, in the Western Hemisphere, advanced, elite body armor was usually made from tightly woven, many-layered cotton, functioning much as Kevlar does in contemporary body armor (Chapter 13).

bond: See **chemical bond**.

bonobo: One of two African great **ape** species that are our closest surviving relatives. The formal scientific name for these animals is Pan paniscus and they are sometimes informally referred to as pygmy chimps. The other ape most closely related to us is the **chimp** or common chimp (Chapters 6 and 7).

bow and arrow: Most readers will be familiar with this weapon (see images in Chapter 12). What concerns us here is the following. First, this weapon has greater range than any previous weapon system, including the **atlatl**. Second, though manufacturing the weapon is relatively complex, requiring specialized expertise, effective use of the weapon is comparatively simple. Thus, becoming an effective archer (bow user) is much easier than becoming an effective atlatl user. Bow use requires less skill and less ongoing practice. As a result, the **opportunity costs** of using the bow (time and effort away from other useful activities) is much lower than for the atlatl. Third, the bolts (arrows) for the bow are lighter than for the atlatl and, thus, many more can be carried in a quiver, for example. The effect of these considerations is that the bow is a much better weapon for projecting the coercive threat that can produce enforced cooperation than any weapons before it. We thus expect the bow to produce local social revolutions (Third Interlude). The **Neolithic** and **agricultural** revolutions in Eurasia and North America apparently resulted from the local introduction of this weapon (Chapter 12).

breach: Refers to the back opening of the barrel of a **handgun**, where the hammer strikes the back of the cartridge casing in a modern **bullet**. Around the same time as **rifled barrels** became popular, technical skills became high enough to produce weapons that could be loaded from the breach (in any of several ways) as more modern weapons continue to be. In contrast, older weapons were loaded from the **muzzle**.

Broca's area: A region in the left cerebral hemisphere of most contemporary humans (see Figure in Chapter 9) along the lower portion of the cerebral hemisphere about a third of the way back from the front of the brain. This region is intimately involved in elite human communication. Broca's area and associated structures are noticeably enlarged in humans compared to the other apes, even in proportion to our overall larger brain size. We can use this enlargement to score when humans evolved language, depending on how we define language (Chapter 9).

bronze: An alloy of copper and tin. Like almost all metals, these two exist mostly as oxidized ores in the Earth's crust. To make them usable for tool production requires chemical reduction in the presence of very high temperatures in specialized furnaces. Moreover, the resulting reduced metals must be melted and poured into molds and/or hammered and worked into final shape. All these activities require extremely specialized expertise (information). They are thus only mastered in the context of relatively large human coalitions. **Body armor** made from bronze and other metals correlates with the social revolutions producing the earliest **archaic states** of Eurasia (Chapter 13).

buffalo hunters of the Great Plains: An adaptation with substantially increased sophistication that arose on the Great Plains of what is now the central United States after the arrival of bow around 600AD. This is one of several examples of increased local social/adaptive complexity associated with a non-agricultural **Neolithic** revolution following the local introduction of the bow (Chapter 12).

bullet ammunition: Refers to projectile rounds for hand-held gunpowder weapons (**handguns**). Their key property is that the projectile, ignition cap, and gunpowder propellant are all together in a single, easily-loaded, self-contained object. Their development in the mid-19th Century further revolutionized the use of handguns (Chapter 15).

by-product mutualism: Refers to the special cases where two non-kin conspecific animals nevertheless cooperate because some benefit can only be achieved in this way and because the two animals have no effective **conflict of interest** for these specific acts at the moment; that is, there is no option to cheat during the cooperative event. Sexual mating is a good example of such a cooperative behavior. Humans engage in by-product mutualism just like other animals. However, unlike other animals, we also engage in substantial cooperation with non-kin even when there are substantial conflicts of interest (Chapter 5).

Calusa: A **Neolithic** culture in southern Florida that emerged between 600-800AD, apparently after local introduction of bow (Chapter 12 and online endnotes). This culture depended on exploitation of the rich marine resources of the estuaries of the Florida coast rather than on agriculture to support its elaborate settlements and social system.

Canela: A **Neolithic** tribe living in Brazil on the savanna-like territory near the edges of the rain forest until aggressive contact with the modern state during the mid-20th Century. At state contact the Canela practiced a high level of sexual promiscuity and matrilocal residence. They also experienced a very high level of adult mortality. The Canela are among the societies illustrating the evolutionary logic of human sexual behavior (Chapter 8).

catalysts/catalysis: Molecules that bind to and influence the chemical reactions of other atoms and molecules without themselves being permanently altered. Their key role in biological systems is to selectively accelerate particular chemical reactions. This process allows cells and organisms to take in simple compounds, like food molecules, and use them for a very few specific reactions that benefit the biological system. These reactions include building larger molecules. These building reactions, in turn, include the replication of **genetic design information**. Most catalysts in biological systems are protein molecules encoded by genes. Such catalysts are referred to as enzymes. Encoding and producing catalysts of this form is an example of the fundamental molecular trick genetic design information uses to build new physical vehicles – encoding the production of tools.

cc: Abbreviation for *cubic centimeters*. This is the volume unit in which most brain sizes are measured in living and **fossil humans** (see **Homo**). This volume corresponds to a cube each of whose sides is one centimeter in length. For comparison, a cubic inch is approximately 6.45 cc's. A cc of water weighs one **gram**. A cc of brain tissue is made up mostly of water and will weigh *roughly* one gram.

cervical tenting: Refers to violent contractions of the vaginal and uterine muscles normally causing the cervix to dip into the seminal pool in the vagina. This process is thought to be the key to the reproductive function of female organism. It apparently has the capacity to allow the female to strongly influence paternity unconsciously. Strong cervical tenting is thought to increase the likelihood of uptake of sperm, biasing subsequent fertilization in favor of males with whom the female has a strong, properly timed orgasm (Chapter 8).

CE: Also see **BCE**. An older dating convention treated a transition point (ostensibly around the birth of Christ, hence AD for Anno Domini, "in the year of our Lord") as the starting of a dating system. At this writing we occupy year 2009 in this dating system. More recently the trend has been to internationalize this dating system by referring to AD as CE corresponding to the "common era." At this writing we occupy 2009 CE under this system.

cheating: We will use this prosaic, everyday term in a narrower, technical sense as professional game theorists often do. For us this term will refer to failure to reciprocate/contribute in a situation of conspecific social cooperation or to pretend falsely to cooperate or to actively mislead or manipulate another in a potentially cooperative situation. Cheating is a convenient description of behavior in pursuit of conflicting individual interests in an otherwise potentially cooperative social encounter (Chapter 5).

chemical bond: Atoms of different elements share and/or exchange electrons according to specific rules. When this happens, the atoms become bonded to one another.

These bonds vary greatly in character and strength, depending on which atoms are involved and what other atoms a multiple bonded atom is bonded to. They can range from very strong, stable bonds referred to as covalent to much weaker, readily reversible bonds. These properties of chemistry and chemical bonds have two important implications for us. First, they mean that atoms can join to produce molecules, even very large macromolecules that have very specific, orderly structures. This results from the capacity of atoms to form specific numbers and kinds of bonds. Second, these molecules can then, in turn, reversibly interact with other atoms and molecules through weaker, reversible chemical bonds. Which other molecules a molecule will interact with is determined by the precise shape and chemical structure of the two molecules. Any two molecules taken at random will, in general, interact very little or not at all, merely colliding with one another and moving on unchanged. However, if molecules are designed, by natural selection, so that their shapes and chemistry are appropriately compatible, they will interact specifically. These specific interactions can result in one or both of the molecules undergoing a new chemical change. In each of a very widely occurring set of cases, a **protein** molecule produced by **genetic design information** encoded in a **DNA** molecule interacts with other sets of molecules to encourage them to undergo one specific type of chemical reaction among the various alternatives that are chemically possible. The protein itself is unchanged and goes on to carry out the same process over and over. A protein acting in this way is called a **catalyst**. Many of the processes organisms carry out, including digesting their food and replicating their genetic design information, are ultimately achieved through the action of specific sets of such catalysts designed by natural selection for that purpose (Chapter 2).

childhood: We will use this term to refer to the time *between* when an ancestral human infant was probably weaned from breast feeding – perhaps around three years of age – and the time the human youngster's brain growth stops and she/he begins to acquire adult dentition (teeth) round six years of age. This particular developmental pattern is a uniquely human **life history** feature. All other animals wean at or after the time that brain growth stops and adult dentition is acquired. The extensive social support necessary to allow the evolution of childhood became available only with the evolution of uniquely human social cooperation (Chapters 6 and 7).

chimps: One of two African great ape species that are our closest surviving relatives. The formal scientific name for these animals is Pan troglodytes. The other ape most closely related to us is the **bonobo** or pygmy chimp (Chapter 7).

chromosomes: These are the physical objects in which **genetic design information** is packaged in contemporary organisms. Chromosomes consist of a single very long **DNA** molecule, together with a large, complex array of protein molecules which package, control, express, and replicate that genetic design information. A typical chromosome in a human or other animal might have one to three thousand genes encoded along its DNA molecule. A typical mammal might have fifteen to thirty such chromosomes making up its entire **genome**. Of course, these animals – including us – are **diploid** so that they actually have two copies of each of these chromosomes (one copy from the organism's mother and one from its father).

Chumash: A **Neolithic** society living in and around the Santa Barbara Channel (present day California) from around 600 CE through European colonial contact. The Chumash made extensive use of marine resources rather than **agriculture**. This culture helps us understand the rise of the Neolithic level of adaptive sophistication (Chapter 12).

coercive threat: Refers to using the credible threat of violence to attempt to require specific behaviors of others. All animals use coercive threat against one another. Humans are unique in our capacity to use the abilities as **remote killers** to project threat conjointly in relatively large social aggregates. This makes our individual **cost of coercion** extremely low under the appropriate conditions. When we project threat conjointly in pursuit of confluent individual interests, **kinship-independent social cooperation** is the inevitable adaptive by-product (Chapter 5). This pattern of **cooperation** is revolutionary and ultimately accounts for all our other unique properties on the theory we are exploring here on our theory (Chapters 8-17).

co-fathers: See **social fathers**. Also see **biological father** for contrast.

coalition: We use this generic term in a very specific, narrow technical sense. For us, coalitions refer specifically to collections of non-kin humans who cooperate in spite of active **conflicts of interest**. We argue that the uniquely human scale of coalitions is central to our biology and results from our access to inexpensive social coercion, allowing management of non-kin conflicts of interest (Chapter 5).

collective action problem: Social scientists independently discovered and analyzed the problem of social cooperation between individuals with conflicts of interest and referred to this problem using this phrase. Thus, the collective action problem is fundamentally the same thing as the non-kin social cooperation problem as discussed in Chapters 3-5.

combinatorial: For our purposes this term refers to the use of combinations of limited numbers of symbols or elements (or their equivalent) to encode information or to build **complexity**. For example, the letters of the alphabet are used to combinatorially encode written language. Also, for example, our phone number system is combinatorial. See Chapter 2 for a discussion of the use of combinatoriality to build complex biological structures and to encode information.

complexity: A characteristic of living organisms that distinguishes them from non-biological objects is their spectacular complexity (Chapter 2). Even a single-celled organism is vastly more complex than a rock, for example. The properties of self-replicating systems also account for and produce this complexity as follows. A successful self-replicating system will come, over time, to be present in many copies in its environment. Thus, the opportunities for one or another copy of such a system to add some new tool or capability supporting reproduction by recruiting cooperating pieces of design information and/or by mutation are enormous. Such added capabilities allow these improved systems to competitively displace earlier, simpler versions of the system. Added capabilities interact **combinatorially** to produce dramatically increased complexity. [The same reasoning applies to our adding tools to a tool kit, combinatorially increasing the number of jobs we can perform. One tool (say, a hammer) might allow us to do a few jobs. However, four different tools (say, hammer, saw, drill, and screwdriver) might allow us to do hundreds of different jobs. Going from one tool to four might increase our capacities not four times but hundreds of times. For the same reason, going from one tool to thousands might increase an organism's capabilities millions or billions of times.] The complex set of tools produced by such cooperative groups of pieces of **genetic design information** are sometimes referred to, collectively, as **vehicles** (organisms) whose evolved job it is to support the replication of the design information that produces them.

A crucial point is that the complexity is actually simple in both its evolutionary origin and it contemporary function. The complex biological world is startling simple once we understand it (Chapter 2).

compound: Refers to the chemical combination of multiple **atoms**. Compounds can be very small, like a salt or sugar, or very large like a **DNA** molecule with millions of bases strung together. However, even these enormous biological molecules are ultimately simple, resulting from **combinatorial** deployment of many copies of relatively simple building block molecules (Chapter 2)

conflicts of interest: Refers specifically to the fact that organisms designed by **natural selection** in a finite (**Malthusian**) world will inevitably come into competition for necessary resources. Since the finite world cannot support both individuals fully, each of the two organisms must attempt to acquire the necessary resources at the expense of the other. More generally, almost any time when two animals or two humans interact, they have potential conflicts of interest and natural selection will shape them to maximize the probability of resolving those conflicts in favor of their own personal copies of **genetic design information** (Chapter 3). This last rule means that close kin animals can often cooperate because they share some unambiguously identical personal design information. **Non-kin conspecific** animals commonly do not share unambiguously identical design information. Management of this problem of conflicts of interest in humans has allowed our unique evolution of vastly expanded **kinship-independent cooperation**.

confluence of interest: Under the appropriate circumstances different individual animals can share common interests. This condition is referred to as a confluence of interest. This can result from close genetic kinship (Chapter 3). However, confluence of interest can also occur between non-kin individuals when the capacity to suppress **conflicts of interest** exists. Under these conditions, non-kin individual can participate in mutually beneficial behaviors and, thereby, would have a confluence of interest (Chapter 5).

conspecifics: Two organisms are said to be conspecifics if and only if they are members of the same species. For example, we are conspecifics because we are all humans. Likewise two chimps are conspecifics, as are two dogs or two apple trees. However, a chimp and a human, for example, are NOT conspecifics even though they have a close evolutionary relationship. Conspecific animals generally have many more **conflicts of interest** with one another than do organisms who are members of different species.

contemporary era: This term specifically describes the current era of pan-global human cooperation (still emerging; Chapters 16 and 17). This contemporary global cooperation involves overtly economic exchange as well as massive movements of people, information, and ideas. On our theory, global cooperation can only be sustained by coercive weapons of global scale. Democratic global cooperation can only be sustained when majority consensus controls the use of these coercive weapons. With **aircraft**, weapons of this scale have become available. We date this era to the development of the first elite military aircraft suitable for routine use in the 1930's. The key initiating coercive event was World War II.

Coolidge Effect: This term refers to a rather remarkable property of the sexual response in many male animals, including humans. Males showing this effect cannot mate repeatedly over short times with the same female, but can successfully mate over that same short time interval with multiple different females. This and related properties of the human sexual response are highly illuminating about our evolution, including our ancestral adaptation to contingent promiscuous mating (Chapter 8). Human females also sometimes show Coolidge-like behavior, with sexual receptivity increased rather than decreased by an individual act of mating.

cooperation: We use this term to refer to two or more organisms acting conjointly so as to produce some outcome that is qualitatively or quantitatively different than these same organisms could produce when acting alone as individuals. Note that this term does not imply anything about the adaptive logic or proximate psychology producing such conjoint behaviors (Chapters 3 and 5).

cost of cooperation: Refers to the expense associated with an act of **cooperation**. For example, cooperative hunting requires time, energy, and risk of injury on the parts of each of the hunters. This is a cost. For a cooperative behavior to evolve, its benefits must exceed its costs, including the cost of cooperation. This is not the only cost that **kinship-independent social cooperation** imposes. There is also the **cost of coercion** necessary to suppress **cheating** on the cooperative enterprise. This second cost is central to the uniquely human story (Chapter 5).

cost of coercion (problem): Refers to the fact that the cost of social coercion that would produce cooperation generally exceeds the net benefit from the resulting co-operation for most animals in the short term. This cost barrier can apparently only be overcome in animals that can kill remotely (from a distance of many body diameters) allowing remarkably inexpensive conjoint projection of violence (and credible threat) by self-interested individuals. Humans are the first animals in the history of the Earth to evolve the capacity to reliably kill adult conspecifics from a distance. This behavior had revolutionary implications for our social evolution (Chapters 5 and 7).

covalent: See **chemical bond**

cranial volume: Refers to the volume inside the brain case of the skull. It can be measured both in living animals and in fossil skulls. It is a very accurate re-flection of living brain size. Human brain size is much greater than chimps, our closest living relative. This brain expansion requires assets that can only be pro-vided by a newly expanded scale of social cooperation (Chapters 6 and 10). Thus, increased cranial volume in the human fossil record allows us to identify the evo-lutionary origin of uniquely human **kinship-independent social cooperation** (Chapters 6 and 7).

cubic centimeters: See **cc**.

culture: See **culturally transmitted design information.**

culturally transmitted design information (**cultural information** or simply **culture**, for short): **Genetic design information** plays a crucial role in building our minds and influencing our behavior. Also important is the culturally transmitted information ab-sorbed by our brains from our social surroundings. This information exists in non-hu-man animals but is vastly larger in scope in humans for reasons predicted by our theory (Chapter 10). Humans understand and look at the world differently than non-human animals, almost entirely because of this vastly enlarged repertoire of culturally transmit-ted information. The contemporary process of science is merely the continuation of our ancient, uniquely human trick of using **kinship-independent social cooperation** to build a vast culturally transmitted informational heritage (Chapter 10).

Some important things about our minds/brains remain unclear, for example, why we experience consciousness or conscious awareness and why reality "feels" so viv-idly sharp. Moreover, many molecular, cellular and histological features of the brain are likewise quite unclear. In spite of our vast ignorance in these areas, we neverthe-less know enough to begin to build some useful pictures of how our minds work and what they do. In particular, their relationship to information is relatively clear (Chapter 10).

It is useful to divide the uniquely human cultural information stream into three functionally distinct components, **pragmatic information, social contract information** and **identifier information**.

Pragmatic information consists of all functional insights into the world and how to manipulate it. All of science and engineering constitute pragmatic information. Likewise the billions of useful skills, from cooking to cleanliness or speech to sewing, we acquire and transmit culturally are pragmatic information.

Social contract information includes legal rules as well as the vast informal mores and norms for all the cooperative elements of our public behavior. Virtually everything we think of as moral and ethical is defined as such by social contract information.

Identifier information consists of information that controls behaviors (including speech, dress, and diet) that mark our social affiliations. Our minds are highly adapted to develop, evince and monitor identifier information because of its central adaptive role in the ancestral past (Chapters 10 and 12). Identifier information includes most public religious ritual, for example. In the contemporary world, narrowly defined identifier information, like wearing a cross, a star of David, or a crescent pendant, say, is generally much more important in hierarchical societies than in democratized ones (Chapters 13 and 17).

Darwinian (or natural) selection: Also sometimes referred to as a Darwinian process. The self-replicating system consisting of a **vehicle** built by **genetic design information** that is a biological organism is susceptible to change (evolution) over time. Replication can never be perfectly accurate and, thus, small changes (**mutations**) in the design information will inevitably occasionally occur. When they do, most will make a less effective self-replicating system and will be lost. However, rarely, such changes will produce a better self-replicating system and this altered system will eventually competitively displace the original system. Thus, these systems continue to undergo improvement and refinement. This same process will mediate adaptation to any new environmental challenges. This is the fundamental process underlying the origin and properties of all biological creatures, including humans (Chapter 2).

democracy effect: Refers to the increased adaptive sophistication of democratically organized human coalitions compared with hierarchically organized coalitions (see **hierarchy**). This effect results from the fact that a larger fraction of democratically organized coalitions experience **confluence of interest**, resulting, in turn, from democratized access to decisive **coercive threat** (Chapters 5, 15 and 17). In contrast, many members of hierarchically organized coalitions have **conflicts of interest** with elite individuals controlling the coalition (Chapters 13 and 17). Among the many consequences of this effect is that members of democratically organized coalitions have access to much larger amounts of **culturally transmitted information** because of the larger numbers of members with confluent interests.

design information: Refers to the information that builds and controls the properties of biological **vehicles** (Chapter 2) or **organisms**. A major type of such information is **genetic design information** (also see **DNA**). However, **culturally transmitted design information**, stored in the brain and replicated through social communication, can also act as design information controlling the behavior of organisms. This second information source is much more quantitatively important in humans than in any other animal (Chapter 10). Notice that all forms of design information are inherently subject to **Darwinian or natural selection**.

diploid: Most large organisms, including all animals and humans, are diploid. The root of this word is *di* meaning *two* and it refers to the fact that the genomes of these animals have two copies of almost every gene. This result is achieved when a sperm cell containing one copy of each gene and an egg cell containing a second single copy of each gene are joined at **fertilization** to produce the diploid zygote (fertilized egg) that will develop into the adult organism. The sperm and egg cell are each said to be **haploid** because they contain only one copy of each gene. The process whereby the diploid organism regenerates the haploid gamete (sperm or egg) is referred to as meiosis which occurs during gametogenesis.

direct access (to culturally transmitted information): This specific, technical term will refer to our capacity to acquire culturally transmitted information stored in the mind/brains of others through language, direct demonstration, or symbolic gesture. Because of the **hostile manipulation problem**, acquiring such information from non-kin requires enforcement of **kinship-independent social cooperation**. It is also possible to have **indirect access** to culturally transmitted information.

Dmanisi: A famous **hominid fossil** site in the Central Asian nation of Georgia. Fossils of early **Homo erectus** have been found at this site. These early humans are around 1.8 million years old. Moreover, the site itself has some useful properties. For these reasons we made extensive use of the results from this site in understanding our early evolution (Chapter 7).

DNA: Abbreviation for the chemical name of the polymer that encodes **genetic design information** (deoxyribonucleic acid). See Chapter 2 for a discussion of the issues summarized here. This polymer is made of a linear (unbranched) string of monomers. These individual monomer units are the DNA bases abbreviated A, T, G and C for the first letters of their chemical names.

The following properties of DNA are relevant to us here. First, because the monomer units making up DNA are slightly different from one another, the sequence of these monomers can be used to encode information, generally analogously to the use of letters to encode words and sentences in written language. The information so encoded is **genetic design information**.

Second, this information can be used to control the sequence of amino acid monomers in the **proteins** encoded by DNA sequences. The twenty amino acids normally used are each different chemically from one another so that the protein polymer is richly complex in structure. Typically the protein polymer folds up into a complex three dimensional object which can function as a molecular machine if so designed by natural selection. These protein molecular machines and structural components are the major class of tool molecules in the contemporary biological **vehicles** or organisms (Chapter 2).

DNA sequence is copied into a similar, but disposable molecule called **RNA** by a process called transcription. This RNA transcript, in turn, is used to direct the synthesis of a protein (according to the genetic code) through a process referred to as translation.

Third, the DNA bases pair with one another in very specific ways so that one chain of DNA can be used as a template to make a new DNA chain of controlled sequence. This constitutes replication of genetic design information. Thus, DNA encodes genetic design information that builds vehicles (organisms consisting of the DNA itself plus all the tools for its replication). Animals like us are very complex vehicles built by genetic design information shaped by **Darwinian or natural selection**.

DNA bases: See **DNA**.

dorsal interosseous: See **muscles**.

early modern state: Refers to the first states consolidated using gunpowder **artillery**. These states grew out of late **archaic states**. As a result of this pre-existing elite concentration of coercive power and the costly, sophisticated nature of gunpowder artillery, early modern states were intensely **hierarchical** (Chapter 14). The France of Louis XIV, the Muscovy of Ivan the Terrible, the Tokugawa Shogunate of Japan, and the other gunpowder empires of Eurasia are all examples. In contrast to archaic states, these early modern states were relatively stable as a result of gunpowder artillery's capacity to support long-term management of elite **conflicts of interest** (Chapter 14).

Earth Mark I: Refers to the earlier proto-planet that collided with the Mars-like planet sometimes called Orpheus to generate the Moon and Earth Mark II, the highly improbable world where complex organisms, including us, evolved over the last roughly 4.5 billion years (Chapter 2).

ecological dominance: Ecology or ecosystem refers to the set of organisms making up a physical locality, ranging in size from a puddle of water to a planet. Ecological dominance refers to the extent to which a **species** controls the resources of an ecosystem for its own uses, generally at the expense of members of other species within the same ecosystem. Humans enjoy a level of ecological dominance never before seen on Earth.

element: Atoms are the most fundamental unit of matter from the point of view of chemistry at temperatures existing on the surface of the Earth. An individual atom contains protons and neutrons in its nucleus (not to be confused with a cell's nucleus) surrounded by electrons in shells with highly orderly substructure. A specific element (carbon, say) consists of one class of atoms with a unique, specific number of electrons. When two atoms undergo a chemical reaction, they exchange and/or share some electrons according to specific rules. The number of electrons possessed by the atoms of any particular element determines the kinds of chemical reactions it will undergo. The number of electrons is responsible for the element's personality or its properties. Thus, for example, gold bars and carbon charcoal are different exclusively because of the number of electrons their component atoms possess. When atoms react chemically, forming **chemical bonds**, they build compounds or molecules. These molecules can be large macromolecules. Such large molecules are central to biology. They encode **genetic design information** and they are the stuff of biological **vehicles**.

elite-dominated polities: See **hierarchy**.

elite throwing: Refers specifically to the capacity humans have to throw repeatedly and rapidly with great force and accuracy. This capacity has required the wholesale redesign of our bodies. Humans throw with elite skill in the same sense that dolphins swim or cheetahs run with elite skill. This elite capability is not displayed by any other known animal alive or extinct, including our closest living relatives, the other **apes**. We argue that this uniquely human property is foundational to all our unique **kinship-independent social cooperation** because it allows us to project **coercive threat** remotely, from a distance, supporting inexpensive management of non-kin **conflicts of interest** (Chapters 5 and 7). This unprecedented social adaptation is responsible, in turn, for all our other unique properties as individual organisms (Chapters 8-10) and for the details of our history (Chapters 11-17) on the theory we explore here.

elite warrior: Refers to the class of individuals who founded and managed **archaic states**. (Also see **armored warrior**.) These individuals occur independently and universally around the world wherever archaic states formed (Chapter 13). These individuals exploited **shock weapons** and **body armor** to achieve coercive dominance. The cost/benefit logic of coercion by these individuals determines the structure and properties of the archaic state. Examples of elite warriors include Roman Legionnaires, Greek hoplites, Japanese Samurai, and Aztec warriors. Control of human social organization by these elite warriors was destroyed by the rise of gunpowder weaponry which ultimately redistributed coercive power and lead to rise of the modern state (Chapters 14 and 15).

empire: We will use this term to apply to the sometimes massive, but unstable and short-lived, human coalitions produced by conquest and originating from an **archaic state** that has a temporary military advantage resulting from innovations in the design, manufacture or use of **armor** and **shock weaponry** (Chapter 13). The Bronze Age Egyptians, Mesopotamians, Japanese Samurai states, Athenian Greeks, Imperial Romans, Mayans, Aztecs, and Inca are all examples of this level of organization.

Enlightenment: The term applied to a period of great intellectual achievement corresponding to the early height of the **Scientific Revolution**. We argue this era is a consequence of the ongoing democratization of the modern state with the attendant **democracy effect** on human **adaptive sophistication** (Chapters 15 and 17).

enzyme: see **catalyst, DNA** and **protein.**

ethnocentrism: See **racism**.

ethnography or ethnology: Refers to the specific comparative analysis of local human cultures. This includes, in principle, every element of cultures from economic practices to marriage, religious, and child-rearing practices, for example. Many anthropologists spend their time collecting ethnographic information and, therefore, are sometimes explicitly called ethnographers or ethnologists.

expanded gestation: Refers to one the new features of human **life history**. These are features we have that our closest living relatives, the other African great apes, do not. (See **babyhood** and **childhood** for related phenomena and Chapter 6 for a detailed discussion.) Gestation refers to the growth of the unborn mammalian fetus in the uterus of its mother. For humans this growth period lasts about nine months. This is one month longer than in chimps, our closest living relative. More importantly, humans invest more resources in the fetus so that our newborns (and their brains) are substantially larger than for the other great apes. This expanded gestation means that human childbirth is among the most difficult of all mammals. Moreover, human females need very high levels of nutrition during the final months of gestation/pregnancy. However, they are physically challenged by the very large, late-term fetus, making personal food procurement difficult for them. Thus, late-term pregnant human females are very dependent on the nutritional assistance of others. The social support necessary to evolve expanded gestation became available only with the evolution of uniquely human **kinship-independent social cooperation** (Chapters 6 and 7).

exponential growth: Refers to a universal property shared by all living organisms (Chapter 2). Specifically, when biological organisms reproduce, they produce multiple new offspring each of which can reproduce in precisely the same way. Thus, consider the simplest case as follows. One organism produces two, but two produces four, four produces eight, and so on ad infinitum! This means that biological populations will grow explosively until checked by some factor. This factor is most commonly competition with other conspecific organisms in a finite **Malthusian** world. Among the many implications of this phenomenon is that conspecific competition is almost always the single most important adaptive challenge a self-interested individual organism faces. Non-kin **conspecifics** have strong **conflicts of interest** as a result (Chapters 2 and 3). This, in turn, means that **cooperation** between non-kin conspecifics can only occur under very special circumstances (Chapters 3 and 5).

extinction: The loss of all the members of a species of organism. Such losses can have many causes. For example, extinction can result from climate changes that alter the local environment faster than organisms can adapt. It can also result from competition with a newly introduced organism of a different species. Extinctions have occurred throughout the history of the Earth and continue today. Current rates of extinction are extremely high, probably as a result of human-generated habitat destruction and climate change.

extra-pair (or EP) matings: Refers to matings by individuals who are members of a pair-bonded couple with someone other than their nominal mate. In some local human cultures such EP matings are socially tolerated or even actively sanctioned. In others, like ours, they are discouraged but continue to occur at some low level. These properties of our mating behavior have very important implications for the evolution of human reproduction (Chapter 8).

femur: The thigh bone. It has a ball joint at the top or head which fits into a socket in the **pelvis**. This joint is very informative about the walking and throwing behaviors of its owner (Chapter 7).

fertilization: Refers to the process whereby a sperm and egg cell unite to produce a fertilized egg/zygote that will develop into an organism. Human mating introduces sperm cells from the male into the reproductive tract of the female where fertilization can occur. A fertilized human egg implants in the uterus (womb) where it undergoes sufficient development to undergo birth. See Chapters 6 and 8 for a discussion of human sexual and reproductive behavior and biology. (Also see **diploid**.)

first metatarsal: The foot bone immediately behind the big toe. It contributes to the ball-like surface we push off on when we throw in the characteristically human way. This bone is also one of the connection points for the peroneus longus **muscle**, apparently involved in controlling parts of the foot's role in the elite human throwing motion. The change in this bone in fossil ancestors helps us pinpoint the evolution of elite human throwing. See Figures in Chapter 7.

fission: Refers to one of two major classes of energy releasing processes that the nuclei of atoms undergo. In this case, certain heavy nuclei decay with the release of energy in various forms that can be rapidly converted to heat. Chain reactions of large numbers of atoms of this form are one of the two bases for nuclear weapons (Chapter 16). Fission weapons are popularly referred to as atomic bombs or atom bombs (also see **fusion**). The first successful test of nuclear explosive involved a fission device code named *Trinity* exploded in the New Mexico desert in the summer of 1945. In its naturally occurring form, nuclear fission is the source of most the endogenous heat that keeps the core of planet Earth molten, thus driving the tectonic activity crucial to the long term survival of life on Earth (Chapter 2).

Five Articles Oath: See **Meiji Restoration**.

flintlock: see **handguns**.

foramen magnum: The large opening at the base of the skull through which the spinal column enters the brain case. The position of this opening in a fossil skull reveals whether the animal stood upright on two legs or walked down on four legs.

fossil: An artifact formed when plant or animal tissues are preserved, often in sediments formed in shallow, slow moving water. Portions of the tissue are replaced by minerals from the surrounding water. Though soft tissues do fossilize, this event is extremely uncommon. Much more common is the fossilization of hard tissues, especially bone. The resulting object is often an extremely accurate, detailed hard stony replica of the original relatively soft bone, consisting of some material from the original object and some mineralized material. Fossils are one of the very few direct clues we have as to the nature and timing of existence of now-extinct ancient species, including our own ancient ancestors (Chapter 7).

fossil record: The sum total of all the fossils that have been left by earlier life on Earth. This record is often organized stratigraphically or in other ways that allow us to infer the temporal sequence of events making up the evolutionary history of contemporary organisms (see **stratigraphy**), to determine their relative ages. We can also determine the absolute age of many fossils using radiological dating techniques (Chapter 7).

free rider: The term was coined by social scientists. This original metaphor was that of a turn-style jumper in a subway system. The idea was that each individual has an interest in riding the subway free but in having everyone else pay so that the system will continue to operate. Because of this **conflict of interest** between the self-interested individual and the larger system, only subway systems in which such free riders are blocked can actually work in practice. This term has now been generalized. Free rider now refers to an individual who pursues self-interest by **cheating** (sensu lato) in the context of any potentially cooperative social enterprise. The crucial question for us is under what circumstances free riding can actually be controlled (Chapter 5).

French Revolution: Initiated in 1789, this revolution toppled the French crown, including the execution of Louis the 16th and Marie Antoinette. As a result of entrenched elite power in France, this revolution was terribly violent and played out over many years, through and including the Napoleonic Wars that engulfed much of Europe (Chapter 15).

fusion: Refers to one of two major classes of energy releasing processes that the nuclei of atoms undergo. In this case, light atoms (hydrogen, for example) are driven together with such high energy that they fuse to form heavier atoms. This is the energy releasing process that powers the Sun through a fusion chain reaction originally ignited by extreme gravitational compression. Fusion can also be driven artificially by compressing light atoms under the influence of a **fission** explosion. This is how human-made fusion bombs, like hydrogen bombs or thermonuclear devices work (Chapter 16).

gamete: See **diploid, haploid** and **sexual reproduction**

GDP: Abbreviation for gross domestic product. This number is one measure of economic output and **adaptive sophistication**. Most commonly this number is normalized to the number of people in a social unit like a state. So, for example, the per capital GDP of the United States at this writing is around $45,000 by one estimate.

gene: A specific piece of **genetic design information** typically encoding a single specific **protein** molecule. The gene is a segment of **DNA** and a mammal like us would have around 23,000 genes. Each gene is a unit of genetic design information encoding one of the tools that build the **vehicle** (organism) whose role is to replicate that design information.

genetic design information Refers to the information encoded in **DNA** sequence, that is capable of replication and that directs the production of tools (often **proteins** acting as **catalysts** or to build structures like organs and skin) that build the **vehicles** that, in turn, support and permit replication of the design information (Chapter 2). The logic of getting this design information replicated in a **Malthusian** world determines how natural selection shapes all animal behaviors, including social behaviors (Chapter 3).

There are several slightly subtle but crucial points here. First, it is the information that is important. It is the information encoded in DNA and not the physical DNA molecule itself that matters. This is quite analogous to saying that Shakespeare is not the paper and ink in which it is encoded but the information Shakespeare intended to convey. Second, genetic design information is encoded in DNA as a result of historical accident. There is nothing unique or magical about DNA. Any physical medium - on another planet, perhaps - that has the necessary properties would work just as well as and analogously to DNA.

genetic relatedness: This term will have a very specific technical meaning for us in the context of evolved social behavior. It refers to the probability that any individual piece of **genetic design information** is identical by very recent descent in two individual organisms. A piece of genetic design information can be identical in this way only if it was inherited from a recent common ancestor. Genetic relatedness falls off sharply with changing pedigree relationship. Thus, for example, the relatedness of two human siblings is 50 percent, two first cousins is 12.5 percent, and of two second cousins is only about 3 percent (see online endnotes to Chapter 3). It is this genetic relatedness that determines the likelihood of kin-selected social cooperation.

It is important to be aware of the following. Genetic relatedness is not the same as **genetic similarity**. Indeed, it is possible, in principle, for two organisms to be genetically identical in the sense of DNA sequence similarity (say greater than 99.9999%) but to have very low genetic relatedness (say, 0.1%). The evolutionary logic of kin-selection and not some rational understanding of genetic identity produce social behavior (Chapter 3). Natural selection can be insane from our perspective.

genetic similarity: Refers to the fraction of **DNA** sequence (**genetic design information**) that is identical in two genomes. For example, if you compare the two copies of the human genome that make you up (the one you got from your mother and the one from your father), they are about 99.9% identical. That is, about one base in one thousand, on average, is different. Further, for example, if you compare any contemporary human genome copy to any contemporary chimp genome copy, they are roughly 98.5% identical. That is about fifteen bases out of one thousand, on average, are different. This last number is very useful because it gives us strong evidence of the time since we last had a common ancestor with living chimps. Putting this information together with the fossil record and other evidence allows us to build a very complete, clear picture of our kinship with chimps. [NOTE: Genetic similarity here is not the same thing as the **genetic relatedness** between conspecifics of differing pedigree relationships. It is genetic relatedness and not genetic similarity that determines the scope and nature of kin-selected conspecific social behavior (Chapter 3).]

genetic variation: See **mutation**.

genome: The technical term for all the **genetic design information** directing the building of a specific organism (**vehicle**). Contemporary genomes include many pieces of design information which interact combinatorially to produce the complexity of the organisms we see. Also see **chromosome**, **DNA**, **haploid** and **diploid**.

genotype: Refers to the specific collection of versions of pieces of genetic design information (**alleles** of **genes**) in a specific individual organism's **genome**. For example, an individual might have brown hair and brown eyes rather than, say, red hair and blue eyes. These differences in our appearance are due to differences in our genotypes, the specific versions of our genes we inherited from our particular parents. The anatomical differences themselves (brown eyes, say) are referred to as our **phenotype**, the consequence of our genotype.

A crucial detail is that genotype can control behavior in startlingly extensive and specific detail. Thus, behavior will evolve to serve the interests of replicating the **genetic design information** that produces that behavior (Chapter 3).

geon: The meaningless subunits into which perceived objects are broken down by our perceptual apparatus. A crescent and a cylinder are each geons, for example. They can be assembled into larger composite units that, in turn, are treated as meaningful or functional objects. A crescent and cylinder can be combined to make a coffee cup or a beer mug, say. This process creates a nested, hierarchical structure for the perception of objects and all of reality. This structure of perception is strikingly analogous to the structure of human speech, suggesting that they have a common evolutionary origin (Chapter 9).

germ line: The cells in an animal's body that will mature into eggs or sperm. This term also applies to eggs and sperm themselves. The germ line is where the **genetic design information** responsible for the building of the class of vehicles we call multicellular animals is recorded, stored and replicated in preparation for sexual reproduction.

gestation: See **expanded gestation**.

gestural communication and the gestural theory of language origins: This theory proposes that all animals, including ancestral humans, communicate most fundamentally by using salient movements of their bodies and parts of their bodies. There is very good reason to believe that animal brains are shaped by natural selection to allow this form of communication. In the human lineage, this mode of gestural communication is proposed to have been dramatically expanded and refined early in our evolution (Chapter 9). Speech communication might then have evolved secondarily or in parallel. The nested, hierarchical structure of spoken grammar is supposed to have been based on the previously existing structure of gestural communication which, in turn, is based on the fundamentally hierarchical nature of animal perception (see **geon**; Chapter 9).

glaciations: On a number of occasions over the last several million years, the Earth's climate has cooled to the point that the polar ice caps and ice deposited on high mountains began to expand. Over time these ice sheets, called glaciers, grew enormously.

During the last such glaciation, ice sheets in the Northern Hemisphere expanded to cover what is now Canada and the Northern US as well as Northern and Central Europe and Asia. This last glacier receded by around fifteen thousand years ago to create the contemporary climate pattern. The large amount of water tied up in these glaciers caused sea levels to drop rather substantially during peak glacial extents. As a result, for example, Tasmania was joined to Australia, Asia to North America (through "Beringia") and England to Europe across what is now the English Channel.

Glorious Revolution: The "popular" revolution in England beginning in 1688 that displaced the last British monarch who claimed divine right to rule. This resolution resulted in the granting of extensive power to Parliament creating an early approximation of the **modern democratic state** (Chapter 15).

gluteus maximus: This sizable human buttock muscle is the largest single muscle in our bodies. It is dramatically enlarged relative to the same muscle in chimps. This has been interpreted as an adaptation to bipedal locomotion in humans; however, its enlargement is probably more likely to be part of the elite adaptation to high momentum throwing in humans (Chapter 7). Also see **muscles**.

gm: Abbreviation for *gram*. See **cc**.

gorilla: One of three African great ape species (also see **ape**) that are our closest surviving relatives. The formal scientific name for these animals is Gorilla gorilla. Gorillas are slightly more distantly related to us than are the other two African great apes – **chimps** and **bonobos** (Chapter 7).

gram: See **cc**.

great apes: (Also see **ape**.) This term refers to a group of living animals descended from a relatively recent common ancestor. There are four African great ape species: gorillas, chimps, bonobos (pygmy chimps), and humans. (The Asian orangutan is the sole remaining Asian great ape species.) The gorilla lineage diverged from the lineage that would give rise to chimps, bonobos and humans about ten to twelve million year ago. The hominid lineage that would give rise to us diverged from the lineage that would give rise to chimps and bonobos about six million years ago (Chapter 7).

group selection fallacy: Because of our evolved **proximate psychologies,** we are highly prone to make a false assumption about how our evolution works. We tend to believe that behaviors that generate benefits shared between non-kin **conspecifics** are sufficiently accounted for by supposing that they evolved solely because of those benefits (Chapters 3 and 10). In other words, we evolved various social behaviors solely *because* they benefit the groups of which we are members. In fact, under very special circumstances this kind of group selection can probably work in principle. However, such group selection is almost never effective under realistic circumstances. We believe that it played virtually no role in the evolution of human uniqueness (Chapters 3, 7 and 10). Our evolved fondness for group selection hypotheses is apparently a persistent psychological illusion, not a reflection of the organization of the world or of the logic of **natural selection** (Box 10.3 in Chapter 10).

guilt: Guilt is the emotional reaction associated with the functioning of a **proximate psychological device** that helps us anticipate and avoid becoming a target for coercive actions by others (Chapters 5 and 10). We generally experience this feeling when we believe we may have done something that others around us will perceive as inappropriate or socially parasitic. This behavior is sometimes provoked or exacerbated in us in response to expressions of **moral outrage** by others directed at us.

gunpowder: Throughout the first roughly two million years of human history, our weaponry was entirely dependent on human muscle power, directly or indirectly. However, very recently, we acquired the capacity to exploit chemical energy to launch projectiles. Black powder, the original gunpowder, was the first of many such chemical strategies. It results from the combination of charcoal (energy-rich reduced hydrocarbon) with nitrate salts and sulfur (endogenous oxidizing agents chemically reacting with the charcoal to release amounts of energy). This was utterly revolutionary. It greatly increased the range and performance (including penetrating power) of our projectile weaponry. As expected, this enhancement, in turn, revolutionized the scale and form of our **kinship-independent social cooperation**, leading to the **modern** condition (Chapters 14 and 15). Gunpowder was widely available throughout Eurasia by around the 13th Century. However, it was first developed into reliable weaponry in the Mediterranean basin. **Artillery** was technically simpler and developed first, apparently producing the **early modern state** (Chapter 14). Sophisticated but inexpensive gunpowder **handguns** were more technically challenging. These handguns lagged behind the development of artillery by several centuries. The ultimate development of the flintlock (and its many successors) returned decisive local coercive threat to large commoner aggregates. This produced the **modern democratic state** or the **modern authoritarian state** depending on the capacity for elite monopolization of access to these weapons (Chapters 15 and 17).

Hawaii: This island archipelago in the mid-Pacific contained a culture that we will argue was an excellent example of an **archaic state**, well preserved into the modern scientific era and contributing substantially to our empirical understanding of these states (Chapter 13).

Hamiltion's Law or Rule: Bill Hamilton proposed that social cooperation between **conspecifics** would evolve when the benefit of that cooperation (B) to the receiver of the behavior exceeded the cost of that cooperation (C) to the provider and when the benefit was discounted according to **genetic relatedness** (Chapter 3). This law is very powerful in understanding many features of animal social behavior (Chapter 3) and some very specific features of human social behavior (Chapter 4). Hamilton's Law predicts the consequences of **kin-selection.** We are primarily concerned with social cooperation beyond that predicted by Hamilton's Law, that is, **kinship-independent social cooperation.** We argue that the **conflicts of interest** between non-kin implied by Hamilton's Law are the fundamental barrier to kinship-independent social cooperation and that humans are the first animal to evolve a solution allowing them to overcome this barrier (Chapter 5).

handguns: Refers to individual **gunpowder** projectile weapons. In the contemporary environment these include all the popularly known individual weapons, like Glocks, Colts, Kalashnikovs (AK-47s), and so on. Historically the earliest handguns were matchlocks and arquebuses (variously spelled), dating to the sixteenth and early seventeenth centuries. These guns were primitive, dangerous, and expensive. They tended to favor elite troops rather than commoner owners. The wheelock was invented later. These weapons were safer and more reliable, but still relatively expensive, again favoring elite infantry over commoners. The key transition to the modern condition of cheap reliability came with the flintlock, developed in the mid seventeenth century. This weapon was robust and reliable, yet simple enough to be made by local gunsmiths. The effective cost of this weapon was low enough that commoner militia became highly credible political threats. See Chapters 15 and 17 for a discussion of the revolutionary social implications of these new coercive tools.

haploid: Refers to genomes of gametes (eggs and sperm). The term refers to the condition of having half of the **diploid** number of genes in the adult organism (and the fertilized egg that will grow into this adult). A haploid gamete has one copy of each piece of **genetic design information** (each **gene**) while the diploid individual has two copies of each piece, one from each of parent.

hierarchical/hierarchy (political): Refers to the human social organizations that arise when decisive coercive threat is held by small elites. These polities predictably serve elite interests selectively and include **archaic states** (Chapter 13), **early modern states** (Chapter 14), and **modern authoritarian states** (Chapters 15 and 17).

hierarchically nested combinatoriality: See **nested combinatoriality.**

hominids: We will use this term to apply collectively to all the members (surviving and extinct) of the bipedal African great apes. Thus, hominids include our relatively recent non-human ancestors (or cousins to those ancestors) including **australopithecines,** but do not include **bonobos, chimps** or **gorillas.** We will also use the term to include all members of the genus **Homo** including our human ancestors and ourselves. We will avoid using the roughly synonymous term hominins, popular among professional taxonomists for reasons that need not concern us.

hominins: See **hominids**.

Homo: The genus name of all humans, ancestral and living, as we will use the term. See specific examples below.

Homo erectus: The name given to some early members of our human genus, Homo. These individuals showed relatively modest, but clearly significant increases in absolute brain size relative to their **australopith hominid** ancestors (Chapters 6 and 7). This term is sometimes construed to apply to early African humans called by others, Homo ergaster. Examples of Homo erectus, as we use the term, include the **Nariokotome boy** fossil and the **Dmanisi** fossils (Chapter 7). Members of Homo erectus generally have brain sizes around 950cc. This is larger than **apes** and **australopiths** (350-450cc), but smaller than contemporary humans (averaging around 1350cc and ranging from 1100-1500cc).

Homo habilis: The name given to one putative very early species of Homo (Chapter 7). These animals lived about two million years ago. This term is somewhat variably used. We will use it to apply specifically to relatively small-brained hominids alive at this time. These animals are sometimes called Australopithecus habilis instead of Homo habilis; however, we prefer Homo as they appear to show brain anatomy, including **Broca's area** enlargement, indicating initial evolution of uniquely human elite information exchange (Chapter 9). These fossils are associated with remains consistent with early **power scavenging** and therefore, presumably, **elite throwing.** Homo habilis individuals often have brains only slightly larger (about 650cc) than the **apes** and **australopiths** (about 350-450cc).

Homo heidelbergensis: This species evolved from **Homo erectus** or a similar earlier human, probably around five hundred thousand to seven hundred thousand years ago. This animal had a still larger brain than erectus and approached modern brain size (around 1200-1400 cc; while modern humans range up to 1500cc with a few individuals having larger brains; see **Homo sapiens**). Homo heidelbergensis is distinguished from later humans largely on the basis of subtle morphological characteristics, including things like facial shape. These animals probably continued to make Acheulean tools for some time. However, they ultimately made slightly different tools, including the so-called Mousterian or **Levallois** stone tool traditions. These animals apparently spread through virtually all of Africa and Eurasia, displacing other human species - including erectus - as they went. They were apparently ancestral to all later humans, including us.

Homo neanderthalensis: One of two major species that probably evolved from **Homo heidelbergensis** around one hundred fifty thousand to two hundred fifty thousand years ago (Chapter 11). They are commonly referred to as Neandertals or Neanderthals. They have a slightly larger brain than heidelbergensis (1400-1700 cc; compared to our brains sizes which are typically 1350cc and range from 1100-1500cc) and their brains are similar to or slightly larger in size than ours. These individuals are distinct from other human species in various ways. Most important are shorter, heavier stature and an elongated, enlarged nasal apparatus. These are probably adaptations to the relatively cold climate across central Europe and western Asia where these individuals lived. Neandertals continued to make **Mousterian** tools throughout their evolutionary history. They ultimately became extinct around thirty thousand years ago, apparently as a result of unsuccessful competition with our immediate ancestors, **behaviorally modern humans** (Chapter 11).

Homo rudolfensis: The name given to one of the earliest known larger-brained hominids (Chapter 7). This human also lived around two million years ago. Based on its larger brain and its postcranial skeleton, this animal is a plausible candidate to have been our direct ancestor. This animal may have been ancestral to **Homo erectus.** Homo rudolfensis individuals may have made the so-called developed **Oldowan** stone tools, mostly simple flakes, and **manuports** (including throwing stones). Homo rudolfensis have brain sizes around 850cc, significantly larger than **apes** and **australopiths** (350-450 cc).

Homo sapiens: The name given to the species represented by all currently surviving humans. This species is probably at least several hundred thousands years old, including both **anatomically modern humans** and **behaviorally modern humans** (Chapter 11). Our brains average around 1350cc and range from 1100-1500cc with a few individuals outside of this range, both larger and smaller.

Hopewell: A Native American culture distributed throughout the Ohio River drainage and existing between about 200 BCE and 500 CE (Chapter 11). This society had a relatively low population density with no large permanent settlements and certainly no cities. It was supported by small-scale farming formally called **horticulture** as well as by extensive gathering of wild foods. We argue that the Hopewell represent a relatively late persistence of the **behaviorally modern human** scale of social cooperation (Chapter 11).

Hoplite: This is the ancient Greek term for an **elite** or **armored warrior** inevitably produced when effective **body armor** becomes available, producing **archaic states** (Chapter 13). These elite warriors had a very different social role than most contemporary soldiers. They were privileged members of the societies in which they lived. Indeed, early democracies, including Athenian Greece and the Roman Republic, had an enfranchised, voting population consisting exclusively of these warriors and their designated functionaries. Elite warrior and citizen were virtually synonymous. Democracies of this faux form are ultimately, predictably unstable (Chapter 13). These elite soldiers had different names in each archaic state including Samurai, legionnaire, hoplite, knight and warrior.

horticulture: This term refers to the domestication and use of plants on a relatively small scale, for example by families or very small local coalitions. Such people are referred to as horticulturalists. Most horticulturalists inhabit **behaviorally modern human** scale social adaptations (Chapter 11). Such people were responsible for the domestication of some of the food plants we now use. Other domesticates were developed by **Neolithic** scale adaptations (Chapter 12). **Agriculture,** as we will use the term, is distinct from horticulture in scale. Agriculture represents a much larger, industrial-scale exploitation of domesticates. For example, in prehistoric North America, the **Hopewell** were horticulturists while their successors, the **Mississippians,** used large-scale agriculture to support their large permanent towns, like Cahokia. The larger scale of Neolithic social cooperation, driven by the **bow** in Eurasia and North America, produced the scale-up of horticulture to agriculture on our theory (Chapter 12).

hostile manipulation problem: Refers to a particular subclass of the **conflict of interest** problem. Specifically, non-kin conspecifics have an incentive to mislead and manipulate one another during the exchange of contingent (potentially false) information. As a result, systematic non-kin exchange of such information cannot evolve as a Darwinian adaptation unless the conflict of interest problem can be managed. Humans are the first animals to achieve the necessary mastery of the hostile manipulation problem (Chapter 9).

human uniqueness problem: Refers to the scientific or theoretical problem presented to us by the fact that humans are merely one of millions of terrestrial animal species, on the one hand, and yet have capabilities and a level of **ecological dominance** orders of magnitude beyond those of any other animal that ever lived on the planet. The central objective of this book is to develop a coherent, **parsimonious** theory accounting for human uniqueness (Preface and Introduction).

humans: We will use the term to apply to all members of our genus **Homo**. All humans defined in this way show post-cranial skeletal adaptations for **elite throwing** and brain expansion reflecting the emergence of **kinship-independent social cooperation** allowing characteristically human **life history** redesign (Chapters 6 and 7). The first humans arose around two million years ago in Africa.

hydrogen bomb: See **fusion**.

hymenopteran: Literally translates *membrane wing* and refers to a group of insects including ants, bees, and wasps. This group will be of special interest to us because its members reveal a great deal about how **kin-selection** works (Chapter 3). These animals have a very unusual sex-determination system. For most organisms – like us, for example – the sex of the fertilized egg (the zygote) is determined by presence or absence of the Y chromosome in the sperm and both sexes are diploid. In contrast, in hymenopterans sex is determined by ploidy so that **diploid** individuals are female and **haploid** individuals are male. In other words, haploid male individuals arise from unfertilized eggs. This has several implications. First, since males are haploid, all their sperm show 100 percent **genetic identity** and **genetic relatedness** to one another rather than being only 50 percent identical/related, as in a human or most other animals. Second, as a result, female hymenopteran full siblings are 75 percent genetically related rather than 50 percent as in our case. [This is because sperm are 100 percent related and eggs are 50 percent related averaging to a total of 75 percent genetic relatedness.] This predicts that female hymenopterans will show some special kin-selected behaviors and they do (Chapters 3 and 5). Third, a mated adult female hymenopteran can control whether she produces females or males by choosing to fertilize or not fertilize her eggs as they are laid, and she exerts this control adaptively. Fourth, virgin hymenopteran females can still reproduce, though they produce only sons. **Worker policing** (Chapter 5) arises when half-sister hymenopteran workers (from queens/mothers who have mated with multiple males) eat one another's male eggs (12.5 percent related), creating the preference to raise their common brothers (25 percent related) as the next best individual option (Chapter 5).

hyoid bone: A small, horseshoe-shaped bone high in the throat just under the tongue. This free-floating bone is anchored to surrounding large structures through ligaments and is thus relatively mobile. It then anchors many of the muscles of the tongue. This allows motion of the tongue to be independent of the lower jaw. This capability evolved originally in ancient animals to support complex mouth/tongue movements during eating, drinking, and the like. In humans, this movement has been expanded to also play a role in control of sound generation during speech (Chapter 9). As a result, the human hyoid bone is bigger, more robust, and somewhat redesigned relative to our pre-human ancestors. This bone is *relatively* small and delicate - compared to a femur or a pelvis - and does not often survive in fossils. However, we have a few fossil hyoids and we can use them to assess some details of the evolution of human speech (Chapter 9).

identifier information: See **culturally transmitted design information**.

ilium and iliac crest: The ilium is the large blade-like portion of the pelvis (see images in Chapter 7). The iliac crest is a thickening along the top of this bone. The ilium represents the attachment point for the gluteus maximus and the iliac crest for the so-called oblique muscles. The placement of these muscle connections has apparently been moved in humans relative to our last pre-human ancestors as part of our adaptation to elite aimed high-momentum **elite throwing** (Chapter 7).

Inca: The enormous **archaic state** (**empire**) that ruled much of present day Ecuador, Peru, and Chile as well as western Argentina and Bolivia. Like all archaic states/empires, the Inca Empire was ruled by a warrior elite (Chapter 13). Like the Aztecs, the Inca state was still in existence at European contact early in the 16th Century.

indirect access (to culturally transmitted information): We will use this phrase in a specific technical sense. It will refer to our capacity to acquire access to the fruits of culturally transmitted information we do not personally hold, but is held by others. For example, when we purchase food we do not know how to grow or make, we have acquired indirect access to culturally transmitted know-how enabling the production of this food. Virtually every transaction in our public economic lives has the purpose of obtaining such indirect access to cultural information (Chapter 10; Third Interlude). Because of the **conflict of interest** problem, acquiring such access requires enforcement of **kinship-independent social cooperation**. Otherwise, goods produced for exchange would be stolen instead (Chapter 5). It is also possible to have **direct access** to culturally transmitted information as know-how we actually hold in our individual minds.

infanticide: The killing of juveniles – most often non-kin - by conspecific adults. This occurs under a variety of conditions for strategic reasons (Chapters 3 and 4) and is a universal animal behavior. Its occurrence and partial suppression in humans is illuminating about our unique social cooperation (Chapter 4).

information: This is the central concept of all of biology. Biological systems are first and foremost the products and the agents of self-replicating information (Chapter 2). Two considerations are important. First, information is encoded in physical entities – in **DNA** sequence or the pages of a book, for example. However, the information is not the physical entity itself. It is merely recorded in its physical instantiation. Though this distinction might seem a little subtle, it turns out to be absolutely crucial. Second, though we all have a robust intuitive understanding of what information is, it is shockingly elusive to produce a rigorous, analytical (scientific) definition. What matters at the moment is that information consists of order reflected in its physical instantiation that, in turn, allows that order to be copied and/or to interact with the external physical world. Notice that the **culturally transmitted information** we think of as knowledge has these properties as does **genetic design information**.

iron: Iron is a metal generally found in an oxidized from in ore that can be reduced back into its metallic form by the application of intense heating (smelting). Metallic iron alone is relatively soft and breakable. However, when alloyed with carefully controlled amounts of carbon (from burning wood or coal, for example), it becomes very useful **steel**. Iron ore (rust) is extremely abundant. However, smelting this ore to produce metallic iron and then alloying it with carbon to produce steel requires even higher temperatures and greater expertise than working with copper and **bronze**. Thus, iron working was mastered relatively late (around 1200 BCE). Moreover, until Bessemer's process was developed in the mid 19th Century, production of strong steel was only done by sophisticated specialists on a small scale. Thus, steel shock weapons, like high quality swords, remained expensive, elite articles throughout the era of **armored warriors** and **archaic states**.

Kalashnikov: See **handgun**.

kin (kinship): We will use this term with a very specific and technical meaning. [We will not use it as we might casually to describe a distant relative or even a political ally as in brother or sister fellow believer.] Kin are individuals with whom the reference animal shares one or more *very recent* common ancestors. As are result, two kin individuals have a specific and predictable probability of having been built by copies of some of the very same pieces of **genetic design information**. This known commonality is vital to understanding the evolution of their social behavior (Chapter 3). See **kin-selection**.

kin-selection: Refers to the process when natural selection favors behaviors in one animal because of their impact on the reproductive success of another, close **kin** animal (Chapter 3). This process works because close kin are built by many pieces of the same **genetic design information**. Thus, when genetic design information builds an animal to help its close kin, that design information is supporting its own replication, discounted by the probability that the close kin individual did not receive this same piece of design information during sexual reproduction (Chapter 3 and online endnotes to Chapter 3). As a result of kin-selection, individual animals will behave as if they have a **confluence of interest** with close kin. In contrast, therefore, non-kin conspecific animals will behave as if they have **conflicts of interest**.

kinship-independent social breeding: This is the uniquely human sexual/reproductive adaptation (Chapters 6 and 8). It is a profoundly new and different method of reproduction in comparison both to our closest living relatives (chimps and other apes) and to all other animals. It is, in fact, merely a specific example of the wider human novelty of **kinship-independent social cooperation**. Specifically, this reproductive adaptation involves many non-kin individuals helping the pregnant female, the young mother and the young child, rather than having the entire responsibility for birth and raising of the youngster fall exclusively on the individual mother or a pair-bonded couple, as in the case for other animals. This help can involve *potentially* related individuals – especially males who have mated with the adult female in question and thus who could be **biological fathers**. It also includes relatives of the female. However, members of the larger kinship-independent coalition in which these individuals are embedded also contribute, especially through economic cooperation. It is also important to recognize that even when human reproduction involves help primarily from (potentially) genetically related individuals, these individuals will often be of much lower **genetic relatedness** to the mother than would characterize such helpers in **kin-selected** social breeding systems in non-human animals. Thus, this human system is very much a kinship-independent reproductive system.

kinship-independent (conspecific) social cooperation: Also sometimes called non-kin cooperation, this cooperation is not based on genetic kinship. This pattern of social cooperation is not produced by **kin-selection**. In principle, two non-kin conspecific animals can cooperate to mutual benefit, "you scratch my back, I scratch yours", so to speak. However, the problem with this logic is that the design information of two non-kin conspecifics will almost always come into conflict. They or their descendants will inevitably compete in the future. Thus, in the absence of other special factors, it is always in the short-term interest of design information to build animals that cheat in non-kin cooperation - who "take the money and run", so to speak. This is the non-kin social cooperation problem and is synonymous with the **collective action problem** in the social sciences. This problem can be solved only when cost-effective means exist to project **coercive threat**, enforcing cooperation by ostracism as a by-product. We argue that humans are the first animals to have access to this cost-effective social coercion (Chapters 5 and 7).

Lucy: A famous **Australopithecus afarensis** fossil discovered by Donald Johansen and his team (Chapter 7). She may have been a young adult female. From her skeletal anatomy she was clearly a biped. **Australopithecines**, including Lucy, had an ape-sized brain and left no material artifacts that would indicate that they had any greater **adaptive sophistication** than living non-human **apes**, like **chimps** or **bonobos**.

lactation: The production of mother's milk in the breasts of mammals, including humans. This is high value food necessary to help the young mammal, including the young human mammal, grow properly. In the special case of humans, the baby needs unusually large amounts of milk to continue its rapid fetal rate of brain growth (Chapters 6 and 7). This high demand means that human mothers must be well provisioned during the infant's life so that sufficient milk can be produced. In light of the extreme infant dependence (see **babyhood**) this can only be achieved by others helping to provide for the nursing female. An extended network of individuals of variable genetic relatedness to the mother makes this possible. This provisioning is part of the uniquely human reproductive strategy – **kinship-independent social breeding**. Also see **life history** and **childhood**.

Laetoli footprints: This is an extensive set of footprints originally made by three **australopithecines** that walked across an area of damp volcanic ash in East Africa about 3.7 million years ago. These were discovered by the famous paleoanthropologist Mary Leakey and are extremely illuminating about how australopithecines walked (Chapter 7).

language: We will use this term to refer specifically to the highly evolved capacity of humans to exchange information using spoken and gestural symbols (Chapter 9). Human language is characterized by extensive semantic content (word/object and word/action relationship) as well as elaborate compositional rules, including phonology (meaningless short sound segments), morphology (combining phonological elements to make meaningful words), and syntax (combining words into complex sentences). As a result of these properties, human language is said to be capable of producing an effectively infinite number of sentences expressing a nearly infinite number of thoughts, ideas and pictures. Thus, language is generative. We will argue that language results from the redeployment and refinement of ancient properties of the universal ancestral animal mind rather than invention of any qualitatively new capabilities (Chapter 9).

language area (of the brain): See **Broca's area**.

larynx: The highly structured, controllable constriction in the wind pipe containing muscular flaps of tissue that can open and close variably so as to be vibrated by the air forced out of the lungs. This is the fundamental vibrator that generates the sounds that are then shaped and controlled by the throat, mouth, and lips to generate speech sounds (see diagrams in Chapter 9).

last common ancestor (abbreviated LCA): This term refers to the last organism who left descendants who gave rise to any two different contemporary organisms in question. Most commonly this is used on the time scale of speciation and extinction events so that we speak about the LCA to two different species. For example, there was a single species of ape-like animal (call it X for the moment) who lived in southern and/or central Africa around six million years ago. This population underwent a speciation event, producing at least two new species. One of those two species ultimately produced chimps and the other produced humans. Thus, species X was the LCA of chimps and humans.

latissimus dorsi: See **muscles**.

Levallois: This term is used, somewhat variably by different specialists, to refer to a series of stone tools that are noticeably different than **Acheulean** tools. Acheulean tools are essentially sculpted from a stone core. The Levallois technique flakes the sharp tool off a (prepared) core. Levallois tools begin being made around roughly four hundred thousand years ago and ultimately throughout Africa and Eurasia. These tools were arguably a little more sophisticated than Acheulean tools. However, if they were, it was not by much. These tools remain unsophisticated by our contemporary standards. They continued to be made up until forty to sixty thousand years ago when they were replaced by the more sophisticated tools of **behaviorally modern humans**. Levallois tools were apparently made by both **Neandertals** and **anatomically modern humans**.

life: We now understand that the transition from chemistry to biology or from non-life to life occurs when a chemical system first arises with the following property. It contains **genetic design information** encoded in some form that can be replicated. Moreover, this design information can direct the production of products - tools - that promote this replication. Thus, such a design information/product system is self-replicating. The earliest of these systems probably used non-biologically produced precursors or building blocks to synthesize copies of the design information and tools (Chapter 2).

life history: Refers to all the details of an animal's entire life from conception through gestation, birth, infancy, childhood, later juvenile stages, adolescence, adulthood, and old age. What will be important to us is that the myriad details of the life history of an animal – including humans and our ancestors and relatives – tell us a great deal about the social life of the animal. Life history is informative about the parts of social life that impinge on sexual behavior and childrearing (Chapters 6 and 8). Moreover, many features of life history can be inferred from skeletal evidence. Since skeletons frequently form **fossils** we can often infer details of the life histories of our extinct ancestors. We will make use of this knowledge to infer many things about human evolution and evolutionary origins (Chapter 7). We will find that **kinship-independent social breeding** is the fundamental human reproductive trick. This adaptation is reflected in some dramatically new, unique features of our life history including **expanded gestation, babyhood,** and **childhood**. Moreover, we will find that this novel, uniquely human reproductive adaptation probably evolved in its initial form at or immediately after the origins of the first members of our genus, **Homo** – the first humans – about two million years ago (Chapters 6 and 7).

ligaments: Elastic fibers that hold jointed bones together. They do not actively contract like muscles but they are like rubber bands, causing joints to snap back into their natural resting positions when not in use or after relaxation of a violent muscle contraction.

macromolecule: See **chemical bond, DNA, genetic design information,** and **catalyst**.

maize: see **Mesoamerican triad**

Malthusian: A world or segment thereof (like a local environment) characterized by its *finite* resources. This limitation in resources, in turn, inevitably limits the growth of **exponentially growing** biological organisms. This word is derived from Thomas Robert Malthus who wrote the famous book *An Essay on the Principle of Population* – published in 1798 - which first explicitly emphasized this inevitable feature of the worlds of all biological organisms. Malthus' thought profoundly influenced Darwin.

manuport: Refers to a stone whose archaeological context indicates that it was moved across the landscape by humans, typically by ancient ancestral humans. A subset of these manuports looks as if they represent ammunition for characteristically human **elite throwing** (Chapter 7).

matchlock: see **handguns**.

materialism: This term is widely and somewhat variously used. We will use it to mean the hypothesis that all things in the universe, including organisms and their minds and behaviors, are the intelligible products of material process. In other words, everything in our world, including human minds and societies, emerge simply from the chemical properties of specific types of **vehicles** replicating in a **Malthusian** world. This book explores a materialist theory of human origins, properties, and history.

maternal investment: See **parental investment**.

matrilocal: This term describes a residence pattern for any animal, including humans. It implies that as animals reach sexual maturity the females stay in place on the home territory and males emigrate. The converse pattern is **patrilocal**. Humans are not strictly either matrilocal or patrilocal, though a local culture may be primarily one, the other, or some mixture of the two. Human matrilocality includes adult males focusing important elements of their economic activity around their maternal households rather than the households of their mates. This form of human matrilocality tends to correlate with promiscuous mating systems (Chapter 8) as maternity is substantially more certain under these conditions than paternity.

maximum policeable unit (MPU): Refers to the maximal size of a non-kin cooperative human social unit that can be cost-effectively (adaptively) policed with the coercive technology available at that time and place. This MPU has increased in size dramatically over the course of human history as new coercive technologies became available (Third Interlude; Chapter 12).

Mayans: An **archaic state** (**empire**) in what is now southern Mexico and Guatemala. The Mayans produced large cities (populations of roughly one hundred thousand or more). The Mayan empire collapsed (in places) around one thousand years ago (Chapter 13).

Meiji Restoration: The name given to a dramatic partial democratization of the Japanese state in 1868 which began with introduction of modern European **handguns** into Japan around 1855 and culminated in the elimination of the **Tokugawa Shogunate** and the restoration of the Emperor who, in turn, signed the Charter or Five Articles Oath as a condition for his restoration to the thrown. This famous oath is reproduced in Box 15.4 in Chapter 15.

menopause: Another unique phase of human **life history**. In other animals, adult females typically continue to ovulate throughout their lives. In humans, females typically cease ovulation around the age of fifty, many years before death in a healthy individual. This is thought to be a reproductive adaptation. As a female ages, her capacity to improve the reproductive output of her offspring and relatives (by helping to pass on the uniquely large amounts of **culturally transmitted information** possessed by human coalitions, Chapter 10) produces greater adaptive returns than further, risky attempts at producing more offspring of her own. As with all the uniquely human life history features, this one dependent on our unique human reproductive strategy, **kinship-independent social breeding** on the theory we are exploring.

mental information: See **culturally transmitted information**

Mesoamerica: An archaeologist's term for what is now Mexico and Central America. It is usually applied to this area especially from the time of the earliest recognizable complex cultures (roughly 1500 BCE) through European contact (1519 CE).

Mesoamerican triad: The triad of domesticates forming the basis of the early agricultural civilizations and subsequent archaic states in present-day Central America (Chapter 13). These three crops were originally domesticated in this area and consist of maize (corn), beans, and squash. These cultigens spread into North America, apparently without the cotton-based weapon systems that produced the Mesoamerican early states (Chapter 13). These domesticates were an element of the North American environment in which bow-based agricultural civilizations like the **Mississippians** and the **Anasazi** subsequently developed (Chapter 12).

metatarsal: See **first metatarsal**.

Minie ball: The first convenient projectile for a **handgun** that allowed guns with **rifled barrels** to be routinely used by infantry in active battle. The projectile was conically shaped with a hollow back. This structure allowed the projectile to be easily rammed down the barrel of muzzle-loading rifled flintlock muskets or *rifle*, for short. When the powder charge exploded behind the Minie projectile, the hallow base in soft lead expanded to fit tightly against the spiral grooves of the rifling, imparting spin to the projectile resulting in gyroscopic stabilization. This stabilization produced a dramatic improvement in accuracy of handguns, extending their effective range out to many hundreds of meters. Rifling continues to be used in contemporary weapons to improve their accuracy, though projectiles are now loaded from the back (from the breach).

MISA: See **mutually informed social aggregate.**

Mississippian: A **Neolithic** Native American society distributed throughout large portions of the Mississippi and Ohio River drainages and derived from the **Hopewell**.

Mississippian culture existed from about 700 CE through at least 1500 CE. This culture supported much higher population densities than the Hopewell and had a number of very large permanent towns, each with perhaps as many as about ten thousand inhabitants. This society arose from a substantial expansion of the scale of the human **kinship-independent social cooperation** very shortly after the introduction of the **bow** into this part of the world as predicted by our theory (Chapter 12). This enlarged scale of cooperation allowed the development of **agriculture** on which Mississippian subsistence was based.

Moche: An **archaic state** culture in the Andes of South America, apparently ancestral to the **Inca**.

modern authoritarian state: Though cheap, reliable **handguns** created the potential for democratization of control of decisive **coercive threat**, such democratic control is not inevitable. Elite concentrations of power can monopolize production and access to **gunpowder** handguns, concentrating coercive power in massive state-level military organizations (Chapters 15 and 17). This has become easy to do as handgun manufacture has become more industrialized and capital-intensive. We argue that this potential for the concentration of elite power in the modern state is the single most terrible threat we face in thinking about the global human future (Chapter 17).

modern economic miracle: Refers to the dramatic, persistent increase in both economic output and sustainable population densities beginning roughly in the 18th Century (Chapters 15 and 17). This economic transformation reflects a radical increase in human **adaptive sophistication**, an **adaptive revolution**. On our theory, this adaptive revolution results from the cooperative social effects of democratization of access to local coercive threat after development of the gunpowder **handgun**. This economic transformation lagged roughly one hundred years behind the first effects of the democratization of the **modern state**, the **Scientific Revolution**. We date the beginning of this modern coercive era, somewhat arbitrarily, to Dutch Revolt against the Spanish crown in the late 16th Century.

modern era: This term is most often used to describe the era consisting of the **Scientific Revolution** and the **modern economic miracle** (Chapters 15 and 17). This era reflects an **adaptive revolution** produced by democratization of access to coercive threat resulting from development of gunpowder **handguns** in the context of the **modern democratic state**. This era begins roughly in the late 16th Century. Sometimes the word modern is expanded to include also the **early modern state** produced by gunpowder **artillery** beginning in the 14th and 15th Centuries (Chapter 14).

modern democratic state: After its effects on the **early modern state, gunpowder** apparently had a second dramatic effect on human **kinship-independent social cooperation**. Several hundred years after reliable **artillery** were developed, robust **handguns** emerged. These had the effect of allowing commoner militia to sometimes defeat or control elite coercive power, including driving **elite warriors** exploiting **body armor** from the field. In the hands of our species, with its ancient adaptation to exploiting the coercive power provided by weaponry, this led to the redistribution of political power - out of the hands of the elite few and back into the hands of the many on our theory (Chapters 15 and 17).

modern state: See **modern democratic state** and **modern authoritarian state**.

molecule: Generally applied to chemical compounds that have a large number of atoms, often atoms of several different elements. Molecules can be very large in biological systems and are often referred to as macromolecules as a result. **DNA** and most **proteins** are macromolecules. Note that these enormous biological macromolecules are, nonetheless, relatively simple in their structure, being built from long linear chains of chemically similar small subunit molecules. These long linear molecules are called, generically, polymers and their component subunits, monomers.

monitoring: See **guilt** and **moral outrage**.

monogamy: A mating/reproductive strategy in which a single male and a single female mate exclusively with one another. A strategy approaching simple monogamy is seen in many bird species. However, it is very rarely seen in mammals. Humans are an exception. We practice a mating system approaching monogamy under some specific circumstances. Those circumstances consist of relatively low rates of adult (mate) mortality. However, under conditions of high mate mortality risks, humans practice promiscuity or polygynandry (Chapter 8). Under these conditions humans still pair-bond, but mate with multiple others in addition to their nominal or social spouses. Under these promiscuous mating conditions, a couple is said to participate in social monogamy. Each is said to be the social spouse of the other. The male member of this promiscuously mating pair is said to the **social father** of his mate's offspring. These offspring often are said to have other co-fathers, usually corresponding to other males who may have fathered them (Chapter 8).

monkeys: Primates more distantly related to humans than are the other **apes**. Many of these animals are smaller in size than apes and highly arboreal (tree climbing). These include the macaques and langur monkeys among others.

monomer: See **molecule** and **DNA**.

Mousterian: See **Levallois**.

moral outrage: This emotion is a product of **proximate psychological devices**. This emotion is produced in response to information about our own behaviors and the behaviors of those around us. We constantly monitor the social behaviors of others, collecting this information, mostly unconsciously. This imposes a strong time, effort, and information storage burden on us. The magnitude of this burden increases as roughly the square of the number of people in our social groups. [The actual formula is $(n)(n1-)/2$ where n is the number of people in our social group.] This means that the size of any intimately cooperative social group in which we participate is ultimately limited by this burden. However, by structuring our economic cooperation as **public market exchange,** we can overcome this monitoring burden and extend **kinship-independent social cooperation** to apparently unlimited scales. Also see **guilt**.

MPU: See **maximum policeable unit**.

muscles: Large bundles of tissue that actively contract when appropriate nervous system commands are received. Most muscles span joints between bones and cause the joint to close (flex), open (extend), or twist. Muscle bundles can only do two things - contract and relax. Relaxation is passive while contraction is active. Thus, when a muscle relaxes, the joint either snaps back to its resting position or is pulled through its resting position and beyond in the opposite direction by the active contraction of opposing muscle bundles. The fine control of complex, opposing sets of muscle bundles by our central nervous systems is what allows us to move with grace, power, and control.

The attachment points of muscles on bones and the joint faces between bones are often preserved in **fossils** allowing us to learn a great deal about the behavioral/movement capabilities of ancient animals, including our own ancestors (Chapter 7). The following individual muscles torque the body (or control its torqueing) during the elite human throwing motion. Their placement and structure can be found in the Figures in Chapter 7: **latissimus dorsi, obliques (internal and external), tensor fascia lata, gluteus maximus** and **peroneus longus**. The **dorsal interosseous** is an important hand muscle involved in throwing (Figures in Chapter 7).

musket: Refers to long **handguns**, especially matchlocks, wheelocks and flintlocks (see **handgun**; Chapters 15 and 17). These individual weapons had smooth surfaces on the inside of the barrel where the projectile exited. This limited the accuracy of these weapons. With the introduction of **rifled barrels** imparting a controlled spin to the exiting projectile, these weapons became dramatically more accurate. With this introduction, these weapons came to be called *rifles* rather than muskets.

mutation: Can be used as either a verb or a noun, depending on context. As a verb, it refers to the process whereby **DNA** sequence is accidentally changed – generally rarely – during replication of a DNA molecule. For example, an A base might be put in where a G base was before. This changes the sequence of the DNA, thereby altering the **genetic design information** it encodes. Such changes generate the differences between different copies of individual pieces of genetic design information or **genes**. These differences constitute the genetic variation on which **Darwinian selection** acts (Chapter 2). As a noun it generally refers to the change in DNA sequence produced by mutation (the verb). Also see **allele**.

mutually informed social aggregate (MISA): This term refers to human non-kin public social aggregates sharing information in pursuit of individual self-interest. These aggregates generally correspond to **maximal policeable units** or some subset of these units.

muzzle: Refers to the mouth or outer opening of the barrel of a **handgun**. Older weapons, like the flintlock **musket**, for example, were loaded from the muzzle, with **gunpowder** and the projectile or bullet being forced down the barrel with a ramrod. As engineering skills increased, weapons were developed that could be loaded from the back of the barrel – from the **breach** – as modern handguns continue to be.

Nariokotome Boy: A famous fossil of a young male member of **Homo erectus** (sometimes called Homo ergaster) from East Africa. This individual lived roughly 1.6 million years ago. He is the most complete skeleton of a very early member of our genus we currently possess and we will make frequent reference to him (Chapter 7).

natural (or Darwinian) selection: Also sometimes referred to as a Darwinian process. The self-replicating system, consisting of a **vehicle** built by **genetic design information** that is a biological organism, is susceptible to change (evolution) over time. This occurs for the following reason. Replication can never be perfectly accurate and, thus, small changes (**mutations**) in the design information will inevitably occasionally occur. When they do, most will make the self-replicating system less effective and will die out. However, rarely, such changes will produce a better self-replicating system and this system will eventually competitively displace the original system. Thus, these systems continue to undergo improvement and refinement. New selection in a novel adaptive context acts in this same way to change the adaptive behaviors and anatomies of organisms, including ancestral humans in new physical and social environments (Chapters 5 and 7).

Neandertals or Neanderthals: See **Homo neandertalensis**.

Neolithic: This term translates literally as "new stone age." We use this term to describe the dramatic adaptive changes associated with a new scale of social cooperation apparently produced by the **bow** in Eurasia and North America (Chapter 12). The **agricultural**

revolutions in these areas are Neolithic revolutions. However, Neolithic revolutions also produce non-agricultural **adaptive revolutions** in areas where agriculture was not the adaptation of choice. For example, see the **Calusa**, the **Chumash** and the **buffalo hunters of the Great Plains**.

nested combinatoriality: The fundamental strategy used to build complexity throughout the biological world (Chapter 2). It refers to the fact that each new organizational level (multicellular animals, say) emerges by simple combination of the units from the level immediately below (cells, in this case). Moreover, the elements or members of this level can combine to form still higher organizational levels (animal social aggregates, in this case). This relatively simple strategy is profoundly powerful. Understanding it is one of the key steps to being able to comprehend the elegance and ultimate simplicity of the biological world.

Nineteenth Amendment to the American Constitution: This amendment granted the right to vote to American women in 1920. This is one generation after technical developments in **handguns** (also see **bullet ammunition**) placed decisive local coercive power in the hands of all adults – not just men. (See **Annie Oakley**, for example). This and several other nearly contemporaneous cases involving **modern democratic states** represented the first time in the 5500 years since the first **archaic states** that women were formally re-enfranchised (Chapters 15 and 17).

non-kin cooperation: see **kinship-independent social cooperation**.

non-kin: See **kin**, **kin-selection** and **kinship-independent social cooperation**.

nuclear: This term will be used in two completely independent ways. First, it will refer to events that go on in the nuclei of cells. These are macromolecular **organelles** containing the **DNA** housed in large **chromosomes** in which contemporary **genetic design information** is encoded. Second, it will refer to events involving the nuclei of **atoms**. These are structures consisting of subatomic particles making up the center of individual atoms. Our primary concern with these processes will be with their subsets – **fission** and **fusion** – which are sources of explosive power in human weapons (Chapter 16) and sources of naturally occurring energy crucial to life on Earth.

obliques (internal and external): See **muscles**.

occult ovulation: Refers to the practice of various mammalian females (including humans) of hiding ovulation so that no surrounding individuals (and often the female herself) know when she is ovulating. Occult ovulation in humans is in stark contrast to the active advertisement of ovulation by our closest living relatives, the chimp. We argue that this is one of many human adaptations to strategic promiscuity in the ancestral human **kinship-independent social breeding** environment (Chapter 8).

Okada, Daijiro: Game theorist and economist. Daijiro was a vital collaborator in working out the fundamental logic of self-interested coercion and social cooperation. His paper in collaboration with author Bingham is one of the most important pieces of work on which this book is based (see Okada and Bingham, 2008).

Oldowan stone tools: This name is derived from the Olduvai Gorge in East Africa where the original examples of these tools were found by Lewis and Mary Leakey. The early forms of these tools are very simple, consisting of stone flakes used for sharp edges and rounded stones probably used for battering and throwing, including **manuports**. Early members of Homo (possibly including **Homo rudolfensis** and/or early **Homo erectus**) may have made early Oldowan tools. Oldowan tools might also have been made by **Australopithecus garhi** and by the **Dmanisi Homo erectus** individuals (Chapter 7).

opportunity cost: This is a slightly subtle but very important principle. Though it is taken from traditional economics, it actually has a much broader applicability. Any activity (call it X) that an animal engages in and requires any amount of time or resources is done at the expense of all other activities that this time and these resources could alternatively have been used for. Among the activities that X displaces during this time are other activities that are adaptively important – child care, feeding or mating, for example. The loss of these displaced activities constitutes an indirect cost referred to as an opportunity cost. For any activity X to be adaptive, it must produce an adaptive advantage that offsets all its costs, including its opportunity cost. Moreover, the sum of all activities X must be such that essential adaptive activities can still occur. While it is often difficult or impossible to precisely quantify opportunity costs, we know they are very important. They are central in considering the difference between the **atlatl** and **bow** (Chapter 12) and the social status of **elite warriors** (Chapter 13).

organelles: Refers to relatively large structures within cells which carry out specific functions and typically consist of many copies of large machines generally containing dozens of **proteins** and, often, other **molecules**. Organelles include the **nucleus** in which **genetic design information** is stored in contemporary animals. These also include a variety of other structures like the mitochondria where key steps in energy metabolism occur or the Golgi apparatus where protein gene products are prepared for secretion to the outside world.

organism: Refers to any individual biological creature, from a human being to a bacterium, a lion or an oak tree. We will be most interested in the **social behavior** of organisms, but occasionally also in their individual and sexual behaviors. All these behaviors are produced by **natural selection** acting on **design information**. This design information can be **genetic design information** or **culturally transmitted design information**. The following insight is crucial to have in mind if we wish to understand the **ultimate causation** of any organism's behavior, including our own human behavior. Organisms are **vehicles** built by design information to behave as if the replication of their design information were their overriding priority or ultimate purpose. These properties are produced by design information because these strategies replicate the design information producing them more successfully over time – natural selection, again.

Orpheus: The names sometimes given to the Mars-sized planet that collided with Earth Mark I to produce the Moon and Earth Mark II, the highly improbable world we occupy (Chapter 2).

paleoanthropology: Anthropology (sensu stricto) is the study of humans. Paleontology is the study of the remains – mostly **fossils** – of ancient organisms. Paleoanthropology is, thus, the study of (mostly) fossil remains of humans and their immediate ancestors. We will use the term to apply to the study of the fossils of **australopithecines** and of all the members of the human genus, **Homo**.

parental investment: Raising a behaviorally complex animal to independence requires a great deal of protection, support, and guidance from parents. This is true of all mammals and birds and is especially true of humans. This, in turn, requires time and effort by adult animals (kin/parents, in most common practice). In most mammals, this care is given exclusively or nearly exclusively by mothers. This results from the fact that maternity is always certain in mammals whereas paternity is always uncertain, at least in principle. However, in humans and in many birds, both parents typically contribute extensively. More specifically, female parental investment is referred to as maternal investment and male parental investment as paternal investment.

Note that humans also contribute significant parental investment – often in the form of economic cooperation – to offspring who are not their own. This uniquely human pattern reflects our unprecedented **kinship-independent social breeding** (Chapters 6 and 8).

parsimony: Refers to a scientific theory or explanation that is both simple, economical on the one hand and powerful – predicting many events within a wide domain – on the other. Parsimonious theories like Newton's Laws or Maxwell's Field Equations can literally be written on the front of a tee shirt. A broad and parsimonious theory is sometimes referred to informally as a theory-of-everything or a TOE, for short, for its domain.

partible paternity: See **monogamy** and **social father**.

paternal investment: See **parental investment.**

patrilocal: We will use this term to describe a particular residence pattern for animals. As animals reach sexual maturity, the males stay in place on the home territory and females emigrate. The converse pattern is **matrilocal**. Humans are not strictly either matrilocal or patrilocal, though a local culture may be primarily one, the other, or some mixture of the two.

pedigree: A way of describing the genetic or family relationships between individuals. There are a number of different ways of diagramming these relationships.

pelvis: The large bowl-shaped hip bone to which the legs (**femur**) and the base of the spinal column attach. This bone is very different in us than in our closest living relatives, the chimp. These differences tell us about the evolution, first, of bipedal locomotion and, second, of uniquely human **elite throwing** (Chapter 7).

peroneus longus: See **muscles**.

phenotype: Refers to the morphology and behavior of an individual organism. For example, one individual might have brown hair and brown eyes whereas another might have black hair and black eyes. These overt features represent elements of the individual's phenotype. Many features of an organism are determined, in turn, by its **genotype**. So a slightly different way of saying this is that individual **vehicles** have different phenotypes because of the differences in the specific **genetic design information** they inherit. If the two vehicles are members of the same species, their genetic design information and phenotypes will differ only subtly, slightly. Conversely, if two individuals they are members of different species, their genetic design information and phenotypes will be very different.

A crucial detail for us here is that the behavioral phenotype of an organism is controlled by its genotype in startlingly extensive and specific detail. Thus, behavioral phenotypes will evolve to serve the interests of replicating the genetic design information that produces those phenotypes (Chapter 3).

phylogenetic analysis: This term has a series of closely related but somewhat different meanings, depending on the specific discipline in which it is used. We will use the following definition: Assessment of the likely behavior, morphology or other property of an extinct ancestor based on the corresponding properties in currently surviving descendants of that ancestor. For example, we will use the properties of the four currently surviving African great ape species - gorillas, chimps, bonobos (pygmy chimps) and humans - to infer the properties of the **last common ancestor** (LCA) of humans and chimps (Chapter 7). This, in turn, will contribute to our understanding of what happened to our more recent ancestors, after the time of this LCA. The logic of this kind of phylogenetic analysis is the following. If a property is shared by gorillas, chimps and bonobos, it is almost certain to have been shared by their common ancestors and by the LCA of chimps and humans. On this basis, we infer that this chimp/human LCA was probably much more like contemporary chimps and gorillas than like contemporary humans. In other words, the **hominid** lineage (leading to us) has "gone off in a new direction" compared to the other great apes. See **phylogenetic tree**.

phylogenetic tree: A convenient way of graphically representing the results of **phylogenetic analysis** and of data from **paleoanthropology** and related disciplines. It depicts the relationships of organisms at any one time by explicitly representing their descent from older species. The usefulness of these trees will become apparent by inspecting specific examples in the Figures in Chapter 7.

polyandry: Mating of one female exclusively with the members of a group of multiple males who, in turn, mate only with this single female (Chapter 8). This is the converse or mirror image of the conventional harem arrangement in **polygyny**, for example.

polygynandry: Both sexes mate with multiple members of the opposite sex. **Promiscuous** mating schemes of this form are seen in humans under conditions of high mate mortality risks (Chapter 8).

polygyny: A single male mates exclusively with the members of a group of multiple females who, in turn, each mates with only with this male (Chapter 8). This is a conventional arrangement idealized as a harem, for example.

polymer: See **molecule** and **DNA**.

postcranial skeleton: Refers to the entire skeleton other than the head. The prefix *post* is used instead of *sub* because most animals are quadrupeds and, thus, their non-head skeleton is behind rather than under their heads. As bipeds we are the exception rather than the rule among mammals. It will be very convenient to talk separately about the evolution of the head and of all the rest of the skeleton – the postcranial skeleton - as two separate units (Chapter 7).

power scavenging: See **scavenging.**

pragmatic information: See **culturally transmitted design information**.

precocial: Refers to young who have relatively high functional competence shortly after birth. These youngsters are, thus, precocious. This life style tends to be characteristic of prey species like the herbivores. Precocial youngsters are high-functioning compared to conspecific adults. However, because they forego most of the post-birth learning of **altricial** animals, their adult behavior tends to be significantly less sophisticated than adult altricial animals, including supremely altricial humans. (The noun form of this adjective is precociality.)

primates: The group of related mammalian species which includes **humans, apes,** and **monkeys**. These animals all share a set of adaptations they originally inherited from a common arboreal (tree-dwelling) ancestor. These include grasping, five-fingered hands and feet. (The grasping feet have been lost in humans in the roughly six million years since we diverged from the other apes.) These primate properties also include high quality color binocular vision.

promiscuity: In daily speech this word sometimes has a negative ethical connotation. However, we will use it in a different and technical sense. We will use this word as synonymous with **polygynandry** – mating systems in which multiple males mate with multiple females and vice versa. Humans practice such mating systems under conditions of high mate mortality risks (Chapter 8).

protein: Also see discussion under **DNA** and **catalyst**. Proteins are linear polymers of diverse amino acid monomers whose sequence is controlled, in turn, by the linear sequence of bases in DNA. Proteins fold into highly idiosyncratic molecular shapes determined by their amino acid sequence. This structure permits them to interact in an enormous variety of ways with other molecules, including DNA and other proteins. Proteins also collaborate to build large sophisticated molecular machines which, in turn, are assembled to build larger structures, including **organelles** and the cells assembled from multiple organelles. Proteins are the tools (including catalysts) that carry out all the myriad processes necessary to ultimately allow the **genetic design information** that produced them to be successfully replicated. They make up the large bulk of the macromolecules in a **vehicle** built by this information.

proximal killing: Refers to those strategies animals use to kill/injure other animals when killing requires direct physical contact. For example, a lion strike or the charge of a buffalo is an example of proximal killing strategies. In fact, all non-human animals use proximal killing strategies against conspecifics. Such strategies can also be thought of as using tooth and claw. Proximal killers do not have access to inexpensive conjoint **coercive threat** that is accessible to **remote killers** (Chapters 5 and 7).

proximate causation: Refers to the immediate internal (often subjective or even unconscious) cause of a behavior in an animal. For example, we eat because we "feel hungry" – a proximate cause. However, to fully understand why we eat we must look beyond this subjectivity to the **ultimate causation** of our feeding behavior. We are non-equilibrium thermodynamic systems requiring input of high energy compounds (food) in order to continue to function. **Natural selection** has shaped our hunger-feeling psychological devices to cause us to act *as if* we understood the Second Law of Thermodynamics, when usually we do not. All human and animal behaviors have this proximate/ultimate duality including in our social behaviors. To truly understand our behaviors, we must look beyond subjective proximate causation – often ignoring it entirely – and focus, instead, on ultimate causation.

proximate psychological device/mechanism: An informational structure in the mind/brain that produces a specific, generally adaptive, goal, behavior, or pattern of behavior. These adaptive devices are produced by **genetic design information** and\or **culturally transmitted design information** shaped by **Darwinian selection**. The behaviors produced by these devices, including **guilt** and **moral outrage,** are often associated with a specific affect or emotional feeling or flavor. See Chapter 10 for a discussion of the evolution of our minds.

psychological device: See **proximate psychological device.**

public market exchange: We will use this term to describe a special kind of cooperative economic exchange between humans. Typically this occurs when we are functioning in cooperative coalitions too large to function on remembered favors and exchanges – the way we deal with a few close friends and colleagues in small coalitions (see **moral outrage**). Instead, public market exchange involves highly formal, public exchange of information, goods, etc. Moreover, this exchange is rigidly, immediately reciprocal and, generally, ostentatiously public. This allows the solution of the monitoring problem associated with very large coalitions. When we buy an apple at the grocery store (or a financial derivative "over the counter") in exchange for money in the contemporary human economy, we are engaging in public market exchange. Barter exchange in a public setting in a non-money social system is also public market exchange.

purpose (biological version): The evolved **proximate psychological devices** making up our minds are intensely focused on the purposes of things, organisms, and behaviors (Chapter 10). From a strictly biological, **materialist** perspective, purpose is a metaphor when applied to the behaviors of organisms (including ourselves). **Natural selection** acts on **genetic design information** (and **culturally transmitted design information**) so that the organisms built by this information will behave *as if* they have specific purposes. These behaviors in no way imply that organisms have any awareness of these purposes or their ultimate origins. Organisms carry out these apparently purposeful behaviors in response to the **proximate causation** of their minds with no awareness of the **ultimate causation** of their behaviors. We argue in Chapter 17 that self-aware humans can aspire to escape the emptiness of this biological kind of purpose and build a future based on a consensually defined higher purpose.

racism: Together with sexism and ethnocentrism, these false, arbitrary beliefs are elements of **social contract** and **identifier** beliefs developed to sustain **hierarchical,** elite **dominated** political systems (Chapter 13). Such beliefs persisting into the contemporary world are symptomatic of attempted elite domination of the state in all cases of which we are aware (Chapter 17).

rare Earth hypothesis: This hypothesis proposes that planets (like Earth) that have all the diverse features necessary to support the evolution of very complex organisms (like us) are very rare (Chapter 2). Among the implications of this hypothesis are that animals with the complex capabilities of humans may exist nowhere else in our galaxy or, perhaps, even in our universe. This is a profoundly sobering reminder of how important human self-understanding may be. Also see **Orpheus** and **Earth Mark I**.

rational actor fiction: Refers to artificially assuming that individual animals or people behave as if they had omniscient access to the strategic consequences of their behavior. In fact, natural selection often shapes the **proximate** psychological mechanisms making up the minds of animals in a way that causes them to behavior *as if* they were rational actors, as if they had this awareness of the **ultimate** causation of their behavior, even though they do not. This fiction is a useful descriptive approach, but its products must be subjected to empirical test before they can be taken to have any analytical value.

reductionism (reductionist explanation): The form or style of explanation characteristic of the scientific method (Chapter 1). It supposes that the universe is organized in levels of complexity, each emerging from a small subset of the properties of the simpler level immediately below. So, for example, we argue that life is merely a special case of chemistry. Biological systems are just a very specific subclass of chemical systems. Likewise we argue that human social behavior emerges from the simple control of a few specific elements of non-human animal social behavior (Chapter 5). The new theory of human origins, properties, and history we propose in this book is a reductionist explanation of these things.

remote killing: Refers to the capacity to kill/injure another animal from a distance of many body diameters. In practice, the only animals that have had the capacity to kill adult conspecifics remotely are the members of the human lineage. This capacity is crucial to the uniquely human access to inexpensive **coercive threat** necessary to allow the evolution of **kinship-independent social cooperation** as an adaptive by-product (Chapter 5).

replication of genetic design information: See **DNA** and **self-replicating information.**

reproductive value: The likely future reproductive output of an individual. This is expected to influence the gene's-eye-view of an individual. For example, a healthy young adult has a higher likely future reproductive output than an aged adult (who has already reproduced in many cases) OR than an infant who still has a significant probability (in ancestral human populations) of dying before reaching reproductive age. Reproductive value is one of the major determinants of how conspecific animals – especially close kin animals – behave toward any individual.

reciprocal altruism: A term invented by the famous 20th Century biologist, Bob Trivers. It refers to two organisms helping one another on the basis of the benefits each receives from that cooperation. Ostensibly, reciprocal altruism is an adaptive strategy open to **non-kin conspecifics,** leading to the evolution of **kinship-independent social cooperation.** We argue that reciprocal altruism between non-kin conspecifics is viable only under rare, narrowly specific circumstances (Chapter 5). Most of the time, conspecific **conflicts of interest** preclude reciprocal altruism. On our view, humans engage in substantial non-kin cooperation NOT because they use reciprocal altruism as an adaptive strategy, but RATHER because they have evolved the capacity to adaptively manage non-kin conflicts of interest (Chapter 5). Our apparent "reciprocal altruism" emerges as an effect of this capacity, not as a cause of the evolution of our social behavior.

rifle: See **rifled barrels.**

rifled barrels or rifling: Refers to the practice of introducing helical or spiral groves on the inside of the barrels of gunpowder **handguns** or **artillery** (Chapter 15). These groves impart a spin to the projectile which stabilizes it in flight and thus greatly increases the accuracy of the weapon. Also see **Minie ball** and **musket.**

RNA: Molecules whose fundamental structure is essentially identical to **DNA.** The subtle chemical differences allow RNA molecules to conveniently serve as temporary disposable copies of DNA information. RNA copies of DNA are produced by the process of transcription. These resulting RNA copies (transcripts) are used to direct the synthesis of the **proteins** ultimately encoded in the DNA sequence through the process of translation.

Romans: The Republican and Imperial Romans of the era from roughly 400BCE through 500CE were a massive and well documented **archaic state,** sometimes dominating the entire Mediterranean basin (Chapter 13).

Samurai: See **armored warriors** and **elite warriors** (Chapter 13).

savanna: Open tropical grasslands with few trees. These include the regions of East Africa where the great herbivore herds (like gazelles and wildebeests) and their predators (like lions and cheetahs) live today. This kind of relatively open territory, together with nearby woodlands, in East Africa is apparently where our ancestors evolved immediately after the divergence of our **hominid** lineage from the chimp/bonobo lineage (Chapter 7).

scavenging: Refers specifically in this context to a life style based on eating the remains of the kills originally made by members of other species who are generally professional carnivores. Many vultures are scavengers, for example. There is reason to believe that our most immediate **australopithecine** ancestors, possibly including **Australopithecus garhi**, were scavengers. If scavengers use threat to scare carnivores off their kills, they are sometimes referred to as *power scavengers*. A very attractive hypothesis for the original selective pressure leading to the evolution of **elite throwing** proto-humans was adaptation to power scavenging by an Australopithecus garhi-like ancestor (Chapter 7).

scientific method: See **reductionism**.

Scientific Revolution: Refers to the period beginning in the 16th or early 17th Century (depending on some partially arbitrary dating choices) and, arguably, extending through the present moment. Isaac Newton's *Principia* was published early in this revolution and the revolution continues through the present in which the human understanding of the physical world continues to increase at an explosive rate. This period is quite different than any earlier phase of human history. It apparently represents a symptom of an **adaptive revolution** on a vast new scale (Chapter 10 and Third Interlude). We argue that the Scientific Revolution is a product of the adaptive power of the **modern democratic state** for very specific and transparent reasons (Chapter 17). This argument, in turn, has profound implications for the human future.

Second Law of Thermodynamics: This is a slightly subtle but profoundly important law of the physical universe. Like all other physical systems, organisms are governed by this law. To understand the world in general and to understand biological organisms in particular, it is essential to have a good understanding of this law. The Second Law can be stated as saying that all physical systems, including organisms, tend to increased disorder with time. This process can only be overcome by the input of energy from the outside. However, the preservation of perfect order requires the (impractical) input of infinite amounts of energy. Thus, orderly physical systems in the real world are always "at war" with the Second Law causing diverse consequences. Among these, for example, is that replication of **genetic design information** is never perfect. Errors or **mutations** always occur because there is not infinite energy to invest in their prevention. Another implication, for example, is that most mutations degrade the performance of a **vehicle** while only a very small fraction improves it.

self-replicating information: This phrase refers to a central concept of all of biology. Biological organisms are **vehicles** built by **genetic design information** shaped by **natural selection** to behave as if generating new copies of that design information were their sole ultimate purpose. Thus, this design information builds tools or devices that assist it in making new copies of itself. This is what makes the information "self-replicating." Also see **life**, **vehicle**, **genetic design information**, **DNA,** and **culturally transmitted design information**.

sensu lato: This is used to mean *in the broad sense* or *broadly defined*.

sensu stricto: This is used to mean *in the narrow sense* or *strictly defined*.

sexism: See **racism**.

sexual dimorphism: This term refers to the ways in which the two sexes of a species, including us, are different, including the reproductive apparatus, of course – vaginas versus penises. However, it also sometimes includes a number of other features. What these other features are and how extreme the sexual dimorphism is are strongly affected by the mating system of the animal. Highly dimorphic animals usually mate polygynously (see **polygyny**), with high levels of choice on the part of females and high levels of competition between males to mate with large numbers of females. This process of female choice is an example of sexual selection. At the other extreme, animals that mate monogamously show only very limited sexual dimorphism in most cases. Humans are between these two extremes.

sexual reproduction: Refers to a particular reproductive or replication strategy whereby **genetic design information** building an individual **organism (vehicle)** is provided in two copies – one from each of two parents. Typically one copy is provided by the female in an egg cell and the other copy by the male in a sperm cell. The sperm and egg are each said to be **haploid** having one copy of each **gene** or element of genetic design information. The fertilized egg (that will go on to grow in to the mature offspring) resulting from the fusion of the sperm and egg at **fertilization** is said to be **diploid** as a result of having two copies of each piece of genetic design information.

sexual selection: See **sexual dimorphism**.

shock weapons: This term applies to any weapon that is used to deliver a powerful blow whose lethality comes from blunt trauma and/or from slashing or stabbing penetration. This class of weapons is intended to be in contrast to a projectile weapon like a thrown stone and a **handgun**. The most well known examples of shock weapons are swords, maces, war clubs, and stabbing spears. Elite shock weapons are those that require unusual expertise to make and/or use. Good examples would be slashing and stabbing swords made from sophisticated metals (**bronze** or **iron**) or by complex composite manufacture from specialized wood and very expensive stone (like obsidian, volcanic glass). Such elite shock weapons were used by **hoplites** and other **armored warriors** of the **archaic state/empire**. A crucial point is that shock weapons dominate projectile weapons *only* when elite **body armor** is available. Otherwise, individuals with shock weapons are felled by projectile fire before they can close to the point that their weapons are effective.

social breeding: Refers to a reproductive strategy in which individuals in addition to the parents of a specific offspring contribute to the rearing of that offspring. Non-human animals sometimes do this. Examples include social **hymenoptera** and various other animals like the African bee eater. However, in non-human animals the non-parental helpers are virtually always very close kin – most commonly older siblings of the youngster or other close relatives of the parents. Humans are the first example ever of a **kinship-independent social breeder** (Chapters 6 and 8).

social contract information: See **culturally transmitted design information**.

social cooperation: See **cooperation, kin-selection** and **kinship-independent social cooperation**.

social father: Refers to the adult male who is nominally considered to be the father of a specific child. In contemporary state-level cultures, the social father is also commonly the **biological father** or genetic father. However, this is not always true in other cultures. There is good reason to believe that it was very commonly not true among our more ancient human ancestors (Chapter 8). In contemporary pre-state cultures, such as the **Canela**, adult females typically mated with multiple males in addition to their nominal husbands. Under these conditions, the social father (husband) is often not the biological father. Many cultures identify all the males who might be the biological father of a specific child as co-fathers of that child. Such cultures are said to believe in partible paternity.

social monogamy: See **monogamy**.

soma: Refers to the body of a sexual organism as distinct from its **germ line**. The body or soma of an animal is the **vehicle** built by its **design information** for the "**purpose**" of getting that information replicated (through reproduction).

spear-thrower: see **atlatl.**

speciation: The process whereby a single **species** gradually, over many generations, splits into two populations whose members can no longer successfully interbreed. This can happen in any of a number of different ways in detail. Once this reproductive barrier is formed, these two populations no longer exchange **genetic design information** and they evolve independently thereafter, inevitably becoming more and more different with time if both survive. Two species that are the descendants of relatively recent speciation events – lions and leopards, for example – are very similar. In contrast, those that are descendants of the products of an extremely ancient speciation event (with many intervening evolutionary changes ensuing) – humans and oak trees, for example – are very different.

species: All sexual organisms belong to a population of organisms that are relatively similar in their anatomy and behavior and whose opposite sex members are able to mate to produce viable, fertile offspring. This interbreeding population is what we will mean by the term species. All people alive today are members of a single human species, for example.

Every known population of interbreeding animals is given a species name. This species is, in turn, a member of a genus (pl. genera) which frequently includes several closely related species. In other words, the different species that make up a genus are almost always descended from a common ancient ancestral species *relatively* recently; however, they have diverged as a result of multiple speciation events since the time of that **last common ancestor**.

It is standard practice to refer to an animal as belonging to a species by naming both the genus and species. So, for example, the reader and the authors are humans and our genus name is **Homo** and our species name is sapiens. This is from the Latin - homo means *man* in the generic sense and sapiens refers to *wise* or wisdom. [Thus, we have named ourselves *wise man* - obviously not necessarily an objective characterization.] We, thus, refer to contemporary humans then as members of **Homo sapiens**.

sperm competition (or warfare): Refers to the situation that arises when females mate polyandrously or polygynandrously (see **polyandry** and **polygynandry**). The sperm from multiple males are simultaneously present in the female's reproductive tract. Sperm that win this competition by fertilizing an egg successfully transmit their design information. Thus, we expect some adaptations in males to be designed to enhance their success in sperm competition.

squash: see **Mesoamerican triad**

stochastic: A process is stochastic if it is accidental rather than causal and predictable. Errors in **DNA** replication producing **mutations** are stochastic events from the point of view of the **genetic design information**. Outcomes of the flips of a coin are stochastic as well, from the perspective of individual people tossing the coins.

stratigraphy: Refers to the fact that deposits from the past are often organized spatially one on top of the other. Thus, each layer of soil deposited by water or wind covers the one that came immediately before. Each layer is a *stratum* (plural strata) and the analysis of such sequences is called stratigraphy. The relative stratigraphic positions of **fossils** can often be used to infer their historical and evolutionary relationships. New fossils are in higher strata whereas older fossils are more deeply buried in older strata.

Tasmania: An island off the southern coast of Australia. Tasmania is separated from Australia by the Bass Strait. This strait is very shallow and it was uncovered when sea levels dropped during the last **glaciation**. As a result, Australia and Tasmania became a single land mass and **behaviorally modern humans** spread across it before the last glacial retreat. When sea level rose again around fifteen thousand years ago, Tasmania became isolated. It supported only a very small human population. This population apparently underwent a decline in **adaptive sophistication** to the point that they made tools no more sophisticated than ancient **Homo erectus** at the time of European contact. However, the contemporary descendants of these people are fully contributing members of the local modern economy. These observations represent powerful support for the theory that human cooperative coalition size ultimately limits and determines human adaptive sophistication (Chapter 10).

tectonic activity: All the diverse consequences of the movement of the molten core of planet Earth. This includes volcanic activity and the movement of the large tectonic plates on which the continents ride. These forces are responsible for the appearance of the contemporary Earth in a variety of obvious and less obvious ways. For example, they are responsible for mountain building that is probably crucial to long-term nutrient cycles that fuel the persistence of complex life. The detailed composition of Earth, as produced by the **Orpheus-Earth Mark I** collision early in the development of our solar system, is necessary for the characteristic tectonic activity of our Earth – Earth Mark II.

tensor fascia lata: See **muscles**.

Teotihuacan: The first large city-state built in **Mesoamerica**. It was constructed over several centuries (from ca. 1 CE to ca. 600 CE) and then was destroyed catastrophically around 750 CE. Teotihuacan was a central element of an early **archaic state** (Chapter 13).

theory-of-everything (TOE): See **parsimony**.

thermonuclear bomb or **device:** See **fusion**.

Tokugawa Shogunate: Refers to the **early modern state** in Japan (Chapter 14). This era was followed rapidly after the introduction of gunpowder **artillery** into Japan through European trade contact.

transcription: See **DNA** and **RNA**.

translation: See **DNA** and **RNA**.

Trinity: See **fission**.

ultimate causation: This refers to the most fundamental reason for or causal origin of a behavior in an animal. For example, we eat because we "feel hungry." However, this feeling is merely the **proximate causation** of our eating behavior. The ultimate causation of this behavior is that the **Second Law of Thermodynamics** requires that we eat and **natural selection** has built animals – including us – designed to behave in ways that satisfy that requirement. The mechanism natural selection has built to cause this to happen is our neurological/physiological "feeling hungry" response. The ultimate causation of a behavior is always illuminating, whereas its proximate causation is often quirky, idiosyncratic, and not necessarily even interesting from an analytical view (Chapter 10).

universal Darwinism: This is a very general and important claim about the universe. It states that all complex adaptive order in the universe results from **Darwinian selection**, that is, Darwinian processes. We know this is true of the complex order of organisms. However, there is good reason to believe that it applies to other things – thinking and ideas, for example. This theoretical postulate – it is not yet an established theory – appears to be profoundly illuminating about otherwise inscrutable processes.

vehicle: We will use this term in a specific technical sense and as it was originally used by Richard Dawkins in the *Selfish Gene* (1976). It refers to the physical entity that is an **organism** (Chapter 2). It can be any organism, including a virus, an oak or a human being, for example. This term emphasizes the status of the organism in the long sweep of evolutionary history. The organism is built by **genetic design information** for the sole teleological **purpose** of replicating that design information through what we think of as biological reproduction. In other words, on an evolutionary time scale an individual organism is merely a transient, ultimately disposable vehicle for the replication of potentially immortal genetic design information. Use of the term vehicle – rather than animal or organism – tends to focus our attention on **ultimate causation** where this attention needs to be focused for authentic understanding.

village: We will use this term with a specific meaning. This term will refer to all the kinship-independent social units constructed by humans since their evolution beginning around two million years ago. The human village refers to any such unit, no matter its physical structure or possible transience. The human village is the social unit – however deployed – in which non-kin adults of both sexes cooperate to carry out **kinship-independent social cooperation**, including **kinship-independent social breeding**. As these units get very large as in more recent eras, we will sometimes use other terms, including **coalition** or **state**, for example. However, the underlying logic of all human social units, including the global "village," remains fundamentally the same (Chapter 5 and Third Interlude).

World Wars: We argue that the underlying coercive logic of World War I and World War II were fundamentally different (Chapters 15 and 16). World War I was dominated by **gunpowder** weaponry (**artillery** and **handguns**) and was fought to an inevitable lack of resolution. World War II was dominated by **aircraft** and fundamentally changed the logic of international coercion and, thus, of international cooperation.

waggle dance: A dance done by honey bees to communicate the location of external food sources to other honey bees while they are still together in the home hive. The details of this dance indicate that non-human animals are capable of abstract symbolic **gestural communication** of contingent or fakable information (Chapter 9).

wheelock: See **handgun**.

Williams, George C.: One of the great evolutionary biologists of the 20th Century. His work, including that embodied in his famous book *Adaptation and Natural Selection* (1966) was one of the crucial foundations on which our work is based.

worker policing (hymenoptera): Refers to the practice of worker females in large-colony bees and ants eating one another's (male) eggs. Females behave this way because they are mostly half-sisters and their half-nephews have a lower level of **genetic relatedness** to them than do their brothers (male eggs laid by the queen; Chapter 7). By acting in this way individual females are actively coercing one another to raise their brothers, allowing the colonies to grow large and live for many years, rather than competing to raise their own sons, destroying the cooperative colony. Thus, this eating of one another's eggs represents *policing* of the cooperative adaptation, resulting in the prevention of **cheating** or **free riding**. Notice that eating one another's eggs is easily, cheaply done and, thus, each female exerts a credible **coercive threat** against her half-sisters.

zygote: See **diploid**, **haploid** and **sexual reproduction**.

❧

About the Authors

Paul M. Bingham earned his Ph.D. from Harvard University in Biochemistry and Molecular Biology, where he also continued to develop his fascination with fundamental unanswered questions about how humans evolved. During his 27-year career on the faculty of Stony Brook University, he has continued to explore human origins while also contributing to fundamental cell and molecular biology, including the discovery of the P element transposon and new approaches to cancer therapy.

Joanne Souza is a successful business industry consultant in health & education trained by AT& T and a faculty member at Stony Brook University. She earned her BA in Psychology, summa cum laude, from Stony Brook University, receiving a Recognition Award for Academics & Research and the University Award for Senior Leadership & Service. In the last six years she has continued to pursue her life-long research interest in human behavior, evolution, and history while earning a Masters of Science in Psychology from Walden University.

Photos courtesy of Christina Luciw

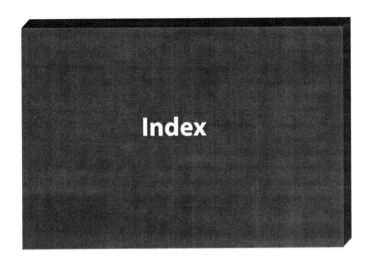

Index

CPSIA information can be obtained at www.ICGtesting.com
Printed in the USA
BVOW021208130912

300301BV00001B/6/P